Molecular Reaction Dynamics and Chemical Reactivity

Oxford University Press

Oxford   New York   Toronto
Delhi   Bombay   Calcutta   Madras   Karachi
Pealing Jaya   Singapore   Hong Kong   Tokyo
Nairobi   Dar es Salaam   Cape Town
Melbourne   Auckland

and associated companies in
Beirut   Berlin   Ibadan   Nicosia

Copyright © 1987 by Oxford University Press, Inc.

Published by Oxford University Press, Inc.,
200 Madison Avenue, New York, New York 10016

Oxford is a registered trademark of Oxford University Press

Library of Congress Cataloging-in-Publication Data
Levine, Raphael D.
Molecular reaction dynamics and chemical reactivity.
Bibliography: p.
Includes index.
1. Molecular dynamics.   2. Chemical reactivity.
I. Levine, Raphael D.   II. Bernstein, Richard Barry, 1923–
Molecular reaction dynamics.   III. Title.
QD461.L66   1987   539'.6   86-8452
ISBN 0-19-504139-9
ISBN 0-19-505395-8 (pbk.)

Printing (last digit): 9 8 7 6 5 4 3 2 1

Printed in the United States of America
on acid-free paper

Molecular Reaction Dynamics and Chemical Reactivity

Raphael D. Levine

The Hebrew University of Jerusalem

Richard B. Bernstein

University of California, Los Angeles

New York   Oxford

OXFORD UNIVERSITY PRESS

1987

Preface

The 12 years that have elapsed since the completion of our book *Molecular Reaction Dynamics*† have been exciting and rewarding. We have witnessed an exponential increase in activity in the field of molecular dynamics, both experimental and theoretical. Our original intention, to write a book that serves as an introduction to the field, remains the same. The field itself, however, has grown to such an extent that our attempt at a second edition resulted in a completely rewritten volume. Furthermore, we were faced with the need to be highly selective with regard to the choice of material discussed. The only surviving aspect of the earlier work is our intent: to provide a primer of molecular reaction dynamics.

This volume is organized not according to methodology or technique; rather, it is intended to help the reader proceed from the simple to the more complex. As we progress through the book, we probe ever deeper into the microscopic details of chemical reactivity.

Our field is growing, but it is far from reaching maturity, let alone middle age. We believe that it is still in its vigorous adolescence, as demonstrated by the many sections that are forced to conclude that much work on the topic in question must yet be done. Some of the subjects are in the midst of their growing period and some, for example, the dynamics of reactions on surfaces, are only recently out of their infancy.

Another manifestation of the tremendous activity in the field of molecular reaction dynamics is the breadth and extent of the review literature. We have thus limited all of our suggested readings to such overviews, most of which have appeared since our earlier work. Moreover, a reference is cited usually only once, at the earliest point at which it is relevant. It must be assumed, therefore, that the reading lists for the more advanced chapters include those of the introductory ones. To help the reader, we have provided titles of all the suggested readings. At the end of Chapter 1, we provide a list of the review volumes containing authoritative articles to which we shall be referring.

Theoretical considerations play a decisive role in our field, both for the interpretation of observations and in the design of new experiments. This introductory volume is not, however, intended as a handbook of theory (or

---

† Japanese translation by Hokotomo Inouye, Tokyo University Press, Tokyo, 1976; Chinese translation by Yusheng Tao, Press of the Academy of Science, Beijing, 1986.

of experiment). No one expects to be able to go into the laboratory and reproduce any one of the experiments that we discuss (without much additional study of the literature). By the same token, we do not provide enough detail for the reader to carry out a computational implementation of the theory. Our aim is only to convey the essence of the ideas and to demonstrate the blend of concepts and experiments that is so characteristic of our field. The reading lists are intended to help those interested (and we hope this includes most of our readers!) to gain sufficient expertise to join the growing community of researchers, both experimental and theoretical, in the field of molecular reaction dynamics.

The good health of the field is evident from the quality and number of its practitioners. We are indeed proud of our collective colleagues all over the world who have made the field what it is today. Unfortunately, we can do no more than list in the Author Index those whose work happens to have been selected for the illustration of specific points in the text.

We thank our families, who have borne with us with their usual good grace. Also understanding were the science funding agencies that have supported our research. We are indebted in particular to the U.S. Air Force Office of Scientific Research, the U.S. National Science Foundation, the U.S.–Israel Binational Science Foundation, the Minerva Gesellschaft, and the Volkswagen Foundation.

We hope that this volume will serve to catalyze the "growth" to maturity of the field of molecular reaction dynamics and the understanding of chemical reactivity.

Jerusalem   R. D. L.
Los Angeles   R. B. B.
December 1985

Contents

Molecular Reaction Dynamics and Chemical Reactivity

1
Understanding molecular collisions

1.1 What is molecular dynamics?

Molecular dynamics involves the study of the molecular mechanism of elementary physical and chemical rate processes. It is concerned with both intramolecular motions and intermolecular collisions, including molecule–surface encounters, and with the means of promoting or interrogating such collisions, say, by molecule–photon interaction. An understanding of the dynamic behavior of a system at the *molecular* level is the key to the interpretation of the "macroscopic" kinetics of the bulk system.

Since the acceptance of the kinetic theory of gases it has been recognized that intermolecular collisions serve as the microscopic mechanism "underneath" all observed rate phenomena involving gases and liquids. It is only fairly recently that progress in both experimental techniques and fundamental theory has brought us to the state where we can probe a rate process *directly* at the microscopic, molecular level. We are entering an era when the intimate details of physical change and chemical reactivity can be observed experimentally and understood theoretically. We are beginning to secure a molecular vantage point. From here we can view the very *process* of chemical change itself, that is, the *chemical dynamics* of a reactive system.

Molecular dynamics has by now become a field in its own right, with chemical dynamics as a major subdiscipline. The latter, in turn, serves not only as the foundation of macroscopic chemical kinetics but also as a rich source of new knowledge of the basic phenomena involved in the elementary chemical act.

Although the technical details of the experimental and theoretical methods of molecular dynamics are indeed complex, the new *concepts* are quite simple. An understanding of these concepts—the ability to read the language—is all that is necessary to understand the new "microlevel" phenomena revealed by molecular dynamics. This book is a primer of the new language.

1.2 An example: Infrared chemiluminescence

Typical of the kind of information that is becoming available from the new techniques of molecular dynamics is the determination of the *energy disposal* in exoergic atom–molecule exchange reactions. One example of such a system is the H-atom transfer reaction

$$Cl + HI \rightarrow I + HCl$$

In the course of this reaction the relatively weak HI bond is broken and replaced by the stronger HCl bond (see Fig. 1.1). The reaction thus liberates "chemical energy": the exoergicity is $-\Delta E_0 = 133\ \text{kJ mol}^{-1}$.† The question is. In what *way* is the energy released following a reactive collision of Cl + HI? Even if both products are formed in their electronic ground states, there are three remaining "modes" of energy disposal: vibration of HCl, rotation of HCl, and relative translation of the recoiling I + HCl. But what is the *distribution* of energy among these three modes?

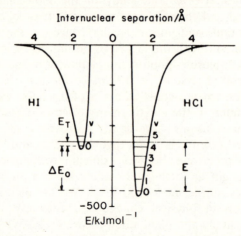

Fig. 1.1. Schematic illustration of the energetics of the reaction $Cl + HI \rightarrow I + HCl$. Starting from ground state HI with translational energy E_T, we can form HCl in any vibrational–rotational state with an energy less than the total available energy E. The ΔE_0 for the reaction is the difference in bond dissociation energies: $\Delta E_0 = D_0(HI) - D_0(HCl) = -133\ \text{kJ mol}^{-1}$.

When the reaction is studied in the "bulk" gas phase, the products collide with other molecules, energy is transferred upon collision (thus becoming effectively partitioned among all molecules), and the overall reaction exoergicity is finally liberated in its most degraded form, that is, heat. In macroscopic terms, the reaction is exothermic. The microscopic approach of molecular dynamics, however, is concerned with the outcome of the *individual* reactive collisions per se. Thus the techniques usually involve

---

† Energy and other conversion factors can be found in the General Appendix.

measurements at very low pressures, often employing *fast-flow* arrangements, or very short times. The newly formed products are thereby prevented from undergoing extensive collisional relaxation. The vibrotationally excited newly formed HCl molecules can then lose their excess energy by radiation (here the emission is infrared). By monitoring this infrared chemiluminescence with a spectrometer we can determine the relative populations of the various excited states and thus the relative rates of their formation.

Alternatively, rather than try to prevent relaxation over the (comparatively long) time scale for the spontaneous emission of radiation in the infrared, we can (1) induce emission or (2) determine the concentration of the molecules in different levels by light absorption. The former is just what happens in a chemical laser, discussed in Section 1.2.2.

1.2.1 Distribution of products' energy states

Figure 1.2 shows a typical experimental result, showing the energy disposal in terms of the distribution of population of the vibrational states of HCl, irrespective of the rotational distribution within a given vibrational state. It is found that a large fraction of the available energy goes into vibrational excitation of the HCl, and thus, by difference, only a small fraction goes into translational recoil of HCl + I.

This vibrational distribution can be compared with that expected when the reaction is run under "bulk" conditions, where a Boltzmann distribution

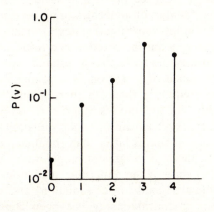

Fig. 1.2. Semilog plot of $P(v)$, the probability of formation of HCl in a given vibrational state v, from the Cl + HI reaction of Fig. 1.1. Note the "population inversion," that is, the excess population of the higher v states. At thermal equilibrium $P(v)$ will decline monotonically as v increases. [Adapted from the experimental infrared chemiluminescence results of D. H. Maylotte, J. C. Polanyi, and K. B. Woodall, *J. Chem. Phys.*, **57**, 1547 (1972).] Since the $v = 0$ state cannot emit infrared, $P(0)$ is not available from such chemiluminescence experiments. However, the ratios $P(0)/P(1)$, $P(1)/P(2)$, and so on, can be deduced from chemical laser data.

would be produced. Here the most probable state is $v = 0$ and the relative populations decline strongly with vibrational quantum number. Of course, the bulk population distribution does not arise from an elementary process but, rather, from a succession of energy-degrading collisions of the energy-rich HCl molecules with various other molecules.

Molecular dynamics in its "purist" approach tries to seek out (and understand) the truly elementary events. Thus it is interested in determining the nascent distribution as shown in Fig. 1.2. It is, however, concerned not only with the primary reactive collision process but also with the nonreactive, inelastic energy transfer steps which take the system from the nascent distribution of Fig. 1.2 to thermal equilibrium.

1.2.2   From population inversion to chemical lasers

The Cl + HI system is not exceptional. Many exoergic reactions release a substantial part of their energy as internal excitation of the nascent products. In Section 1.4 and more in detail in Chapter 4 we shall argue that this is due to the energy being released already as the reagents approach one another. A well-studied example with practical implications is the elementary reaction

$$F + H_2 \rightarrow HF + H$$

We can generate the reactive F atoms *in situ* by photolysing some volatile fluorine compound, for example, CF_3I. Through flash photolysis a high initial concentration of F atoms is rapidly converted to a high concentration of vibrationally and rotationally excited HF molecules. This nascent HF population (like that of HCl; see Fig. 1.2) deviates from equilibrium in an extreme manner: The fraction of molecules in upper (initial) states is much greater than the fraction in lower (final) states for many allowed transitions in the infrared. By carrying out the reaction in a resonant cavity, the system can amplify the infrared radiation. An excited HF molecule can be stimulated by the presence of a photon of a proper frequency to emit a photon of the same frequency that can then stimulate a second excited molecule, and so forth. The radiation is coherent (and thus well collimated), since the stimulated emission synchronizes the radiation of different excited molecules. Stimulated emission is faster than spontaneous emission and will continue as long as there is a greater population in the emitting (as compared to the final) state of the molecule. Such a device is a laser (*l*ight *a*mplification by *s*timulated *e*mission of *r*adiation) and specifically, here, a chemical laser, where chemical energy has been converted to coherent radiation.

1.3 Why molecular dynamics?

Obtaining the details of *energy partitioning* in an elementary reaction, as an example, requires considerable experimental ingenuity as well as sophisticated (and often expensive) apparatus. Yet many laboratories around the world are probing the field of molecular dynamics. Why? What can we learn from the study of the elementary processes? There are, in fact, two justifications for our interest in such a fundamental approach: the "pragmatic" and the "purist." Let us examine the arguments.

1.3.1 The pragmatic approach

The study of elementary processes has already proved to be of practical utility in many areas of physics and chemistry. One example is the development of the *chemical laser,* where the very *specific* energy release, that is, *population inversion,* found in many exoergic reactions (such as those discussed, $Cl + HI$ and $F + H_2$, among others) is responsible for this intense source of light. For efficient conversion of chemical energy to radiant energy one needs not only rapid generation of the excited states but also the suppression of energy-degrading collisional depopulation processes and removal of the lower (final) level of the laser transition (to maintain the required population inversion).

An ingenious solution to the problem is responsible for the excimer lasers currently available in the chemically important ultraviolet range of the spectrum. The key step is the formation of a halide of an electronically excited rare-gas atom, for example, Kr^*F. Since the outer electron in Kr^* is loosely bound, it is similar to an alkali atom and Kr^*F is a fairly strongly bound molecule. Upon making a downward electronic transition, it forms KrF, which dissociates. Many steps are, however, required in the real laser to form Kr^*F. Thus for the intelligent design of a high-powered laser, that is, for *purely technological* progress, we need detailed information on many different elementary processes, all of which lie within the domain of molecular dynamics.

Relaxation processes are not entirely avoidable and hence laser photons are expensive. Not only their generation but also their application needs to be done with due regard to the processes on the molecular level. One exciting possibility is the laser catalysis of chemical reactions. A more immediate application to chemical synthesis is the use of lasers to initiate a chain reaction. An example is the use of a KrF laser to dissociate Cl_2 in the presence of ethylene and thereby lead to the formation of the $\cdot CH_2CH_2Cl$ radical in the chain reaction for the formation of $CH_2{=}CHCl$. (The vinyl chloride monomer, used in the production of polyvinylchloride, or PVC, is a key industrial chemical with an annual production in megatons.)

1.3.2 Disequilibrium

An understanding of elementary processes is also important if we wish to enhance our practical ability to manipulate a variety of environmental phenomena (e.g., atmospheric and ionospheric chemistry, combustion, and air pollution). These involve a complex web of series and parallel elementary steps comprising the overall system of reactions. Many of the elementary processes are fast reactions, yielding products in a state of *disequilibrium*. We know that the nascent products of such (exoergic) elementary reactions are often disproportionately *energy-rich*. We must therefore consider the influence of their excitation upon their (subsequent) reactivity. In particular, if one of the "slow steps" in the overall scheme has an activation barrier, its rate may be very much enhanced if one of the reactants has an abnormally large population of excited states. In many cases such reactions are highly *selective* in their energy requirements.

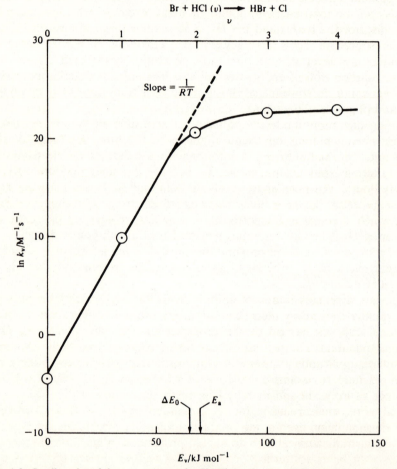

Fig. 1.3. Semilog plot of the rate constant (at 298 K) for the Br + HCl reaction as a function of the vibrational energy of the HCl. [Data from D. J. Douglas, J. C. Polanyi, and J. J. Sloan, *J. Chem. Phys.*, *59*, 6679 (1973); D. Arnoldi and J. Wolfrum, *Ber. Bunsenges. Phys. Chem.*, *80*, 892 (1976).] Shown are the endoergicity and barrier energies. The dashed line has a slope $1/RT$ predicted from general theoretical considerations [R. D. Levine and J. Manz, *J. Chem. Phys.*, *63*, 4280 (1975)].

Frequently vibrational energy is more effective than relative translational energy in overcoming the barrier.

To see this, imagine a detailed slow-motion movie of an exoergic reactive collision. Assume that the reaction exoergicity is released in a very specific fashion. Now let us run the movie backward. The (former) "products" converge, collide, and emerge as the "reactants." In the process, the disposed exoergicity is being consumed to overcome the energy barrier of the reverse, endoergic reaction. It is clear that if the energy disposal in the exoergic reaction is specific, the energy consumption in the reverse, endoergic reaction must be selective. In other words, for a given total energy, the endoergic reaction will be more "efficient" when a larger fraction of the total energy is initially present as the internal energy of the reactants, for the reaction can draw on the internal energy to surmount the barrier. Once the excess internal energy is higher than the barrier, however, the effect may level off.

The reaction of Br + HCl is endoergic by 65 kJ mol^{-1}. The dependence of the reaction rate constant on the initial vibrational excitation of HCl is shown in Fig. 1.3. The exponential increase in the rate constant up to $v = 2$ and the much more moderate increase past that point are clearly evident.

Nor need we limit ourselves to scalar aspects. We shall derive considerable information from the distribution of the reaction products in space, which need not be, and often is not, isotropic. Also, the relative orientation of the reagents is expected to have a dominant effect on their reactivity. The deviance from the "uniformity" of an equilibrium-type behavior is a prime characteristic of real processes when examined at the molecular level.

1.3.3 The purist approach

Molecular dynamics is a natural outgrowth of long-standing attempts to explore the *elementary chemical act*. In the pursuit of this grail, classical chemical kinetics has probed ever-shorter time scales, toward the femtosecond domain. The realization that elementary reactive events can really be studied only under "single collision" conditions has led to the development of modern molecular dynamics. It is simply the next step in our attempt to probe the *molecular mechanism* of chemistry.

Of course, nuclear and high-energy particle physics have been looking at individual collision events since Rutherford, and for half a century workers in atomic, molecular, and electron physics have been probing interparticle forces with beam-scattering techniques—as well as liberal applications of Schrödinger's equation! It is only natural that the physical chemist has taken up these tools to study the elementary chemical act. With the more recent introduction of laser techniques we can study real-time dynamics (even without the use of particle beams).

The dynamics of single molecular collisions are of interest not only for understanding chemical reactivity, of course. They are involved in all systems involving gases and fluids in a state of disequilibrium, whether chemical or physical. Just as reactive collisions can produce disequilibrium, nonreactive collisions provide the mechanism for bringing about the return to equilibrium.

As we learn to deal with polyatomic dynamics and larger aggregates, we can hope to contribute to the understanding of the dynamics in the liquid state, and eventually to attack the problem of the disequilibrium condition so characteristic of biological systems. Ultimately, there too a molecular dynamical interpretation will follow.

1.4 A simple model of energy partitioning

In Section 1.3 we introduced the related subjects of *specificity* of the energy disposal in elementary exoergic reactions and *selectivity* of energy consumption in elementary endoergic reactions, or those involving an activation energy barrier. Obviously we will require a considerable development of the "fundamentals," both experimental and theoretical, before we can properly deal with these phenomena. However, it may be instructive to take a shortcut, deferring until later the more rigorous background material.

Consider the H-atom transfer reaction referred to in Section 1.2, $Cl + HI \rightarrow I + HCl$. Can we provide a simple interpretation of the observed energy disposal? Can we construct a model—oversimplified, of course—that will at least cast some light on the experimental results? Let us try to take advantage of any "handle" that may help us simulate the dynamics of the problem.

1.4.1 The spectator

One aspect of the collision is that it involves the transfer of a very light atom (H) between two much heavier atoms. The ClHI triatomic system is similar in this respect to the H_2^+ molecule-ion, where the lightweight electron "mediates" between the two heavy protons.

When a molecule undergoes an electronic transition, we obtain qualitative (and often semiquantitative) insight into the distribution of the final states from the Franck–Condon principle, which says that during the very fast *electronic* transition the *heavy nuclei* do not change their momentum appreciably. The nuclei merely act as *spectators* during the rapid electronic rearrangement. (A spectator is someone who is not involved, i.e., does not feel an impulse. From Newton's second law it thus follows that a spectator has a constant momentum.)

To apply similar ideas to the present problem we assume that the transfer of the light H atom is fast, that the transfer reaction is over in a time short compared to the time required for the heavier nuclei to move substantially.

The model is then one where the heavy iodine atom acts as a spectator during the (rapid) transfer of the light hydrogen atom to the heavy chlorine. This means that the final momentum of the I atom after the collision, $\mathbf{p}'_I$, is essentially the same as the initial momentum of the I atom, $\mathbf{p}_I$:

$$\mathbf{p}'_I = \mathbf{p}_I \qquad (1.1)$$

Because the H atom is so light compared to the I atom, the spectator model of Eq. (1.1) implies that the kinetic energy of HI before the collision is very nearly that of the I atom after the collision. For thermal energy reagents the initial relative kinetic energy is small so that the final translational energy of the products is low and the reaction exoergicity must show up in the form of internal energy of the HCl. Before we make the argument more quantitative, let us look at the experimental evidence that supports the model.

1.4.2 Product angular distribution

Using the technique of crossed molecular beams (see Sec. 5.3), it is possible to measure the angular distribution of the reaction products. According to the spectator model, not only the magnitude but also the direction of the momentum of the iodine atom is unchanged in the collision. (Newton's second law requires an external force to change the direction of the momentum.) Hence the product iodine atom should appear in the same direction as that of the incident HI, while the product HCl should appear in the direction of the incident chlorine atom. Since the iodine is a spectator, the chlorine merely picks up the light hydrogen, both heavy atoms continuing in their original direction. This is qualitatively the behavior found experimentally, via the crossed molecular beam scattering technique. The product HCl appears mainly in the "forward" direction (i.e., in the direction of the incident chlorine atom).

The very fact that the angular distribution of the product is highly *anisotropic* implies that there is *no long-lived* ClHI intermediate "complex." If the reaction duration were long compared to the period of rotation of such an intermediate (i.e., several picoseconds), all memory of the initial directions would be lost and the angular distribution would have a forward–backward symmetry.† This is not the case for the Cl + HI reaction. Indeed, if there were a long-lived intermediate, we would not expect a very specific energy disposal, since then there would be time for the energy disposal to be "equipartitioned" among the different modes, that is, vibration, rotation, and translation.

---

† Such *symmetric* product angular distributions *have* been observed, for example, for the exchange reaction $RbCl + Cs \rightarrow CsCl + Rb$. Here there is a CsClRb intermediate whose lifetime exceeds several picoseconds.

1.4.3 A quantitative detour

Having discussed the qualitative features, we should try to go a step further and consider a more quantitative treatment of the model. For the purpose of simplicity we shall neglect the zero-point vibrational energy of the initial state (ground-state) HI molecules. We shall work out the results for the more general case, that is, for the "B"-atom transfer reaction $A + BC \rightarrow AB + C$, remembering (at the end) that A, B, and C represent Cl, H, and I, respectively.

Assume that A and BC have initial velocities $\mathbf{w}_A$ and $\mathbf{w}_{BC}$. (We take all velocities to be with respect to the *center of mass* of the ABC system, which itself moves at a fixed velocity in the laboratory framework.) Conservation of linear momentum, $\mathbf{p}_A = -\mathbf{p}_{BC}$, means that $m_A\mathbf{w}_A = -m_{BC}\mathbf{w}_{BC}$, and so the relative translational energy, E_T, is

$$E_T = \tfrac{1}{2}m_A w_A^2 + \tfrac{1}{2}m_{BC} w_{BC}^2 = \tfrac{1}{2}M(m_{BC}/m_A)w_{BC}^2 \qquad (1.2)$$

where $M = m_A + m_B + m_C$.

After the reaction $\mathbf{p}_{AB} = -\mathbf{p}_C$, so $m_{AB}\mathbf{w}_{AB} = -m_C\mathbf{w}_C$, where $\mathbf{w}_{AB}$ and $\mathbf{w}_C$ are the velocities of the products AB and C with respect to the center of the mass. Then the final relative translational energy is

$$E_T' = \tfrac{1}{2}M \frac{m_C}{m_{AB}} w_C^2 \qquad (1.3)$$

and the ratio of final to initial relative translational energies is

$$\frac{E_T'}{E_T} = \frac{m_A m_C}{m_{AB} m_{BC}} \left(\frac{w_C}{w_{BC}}\right)^2 \qquad (1.4)$$

At this point we introduce the *assumption* of the spectator model that C is nothing but a spectator, that is, $\mathbf{p}_C' = \mathbf{p}_C$. Since initially BC has no vibrational energy, the initial velocity of C is $\mathbf{w}_{BC}$, so that in the spectator model $m_C\mathbf{w}_C = m_C\mathbf{w}_{BC}$, or $\mathbf{w}_C = \mathbf{w}_{BC}$. Then from Eq. (1.4) the ratio of final to initial translational energies is simply a mass factor,

$$E_T' = \frac{m_A m_C}{m_{AB} m_{BC}} E_T = (\cos^2 \beta)E_T \qquad (1.5)$$

(For later use, the mass factor is written as the square of the cosine of a "skew angle" β.) For the Cl + HI system $\cos^2 \beta = 0.96$, confirming our qualitative conclusion that for a light-atom transfer there is almost a conservation of kinetic energy, that is, most of the reactants' relative translational energy is "converted" into the relative translational energy of the products. For the *isotopic* reaction of Cl with DI, $\cos^2 \beta = 0.92$, so the final E_T' should be similar to that for the hydrogenic reaction.

It is sometimes useful to consider a quantity Q, defined as $Q = E_T' - E_T$, the "collisional" or "translational" exoergicity, and express it also in terms of β, using Eq. (1.5):

$$Q = (-\sin^2 \beta)E_T \qquad (1.6)$$

Of course, as the initial kinetic energy is increased, and if $\sin^2 \beta$ is not small, more and more of the reactants' translation is converted into internal excitation of the nascent product. Ultimately, its energy content will be so high that it will dissociate. Collision-induced dissociation via incipient formation of products is thus expected to be an efficient process, particularly if the transferred atom is heavy ($\cos^2 \beta < \sin^2 \beta$).

The present model is very oversimplified. Both experiment and more detailed theoretical work show that there is a *distribution* of final E'_T values around the most probable one, as well as a distribution of scattering angles (albeit mainly in the "forward" hemisphere). Nevertheless, the simple model is a good beginning.

1.4.4 The reverse reaction

Finally, we can ask the question: Is our model predictive? Can it deal with the energy consumption in the reverse, endoergic reaction

$$I + HCl \rightarrow Cl + HI ?$$

The reaction endoergicity is to be supplied by the initial energy of the reactants. This can be provided by the relative translational energy of the colliding pair and/or by the internal vibrotational energy of the HCl. When the energy of the reactants just exceeds the endoergicity, there will not be enough energy to form vibrationally excited HI, and, of course, the final momentum of the I atom will be very small. If this momentum were to be nearly unchanged during the collision, the necessary energy for the reaction could not be provided by the relative translational energy of the reactants, for this would require a high initial momentum of the I atom relative to the center of mass. The reaction endoergicity, at least just above the energy threshold, must therefore be provided by the initial *internal* energy of the HCl.

This tentative conclusion can be strengthened by consideration of the principle of microscopic reversibility, which we already invoked in connection with endoergic reactions. Recall that the experiments on Cl + HI showed that at low energies vibrationally *cold* HI leads mainly to the formation of the vibrationally *hot* HCl, with only a small fraction of the energy released as translation. Since vibrationally cold HCl is formed with a low probability in the forward reaction, it follows that for the reverse reaction involving the collision of vibrationally cold HCl with an I atom at high translational energies, most collisions will prove to be *nonreactive,* that is, the reaction will occur only rarely on such collisions. In contrast, collisions of vibrationally *hot* HCl molecules with I atoms will be fruitful even at low translational energies. The results shown in Fig. 1.3 for HCl(v) + Br fully confirm these expectations.

Implicit in our model is the assumption that the reaction exoergicity is released early, during the approach of the reagents. If, instead, the energy is released only as the products begin to separate, then it will be channeled preferentially into the relative translation of the two products. (In the reverse reaction it will be the initial translation that will be more inducive to reaction.) Since the population in the excited internal states of the products will be small, it is sometimes more convenient to measure the recoil velocity, for example, via the time of flight of the products. To induce the reverse reaction one can prepare translationally fast reagents using the technique of seeded supersonic molecular beams. We shall discuss both these techniques in Chapter 5. Here we will comment on another approach, made possible by the availability of lasers.

1.4.5 Lasers in molecular dynamics

Lasers play a key role in the selective preparation of reagents in nonequilibrium distributions, as well as in the probing of the nonequilibrium distributions in newly formed products. As an example we consider the hydrogen exchange reaction

$$H + D_2 \rightarrow HD + D$$

which is the simplest nontrivial chemical reaction. It has an activation barrier of 41 kJ mol^{-1}. To overcome the barrier, fast H atoms are generated by photolysis of HI using a short ultraviolet laser pulse.† A second "probe" laser provides a pulse at another wavelength that is delayed with respect to the photolysis laser. It reveals the concentration of the HD molecules in different vibrational states. By using a short time interval between the two laser shots it is possible to interrogate the newly formed HD molecules before they undergo secondary collisions. (How short is "short"? See Sec. 2.1.5.) Figure 1.4 shows the observed results for the HD vibrotational distributions.

The initial kinetic energy of the reagents is converted, in part, to internal excitation of the HD product. The fraction converted is clearly less than implied by Eq. (1.5) ($\cos^2 \beta = \frac{1}{6}$), and we need to develop additional concepts before we can qualitatively understand this trend. The quantitative route is, however, possible. The forces between the atoms can be (and have been) determined by quantum chemists. Assuming the applicability of classical mechanics, all we need to do is solve Newton's equation of motion. When this is done, the results (as shown in the right panel of Fig. 1.4) are in good accord with the observations. But does quantum mechanics play no role other than determine the forces? And if we do know the forces, can we not draw better qualitative inferences before we embark on a full-scale computation? We clearly have some way to go.

---

† Changing the wavelength of the photolysis laser will change the kinetic energy of the H atoms, determined by the photon energy minus the dissociation energy of HI. Since H is so much lighter than I, its kinetic energy is practically all the excess energy.

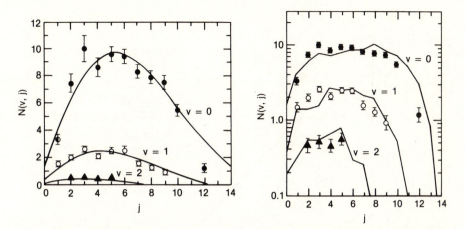

Fig. 1.4. Rotational and vibrational state population distributions for the nascent HD product of the reaction $H + D_2$ at a collision energy $E_T = 1.3$ eV. Left: Plot of experimental results; the solid line is the so-called "linear surprisal fit" to the data (see Sec. 5.5). Right: Semilog plot of same data compared with quasi-classical trajectory calculated results. [Adapted from D. P. Gerrity and J. J. Valentini, *J. Chem. Phys., 81,* 1298 (1984); *82,* 1323 (1985); quasi-classical trajectory results of N. C. Blais and D. J. Truhlar, *Chem. Phys. Lett., 102,* 120 (1983); see also E. E. Marinero, C. T. Rettner, and R. N. Zare, *J. Chem. Phys., 80,* 4142 (1984).]

Earlier we stressed the deviance from equilibrium characteristic of dynamical processes. Implicit in the fits to the data in the left-hand panel of Fig. 1.4 is a quantitative measure of such deviance for the $H + D_2$ reaction. It is based on the notion that at equilibrium, at a sharp value of the total energy, all quantum states of the system are equally probable. This is *not* the case for the results shown in Fig. 1.4 or for many other systems.

Taking advantage of their high frequency resolution, lasers can selectively pump and probe polyatomic molecules, providing a better understanding of intramolecular dynamics. Even such a simple problem as the relaxation of a vibrotationally excited aromatic molecule by a He atom (Fig. 1.5) turns out to be quite complex under the high resolution provided by the laser.

Our ability to develop simple models and to predict trends is, of course, based on familiarity with a large body of experimental results and more elaborate theoretical developments. The aim of the present book is to provide an introduction to the necessary background to be able to understand key results in the field of molecular dynamics. Let us now consider what it is that we *need to know*.

1.5 The need to know

We have already discussed the subject of population inversion in elementary exoergic reactions, more specifically the relative probabilities of production of different vibrational states of HCl in the Cl + HI reaction. This is a fairly

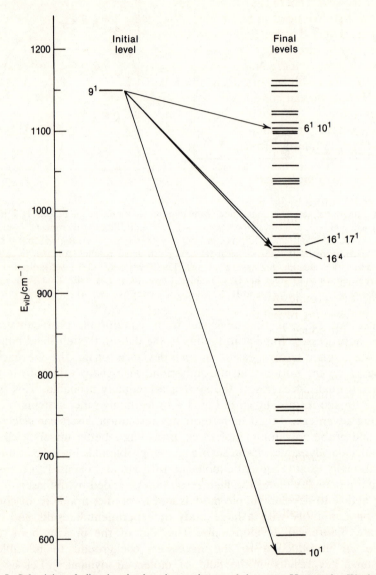

Fig. 1.5. Selectivity of vibrational relaxation pathways. A benzene–He van der Waals bound adduct (see Sec. 7.4) is laser excited to the 9^1 vibrational level. The vibrational redistribution occurs during the photodissociation of the complex and is monitored by analysis of the dispersed fluorescence spectrum. Only a small subset of the many energetically available states are populated. The vibrational quantum number assignment is given for those states that are significantly accessed. [Adapted from T. A. Stephenson and S. A. Rice, *J. Chem. Phys.*, *81*, 1083 (1984).]

sophisticated kind of inquiry, albeit of great theoretical and practical interest. However, there is a far more obvious question that really should be answered first. We need to know how efficient the *overall* reactive collision process is. Does the reaction occur every time Cl and HI collide or is reaction a very rare event? Perhaps in most collisions the reactants simply rebound without undergoing chemical reaction, as happens commonly in many reactions involving an activation barrier.

Even *before* we can consider this question we need to know how often a collision (reactive or nonreactive, as the case may be) between two molecules takes place, and, more important, what are the intrinsic and extrinsic factors that determine the dynamics of a collision.

1.5.1 *The size of molecules*

We must therefore begin with a clarification of the concept of the "size" of molecules. We will introduce a quantitative measure for size, namely, the *collision cross section,* and then find that it is actually *energy dependent.* We trace its connection to the distance-dependent forces between the colliding molecules. We seek out some general dynamical principles that are applicable to all types of molecular collisions. These will provide us with sufficient background to understand how experimental observations of *molecular scattering* reveal the *collision dynamics.*

We next introduce the concept of the *reaction probability* and then the *reaction cross section.* We will learn how to characterize the *size* of reacting pairs of molecules with respect to their propensity to undergo chemical change. We will see why this size is large or small, that is, the factors governing the magnitude of the reaction cross section. Chapter 2 concludes with a discussion of the energy dependence of the cross section for reactions with and without a "barrier."

1.5.2 *Angular distribution of elastic and reactive scattering*

In Chapter 3 we probe deeper into the dynamics and consider the *angular distribution* of the scattering. First we establish the role of the potential in determining the so-called *differential elastic scattering cross section* and then see what we can learn about the intermolecular force field from detailed elastic scattering experiments. We shall find that classical mechanics is not always adequate and that we require a quantal approach to interpret such experiments. As an example of the results of such studies we present an overview of the main features of *interatomic forces,* as determined from both *experiment and theory.*

Now we are ready to get down to business. We look at the *angular distribution of the products of reactive collisions* and learn how to tell whether a reaction occurs by a *direct* route or via a *long-lived complex.* Within the family of direct reactions we compare two antithetical examples; the *forward and rebound* modes of reactive scattering.

As an example of the microscopic details becoming available, we shall consider the steric effect in the reaction of methyl iodide with alkali atoms. Direct colliding beam experiments with *oriented* methyl iodide molecules have shown that the reaction probability for the "unfavorable" configuration $Rb + H_3CI$ is negligible compared to that for the "favorable" orientation $Rb + ICH_3$. Thus there is a "cone of acceptance" around the axis of the CH_3I molecule such that a reaction will occur only if the incident Rb atom strikes within that cone.

We shall see many other illustrations of the new microlevel experiments and their theoretical concomitants. The goal is to provide insight into the *intimate details* of the *elementary chemical act*.

1.5.3 Energy and chemical change

Chapter 4 introduces us to the many-body approach to reaction dynamics. We begin with more examples of the role of energy in chemical reactivity. To discuss the dynamics behind such specificity we start with the construct of the *potential energy* hypersurface that "interpolates" between reagents and products. The qualitative features of this surface are discussed, but to arrive at quantitative implications we consider the *classical trajectory* method for describing the course of a reactive collision. We are now in a position to appreciate the various heuristic models and approximate theories of the overall reaction rate.

The *transition state* approximation has played a significant role in the development of theories of reaction rates. We will derive this model and see its implications for macroscopic (bulk) kinetics. The central assumption of the approximation, known as the configuration of no return, will be discussed from a dynamical point of view. The two distinct modes in which this assumption is used for "direct" and for "compound" collisions will be discussed.

We shall not be content with considering the overall rate alone. "State-to-state" rate constants will be discussed in some detail, and we shall find the principle known as "detailed balance" to be quite useful in our considerations.

1.5.4 The practice of molecular reaction dynamics

Chapter 5 is a probe into the practice of our field: what it is that we wish to do, what we can do, and what the typical results are.

We turn first to the use of lasers as promoters and as detectors of chemical reactions, to chemical lasers and laser-induced chemistry, and to photofragmentation. Then we examine the molecular beam technique and the information it provides on the angular, velocity, and state distributions of the scattered products from reactive collisions.

We recall the old adage in chemical kinetics that a postulated mechanism can never be proved, only disproved. This is no longer true, now that we can make observations of individual, elementary reactive steps. Consider the formerly "well-known" bimolecular reaction $2HI \rightarrow H_2 + I_2$. Molecular beam data indicate that the direct reaction simply does not take place in a single collision. The colliding molecules simply rebound without any chemical change. A similar example of a four-center reaction is a mixture of Br_2 and Cl_2, where in the bulk the net process is partial formation of BrCl. Yet in molecular beams the reaction $Br_2 + Cl_2 \rightarrow 2BrCl$ does not take place. The remarkable observation is that through use of a beam of dimers $(Cl_2)_2$ (formed by expansion of Cl_2 in a supersonic jet), the single-collision process $(Cl_2)_2 + Br_2 \rightarrow 2BrCl + Cl_2$ readily occurs.

An analysis of the role of *energy* in promoting chemical change and the complementary problem of *energy disposal* (in exoergic reactions) and the quantum mechanical description of molecular collisions are the final subjects of the chapter.

1.5.5 Energy transfer collisions

By now, at the start of Chapter 6, we will have encountered considerable evidence, both experimental and theoretical, for the *enhanced reactivity* of excited-state reagents, as well as the *specificity of formation* of excited product species. In order to take practical advantage of such knowledge, we need to know something about the rate of *depletion of excited-state species*. How long do excited molecular states "survive" in the gas phase? Only after we know something about the *radiative and collisional lifetimes* can we consider such intriguing questions as how to *prepare and maintain* steady-state *population inversions* to provide a source of *energy-rich reagents,* or how to make use of the often *specific* pattern of *energy release* in exoergic reactions. The efficiency of any gas-phase laser is governed to a large extent by such relaxation processes.

One of the problems in the study of these molecular *energy transfer processes* is the wide range of rates involved. An important measure of understanding is achieved even when we can distinguish between the more efficient and less efficient energy transfer modes. Typical collisional relaxation rates for different gas-phase processes, even at room temperature, cover a range of more than nine orders of magnitude in rate constants. To understand the "propensity rules" for intermolecular energy transfer we shall deal both with simple models and with the fundamental dynamical theory. We shall also pay attention to the collisional transfer of electronic energy.

An important practical example of the relaxation process "at work" is the supersonic jet, that is, the substantial cooling of the internal degrees of freedom of molecules that can occur in an expanding jet. Supersonic beams of "cold" molecules can thereby be produced that can then be studied by laser spectroscopy, revealing much new structural and dynamical information.

Energy transfer in molecule–surface collisions and the energy distribution in molecules desorbed from surfaces are also discussed in Chapter 6. Laser probing of such molecules shows that here too equilibrium is not necessarily the case.

Spectroscopic manifestations of energy transfer collisions, laser-assisted collisions, and the spectroscopy of the very act of the collision conclude Chapter 6. Indeed, the "spectroscopy of the transition state" is likely to add significantly to our understanding of the dynamics.

1.5.6 Molecular reaction dynamics

Chapter 7 is devoted to a deep probe into the details of the dynamics of chemical change. As a *case study of a well-characterized elementary reaction* we examine all the available dynamical information on the "direct" $F + H_2$ reaction. Next we learn about the intimate details of the *formation and decay of long-lived complexes* from a theoretical as well as experimental viewpoint. We note the host of activation processes made possible by high-power lasers. These include such aspects as the direct overtone pumping of molecules to high vibrational states. At very high powers it is also possible for a molecule to absorb several successive photons within a very short time span. Such multiphoton processes are possible using both infrared and visible–ultraviolet photons.

Increasing the molecular complexity, we turn to the structure, reactivity, and dynamics of van der Waals bound molecules and clusters. The introduction of new techniques for the formation and observation of such cluster molecules has provided intermediate steps between elementary molecule–molecule processes, molecule–solvent, and molecule–surface interactions, to which we turn next. The ultimate aim here is to understand the molecular-level basis of catalysis. Many interesting and several unexpected details are revealed along the way. Finally we examine the current understanding of the role of the mutual orientation of the reagents on chemical reactivity and of the preferred relative spatial orientation of the separating products. We conclude with a look at our new frontiers.

Also discussed throughout this volume is the strong interplay of basic molecular theory with experiment at the molecular level. The *ab initio* quantum mechanical calculations of interatomic and intermolecular interactions are beginning to produce potential-energy surfaces of "chemical" accuracy. For the interaction of two (simple) atoms this means that we can now predict the observed *elastic scattering* behavior, that is, the angular distribution and energy dependence of the scattering. This implies that we can also account for the macroscopic transport properties and many other bulk observables for the system. For interacting diatomic or polyatomic molecules, accurate molecular quantum-mechanics computations are feasible, though costly, but various approximation schemes can provide useful estimates of potential energy surfaces. Given the surface, we invoke scattering theory (quantal when feasible, and classical or semiclassical approximations more commonly) and calculate the inelastic and/or reactive scattering cross sections and then state-to-state collisional relaxation rates, and/or detailed reaction rate coefficients to compare with experiment. In certain cases we must consider curve crossing and/or multiple potential surfaces to account for observations such as charge transfer, collisional ionization, unimolecular breakdown of excited molecules, and the quenching of fluorescence. Just as theory (both quantal and classical) plays a central role in the interpretation of molecular spectra in terms of structure, so it is that the newer developments in collision theory are intimately related to and interwoven with the new kinds of experimental observations that characterize the field of molecular dynamics.

In the chapters that follow we cannot do more than touch upon a few of the myriad of interesting and significant findings of recent research in molecular reaction dynamics. What we hope to do, however, is introduce the *language and spirit* of the subject to a wider audience.

Suggested reading

Review Articles

Bernstein, R. B., "Introduction to Atom–Molecule Collisions: The Interdependency of Theory and Experiment," in Bernstein (1979).

Kuppermann, A., and Greene, E. F., "Chemical Reaction Cross Sections and Rate Constants," *J. Chem. Educ.*, *45*, 361 (1968).

Lawrence, W. D., Moore, C. B., and Petek, H., "Understanding Molecular Dynamics Quantum-State by Quantum-State," *Science*, *227*, 895 (1985).

Lee, Y. T., and Shen, Y. R., "Studies with Crossed Lasers and Molecular Beams," *Phys. Today*, *33*, 52 (1980).

Leone, S. R., "Laser Probing of Chemical Reaction Dynamics," *Science*, *227*, 889 (1985).

Letokhov, V. S., "Laser-Induced Chemical Processes," *Phys. Today*, *33*, 34 (1980).

Levy, D. H., "The Spectroscopy of Supercooled Gases," *Sci. Am.*, *245*, 68 (1984).

Ronn, A. M., "Laser Chemistry," *Sci. Am.*, *240*, 103 (1979).

Zare, R. N., and Bernstein, R. B., "State-to-State Reaction Dynamics," *Phys. Today*, *33*, 43 (1980).

Zewail, A. H., "Laser-Selective Chemistry—Is It Possible?" *Phys. Today*, *33*, 27 (1980).

List of edited volumes of relevant review literature

Almoster, M. A. (ed.), *Ionic Processes in the Gas Phase*, Reidel, Boston, 1984.

Ausloos, P. (ed.), *Interaction Between Ions and Molecules*, Plenum, New York, 1975.

Ausloos, P. (ed.), *Kinetics of Ion–Molecule Reactions*, Plenum, New York, 1979.

Baer, M. (ed.), *The Theory of Chemical Reaction Dynamics*, CRC Press, Boca Raton, Fla. 1985.

Bamford, C. H., and Tipper, C. F. H. (eds.), *Chemical Kinetics*, Vol. 24, Elsevier, Amsterdam, 1983.

Bederson, B., and Fite, W. L. (eds.), *Methods of Experimental Physics*, Vol. 7, Academic, New York, 1968.

Benedek, G., and Valbusa, U. (eds.), *Dynamics of Gas–Surface Interaction*, Springer–Verlag, Berlin, 1982.

Bernstein, R. B. (ed.), *Atom–Molecule Collision Theory. A Guide for the Experimentalist*, Plenum, New York, 1979.

Birks, J. B. (ed.), *Organic Molecular Photophysics*, Vols. 1 and 2, Wiley, New York, 1973, 1975.

Birnbaum, G. (ed.), *Phenomena Induced by Intermolecular Interactions*, Plenum, New York, 1985.

Bowers, M. T. (ed.), *Gas Phase Ion Chemistry*, Vols. 1–3, Academic, New York, 1979–1984.

Bowman, J. M. (ed.), *Molecular Collision Dynamics*, Springer-Verlag, Berlin, 1983.

Brooks, P. R., and Hayes, E. F. (eds.), *State-to-State Chemistry*, American Chemical Society, Washington, D.C., 1977.

Clary, D. C. (ed.), *The Theory of Chemical Reaction Dynamics*, Reidel, Boston, 1986.

Crosley, D. R. (ed.), *Laser Probes for Combustion Chemistry*, American Chemical Society, Washington, D.C., 1980.

Davidovits, P., and McFadden, D. L. (eds.), *The Alkali Halide Vapors*, Academic, New York, 1979.

Ehlotzky, F. (ed.), *Fundamentals of Laser Interactions*, Springer-Verlag, Berlin, 1985.

Eichler, J., Hertel, I.V., and Stolterfot, N. (eds.), *Electronic and Atomic Collisions*, North-Holland, Amsterdam, 1984.

El-Sayed, M. A. (ed.), *J. B. Fenn-Dedicated Issue, J. Phys. Chem., 88*, No. 20 (1984).

El-Sayed, M. A. (ed.), *Chemical Kinetics Issue, J. Phys. Chem., 90*, No. 3 (1986).

Eyring, H., and Henderson, D. (eds.), *Theoretical Chemistry: Advances and Perspectives*, Academic, New York, 1981.

Eyring, H., Jost, W., and Henderson, D. (eds.), *Physical Chemistry—An Advanced Treatise*, Vol. 6, Academic, New York, 1974–1975.

Faraday Discussions of the Chemical Society, 55, Molecular Beam Scattering, (1973).

Faraday Discussions of the Chemical Society, 62, Potential Energy Surfaces, (1977).

Faraday Discussions of the Chemical Society, 72, Selectivity in Heterogeneous Catalysis (1981).

Faraday Discussions of the Chemical Society, 73, van der Waals Molecules (1982).

Faraday Discussions of the Chemical Society, 75, Intramolecular Kinetics (1983).

Ferreira, M. A. A. (ed.), *Ionic Processes in the Gas Phase*, Reidel, Dordrecht, 1982.

Fontijn, A. (ed.), *Gas-Phase Chemiluminescence and Chemi-Ionization*, North-Holland, Amsterdam, 1985.

Fontijn, A., and Clyne, M. A. A. (eds.), *Reactions of Small Transient Species*, Wiley, New York, 1983.

Gardiner, W. C., Jr. (ed.), *Combustion Chemistry,* Springer-Verlag, New York, 1984.

George, T. F. (ed.), *Theoretical Aspects of Laser Radiation and Its Interaction with Atomic and Molecular Systems,* University of Rochester Press, Rochester, N.Y., 1978.

Gerber, R. B., and Nitzan, A. (eds.), *Dynamics of Molecule–Surface Interactions, Isr. J. Chem, 22,* No. 4 (1982).

Gianturco, F. A. (ed.), *Atomic and Molecular Collision Theory,* Plenum, New York, 1982.

Glorieux, P., Lecler, D., and Vetter, R. (eds.), *Chemical Photophysics,* Editions du CNRS, Paris, 1979.

Gole, J., and Stwalley, W. C. (eds.), *Metal Bonding and Interactions in High Temperature Systems,* American Chemical Society, Washington, D.C., 1982.

Gomer, R. (ed.), *Interactions on Metal Surfaces,* Springer-Verlag, Heidelberg, 1975.

Gross, R. W. F., and Bott, J. L. (eds.), *Handbook of Chemical Lasers,* Wiley, New York, 1976.

Haas, Y. (ed.), *Multiphoton Excitation, Isr. J. Chem., 24,* No. 3 (1984).

Hall, R. B., and Ellis, A. B. (eds.), *Chemistry and Structure at Interfaces: New Laser and Optical Techniques,* VCH Publishers, Deerfield Beach, FL, 1986.

Harper, P. G., and Wherret, B. S. (eds.), *Nonlinear Optics,* Academic, New York, 1977.

Henderson, D. (ed.), *Theoretical Chemistry: Theory of Scattering,* Academic, New York, 1981.

Hinkley, E. D. (ed.), *Laser Monitoring of the Atmosphere,* Springer-Verlag, Berlin, 1976.

Hinze, J. (ed.), *Energy Storage and Redistribution in Molecules,* Plenum, New York, 1983.

Hirschfelder, J. O. (ed.), *Intermolecular Forces, Adv. Chem. Phys., 12* (1967).

Hochstim, A. R. (ed.), *Kinetic Processes in Gases and Plasmas,* Academic, New York, 1969.

Jackson, W. M., and Harvey, A. B. (eds.), *Lasers as Reactants and Probes in Chemistry,* Howard University Press, Washington, D.C., 1985.

Jortner, J., and Pullman, B. (eds.), *Intramolecular Dynamics,* Reidel, Dordrecht, 1982.

Jortner, J., Levine, R. D., and Rice, S. A. (eds.), *Photoselective Chemistry, Adv. Chem. Phys., 47* (1981), Parts 1 and 2.

King, D. A., and Woodruff, D. P. (eds.), *The Chemical Physics of Solid Surfaces and Heterogeneous Catalysis. Adsorption at Solid Surfaces,* Vol. 2, Elsevier, Amsterdam, 1983.

Kleinpoppen, H., and Williams, J. F. (eds.), *Coherence and Correlation in Atomic Collisions,* Plenum, New York, 1980.

Kleinpoppen, H., Briggs, J. S., and Lutz, H. O. (eds.), *Fundamental Processes in Atomic Collision Physics,* Plenum, New York, 1985.

Kliger, D. S. (ed.) *Ultrasensitive Laser Spectroscopy,* Academic, New York, 1983.

Kompa, K. L., and Smith, S. D. (eds.), *Laser-Induced Processes in Molecules,* Springer-Verlag, Berlin, 1979.

Kompa, K. L., and Walter, H. (eds.), *High-Power Lasers and Applications,* Springer-Verlag, Berlin, 1979.

Lahmani, F. (ed.), *Photophysics and Photochemistry Above 6 eV,* Elsevier, Amsterdam, 1985.

Laubereau, A., and Stockburger, M. (eds.), *Time-Resolved Vibrational Spectroscopy,* Springer-Verlag, Berlin, 1985.

Lawley, K. P. (ed.), *Molecular Beam Scattering, Adv. Chem. Phys., 30* (1975).

Lawley, K. P. (ed.), *Photodissociation and Photoionization, Adv. Chem. Phys., 60* (1985).

Leach, S. (ed.), *Molecular Ion Studies, J. Chim. Phys., 77,* 7/8 (1980).

Lester, W. A., Jr. (ed.), *Potential Energy Surfaces in Chemistry,* IBM, San Jose, Calif., 1970.

Letokhov, V. S., and Ustinov, N. D. (eds.), *Power Lasers and Their Applications,* Harwood, New York, 1983.

Levine, R. D., and Jortner, J. (eds.), *Molecular Energy Transfer,* Wiley, New York, 1976.

Levine, R. D., and Tribus, M. (eds.), *Maximum Entropy Formalism,* MIT Press, Cambridge, Mass., 1979.

McGowan, J. W. (ed.), *The Excited State in Chemical Physics,* Parts 1 and 2, Wiley, New York, 1975, 1981.

Miller, W. H. (ed.), *Dynamics of Molecular Collisions,* Plenum, New York, 1976.

Mooradian, A., Jaeger, T., and Stockseth, P. (eds.), *Tunable Lasers and Applications,* Springer-Verlag, Berlin, 1976.

Moore, C. B. (ed.), *Chemical and Biochemical Applications of Lasers,* Vols. 1–5, Academic, New York, 1974–1980.

Pao, Y. H. (ed.), *Optoacoustic Spectroscopy and Detection,* Academic, New York, 1977.

Picosecond Phenomena, Vol. I, Shank, C. W., Ippen E. P., and Shapiro, S. L. (eds.) (1978); Vol. II, Hochstrasser, R. M., Kaiser, W., and Shank, C. V. (eds.) (1980); Vol. III, Eisenthal, K. B., Hochstrasser, R. M., Kaiser, W., and Laubereau, A. (eds.) (1982); Vol. IV, Auston, D. H., and Eisenthal, K. B. (eds.), Springer-Verlag, Berlin (1984).

Pimentel, G. C. (ed.), *Opportunities in Chemistry,* National Academy Press, Washington, D.C., 1985.

Pullman, B. (ed.), *Intermolecular Interactions from Diatomics to Biopolymers,* Wiley, New York, 1978.

Pullman, B. (ed.), *Intermolecular Forces,* Reidel, Boston, 1981.

Pullman, B., Jortner, J., Nitzan, A., and Gerber, R. B. (eds.), *Dynamics on Surfaces,* Reidel, Boston, 1984.

Radziemski, L. J., Solarz, R. W., and Paisner, J. A. (eds.), *Applications of Laser Spectroscopy,* Dekker, New York, 1986.

Rahman, N. K., and Guidotti, C. (eds.), *Collisions and Half-Collisions with Lasers,* Harwood, Utrecht, 1984.

Rahman, N. K., and Guidotti, C. (eds.), *Photon-Assisted Collisions and Related Topics,* Harwood, New York, 1984.

Rhodes, C. K. (ed.), *Excimer Lasers,* 2nd ed., Springer-Verlag, Berlin, 1984.

Rhodin, T. N., and Ertl G. (eds.), *The Nature of the Surface Chemical Bond,* North-Holland, Amsterdam, 1979.

Ross, J. (ed.), *Molecular Beams, Adv. Chem. Phys., 10* (1966).

Schlier, Ch. (ed.), *Molecular Beams and Reaction Kinetics,* Academic, New York, 1970.

Scoles, G., and Buck, U. (eds.), *Atomic and Molecular Beam Methods,* Oxford University Press, New York, 1986.

Setser, D. W. (ed.), *Gas Phase Intermediates, Generation and Detection,* Academic, New York, 1980.

Shen, Y. R. (ed.), *Nonlinear Infrared Generation,* Springer-Verlag, Berlin, 1977.

Smith, I. W. M. (ed.), *Physical Chemistry of Fast Reactions: Reaction Dynamics,* Plenum, New York, 1980.

Steinfeld, J. I. (ed.), *Electronic Transition Lasers,* MIT Press, Cambridge, Mass., 1976.

Steinfeld, J. I. (ed.), *Laser and Coherence Spectroscopy,* Plenum, New York, 1981a.

Steinfeld, J. I. (ed.), *Laser-Induced Chemical Processes,* Plenum, New York, 1981b.

Truhlar, D. G. (ed.), *Potential Energy Surfaces and Dynamics Calculations,* Plenum, New York, 1981.

Truhlar, D. G. (ed.), *Resonances in Electron Molecule Scattering, van der Waals Complexes, and Reactive Chemical Dynamics,* ACS Symp. Ser. No. 263 (1984).

Vetter, R., and Vigué, J. (eds.), *Recent Advances in Molecular Reaction Dynamics,* Editions CNRS, Paris, 1986.

Walther, H. (ed.), *Laser Spectroscopy of Atoms and Molecules,* Springer-Verlag, Berlin, 1976.

Wilson, L. E., Suchard, S. N., and Steinfeld, J. I. (eds.), *Electronic Transition Lasers II,* MIT Press, Cambridge, Mass., 1977.

Woolley, R. G. (ed.), *Quantum Dynamics of Molecules,* Plenum, New York, 1980.

Zewail, A. H. (ed.), *Advances in Laser Chemistry,* Springer-Verlag, Berlin, 1978.

Zewail, A. H. (ed.), *Photochemistry and Photobiology, Proceedings of the International Conference, Alexandria, Egypt,* Vols. 1 and 2, Harwood, Schur, Switzerland, 1983.

2
Molecular collisions

A bimolecular chemical reaction involves an exchange or rearrangement of the atoms of the reactants. The prerequisite for reaction is of course the encounter itself (i.e., the collision) of the two reactants. Thus we first need to understand the factors determining the relative (*inter*molecular) motion of the reactants. Only then can we begin to deal with the dynamics of the *intra*molecular motions that take place during the reactive collision. Our first task is to understand the *dynamics* of simple molecular collisions. Then we will be in a position to inquire into the *conditions* that must obtain so that a *collision can lead to reaction*.

In Section 2.3 we shall get down to the business of chemical reaction dynamics. (This is not to say that skipping Secs. 2.1 and 2.2 is recommended. Chemical reactions *also* require overcoming an activation barrier!) The concepts and quantitative machinery that we shall set in motion in these somewhat technical sections will stand us in good stead throughout our entire journey. In reading Sections 2.1 and 2.2 one should pay attention to the concepts. Neither the (simple) derivations nor the precise form of the quantitative results need to be mastered, but an understanding of the physical implications is required. In later chapters we shall find that the very same ideas enable us to understand other aspects of molecular dynamics, such as the way collisions sample the intermolecular force or the propensities for different types of energy transfer collisions.

2.1 Molecular collisions and free-path phenomena

2.1.1 The concept of molecular size

We recall that many of the properties of low-density gases can be readily understood in terms of the simplest molecular model of *noninteracting, point-mass molecules*. These include mainly *equilibrium* properties, including the ideal-gas equation of state and the specific heat of atomic gases, as well as the distribution of molecular velocities and gaseous effusion. The laws of classical mechanics were successfully applied a century ago to such problems, involving "structureless" particles. To account for the equilibrium specific heat of *molecular* gases one must also take cognizance of their "internal" structure, but the influence of the *size* (and shape) of the molecules per se becomes important when we try to explain the properties of gases in *disequilibrium.*

As soon as we perturb the spatial homogeneity characteristic of a gas at equilibrium we induce a response in the gas that tends to remove the disturbance. The mechanism of this *relaxation* is the ever-present process of *molecular collisions.*

2.1.2 Transport phenomena

Consider a temperature gradient imposed upon a gas by means of a pair of opposing plates at different temperatures. By collisional transfer, *energy* will be *transported* from the hot plate, through the gas, and deposited at the cold plate, tending to equalize their temperature and thus wipe out the temperature gradient. Similarly, a pressure gradient impressed upon a gas tends to be dissipated by the resulting flow of the gas, which involves *momentum*-transferring molecular collisions. In general, the response of the gas is in the direction of restoring equilibrium. The observable rate of this response is governed by the rate of intermolecular collisions, and therefore by the molecular size. This is because the molecular size determines the mean free path λ, or the average distance traveled by a molecule between successive collisions. Measurements of the transport coefficients therefore provide a route to the characterization of the size of molecules. However, rather than follow this route we shall consider a direct experimental determination of the mean free path by the use of *molecular beams.*

2.1.3 Direct determination of the mean free path: A simple scattering experiment

The essential ingredients in such an experiment are a collimated beam of molecules (of type A) directed along the x axis, which passes through a *scattering chamber* of length l that contains gas (of type B) consisting of the so-called target molecules. The intensity of the beam is thereby reduced as some of the A molecules are deflected out of the beam due to scattering (collisions) by the B molecules. By means of an appropriate molecular detector we measure the fraction of A molecules transmitted as a function of the number density n_B of the target gas. To determine the mean free path we argue as follows: Let $I(x)$ be the *flux* of A molecules in the beam. The flux is defined as the number of molecules crossing a unit area (perpendicular to the direction of the beam) per unit time. Note that in terms of the velocity v of the beam molecules,

$$I(x) = v n_A(x) \tag{2.1}$$

where $n_A(x)$ is the number density of A molecules in the beam at position x.

Owing to collisions with molecules of the target gas, A molecules are deflected out of the beam and so the beam flux along the x direction decreases down the length of the scattering cell.

The mean distance λ that a beam molecule travels in the target gas before it is deflected out of the beam by collision is related to the fractional loss in beam intensity by

$$-\frac{\Delta I}{I} = \frac{I(x) - I(x + \Delta x)}{I(x)} \approx \frac{\Delta x}{\lambda} \tag{2.2}$$

that is, the fractional decrease in beam flux, for sufficiently small Δx, is just $\Delta x/\lambda$. In this limit Eq. (2.2) yields

$$-\frac{dI}{I\,dx} = -\frac{d\ln I}{dx} = \lambda^{-1} \tag{2.3}$$

Integrating Eq. (2.3) between 0 and the length l gives the expected result

$$I(l) = I(0)\exp(-l/\lambda) \tag{2.4}$$

where $I(0)$ is the initial flux of the beam at $x = 0$ and $I(l)$ is the flux going out of the scattering chamber.

The beam flux is thus an exponentially decreasing function of the length of the scattering path, a result similar to that for a beam of light attenuated by absorption (where it is known as the Beer–Lambert law).

Figure 2.1 shows results obtained for a beam of CsCl scattered by two target gases, Ar and CH_2F_2. Plotted is the (logarithm of the) fractional transmission $I(l)/I(0)$ versus the scattering gas pressure, showing that the mean free path is an inverse function of the target gas pressure, $\lambda \propto P^{-1}$.

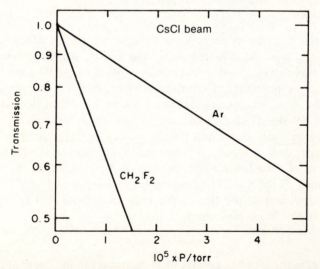

Fig. 2.1. Attenuation data for the scattering of a thermal beam of CsCl ($T \approx 1100$ K) by Ar atoms and by the polar molecule CH_2F_2 (both at 300 K) in a 44-mm cell. The logarithm of the transmission decreases linearly with target gas pressure P (and thus n_B). [Adapted from H. Schumacher, R. B. Bernstein, and E. W. Rothe, *J. Chem. Phys.*, *33*, 584 (1960).]

This is just what we should expect if the beam flux is attenuated by scattering collisions with target gas molecules.

Inspection of Fig. 2.1 shows that at each density of the target gas, the attenuation of the CsCl beam is higher (and hence the mean free path is shorter) for the polar CH_2F_2 molecules than for Ar. It is also evident from Fig. 2.1 that the mean free path declines with increasing density of the target gas. In Section 2.1.4 we factor out this density dependence to arrive at a truly "molecular" characterization of the size: the cross section for A–B collisions.

2.1.4 The collision cross section

The probability of a beam molecule undergoing a collision in the interval x to $x + \Delta x$ is given by $\Delta x / \lambda$; it should be proportional to the number density n_B ($n_B = P_B / kT$) of the target gas. Thus $\Delta x / \lambda \propto n_B \, \Delta x$. We define the *collision cross section* σ as the proportionality constant such that $1/\lambda = \sigma n_B$, that is,

$$\lambda = (n_B \sigma)^{-1} \tag{2.5}$$

With Eq. (2.5) in Eq. (2.3), Eq. (2.4) becomes

$$I(l) = I(0) \exp(-n_B l \sigma) \tag{2.6}$$

The collision cross section is a measure of the "size" of the colliding molecules. The *larger the cross section, the smaller the mean free path* and hence the *more frequent the molecular collisions*.

From the attenuation data one can immediately obtain the effective† collision cross section:

$$\sigma = (n_B l)^{-1} \ln \left[\frac{I(0)}{I(l)} \right] \propto -\frac{d \ln I(l)}{dP} \tag{2.7}$$

where the pressure P in the scattering cell ($P = nkT$ is a measure of the number density of the scattering B molecules). Note from Fig. 2.1 that the slope is steeper for the CH_2F_2 target molecules than for Ar atoms, so the cross section for the collision of CsCl with CH_2F_2 is appreciably greater than for Ar. In Section 3.2.4 we shall attribute this to the stronger long-range forces between the two polar molecules.

With the molecular beam technique it is also possible to determine a possible *velocity dependence* of the collision cross section.‡ Since molecules

---

† The "practical" cross section is determined by the minimal angle of deflection that can be experimentally resolved.

‡ One way to accomplish the velocity measurement is by the *time-of-flight* method. Here the beam is pulsed by a mechanical shutter before entering the scattering cell. (The switching on of the beam also triggers the time base for detection.) The beam intensity at the detector at a later time t is due to molecules of speed $v = l/t$, where l is the path length from the beam shutter to the detector. Comparing the transmitted intensity (for a given speed v) in the absence and presence of the target gas in the cell, one can determine the dependence of the attenuation factor upon velocity. This tells us how the collision cross section depends upon the collision energy.

are *not* hard spheres, one generally observes a decrease in σ with increasing collision energy (see Sec. 2.1.7).

2.1.5 The collision frequency

Before we close this subject it is instructive to note a few simple relationships between the mean free path, the collision frequency, and the binary rate constant for molecular collision.

For a beam molecule of speed v traversing an average distance λ between collisions, the number of collisions ω per unit time is, on the average,

$$\omega = v/\lambda = v n_B \sigma \qquad (2.8)$$

making use of Eq. (2.5). The quantity ω is known as the *collision frequency*.

Another quantity of interest is Z, the rate of bimolecular collisions per unit volume. The quantity Z is a bimolecular rate coefficient (with the same units, e.g., $1\,\text{mol}^{-1}\,\text{s}^{-1}$, as the familiar second-order rate coefficient, except that it refers to *all* A–B collisions and not only to those that lead to reaction). Since a given molecule A in a gas B of number density n_B undergoes ω collisions per unit time,

$$Z = v\sigma \qquad (2.9)$$

Note that $Z = Z(v)$ is not quite a conventional bimolecular rate constant. It refers to collisions for which the relative velocity (here the beam velocity) is well defined in both magnitude and direction. The usual "thermal" rate coefficient, say, $Z(T)$, is an average of $Z(v)$ over the relative velocity distribution for the gases at thermal equilibrium,

$$Z(T) = \langle Z(v) \rangle = \langle v\sigma \rangle \qquad (2.10)$$

Here the brackets denote an average over the velocity distribution in a gas at a temperature T.

Already at this early point we wish to comment that a bimolecular reaction rate coefficient $k(T)$ will also be given by the same type of expression,

$$k(T) = \langle v\sigma_R \rangle \qquad (2.11)$$

Here, however, σ_R is the reaction cross section in A–B collisions. Since at most all A–B collisions can lead to reaction, $\sigma_R \leq \sigma$; thus $k(T)$ is bounded by $Z(T)$. Because of the extensive averaging hidden by the "innocent" brackets in Eq. (2.10) or (2.11), it is difficult to glean much information about $\sigma_R(v)$ from $k(T)$.

Having established the *method* for direct *measurement* of λ (and thus σ), we shall try to determine the physical interpretation of the collision cross section and of its velocity dependence.

2.1.6 Molecules as rigid spheres

To interpret the physical significance of the cross section we begin with an idealized model of the molecules as rigid spheres. When such a molecule of type A approaches one of type B there is assumed to be no interaction until their relative separation decreases to some definite value, say, d, which we can interpret to be the sum of the radii of the two spheres. In other words, collision occurs whenever the center of one molecule approaches to within an "excluded volume" sphere of radius d about the second molecule (Fig. 2.2).

Imagine that this sphere is centered around each beam molecule. Corresponding to this sphere is a circle of radius d in the plane perpendicular to the beam velocity. Thus the beam molecule sweeps out a cylinder of volume $\pi d^2 \Delta x$ as it traverses a distance Δx through the target gas. If the center of a target molecule lies within that volume, a collision will occur and the beam molecule will be deflected off the x axis and therefore lost to the detector.

The probability that a beam molecule will suffer a collision during the

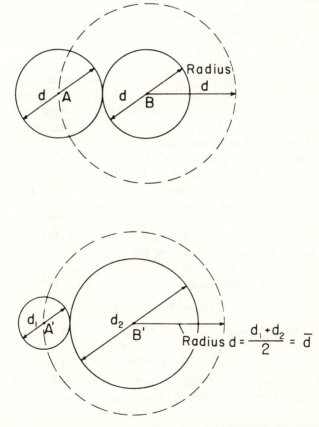

Fig. 2.2 Collision of rigid spheres. Top: Spheres of equal diameter d centered at A and B. Bottom: The dashed "equivalent" sphere is of radius $\bar{d} = \frac{1}{2}(d_1 + d_2)$ equal to d in the case above.

traverse of Δx [i.e., $\Delta x/\lambda$; see Eq. (2.2)] is simply the probability of finding a target molecule in the volume $\pi d^2 \, \Delta x$, or

$$\Delta x/\lambda = n_B(\pi d^2 \, \Delta x) \tag{2.12}$$

where n_B is the number density of the target gas, as before.

Using Eq. (2.5), we obtain for the hard-sphere model (only)

$$\sigma = \pi d^2 \tag{2.13}$$

so the collision cross section is just the area of a beam molecule as seen by a target molecule (and vice versa).

Chemists have long used a refined version of the simple hard-sphere model in which each atom in the molecule is represented as a (hard) sphere with a radius d consistent with its size as revealed by a variety of transport phenomena and deviations from ideal-gas behavior. Such models emphasize an important aspect still missing in our simple description: The inter*molecular* force will depend not only on the separation but also on the mutual orientation of the molecules. This orientation dependence will lead (see Secs. 2.4.4 and 4.2.6) to the concept of a *steric factor* in reactive molecular collisions.

2.1.7 Realistic intermolecular potentials

Of course real molecules are not rigid spheres, yet we can interpret their collision cross sections in the same way: The cross section σ is simply the *effective* (or equivalent) area, such that $(n\sigma)^{-1}$ is equal to the experimentally determined mean free path λ, that is, Eq. (2.5). To make any use of the measurements, however, we must learn how this effective *cross section* tells us about the real *intermolecular potential*. We shall not be fully able to do so until Chapter 3. Here we will summarize the most relevant findings.

First we note that the short-range repulsive force $F(R)$ between real molecules is not "infinitely hard," that is, the potential $V(R)$ is not a vertical "wall" at some critical value of the center-to-center separation R of the colliding particles (e.g., $R = d$), rather, there is an essentially *exponential* repulsion of the form

$$V(R) = A \exp(-R/\rho) \tag{2.14}$$

so that the force is given by

$$F(R) = -(A/\rho) \exp(-R/\rho) \tag{2.15}$$

where ρ is a characteristic length or *range parameter* (typically $0.1 \le \rho \le 0.4$ Å for simple molecules) and A is a *strength parameter* (with units of energy).

The functional form of the repulsive forces is deduced from an analysis of *high-velocity* beam-scattering measurements that show that the *cross section decreases* (in a predictable way; see Chap. 3) with *increasing relative velocity*

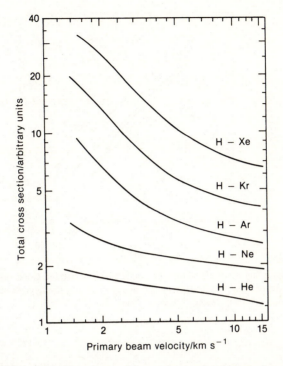

Fig. 2.3 Typical high-velocity dependence of the total scattering cross section. At high collision energies, the scattering is sensitive mainly to the repulsive part of the potential. Shown here is a log–log plot of $\sigma(v)$ for the elastic scattering of H by rare gases. [Adapted from R. W. Bickes, B. Lantsch, J. P. Toennies, and K. Walaschewski, *Faraday Discuss. Chem. Soc. 55,* 167 (1973).]

and thus collision energy (Fig. 2.3). As seen in Fig. 2.4, the exponential form of the repulsive potential is found also for ion–atom systems.

Even more interesting is the evidence from collisional studies at *low* (thermal-range) *energies*. These collisions are sensitive mainly to the *long-range* part of the intermolecular potential (see Sec. 3.1.4). It will be seen in Section 3.1.8 that the overall energy dependence of the cross section demonstrates that a realistic intermolecular potential (Fig. 2.5) has an attractive, long-range contribution usually of the form

$$V(R) \approx -C_6/R^6 \tag{2.16}$$

The R^{-6} dependence is typical of nonpolar closed-shell molecules. Longer-range forces (discussed in detail in Sec. 3.2.4) are involved in the attraction of, for example, polar molecules. For the interaction of ions with molecules the long-range potential is $-C_4/R^4$.

The presence of an attractive "well" in the potential is necessary if matter is to form a condensed phase. This potential *well* is always present,

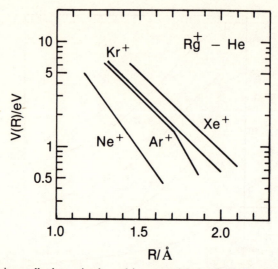

Fig. 2.4. Experimentally determined repulsive potentials for the indicated rare gas ions (Rg$^+$) with He, obtained from the energy dependence of ion beam scattering cross sections. [Adapted from H. Inouye and K. Tanji-Noda, *J. Chem. Phys.*, *77*, 5990 (1982).]

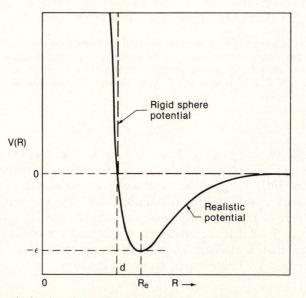

Fig. 2.5. Schematic drawing of a realistic intermolecular potential (solid curve). The minimum of the attractive well is located at R_e; $V(R_e) = -\epsilon$. For comparison, the potential for the rigid-sphere model (radius d) is shown by a dashed line.

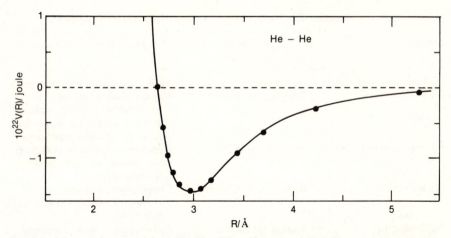

Fig. 2.6 Interatomic potential of He–He. The solid curve represents experimental data; points are theoretical. [Adapted from A. L. Burgmans, J. M. Farrar, and Y. T. Lee, *J. Chem. Phys.*, *64*, 1345 (1976); unpublished computational results by B. Liu and A. D. McLean.] See Chapter 3 for the relation between the potential and scattering data.

even in the absence of chemical binding forces (see Fig. 2.6).† It results from the balance between the ever-present (see Sec. 3.2.3) *long-range attraction* and *short-range repulsion*.

2.1.8 The concept of molecular size

The evidence from collisional studies corroborates that found from studies of deviations from ideal-gas behavior and of transport phenomena.

A particular example is the second virial coefficient $B(T)$ (in units of molar volume), which has the physical interpretation of an average of the fractional difference between the actual volume per mole of molecules (around some reference molecule) and that corresponding to a uniform distribution. For hard spheres, consideration of the "excluded volume" yields $B(T) = \frac{2}{3}N_A\pi d^3 > 0$, where N_A is Avogadro's number. Experimentally $B(T)$ does tend to a positive T-independent asymptote at high temperature, but at low temperature its value is often negative owing to a net overall weak *attraction* between molecules. Other "macro" properties, such as those characterized by the transport coefficients, are also sensitive to the intermolecular forces and have therefore been used to help establish the potentials.

How *sensitive* are the different macrophenomena to the degree of detail of our molecular model? In Table 2.1 we list the type of macroscopic phenomenon that can be accounted for at different levels of description of

† For He–He the depth of the attractive well is only 0.9 meV; yet it is known with an accuracy of better than 5%.

Table 2.1. Concept of the molecule and the phenomena thereby explainable

Concept	Type of interaction explainable	Macroscopic phenomena
(a) Point mass particles	No interaction	Ideal-gas behavior, effusion, etc.
(b) Rigid spheres	Repulsive, on contact	Van der Waals *b* constant (excluded volume); approximate transport coefficients of *spherical* molecules at high temperatures
(c) Rigid convex bodies	Repulsive, as in (b)	Approximate transport coefficients of *aspherical* molecules at high temperatures
(d) Soft spheres	Repulsive, compressible	Temperature dependence of transport coefficients of *spherical* molecules
(e) Same as (d)	Same as (d) but also long-range attraction	Full equation of state of *spherical* molecules; condensation of gases
(f) Soft "ovaloids"	Same as (e) but orientation-dependent (as well as distance-dependent) forces	"Macro" behavior of real fluids; equilibrium and transport properties of anisotropic molecules

the molecules and their mutual interactions. We see that the concept of an (orientation-dependent) intermolecular potential with short-range repulsion and long-range attraction is adequate to describe all the macrophenomena (at least for ground-state molecules). But, unfortuantely, we cannot *determine* the potential uniquely from such properties. The "leverage" is not as strong as we would like. We therefore turn to the *micro,* or collisional, approach, that is, the *direct measurement* of the angular distribution of the *scattering.* Such experiments, carried out with molecular beams, have proven to be the most sensitive means of determining intermolecular forces.

But what is the route from the observed scattering cross sections to the potential? Before we can answer this question we will first have to take a detailed look at the *role* of the *potential* in determining the *dynamics of molecular collisions.*

2.2 Dynamics of elastic molecular collisions

2.2.1 Elastic collisions of "structureless particles"

We cannot hope to deal with reactive (or even energy-transferring, inelastic) collisions until we learn how to deal with the simplest category of molecular encounters: the *elastic collision of structureless particles.* Here the only outcome of the collision is that the particles deflect one another from their original straight-line trajectories.

At low energies, collisions between rare-gas atoms are elastic, but for molecules (which possess internal structure), the collisions may also induce changes in internal energy or even bring about chemical rearrangement. Thermal unimolecular reactions are a familiar example where the energy required for isomerization or dissociation is provided by bimolecular, inelastic collisions. Even at lower temperatures inelastic collisions are quite important. This is particularly so for larger polyatomic molecules, with their closely spaced internal energy levels (see Fig. 1.5). One result of these energy transfer collisions is that the room-temperature absorption spectrum of such molecules is quite complex owing to the many initial states that are populated. (The solution is to cool down these molecules using supersonic jets; see Appendix 5B). In the present section we will confine our attention to the simple elastic scattering of structureless particles, the key features of which are essential to our understanding of the more complicated classes of collision phenomena.

2.2.2 The center-of-mass system

A complete description of the paths of two particles undergoing a collision in space requires a specification of $2 \times 3 = 6$ scalar coordinates as a function of time. The interparticle force depends, however, only upon the *relative separation R* between the collision partners. It is thus advantageous to describe the collision not in terms of the motion of each particle but, rather, in terms of their relative separation and the motion of their center of mass (c.m.). There are usually no external forces operating on the system, and so the motion of the c.m. is unperturbed and its kinetic energy is constant throughout the collision. It is therefore convenient to subtract this constant from the total energy and deal with the remainder, which is the kinetic energy of the *relative* motion. Thus we work in the so-called *center-of-mass system*, that is, a coordinate system in which the c.m. of the colliding molecules is at rest.

In the c.m. system conservation of linear momentum tells us that†

$$m_1\mathbf{w}_1 + m_2\mathbf{w}_2 = 0 \qquad (2.17)$$

where m_1 and m_2 refer to the masses and $\mathbf{w}_1$ and $\mathbf{w}_2$ to the velocities with respect to the c.m. of particles 1 and 2, respectively. The *relative velocity* $\mathbf{v}$ is the difference

$$\mathbf{v} = \mathbf{w}_1 - \mathbf{w}_2 \qquad (2.18)$$

and the *relative kinetic energy* in the c.m. is

$$T = \tfrac{1}{2}m_1\mathbf{w}_1^2 + \tfrac{1}{2}m_2\mathbf{w}_2^2$$
$$= \tfrac{1}{2}\mu\mathbf{v}^2 \qquad (2.19)$$

† The velocity of molecule i, $i = 1, 2$, in the laboratory is given by $\mathbf{v}_i = \boldsymbol{v}_{cm} + \mathbf{w}_i$, where $\mathbf{v}_{cm}$ is the velocity of the center of mass, $\mathbf{v}_{cm} = (m_1\mathbf{v}_1 + m_2\mathbf{v}_2)/(m_1 + m_2)$.

where μ is the *reduced mass*

$$\mu = m_1 m_2 / (m_1 + m_2) \tag{2.20}$$

and we have used Eqs. (2.17) and (2.18) (which imply $m_1 \mathbf{w}_1 = \mu \mathbf{v} = -m_2 \mathbf{w}_2$) in obtaining the second line of Eq. (2.19).

In the absence of any force acting on the particles, their velocities (and hence T) are constant in time. However, as the particles approach to within the range of their mutual interaction, their relative kinetic energy T changes. Only the *total* collision energy is conserved:

$$E_T = \tfrac{1}{2}\mu \mathbf{v}^2 = T + V(R) = \text{const.} \tag{2.21}$$

where $\mathbf{v}$ is the initial value of the relative velocity when the particles are well separated and $V(R) = 0$. The quantity E_T is the initial (and, of course, the final) value of the relative kinetic energy and is conserved throughout the collision. The actual relative kinetic energy at the separation R (i.e., T), with $T = E_T - V(R)$, *does* depend on R when the particles are within the range of their interaction [i.e., when $V(R) \neq 0$].

Equation (2.21) determines the partitioning of the available energy E_T between kinetic and potential energy terms in the course of the collision. In Sections 2.2.3 and 2.2.4 we will examine the implications of this relation.

2.2.3 The centrifugal energy

The relative motion of two particles under the influence of a force that depends only upon their mutual separation is similar to the problem of the relative motion of the sun and a planet. As observed by Kepler and rationalized by Newton, such motions are confined to a plane. It is therefore sufficient to use two coordinates to specify the relative motion. The relative position vector $\mathbf{R}$ can then be specified in terms of its *length* R and its *orientation* ψ with respect to a fixed direction (Fig. 2.7).

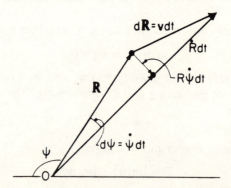

Fig. 2.7. Resolution of the vector $\mathbf{v}\,dt = \mathbf{R}(t + dt) - \mathbf{R}(t)$ into its radial component $\dot{R}\,dt$ and its tangential component $R\,d\psi = R\dot{\psi}\,dt$, perpendicular to $\mathbf{R}(t)$, representing the change of direction of $\mathbf{R}(t)$ at fixed R. The vector $\mathbf{L}$, not shown, is normal to the plane defined by $\mathbf{R}(t)$ and $\mathbf{R}(t + dt)$.

In the course of the collision, both the magnitude of **R** (i.e., the interparticle relative separation R) and its orientation angle ψ will vary with time. We use a dot to designate time rate of change (the time derivative). Thus $\dot{R}$ is the speed with which the colliding particles approach (or recede) from one another and $\dot{\psi}$ is the (angular) speed for the rotation of **R** (Fig. 2.7) during the collision. We will obviously need to know R as a function of time, since $R(t)$ describes the approach and the separation of the colliding particles during the collision. Why, however, do we need $\psi(t)$?

There are two reasons for our concern with the orientation of the interparticle separation. The first is that the rotation of **R**, that is, the change of ψ with time, *takes up energy*. The closer the atoms get, the smaller their moment of inertia $(I = \mu R^2)$ and the higher the energy due to the rotation $[E_R = \frac{1}{2}I\omega^2 = (I\omega)^2/2I = \frac{1}{2}\mu R^2\dot{\psi}^2]$. We need to find an alternative measure for the "closeness" of the collision, that is, for the amount of energy taken up by the rotation of **R**. This energy is not available for overcoming the interparticle potential $V(R)$, that is, for the decrease of R. Rotation prevents the two particles from getting as close as would be possible if $\dot{\psi} = 0$. We address the question of how near the particles do get together and how much energy is available for their approach in Section 2.2.4.

The second reason for our concern with $\psi(t)$ is so important to our theme that Chapter 3 is devoted to it almost exclusively. Before and after the collision, when the particles are far apart and no forces act between them, they travel in a straight path. Hence before or after the collision $\dot{\psi}(t)$ is zero; ψ only changes when the particles get close to one another. By *measuring the overall change in* ψ due to the collision, we can obtain considerable insight into the interparticle forces acting during the collision. Let us postpone such considerations, however, to Chapter 3.

The amount of energy that goes into the rotation, the so-called *centrifugal energy*, increases as the colliding particles approach one another. Hence as R decreases, less of the original kinetic energy is effectively available to the colliding particles. This uptake of energy by the rotational motion plays a key role in molecular collisions. It is for this reason that Section 2.2.5 provides a detailed quantitative treatment of the subject. In Sections 2.2.4 and 2.2.6 we shall confine ourselves to a more qualitative discussion.

2.2.4 The impact parameter

We now seek a measure of the "closeness" of the collision that will enable us to eliminate ψ from the expression for the centrifugal energy. In so doing we shall characterize (and quantify) the "intimacy" of a collision. We begin with a simple case first.

In the *absence* of an interparticle force, each particle travels in a straight line. The relative separation *vector* will trace a straight line as time varies and can therefore be represented as

$$\mathbf{R} = \mathbf{R}_1 - \mathbf{R}_2$$
$$= \mathbf{v}_1 t + \rho_1 - \mathbf{v}_2 t - \rho_2$$
$$= \mathbf{v}t + \mathbf{b} \tag{2.22}$$

where $\mathbf{v}$ is the relative velocity of the two particles $(\mathbf{v} = \mathbf{v}_1 - \mathbf{v}_2 = \mathbf{w}_1 - \mathbf{w}_2)$ and the vectors $\mathbf{v}$ and $\mathbf{b}$ specify the plane of the collision.

The *impact parameter* b is the distance of closest approach of the two particles in the *absence* of an interparticle force. Note, however, that even in the presence of a force we still define b as the magnitude of $\mathbf{b}$, the vector perpendicular to the *initial* direction of the relative velocity and whose magnitude is such that Eq. (2.22) holds at very large separations (i.e., prior to collision). Although b is no longer the distance of closest approach, if a force is present, it is still an indicator of the closeness of the collision. At low b values the collision will be practically head-on, whereas at higher b values it will be a "glancing blow." (For the rigid-sphere model, e.g., if $b > d$, the two particles will not come into contact and no collision will take place.)

Figure 2.8 shows the trajectory of a collision in the *absence* and *presence* of an interparticle force field. The "angle of deflection" χ is that between the final relative velocity vector $\mathbf{v}'$ and the initial value $\mathbf{v}$.

At this point we are ready to begin the consideration of the relative motion of two particles. As we proceed to study chemically reactive collisions we shall find that the simple concepts introduced here provide a

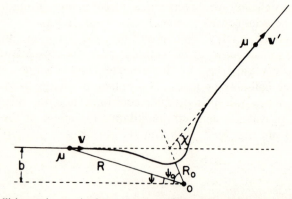

Fig. 2.8. A collision trajectory in the c.m. system. The solid curve represents a trajectory with initial velocity $\mathbf{v}$ and impact parameter $\mathbf{b}$. The relative separation $\mathbf{R}(t)$ is uniquely defined in terms of the distance R and the orientation angle ψ. The final deflection angle is $\chi = \pi - 2\psi_0$, where ψ_0 is the value of ψ at the apse of the trajectory. In the absence of a potential, a trajectory would follow the straight dashed line. The trajectory $\mathbf{R}(t)$ is identical to that traced by a particle of mass μ deflected by a potential $V(R)$ centered at the origin O. We can thus imagine the trajectory $R(t)$ as that of a projectile molecule A (of mass μ) as seen by an observer (usually a task assigned to a beginning graduate student) placed on molecule B.

key to the understanding of the more complicated dynamics of molecular encounters.

During the relative motion of the two particles, their relative position vector $\mathbf{R}(t)$ changes with time t in two ways: (1) Its *magnitude* $R(t)$ changes, bringing the two particles closer (or farther away), and (2) its *direction*, $\psi(t)$ changes. There are correspondingly two terms in the kinetic energy of the relative motion:

$$T = \tfrac{1}{2}\mu\dot{\mathbf{R}}^2 = \tfrac{1}{2}\mu(\dot{R}^2 + R^2\dot{\psi}^2) \tag{2.23}$$

Here we have resolved $d\mathbf{R}$ into its components (see Fig. 2.7), and the two terms in Eq. (2.23) represent the kinetic energy due to the change in magnitude and direction of $\mathbf{R}$, respectively. The first term in Eq. (2.23) is the radial kinetic energy, that is, the kinetic energy due to the component of the velocity along the line of the centers of the collision partners. The second term is the centrifugal energy, that is, the kinetic energy due to the component of velocity perpendicular to the line of the centers of the collision partners.

The impact parameter b is the *measure of the closeness of the collision.* Hence it is to be expected that the centrifugal energy is determined by b. Indeed, we shall see in Section 2.2.5 that the angular velocity ω is

$$\omega \equiv \dot{\psi} = bv/R^2 \tag{2.24}$$

where v is the initial relative velocity of the colliding particles (whose magnitude is given by $E_T = \mu v^2/2$). Hence, $(R\dot{\psi})^2 = b^2v^2/R^2$ and we can rewrite Eq. (2.23) as

$$T = \tfrac{1}{2}\mu\dot{R}^2 + E_T b^2/R^2 \tag{2.25}$$

In this way, the kinetic energy is expressed as a function of only $R(t)$, for a given value of the collision energy and impact parameter.

Having obtained Eq. (2.25), we have completed our reduction of the number of coordinates necessary to specify a collision trajectory. In the c.m. system, for given initial values of E_T and b, the trajectory is uniquely specified by a *single function* of time, $R(t)$, the (magnitude of the) center-to-center distance of the colliding particles.

2.2.5 Conservation of angular momentum; relation to impact parameter

We now turn to a quantitative discussion of the equations of motion for a collision between two structureless particles.

The interparticle force is directed along the lines of the centers of the two particles so that

$$m_1 \frac{d^2\mathbf{R}_1}{dt^2} = F(R)\hat{\mathbf{R}}, \qquad m_2 \frac{d^2\mathbf{R}_2}{dt^2} = -F(R)\hat{\mathbf{R}}$$

where the force on one particle is equal in magnitude but opposite in direction to the force on the other particle, $\hat{\mathbf{R}}$ is a vector of unit length along the line of centers, and $\mathbf{R} = \mathbf{R}_1 - \mathbf{R}_2$. Thus,

$$\frac{d^2\mathbf{R}}{dt^2} = \frac{1}{\mu} F(R)\hat{\mathbf{R}} \tag{2.26}$$

Newton pointed out that Eq. (2.26) implies that the *angular momentum* of the relative motion, $\mathbf{L}$, defined as the vector product

$$L = \mathbf{R} \times \mu\dot{\mathbf{R}} \tag{2.27}$$

is *conserved*. The chain rule shows that

$$\dot{\mathbf{L}} = \dot{\mathbf{R}} \times \mu\dot{\mathbf{R}} + \mathbf{R} \times \mu\ddot{\mathbf{R}} = 0 \tag{2.28}$$

Both terms in Eq. (2.28) vanish, since in both terms the two vectors are parallel.

The conservation of the *direction* of the vector $\mathbf{L}$ implies that the collision trajectory is confined to a plane. To see this, consider the plane normal to the direction of $\mathbf{L}$ defined by the vectors $\mathbf{R}(t)$ and $\mathbf{R}(t + dt)$, where $\mathbf{R}(t + dt) = \mathbf{R}(t) + \dot{\mathbf{R}}\, dt$. Since the direction of $\mathbf{L}$ is constant, $\mathbf{R}(t)$ and $\mathbf{R}(t + dt)$ and hence $\mathbf{R}(t)$ and $\mathbf{R}(t + \Delta t)$ are always in the same plane. It is thus possible to specify the trajectory using only two variables (R and ψ).

We have thus far used the conservation of the *direction* of $\mathbf{L}$ to show that the collision trajectory is confined to a plane. We shall now use the conservation of the *magnitude* of $\mathbf{L}$ to find an expression for $\psi(t)$ in terms of $R(t)$ and the impact parameter.

First consider $\mathbf{L}$ before the collision. Since $\mathbf{b}$ is perpendicular to the initial velocity, via Eq. (2.22), $\mathbf{L} = \mu\mathbf{v} \times \mathbf{b}$, so

$$L = \mu v b \tag{2.29}$$

During the collision, the component of $\dot{\mathbf{R}}$ that is perpendicular to $\mathbf{R}$ is (see Fig. 2.7) $R\dot{\psi}$ so that

$$L = \mu R^2 \dot{\psi} = I\omega \tag{2.30}$$

These two values of L are necessarily the same, since L is constant throughout the collision. Thus we obtain Eq. (2.24), $\dot{\psi} = bv/R^2$.

In quantum mechanics L has only discrete values, $L = [l(l + 1)]^{1/2}\hbar$, where the "orbital angular momentum" quantum number l has integer values, $l = 0, 1, 2, \ldots$. From Eq. (2.29), $b = L/\mu v = [l(l + 1)]^{1/2}\hbar/\mu v$ or $b \approx (l + \frac{1}{2})/k$, where $k = 2\pi/\lambda = \mu v/\hbar$ is the de Broglie wave number. For heavy-particle collisions at all but very low velocities the wavelength λ is quite short compared to the radii of the molecules; so $kd \gg 1$. It follows that even modest b values correspond to fairly high values of l (in the tens or even hundreds). Later we shall present a fully quantal treatment of the elastic collision process, but it should already be evident that for many aspects the classical approximation is adequate.

2.2.6 The centrifugal barrier

During the collision, as R decreases, the kinetic energy, which is initially wholly translational [second term in Eq. (2.25) negligible], is being converted to centrifugal energy, $E_T b^2/R^2$, at the expense of the first (radial kinetic energy) term. In Eq. (2.25) there is no longer any reference to $\psi(t)$, and hence we can determine the collision trajectory, at a given E_T and b, by solving the equation of conservation of energy,

$$E_T = T + V(R) = \tfrac{1}{2}\mu\dot{R}^2 + E_T b^2/R^2 + V(R) = \tfrac{1}{2}\mu\dot{R}^2 + L^2/2\mu R^2 + V(R) \tag{2.31}$$

for R as a function of time. For the purpose of our discussion later, it is convenient to rewrite Eq. (2.31) as

$$E_T(1 - b^2/R^2) = \tfrac{1}{2}\mu\dot{R}^2 + V(R) \tag{2.31'}$$

The approach motion of the two particles is thus equivalent to that of a particle of mass μ moving in a potential $V(R)$ with an effective kinetic energy (sometimes called the energy along the line of centers) equal to $E_T(1 - b^2/R^2)$. The higher b is, the smaller is this "effective" kinetic energy.

An equivalent point of view is to consider the centrifugal energy and the potential energy together, as an "effective" potential (Fig. 2.9),

$$V_{\text{eff}}(R) = V(R) + E_T b^2/R^2 \tag{2.32}$$

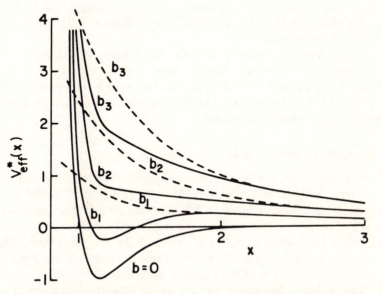

Fig. 2.9. The effective potential V_{eff} at a given E_T for a realistic interaction potential at several values of b. The abscissa and ordinate scales are in reduced (dimensionless) units. The dashed curves are for $V = 0$. Here $b_3 > b_2 > b_1 > 0$. There is a hump in $V_{\text{eff}}(R)$ for $0 < b < b_2$.

so that

$$E_T = \tfrac{1}{2}\mu\dot{R}^2 + V_{\text{eff}}(R) \tag{2.33}$$

The centrifugal energy acts as a repulsive contribution to V_{eff}, often known as the *centrifugal barrier*,† in that it prevents the too-close approach of the colliding particles. The origin of the barrier is as follows. As the two particles approach each other, their centrifugal energy (due to the rotation of the relative separation) $E_T b^2/R^2$ increases. Since the total energy is conserved, this increase is at the expense of the potential and the radial kinetic energy. Eventually one reaches the smallest separation R_0, the *turning point*, or *distance of closest approach*, which is the point where $\dot{R} = 0$. From Eq. (2.31), the turning point is determined as the solution of the implicit equation

$$E_T = V(R_0) + E_T b^2/R_0^2 \tag{2.34}$$

In the absence of a potential, $R_0 = b$, as mentioned earlier. For the hard-sphere model, Eq. (2.34) yields the expected result

$$R_0 = \begin{cases} b, & b > d \\ d, & b \le d \end{cases} \tag{2.35}$$

In other words, if $b > d$, the two spheres will not get near enough to "feel" the hard repulsion. They will pass one another without any "interaction." At their point of nearest approach, $R = b$, the entire initial kinetic energy has been converted to the rotational energy $E_T b^2/R^2$ of their relative separation. For $b < d$, the energy along the lines of centers, $E_T(1 - b^2/R^2)$, suffices for the two spheres to reach $R = d$. Figure 2.10 compares the trajectories in those two cases.

For realistic intermolecular potentials the distance of closest approach depends upon *both b and E_T*, but for large enough values of the impact parameter it can be seen from Eq. (2.34) that $R_0 \to b$, independent of the total energy. (This is sometimes a useful approximation when dealing with collisions of large impact parameter.)

2.2.7 The collision cross section

At a given initial energy and impact parameter we have seen that the collision trajectory is uniquely specified. Consider now collisions with impact parameters in the range b to $b + db$. Since all initial orientations of **L** and hence of **b** are equally probable, the trajectories of these collisions have to pass within an annulus of radius b and width db in a plane perpendicular to the initial velocity **v** (Fig. 2.11). All these trajectories are equivalent, aside from the orientation of the plane of the collision.

We have defined the cross section as an effective area, in a plane

---

† The centrifugal barrier $L^2/2\mu R^2$ is, in quantum mechanical form, $\hbar^2 l(l + 1)/2\mu R^2$.

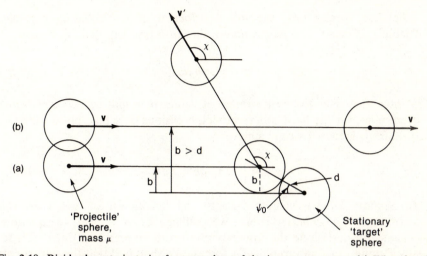

Fig. 2.10. Rigid-sphere trajectories for two values of the impact parameter. (a) When $b \leq d$, there is a collision with deflection χ. Note that $\psi_0 = \arccos(b/d)$ and $\chi = \pi - 2\psi_0$. The energy along the line of centers $E_{LC} = \frac{1}{2}\mu(v \cos \psi_0)^2 = E_T(1 - \sin^2 \psi_0) = E_T(1 - b^2/R^2)$. (b) When $b > d$, there is no collision, $\mathbf{v}' = \mathbf{v}$, and $\chi = 0$.

perpendicular to the initial velocity $\mathbf{v}$, such that the relative separation vector $\mathbf{R}$ has to be within that area for a collision to take place. Hence, when the impact parameter is in the range b to $b + db$, the collision cross section is (see Fig. 2.11)

$$d\sigma = 2\pi b \, db \qquad (2.36)$$

that is, the area of the annulus in b within which the trajectories pass.

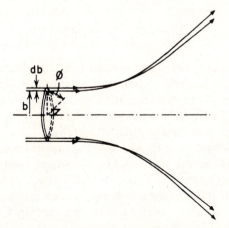

Fig. 2.11. Schematic representation of trajectories within a small range of impact parameters (between b and $b + db$). All initial values of the azimuthal angle ϕ are equally probable, so all trajectories entering the annular ring $(b, b + db)$ are scattered within a given small range of cylindrically symmetrical cone angles.

When we specify only the initial *velocities* of the colliding particles, collisions will occur with *all possible* values of b [see Eq. (2.22)]; hence

$$\sigma = \int 2\pi b \, db \tag{2.37}$$

The integral is over the entire range of values of b *that lead to a collision.* For example, for the hard-sphere model, collisions only take place for $b \le d$ [Eq. (2.35)]. Hence for rigid spheres

$$\sigma = \int_0^d 2\pi b \, db = \pi d^2 \tag{2.38}$$

In Section 3.1 we shall return to the evaluation of σ for *realistic intermolecular potentials*. There we will have to appeal to quantum mechanics to determine the upper bound on the integral of Eq. (2.37) for realistic, long-range intermolecular forces. We shall find that the uncertainty principle ensures that the cross section is finite. At this point, however, we have amassed sufficient background (at last!) to begin our considerations of *reactive molecular collisions*.

2.3 The reaction cross section

The rate of an elementary gas-phase bimolecular reaction, say, $F + HCl \rightarrow Cl + HF$, is characterized by a "thermal" rate constant $k(T)$ that is a function of the temperature only. This rate constant is a measure of the rate of depletion of the reactants or the rate of appearance of the products, for example,

$$-\frac{d[F]}{dt} = \frac{d[Cl]}{dt} = k(T)[F][HCl] \tag{2.39}$$

at a given temperature. On the molecular level the overall chemical change is the result of many molecular collisions, where the colliding partners differ both in their state of internal excitation and in their translational velocity. Similarly, we implicitly sum over all possible energy distributions of the products as well as over their angular distribution.

In this volume we propose to go *below* the macroscopic to the micro-level of description. Among the features that we would want to explore are (1) the "energy requirements" of the chemical reaction—in particular, the *threshold energy* or the minimum energy required for the reaction to occur† and the variation of the reactivity with the reactants' translational (and

---

† The threshold energy is not the same as the Arrhenius activation energy. The latter is the average energy of those collisions that lead to reaction, minus the average energy of all collisions (see Sec. 4.4).

internal) energy; (2) the *steric effect* or the dependence of the reactivity upon the relative orientation of the reactants; (3) the energy disposal of the products; and (4) the angular distribution of the products after they have separated from the region of interaction.

We have made a start on items (3) and (4) in Chapter 1. But before we continue on this road we need to know how to characterize quantitatively the reaction rate under nonthermal conditions.

2.3.1 Definition of the reaction cross section

We define the *reaction cross section* σ_R in a way suggested by the definition of the total collision cross section (see Sec. 2.1). For molecules colliding with a well-defined relative velocity $\mathbf{v}$, the reaction cross section is defined such that the elementary, velocity-dependent reaction rate constant $k(v)$ is given by

$$k(v) = v\sigma_R \tag{2.40}$$

Hence, when we pass a beam through a scattering cell, the loss of the beam flux due to *reactive* collisions is given by

$$-\left(\frac{dI}{dx}\right)_R = k(v)n_A n_B = I(x)n_B\sigma_R \tag{2.41}$$

Here, as in Eqs. (2.1)–(2.4), $I(x)$ is the flux of beam molecules at position x and n_B is the number density of the target molecules. However, *not all* collisions need lead to reaction. While we can conclude that $\sigma_R \leq \sigma$, it is not enough to measure the attenuation of the parent beam; we must also know how many of the collisions leading to attenuation are reactive, that is, we must also determine the fraction of the incident flux converted into reaction products.

For reactions producing ions it is easy to measure the product flux: The ions are simply collected by the application of an electric field. One interesting class of reactions of this sort is that of the endothermic *collisional ionization* type, where two neutral molecules collide to form ions, for example,

$$K + Br_2 \rightarrow K^+ + Br_2^-$$

without atom exchange, or

$$N_2 + CO \rightarrow NO^+ + CN^-$$

with concurrent atom exchange.

Another type of reaction involving *both* reagent and product ions is the so-called ion–molecule class, for example,

$$H_2^+ + He \rightarrow HeH^+ + H$$

Of course, it is not enough just to collect the ionic or neutral products; it is also necessary to identify their chemical nature. This is often achieved by

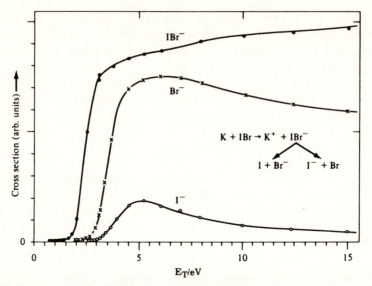

Fig. 2.12. Cross sections for the formation of the indicated negative ions from the collision of fast K atoms by IBr. Here E_T ranges up to 15 eV. [Adapted from D. J. Auerbach, M. M. Hubers, A. P. Baede, and J. Los, *Chem. Phys. 2*, 107 (1973).]

mass spectrometric methods. Such identification is essential when several different reaction paths are possible, for example,

$$K + IBr \rightarrow \begin{cases} KI + Br \\ KBr + I \\ K^+ + IBr^- \\ K^+ + Br^- + I \\ K^+ + I^- + Br \end{cases}$$

and one needs to determine the *branching ratio* or the relative contribution of each process to the total reaction cross section (see Fig. 2.12). Furthermore, a detailed analysis of product *angular* and energy state distribution requires the use of a crossed-beam arrangement, as will be discussed in Sections 3.3 and 5.4.

2.3.2 *The energy threshold for reaction*

We embark on our study of the role of energy in chemical dynamics by examining the dependence of the reaction cross section on the relative translational energy of the colliding partners. Figure 2.13 shows the translational energy dependence of the $H_2^+ + He$ reaction cross section, which is typical of many endothermic reactions. The reaction cross section is

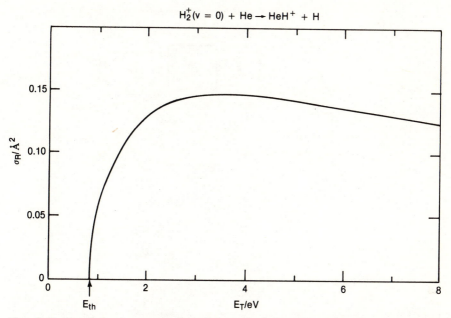

Fig. 2.13. Translational energy dependence of the cross section for the reaction $H_2^+(v=0) + He \rightarrow HeH^+ + H$ from threshold $(E_{th} = 0.8\,\text{eV})$ to 8 eV. [Adapted from T. Turner, O. Dutuit, and Y. T. Lee, *J. Chem. Phys.*, *81*, 3475 (1984).]

zero up to a *threshold energy*† $E_{th} \approx \Delta E_0$,

$$\Delta E_0 = D_0(H_2^+) - D_0(HeH^+) = 2.65 - 1.84 = 0.81\,\text{eV} \qquad (2.42)$$

and increases rapidly as the translational energy increases above ΔE_0.

The same kind of energy dependence as shown in Fig. 2.13 is often observed in reactions that have an *activation barrier*. Here σ_R is effectively zero below some threshold energy, even though the reaction is "allowed" on energetic grounds. Examples include the exoergic reaction

$$O + H_2 \rightarrow OH + H$$

or the nearly thermoneutral reaction

$$H + D_2 \rightarrow HD + D$$

with threshold energies of about 4 and 41 kJ mol^{-1}, respectively (Fig. 2.14). Thus, whereas *all* endoergic reactions necessarily have an energy threshold, many exoergic reactions also have an effective energy threshold. The reaction energy threshold E_{th} can be no lower than the endoergicity (the minimum energy ΔE_0 thermochemically required for the reaction) but may be much higher.

† For this reaction the threshold energy is found to be equal to the minimal value expected on thermochemical grounds, that is, the difference between the binding energies of the reactants and products. Thus there is a negligible activation barrier for the reaction.

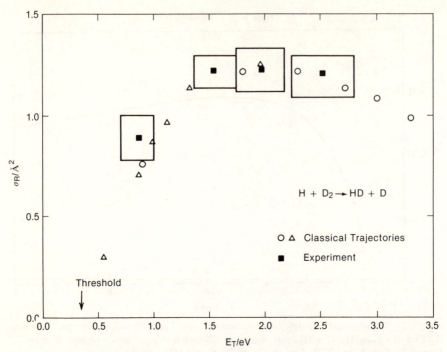

Fig. 2.14. Translational energy dependence of the reaction cross section for the $H + D_2 \rightarrow$ HD + D reaction. The triangles are classical trajectory computations [by N. C. Blais and D. G. Truhlar, unpublished]; circles, independent calculations [by I. Schechter, unpublished]. The rectangles are experimental results obtained using a two-laser arrangement (see Sec. 1.4.5), with one to generate fast H atoms and one to monitor the H- or D-atom concentrations. [Adapted from K. Tsukiyama, B. Katz, and R. Bersohn, *J. Chem. Phys.*, *84*, 1934 (1986), with additional unpublished data.]

An important class of reactions, of particular interest in atmospheric chemistry (aeronomy), is that of *exoergic* ion–molecule reactions, for example,

$$N^+ + O_2 \rightarrow \begin{cases} NO^+ + O \\ N + O_2^+ \end{cases}$$

Such reactions often show no threshold energy, and the reaction cross section at all but the lowest energies is found to be a decreasing function of the translational energy, roughly as

$$\sigma_R(E_T) \approx A E_T^{-1/2} \tag{2.43}$$

(Fig. 2.15). A decreasing cross-section functionality is found also for those neutral–neutral collisions that proceed without any threshold energy, for example,

$$K + I_2 \rightarrow KI + I$$

In Fig. 2.16 we contrast the translational energy dependence of the cross

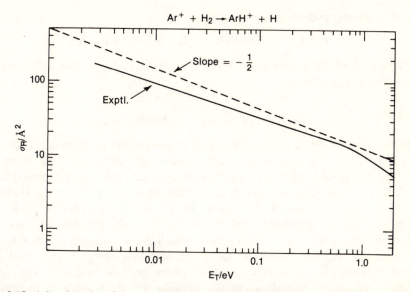

Fig. 2.15. A log–log plot of the translational energy dependence of an ion–molecule reaction with no energy barrier (and no threshold energy). The solid curve is the experimental result for the system $Ar^+ + H_2 \rightarrow ArH^+ + H$. The dashed curve has the slope $-\frac{1}{2}$ demanded by Eq. (2.43); it corresponds to the simple capture model of Eq. (2.59). [Adapted from K. M. Ervin and P. B. Armentrout, *J. Chem. Phys.*, *83*, 166 (1985).]

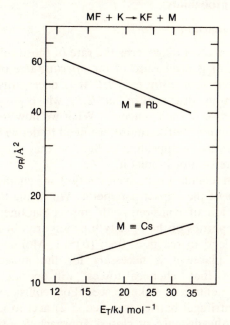

Fig. 2.16. Translational energy dependence of the cross sections for the endoergic reaction of CsF + K and the exoergic reaction of RbF + K. [Adapted from S. Stolte, A. E. Proctor, and R. B. Bernstein, *J. Chem. Phys.*, *65*, 4990 (1976).]

section for the two exchange reactions

$$K + CsF \rightarrow KF + Cs, \qquad K + RbF \rightarrow KF + Rb$$

the first of which is endoergic and the second exoergic. The difference in behavior is self-evident.

2.3.3 Translational energy requirements of chemical reactions

On the basis of the translational energy requirements of chemical reactions, we can make the following rough correlation: Reactions that have an energy threshold (and this necessarily includes all endoergic reactions) have a reaction cross section which is an increasing function of the translational energy in the post-threshold region. Reactions that proceed without any apparent energy threshold (and this includes some, but not all, exoergic reactions) have a reaction cross section which is a decreasing function of the translational energy. However, as the translational energy is increased and other, previously endoergic reaction paths become allowed, the cross section for the previously accessible process will decrease with increasing energy. To make these correlations somewhat more quantitative, we turn next to the concept of the reaction probability.

2.4 The reaction probability

2.4.1 The opacity function

The *reaction cross section* σ_R governs the rate of those collisions that lead to *chemical reaction*. It is a measure of the *effective size* of the molecules as determined by their *propensity to react*. It is necessarily smaller than the collision cross section (discussed in Sec. 2.2), which governs the rate of *all* collisions, irrespective of their outcome. What we now seek is a measure of the reaction probability. Furthermore, we need to define this probability for reaction in an intuitively appealing fashion so that simple models can be readily used to interpret or predict it.

As the reactants collide (at a given energy) we characterize their initial approach in terms of the impact parameter. We define the *opacity function* $P(b)$ as the fraction of collisions with impact parameter b that lead to reaction. Two properties of the opacity function are obvious. Since at most all collisions can lead to reaction, $0 \le P(b) \le 1$. Moreover, for a chemical reaction to take place, it is necessary for the molecules to approach sufficiently close so that "chemical forces" will operate and the necessary atomic rearrangements can take place. For collisions of high impact parameter the centrifugal barrier (see Sec. 2.2) acts to keep the molecules apart (recall that the distance of closest approach $R_0 \rightarrow b$ for large b; see Sec. 2.2.6). We therefore expect that reaction will take place only when b is "small," that is, of the order of the range of the intermolecular force, and hence that the reaction will fail to occur [i.e., $P(b) = 0$] for higher values of b.

2.4.2 Reaction cross section

In terms of the opacity function, we can write the reaction cross section as

$$d\sigma_R = 2\pi b P(b)\, db \qquad (2.44)$$

so that

$$\sigma_R = \int_0^\infty 2\pi b P(b)\, db \qquad (2.45)$$

where $2\pi b\, db$ is the element of the total collision cross section (see Sec. 2.2.7) and $P(b)$ is that fraction of collisions at impact parameter b leading to reaction. Owing to the possibility of reaction, the total cross section is partitioned between the cross section for *reactive* [see Eq. (2.44)] and *nonreactive* collisions,† given by

$$d\sigma_{NR} = 2\pi b [1 - P(b)]\, db \qquad (2.46)$$

where $1 - P(b)$ is the fraction of collisions that are nonreactive at the impact parameter b. We note that while the partitioning between reactive and nonreactive collisions depends in an intimate fashion on the collision dynamics, the sum of their cross sections is always $2\pi b\, db$, irrespective of the fine details. Similarly, if several different reaction paths are energetically possible, their cross sections can increase only at the expense of one another or of the nonreactive cross section. There is never more (or less) than $2\pi b\, db$ to partition among all possible final outcomes of collisions with b in the range b to $b + db$.

True to the principle of "he who hath, receiveth," σ_R is a *weighted* average of $P(b)$, where the contributions at higher b values are more heavily weighted. Hence, if a reaction has an opacity function that extends to higher b values, it will have a particularly large reaction cross section.

In quantum mechanics the reaction probability is defined for each discrete value l of the angular momentum quantum number. The reaction cross section is given by the discrete sum

$$\sigma_R = \frac{\pi}{k^2} \sum_{l=0}^\infty (2l + 1) P(l) \qquad (2.47)$$

where $P(l)$ is the probability of reaction for a given l. When many values of l contribute to the reaction, we can use the semiclassical correspondence $bk \simeq l + \tfrac{1}{2}$ to convert the sum over l to an integral over b, wherein we recover Eq. (2.45).

---

† Strictly speaking, $d\sigma_{NR}$ contains an additional "shadow" term, quantal in origin, which is always present, even when $P(b) = 1$.

2.4.3 A simple opacity function

A determination of the reaction cross section does not uniquely specify†
$P(b)$, but only its b-weighted average, Eq. (2.45). Hence, rather than try to
determine $P(b)$ in detail, let us assume a simple functional form, with very
few parameters, and determine these parameters from the observed σ_R or
from dynamical models. An oversimplified, one-parameter representation is

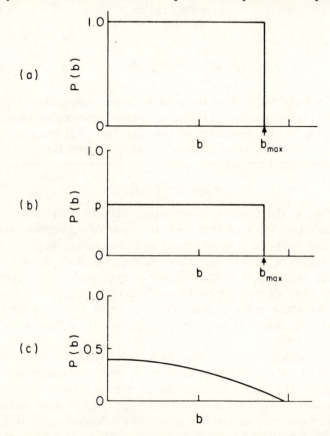

Fig. 2.17. Stylized opacity functions $P(b)$ for a given energy E. The maximum impact
parameter leading to reaction is denoted b_{max}: (a) simplest step function [Eq. (2.48)], (b) step
function with steric factor p [Eq. (2.50)], and (c) realistic function for a reaction with a steric
factor.

a unit step function (Fig. 2.17),

$$P(b) = \begin{cases} 1, & b \leq b_{max} \\ 0, & b > b_{max} \end{cases} \qquad (2.48)$$

where b_{max} is the cutoff impact parameter, the highest b for which a reaction

† We shall see in Section 3.3 that *angular distribution* studies *can* provide information on the
b dependence of the opacity.

occurs. The reaction cross section is then simply

$$\sigma_R = 2\pi \int_0^{b_{max}} b \, db = \pi b_{max}^2 \qquad (2.49)$$

so that b_{max} (and its possible energy dependence) can be determined from the measurements of σ_R.

The simple functional form of Eq. (2.48) and its interpretation in terms of simple dynamical models (see Sec. 2.4.6) have played an important role in the early development of our field. At the present level of our experimental and theoretical understanding, it is no longer accurate enough.

2.4.4 The steric factor

There are usually other requirements, beside the closeness of the approach, for a reaction to take place. In particular, there may be steric requirements, in that some configurations of the colliding molecules may be more conducive to reaction. As an example, consider the reaction $H + D_2 \rightarrow HD + D$. The energy of the three-atom system is such that reaction is favored for a more nearly collinear approach of H to D_2. Yet the impact parameter b is only the "miss distance" from the H atom to the c.m. of the D_2 molecule. At each value of b, *all* possible reactant orientations contribute. Due to collisions with unfavorable orientations, $P(b)$ will fail to reach the maximal value of unity. To account for such steric (and any other) requirements, one can modify the simple unit step function by the introduction of a *steric factor p* (<1), such that

$$P(b) = \begin{cases} p, & b \le b_{max} \\ 0, & b > b_{max} \end{cases} \qquad (2.50)$$

This leads to

$$\sigma_R = 2\pi p \int_0^{b_{max}} b \, db = \pi p b_{max}^2 \qquad (2.51)$$

However, a measurement of σ_R alone can only determine the value of the *product* $p b_{max}^2$. Additional information is required to determine the magnitude of p.

2.4.5 Reactive asymmetry

A direct experimental verification of the steric requirement that provides a measurement of p can be obtained by using *oriented* reactant molecules. The experiment allows the determination of the *reactive asymmetry* or the reactivity as a function of the mutual orientation of the reagents. For example, by orienting CH_3I molecules by a specially designed configuration of electrical fields, we obtain the asymmetry in the reaction probability shown in Fig. 2.18 for a predominantly head-on ($b \approx 0$) collision of CH_3I

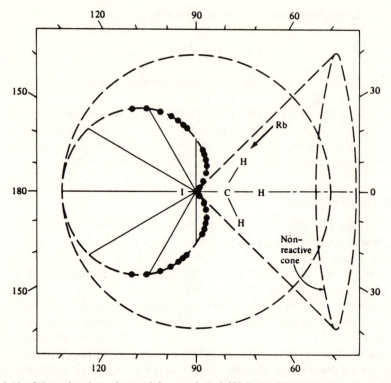

Fig. 2.18. Orientation dependence of the reaction of CH_3I with Rb. Shown here is a polar plot of the reaction probability (for backward scattering in the c.m. system, i.e. the "rebound" direction) as a function of the initial orientation angle between the C–I bond and the incoming Rb atom. The points represent experimental data, with a dashed extrapolation curve. [Adapted from D. H. Parker, K. K. Chakravorty, and R. B. Bernstein, *J. Phys. Chem., 85,* 466 (1981).]

with Rb (yielding $RbI + CH_3$). The reactivity is highest for an approach from the I side of the molecular axis and drops essentially to zero within a cone of nonreaction spanned by the CH_3 group, with an opening angle of about 45°. Integrating over all orientations of CH_3I leads to a directly experimentally determined steric factor (at $b \approx 0$) of about 0.4.

Even in a gas "bulb" experiment it is possible to create a population of aligned excited molecules by absorption of linearly polarized light. The probability of absorption is proportional to the square of the scalar product of the molecular transition dipole moment and the electrical field. Hence the population of the excited states declines sharply for those molecules whose transition moments are oriented away from the plane of polarization. Of course, such an alignment can be very rapidly randomized by collisions.

For reactions with an activation barrier only the excited state reagents can undergo chemical reaction. A polarized laser pulse is used to produce these excited reagents. The reaction products can be detected by probing them with a second, pulsed laser at an appropriate wavelength. If the second pulse is delayed by a very short time [on the order of ω^{-1}; see Eq. (2.8)] after the first one, then the excited molecules have, on the average, undergone less than one collision. Any reactive collisions that did take place did so with aligned reactants (in the laboratory frame). Initiating a reaction with a pump laser with a very short time delay between the pump and probe lasers enables us to study single collisions under otherwise bulk conditions. As we shall see, probe lasers can interrogate not only the internal states of the molecules but also (via the Doppler shift of the laser-induced fluorescence) their direction of motion in space. Since the product molecule–probe laser interaction is also dependent on the orientation of the transition dipole with respect to the laser field, monitoring the reaction products by a polarized laser yields information on any preferred spatial orientation of the products.

With the two-laser pump–probe technique, a great deal of dynamical information has become available. With much detail being deferred for later consideration, Fig. 2.19 shows the probe spectrum of SrF from the reaction

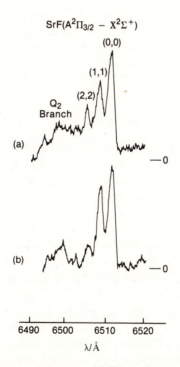

Fig. 2.19. Laser-induced fluorescence spectra of SrF from the reaction of Sr + HF ($v = 1, j = 1$) with the HF molecule aligned (a) perpendicular and (b) parallel with the direction of approach of the Sr atom. Computer simulations of these spectra yielded vibrational state relative populations $v_0':v_1':v_2'$ of (a) 1:0.63:0.29 and (b) 1:0.72:0.13. Broadside attack thus favors the $v_2' = 2$ level in the SrF product. [Adapted from Z. Karny, R. C. Estler, and R. N. Zare, *J. Chem. Phys.*, *69*, 5199 (1978).]

of Sr with vibrationally excited† aligned HF:

$$Sr + HF \ (v = 1, j = 1) \rightarrow SrF \ (v', j') + H$$

2.4.6 *The centrifugal barrier to reaction*

A simple quantitative model for the translational energy dependence of b_{max}, and hence σ_R, is obtained by adopting the following criterion for reaction: If the colliding molecules have sufficient translational energy to overcome the energy barrier for reaction and reach the region of chemical reaction, reaction occurs with unit probability. All collisions that clear the barrier are thus captured by the strong chemical forces. Consider first reactions without an energy threshold. The only barrier to the approach motion of the reactants is the centrifugal barrier (see Sec. 2.2.6). For a given translational energy E_T and impact parameter b the effective potential (see Fig. 2.9)

$$V_{eff} = V(R) + E_T b^2 / R^2 \tag{2.52}$$

will have a maximum at some $R = R_{max}$ (where R_{max} is, in general, a function of b and E). This maximum is due to the combined influence of the very long-range centrifugal repulsion (that dominates at very large R values) and the long-range attractive part of $V(R)$ (which can dominate at a somewhat smaller R). It is defined by

$$\frac{d}{dR} [V_{eff}(R)]_{R_{max}} = 0 \tag{2.53}$$

Only when the molecules can approach to within R_{max} can they enter the region of the chemical forces.‡ Hence the criterion for reaction is that the molecules reach $R = R_{max}$ with at least *some* kinetic energy left. Making use of Eq. (2.31), we can write this criterion as

$$\tfrac{1}{2}\mu(\dot{R}^2)_{R_{max}} = E_T - V_{eff}(R_{max}) \geq 0 \tag{2.54}$$

At a given E_T as b increases, the kinetic energy at R_{max} decreases. The b_{max} is defined as the largest value of b for which the molecules can cross the barrier in the effective potential (see Fig. 2.9), that is, the solution of the implicit equation

$$[E_T - V(R) - E_T b^2 / R^2]_{R_{max}} = 0 \tag{2.55}$$

Sections 2.4.7 and 2.4.8 illustrate the applications of this reaction criterion [from Eqs. (2.53) and (2.55) for R_{max} and b_{max}; we obtain the reaction cross section directly via Eq. (2.49)].

---

† The reaction does not occur with ground state HF molecules.

‡ An exceptional case occurs if the long-range part of $V(R) \propto R^{-s}$ has an inverse power $s \leq 2$. This includes the case of reactions of ions of opposite charge, where $s = 1$.

2.4.7 Cross sections for reactions without an energy barrier

The intermolecular potential between the reactant molecules is, in general, unknown. If we approximate it by its long-range form†

$$V(R) = -C_s/R^s \tag{2.56}$$

we obtain from Eq. (2.53), for $s > 2$,

$$R_{max}^2 = (sC_s/2E_T b^2)^{2/(s-2)} \tag{2.57}$$

Evaluating $V_{eff}(R_{max})$ from Eqs. (2.53) and (2.57), we find that $V_{eff}(R_{max}) \propto (b^2 E_T)^{s/(s-2)}$. At b_{max}, $V_{eff}(R_{max}) = E_T$ [see Eq. (2.55)], we finally have

$$\sigma_R(E_T) = \pi b_{max}^2 = \pi q(s)(C_s/E_T)^{2/s} \tag{2.58}$$

where $q(s) = (s/2)[(s-2)/2]^{-(s-2)/s}$.

For the particular case of singly charged ion–molecule reactions where $s = 4$ and $C_s = \alpha/2$ (see Sec. 3.2.4), we obtain [see Eq. (2.43)]

$$\sigma_R(E_T) = \pi(2\alpha/E_T)^{1/2} \tag{2.59}$$

This simple "Langerin" model accounts for the moderate decrease of σ_R with E_T [Eq. (2.58)] for reactions without an energy threshold. In the present case of ion–molecule reactions it even accounts roughly for the magnitude of $\sigma_R(E_T)$ as well (see Fig. 2.15). Since, from Eq. (2.59), $\sigma_R \propto v^{-1}$, where v is the relative speed, the rate constant $k(v) = v\sigma_R$ is independent of the translational energy. It will be shown in Section 4.4 that the thermal rate constant $k(T)$ is obtained by *averaging* $k(v)$ over the Maxwellian distribution of velocities, $k(T) = \langle v\sigma_R \rangle$. It follows that $k(T)$ will be *independent* of the temperature (as is often found for ion–molecule reactions with no energy threshold).

2.4.8 Reactions with an energy threshold

For reactions *with* an energy threshold, we must take account of both the centrifugal barrier *and* the threshold energy barrier (the "permanent" barrier for reaction). We now replace Eq. (2.52) by the condition that at some separation d, the energy available for the motion along R exceeds the threshold energy‡ $E_{th} = E_0$, or

$$[E_T - E_0 - E_T b^2/d^2] \geq 0 \tag{2.60}$$

Here $E_T(1 - b^2/d^2)$ is the kinetic energy for motion along R evaluated at $R = d$ [see Sec. 2.3 and Eq. (2.25)]. The reaction criterion of Eq. (2.60) can be formulated as follows: *reaction occurs if the kinetic energy along R, the*

---

† This is permissible only if R_{max} (and hence b_{max}) is large enough so that it is determined by the long-range tail of the potential.

‡ We can think of E_0 as the value of $V(R)$ at $R = d$, provided that $V(R)$ is repulsive at that point.

line of centers, exceeds E_0. Note again the complementary views about the role of the "rotational" kinetic energy $E_T b^2/R^2$. Either we can add it to the static potential $V(R)$ to give $V_{\text{eff}}(R)$, whence it plays the role of an additional repulsive potential, or we can take it as part of the kinetic energy, whence $E_T(1 - b^2/R^2)$ is that part of the kinetic energy for motion along the line of centers.

Just as for the previous case, we take $b_{\max}$ as the largest value of b that satisfies Eq. (2.60), that is, $b_{\max} = d(1 - E_0/E_T)^{1/2}$, so that, from Eq. (2.49),

$$\sigma_R(E_T) = \pi b^2_{\max} = \begin{cases} 0, & E_T \leq E_0 \\ \pi d^2(1 - E_0/E_T), & E_T > E_0 \end{cases} \qquad (2.61)$$

In this case, as well, the increase of σ_R with E_T is in qualitative accord with the results often found for reactions with an energy threshold. Figure 2.20

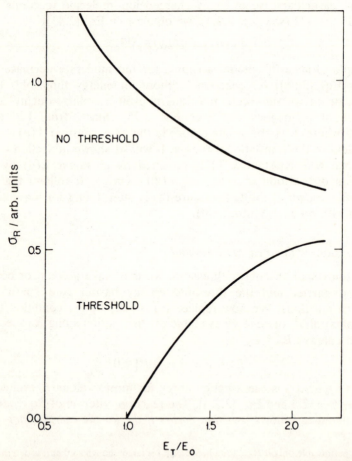

Fig. 2.20. The translational energy E_T dependence of the reaction cross section, σ_R for reactions with and without energy threshold $E_{\text{th}} = E_0$. [Adapted from R. D. Levine and R. B. Bernstein, *J. Chem. Phys.*, **56**, 2281 (1972).]

shows the form of the cross section Eq. (2.61). The energy dependence of Eq. (2.61) is sometimes termed an Arrhenius-like cross-section functionality, or a line-of-centers model dependence. The reason is that upon thermal averaging, it leads to a rate constant of the traditional Arrhenius form, $k(T) = A \exp(-E_0/kT)$. The pre-exponential factor A is given by $A = \langle v \rangle \pi d^2 = (8kT/\pi\mu)^{1/2}\pi d^2$. Indeed, the very introduction of a steric factor resulted from this A being typically an overestimate for most elementary bimolecular chemical reactions. Had we used Eq. (2.51) rather than Eq. (2.49) to derive Eq. (2.61), the pre-exponential factor would have been reduced by a factor of p from its previous value. But this is an *ad hoc* imposition of the steric factor—almost an afterthought. Can we really justify the concept of a steric factor?

2.4.9 A simple model of steric requirements

A hidden assumption in the derivation of Eq. (2.61) was that the height of the activation barrier was independent of the orientation of the reagents. The experimental results show that at a given energy and impact parameter ($b \approx 0$ in Fig. 2.18), the reaction probability depends very much on the initial orientation angle. Equally, quantum mechanical computations of the barrier height (discussed in detail in Chap. 4) show a significant dependence on reagent orientation. The results of such computations for the simplest exchange reaction, $H + H_2$, are shown in Fig. 2.21.

To incorporate these results in the previous model, we modify Eq. (2.60) by allowing the barrier height to depend on the orientation angle† γ [*i.e.*, E_0 becomes $E_0(\gamma)$]. The condition for overcoming the barrier is now written as

$$E_T - E_0(\gamma) - E_T b^2/d^2 \geq 0 \qquad (2.62)$$

For such orientation angles γ that satisfy Eq. (2.62) reaction is possible. Consistent with the previous case, we take it to occur with unit probability.

The reaction cross section for oriented reagents (i.e., for a given value of γ) is obtained by integrating Eq. (2.49) over the range of impact parameters consistent with Eq. (2.62). Just as in the derivation of Eq. (2.61),

$$\sigma_R(\gamma) = \begin{cases} 0, & E_T \leq E_0(\gamma) \\ \pi d^2[1 - E_0(\gamma)/E_T], & E_T \geq E_0(\gamma) \end{cases} \qquad (2.63)$$

The result, Eq. (2.63), as a function of γ, is compared to actual dynamical computations for the $H + D_2$ reaction in Fig. 2.22. The barrier height used in Eq. (2.63) is that shown in Fig. 2.21.

---

† At each impact parameter, the entire range of values of γ, from 0 to π, is, in principle, possible. Only such values that are consistent with Eq. (2.62) can be realized at $R = d$. Note that γ is not the same angle as ψ of Section 2.2. The latter is the angle between **R** and the initial direction of the velocity. For a hard-sphere collision at $R = d$, ψ has a definite value, $1 - b^2/d^2 = \cos^2 \psi$; see Fig. 2.10.

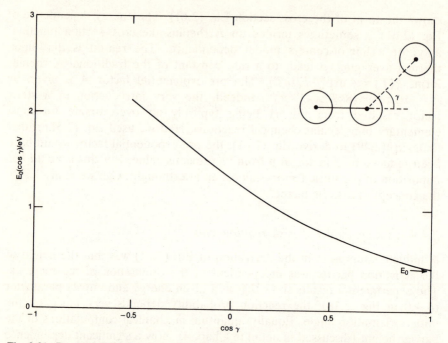

Fig. 2.21. Barrier height E_0 versus $\cos \gamma$ for the *ab initio* H_3 potential surface. Here γ is the "bend angle", defined in the inset. The minimum barrier corresponds to the collinear attack, indicated by the arrow at $E_0 = 0.425$ eV. [*Ab initio* computations by P. Siegbahn and B. Liu, *J. Chem. Phys.*, *68*, 2457 (1978); parametrized surface by D. J. Truhlar and C. J. Horowitz, *J. Chem. Phys.*, *68*, 2466 (1978); *71*, 1514 (1979). Adapted from R. D. Levine and R. B. Bernstein, *Chem. Phys. Lett.*, *105*, 467 (1984).]

For reactions of randomly oriented reactants we need to integrate over the range of values of γ:

$$\sigma_R = \frac{1}{2} \int_{-1}^{1} d(\cos \gamma) \sigma_R(\gamma) \tag{2.64}$$

Care must be exercised, since $\sigma_R(\gamma)$ is nonzero only for such orientations satisfying $E_T \geq E_0(\gamma)$ [see Eq. (2.63)]. In general, the integration needs to be done numerically; however, near the very threshold for reaction, one can expand $E_0(\gamma)$ in a Taylor series and retain only the linear term,

$$E_0(\gamma) \approx E_0 + E_0'(1 - \cos \gamma) \tag{2.65}$$

In Eq. (2.65), E_0' is the negative of the slope of the plot of $E_0(\gamma)$ versus $\cos \gamma$ at $\cos \gamma = 1$ (the barrier in Fig. 2.21 is lowest for a collinear approach). Using Eq. (2.65) in Eq. (2.63) leads to an analytic integral in Eq. (2.64):

$$\sigma_R = \pi d^2 (E_T - E_0)^2 / 2E_0' E_T \tag{2.66}$$

The labor of performing the analytic integration turns out to be worthwhile: Near the threshold [at E_0, where Eq. (2.66) is valid], it predicts a

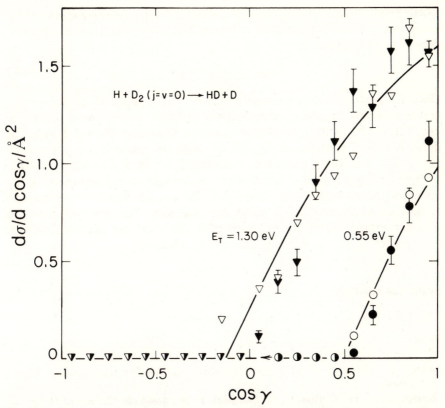

Fig. 2.22. Orientation dependence of the cross section for the reaction $H + D_2$ $(v = j = 0) \rightarrow$ $HD + D$ at the two indicated values of the collision energy E_T. The ordinate is $d\sigma_R/d\cos\gamma = 2\sigma_R(\gamma)$ of Eq. (2.63). The solid curves were calculated from the angle-dependent line-of-centers model, Eq. (2.63) and the points represent quasi-classical trajectory computations (with statistical error bars) on the *ab initio* potential surface referred to in Fig. 2.21. [Adapted from N. C. Blais, R. B. Bernstein, and R. D. Levine, *J. Phys. Chem.*, **89,** 10 (1985)].

concave-up (quadratic) increase of σ_R with the collision energy. This is in contrast to the concave-down functionality of the simpler Arrhenius-like form of Eq. (2.61).

Experimental results often show the concave-up dependence, but near the threshold, where the cross section is small, the unavoidable finite spread in E_T induces large uncertainties in the σ_R values. Dynamical computations do, however, suggest that the initial rise of σ_R versus E_T is indeed typically concave up.

What of the steric factor itself? Assuming that reaction occurs whenever Eq. (2.62) is satisfied and integrating over the allowed range of γ leads to

$$P(b) = \frac{1}{2}\int_{-1}^{1} d(\cos\gamma) H\left[E_T\left(1 - \frac{b^2}{d^2}\right) - E_0(\gamma)\right] = \tfrac{1}{2}(1 - \cos\bar{\gamma}) \quad (2.67)$$

where $H(x)$ is the unit step function, $H(x) = 1$ for $x \geq 0$ and $H(x) = 0$ for $x < 0$, and $\bar{\gamma}$ is the b-dependent maximal angle of the cone of acceptance, defined as the solution of

$$E_T(1 - b^2/d^2) = E_0(\bar{\gamma}) \tag{2.68}$$

Only if the barrier has a constant height E_0 for all approach angles $\gamma \leq \gamma_0$, and is infinitely repulsive otherwise, do we have $\bar{\gamma} = \gamma_0$ and b-independent. This model for $E_0(\gamma)$ is simpler than Eq. (2.65) but also less realistic. It does, however, conform to Eq. (2.50), with the steric factor p given by $(1 - \cos \gamma_0)/2$ and b_{max} defined as in Section 2.4.8, by $E_T(1 - b_{max}^2/d^2) = E_0$.

The seemingly simple concept of the centrifugal energy representing the energy of the rotation of the interparticle distance has thus taken us partway toward an understanding of molecular collisions. We are now ready to consider a rather direct manifestation of this rotation: the angular *deflection* (i.e., *scattering*) of the particles induced by collision.

Suggested reading

Section 2.1

Books

Golden, S., *Elements of the Theory of Gases*, Addison-Wesley, Reading, Mass., 1964.
Hirschfelder, J. O., Curtiss, C. F., and Bird, R. B., *Molecular Theory of Gases and Liquids*, Wiley, New York, 1964.
Jordan, P. C., *Chemical Kinetics and Transport*, Plenum, New York, 1979.
Kauzmann, W., *Kinetic Theory of Gases*, Benjamin, New York, 1966.
Present, R. D., *Kinetic Theory of Gases*, McGraw-Hill, New York, 1958.

Section 2.2

Books

Child, M. S., *Molecular Collision Theory*, Academic, New York, 1974.
Goldstein, H., *Classical Mechanics*, Addison-Wesley, New York, 1950.
Johnson, R. E., *Introduction to Atomic and Molecular Collisions*, Plenum, New York, 1982.
Lawley, K. P. (ed.), *Molecular Beam Scattering*, Adv. Chem. Phys., *30* (1975).
Massey, H. S. W., *Atomic and Molecular Collisions*, Taylor and Francis, London, 1979.
Mott, N. F., and Massey, H. S. W., *The Theory of Atomic Collisions*, Clarendon, Oxford, 1965.
Schlier, Ch. (ed.), *Molecular Beams and Reaction Kinetics*, Academic, New York, 1970.

Review Articles

Buck, U., "Elastic Scattering," *Adv. Chem. Phys., 30,* 313 (1975).
Pauly, H., "Elastic Scattering Cross Sections: Spherical Potentials," in Bernstein (1979).
Toennies, J. P., "Molecular Beam Scattering Experiments on Elastic, Inelastic and Reactive Collisions," in Eyring et al. (1974–5).

Section 2.3

Books

Amdur, L., and Hammes, G., *Chemical Kinetics: Principles and Selected Topics,* McGraw-Hill, New York, 1966.
Bunker, D. L., *Theory of Elementary Gas Reaction Rates,* Pergamon, Oxford, 1966.
Eyring, H., Lin, S. H., and Lin, S. M., *Basic Chemical Kinetics,* Wiley, New York, 1980.
Massey, H. S. W., Burhop, E. H. S., and Gilbody, H. B., *Electronic and Ionic Impact Phenomena,* Vol. 3, Clarendon, Oxford, 1971.
Moore, J. W., and Pearson, R. G., *Kinetics and Mechanism,* Wiley, New York, 1981.
Nikitin, E. E., *Theory of Elementary Atomic and Molecular Processes in Gases,* Clarendon, Oxford, 1974.
Pilling, M. J., *Reaction Kinetics,* Oxford University Press, Oxford, U.K., 1975.
Smith, I. M. W., *Kinetics and Dynamics of Elementary Gas Reactions,* Butterworths, London, 1980.
Weston, R. E., Jr., and Schwarz, H. A., *Chemical Kinetics,* Prentice-Hall, Englewood Cliffs, N.J., 1972.

Review Articles

Grice, R., "Reactive Scattering of Ground-State Oxygen Atoms," *Acc. Chem. Res., 14,* 37 (1981).
Lacmann, K., "Collisional Ionization," *Adv. Chem. Phys., 42,* 513 (1980).
Levy, M. R., "Dynamics of Reactive Collisions," *Prog. React. Kinet., 10,* 1 (1979).
Lin, M. C., "Dynamics of Oxygen Atom Reactions," *Adv. Chem. Phys., 42,* 113 (1980).
Mascord, D. J., Gorry, P. A., and Grice, R., "Alkali Atom–Dimer Exchange Reactions: Na + Rb$_2$," *Faraday Discuss. Chem. Soc., 62,* 255 (1977).
Menzinger, M., "The M + X$_2$ Reactions: A Case Study," in Fontijn (1985).
Steinfeld, J. I., and Kinsey, J. L., "The Determination of Chemical Reaction Cross-Sections," *Prog. React. Kinet., 5,* 1 (1970).
Weber, J. N., and Berry, R. S., "Collisional Dissociation and Chemical Relaxation of Rubidium and Cesium Halide Molecules," *Adv. Chem. Phys., 58,* 127, (1984).
Wexler, S., and Parks, E. K., "Molecular Beam Studies of Collisional Ionization and Ion-Pair Formation," *Annu. Rev. Phys. Chem., 30,* 179 (1979).

Section 2.4

Review Articles

Brooks, P. R., "Reactions of Oriented Molecules," *Science, 193,* 11 (1976).

Bunker, D. L., "Simple Kinetic Models from Arrhenius to the Computer," *Acc. Chem. Res., 7,* 195 (1974).

Chesnavich, W. J., Su, T., and Bowers, M. T., "Ion–Dipole Collisions: Recent Theoretical Advances," in Ausloos (1979).

Clary, D. C., "Rates of Chemical Reactions Dominated by Long-Range Intermolecular Forces," *Mol. Phys., 53,* 3 (1984).

Friedman, L., and Reuben, B. G., "A Review of Ion–Molecule Reactions, *Adv. Chem. Phys., 19,* 33 (1971).

Polanyi, J. C., and Schreiber, J. L., "The Dynamics of Bimolecular Reactions," in Eyring et al. (1974–1975).

Smith, I. W. M., "A New Collision Theory for Bimolecular Reactions," *J. Chem. Educ., 59,* 9 (1982).

Stolte, S., "Reactive Scattering Studies on Oriented Molecules," *Ber. Bunsenges. Phys. Chem., 86,* 413 (1982).

Su, T., and Bowers, M. T., "Classical Ion–Molecule Collision Theory," in Bowers (1979).

Talrose, V. L., Vinogradov, P. S., and Larin, I. K., "On the Rapidity of Ion–Molecule Reactions," in Bowers (1979).

Truhlar, D. G., and Dixon, D. A., "Direct-Mode Chemical Reactions: Classical Theories," in Bernstein (1979).

3

Scattering as a probe
of the collision dynamics

A direct probe of collision dynamics is the observation of the scattering (i.e., the deflection) of the colliding particles as a result of their encounter. From the observation of the particles' deflection after their collision we try to reconstruct the trajectory during the collision and hence determine the interaction forces between the colliding species.

The observation of large deflections in α-particle collisions with atoms led Rutherford, early in this century, to suggest the model of the atom wherein the positive charge is concentrated in a small, central nucleus. Ever since, the angular distribution after collision has been a primary diagnostic tool in the attempt to understand the interactions during the collision.

We begin this chapter with an introduction to this method as applied to elastic collisions and examine the derived intermolecular potentials. Combining our newly acquired knowledge with the considerations of Chapter 2, we can make an inroad into the more complicated problem of the dynamics of reactive collisions. Compensating for the greater complexity is the fact that numerous additional types of probes are available once internal degrees of freedom are coupled with the relative motion: Some of these are scalar in nature, such as the distribution of internal energy states mentioned in Chapter 1; others are vector quantities, such as the orientation dependence of the reactivity or the products' polarization. Furthermore, these different distributions are not independent in that they can be correlated. For example, since real molecules are not perfectly spherical (i.e., since the potential depends not only on relative separation but also on relative orientation), then, corresponding to every deflection angle, the product molecules can have an entire distribution for their orientation. The distribution of the molecular alignments (and also of the internal energy states) will, in general, depend upon the scattering angle. Only by Chapter 7 will we begin to do full justice to these aspects. Here we shall discuss mainly elastic collisions and then try to extend some of the ideas to more complex cases.

3.1 Elastic scattering as a probe of the interaction potential

3.1.1 The angular deflection

We saw in Chapter 2 that the role of the potential is to deflect the colliding molecules from their original, precollision straight-line paths. Section 2.2.4 showed that, for a spherically symmetric potential $V(R)$, the initial relative velocity and impact parameter specify a unique collision trajectory. The deflection of the colliding particles due to the collision is defined as the angle between the final (postcollision) direction of the interparticle distance and its initial (precollision) direction.† It follows that the deflection is a unique function of the initial collision energy and impact parameter. Given the interaction potential, the deflection angle can be determined (see Sec. 3.1.5) once we have computed the collision trajectory $\mathbf{R}(t)$.

As will be discussed shortly, the observable deflection angle is restricted to lie between 0 and π, yet the trajectory itself can result in a total deflection angle of more than π rad. Hence several distinct trajectories can lead to the same observable deflection (see Sec. 3.1.2).

3.1.2 The deflection function

Before we turn to more quantitative considerations, let us anticipate the qualitative dependence of the deflection upon the impact parameter.

We recall that the dependence of the interaction between two molecules upon their relative separation $V(R)$ is typically attractive at long range and steeply repulsive at short range. The long-range attraction is always present even for interactions between rare-gas atoms. The steeply repulsive short-range interaction dominates at separations where there is considerable overlap of the electronic clouds of the two molecules.

The potential well at intermediate separation may be shallow, reflecting the balance between these two physical (i.e., not very specific‡) forces, or it may be deeper, owing to the participation of chemical interactions. Either way, a well is always present. The stronger the attraction, the deeper the well.

To follow the consequences of the presence of the attractive and repulsive parts of $V(R)$, Fig. 3.1 shows schematically various collision trajectories for different values of the impact parameter at a given collision energy E_T. (It is convenient to define a "reduced" impact parameter $b^* = b/R_e$, where R_e is the separation at the minimum of the potential well.) For large b the molecules sample only the long-range attractive force and the trajectories deviate only slightly toward the axis, leading to a small (negative) deflection

---

† This also implies that the deflection is the angle between the final and initial relative velocity vectors, that is, $\chi = \arccos(\hat{\mathbf{v}} \cdot \hat{\mathbf{v}}')$.

‡ Then it is often termed the van der Waals well.

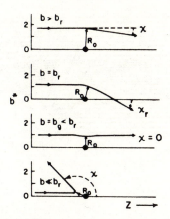

Fig. 3.1. Collision trajectories at various impact parameters for a given collision energy. The "projectile" of mass μ is directed along the z axis with velocity $\mathbf{v}$ and impact parameter b. The ordinate is the reduced impact parameter, $b^* = b/R_e$, χ is the angle of deflection; b_r is the impact parameter at the rainbow angle χ_r, and R_0 is the distance of closest approach.

angle χ. As b decreases, the deflection angle becomes increasingly negative until it reaches its most negative value at the so-called *rainbow angle* χ_r at the impact parameter b_r. As b is decreased further, the influence of the repulsive force becomes dominant, and the deflection angle increases, passes through zero when $b = b_g$, the so-called *glory impact parameter*, and becomes progressively more positive. As $b \to 0$ and the collisions become more nearly head-on, and the molecules rebound, in the backward direction $(\chi \to \pi)$.

The correlation between deflection angle and initial impact parameter (called the deflection function) is so central to our understanding that in Fig. 3.2 we repeat the same information for a swarm of trajectories with different values of the impact parameter, based on a computation for a realistic potential.

Owing to the cylindrical symmetry about the $b = 0$ axis (see Fig. 2.11) the sign of the deflection angle is not experimentally meaningful. The *observable* deflection angle θ is thus the absolute value of the *computed* deflection angle, $\theta = |\chi|$.

The qualitative relation between χ and b at a given E_T for a realistic intermolecular potential is summarized in Fig. 3.3. For $\chi > \chi_r$ there is a unique correspondence between b and χ; for $\chi < \chi_r$ three different values of b correspond to the same deflection angle $\theta = |\chi|$. This multivalued relation between b and θ is *not* found for a purely *repulsive* potential, for which $\chi(b)$ is a monotonically decreasing function of b.

Figure 2.10 showed a hard-sphere collision. It is evident from the construction there that $\chi = \pi - 2\psi_0$, where $b/d = \sin \psi_0$. Thus for the rigid-sphere potential,

$$\chi = \begin{cases} 2\arccos(b/d), & b \leq d \\ 0, & b > d \end{cases} \tag{3.1}$$

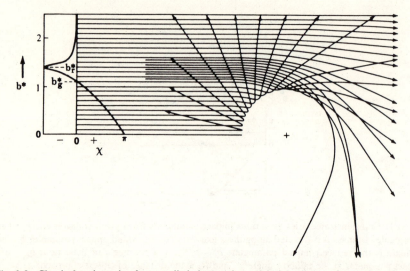

Fig. 3.2. Classical trajectories for a realistic interaction potential, all at a given value of v but with different values of the reduced impact parameter b^*. The graph at the left summarizes the dependence of the deflection function χ upon b^*. Indicated are the glory and rainbow (reduced) impact parameters b_g^* and b_r^*. Note the negative deflection for $b > b_g$; the most negative deflection, say $\chi_{\min}$, at b_r, defines the rainbow angle $\theta_r = |\chi_{\min}|$. [Adapted from H. Pauly, in Bernstein (1979).]

If we could sample the b dependence of the deflection function, we would thus have a very sensitive probe of the intermolecular potential. In fact, we shall see that, at least for *high b*, the functional dependence of $\chi(b)$ is that of $V(b)$, the potential evaluated at $R = b$. At small b, the existence of a minimum in $\chi(b)$ is an indication of the well of the realistic potential, as is evident in Figs. 3.2 and 3.3.

3.1.3 Angular distribution in the center-of-mass system

Our simplistic aim is to ascertain, experimentally, the b dependence of the deflection function. We cannot hope to do this directly, since (even in a collision between molecules of well-defined velocities) *all values* of the microscopic impact parameter are possible (see Sec. 2.2.4). What we can do however, is measure the *angular distribution* of the products (i.e., the scattered flux) as a function of the deflection angle θ and thereby establish the relation between b and θ.

Consider a bunch of collision trajectories with impact parameters in the range b to $b + db$. If b is small enough, the molecules will suffer a deflection. As shown in Fig. 3.4, the final deflection angle lies in the range θ to $\theta + d\theta$, where $\theta = |\chi(b)|$. For a spherically symmetrical "central" potential there will be no dependence on the azimuthal angle ϕ. Thus all

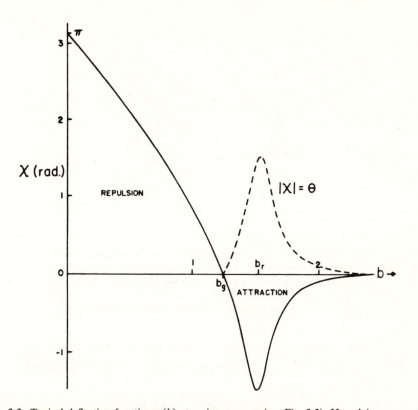

Fig. 3.3. Typical deflection function $\chi(b)$ at a given energy (see Fig. 3.2). Here b is expressed in reduced units. At large b, the angle of deflection is negative (but very small) owing to the long-range intermolecular attraction. As b is decreased, χ becomes more negative, and eventually at $b = b_r$ the greatest negative deflection χ_r is obtained. As b is further reduced, the influence of repulsive forces becomes increasingly evident and, at $b = b_g$, the long-range attraction has been compensated for by the short-range repulsion, so that the net (overall) deflection is zero. At still smaller b, the repulsion dominates and backward scattering results, with $\chi \to \pi$ as $b \to 0$. The observable deflection angle is $\theta \equiv |\chi|$.

the trajectories that entered through an annulus of radius b and width db will exit via an annular cone into a solid angle $d\omega$ with ϕ in the range 0 to 2π and θ in the range θ to $\theta + d\theta$ so that $d\omega = 2\pi \sin \theta \, d\theta$.

We can measure (see Appendix 3A) the number of molecules $d\dot{N}(\theta)$ deflected per unit time into the solid angle $d\omega = 2\pi \sin \theta \, d\theta$. In the same way that the collision cross section is a measure of the rate of all collisions [see Eq. (2.9)], we introduce the *differential collision cross section* $I(\theta)$ as a measure of the rate of collisions leading to deflections in a narrow angular range near θ, that is, $d\dot{N}(\theta) \propto I(\theta) \, d\omega$.

All the trajectories that lead to deflections in the range θ to $\theta + d\theta$ have impact parameters in the range b to $b + db$. The incremental cross section

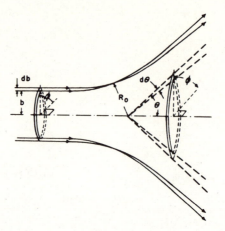

Fig. 3.4. A construction showing the equivalence between the trajectories with initial impact parameters in the range b to $b + db$ and those with final deflections in the range θ to $\theta + d\theta$, where $\theta = |\chi(b)|$. In the usual case, there is no dependence on the azimuthal angle ϕ, and all initial values of ϕ are equally probable, leading to cylindrical symmetry. Trajectories with the same final deflection angle θ will thus form a scattered cone of equi-intensity.

$d\sigma$ for these collisions is [see Eq. (2.36)] $2\pi b\, db$. Thus $d\sigma = I(\theta)\, d\omega = 2\pi b\, db$ so that the differential cross section (at a specified E_T) becomes[†]

$$I(\theta) = \frac{b}{\sin\theta\,(d\theta/db)} = \frac{b}{\sin\theta\,|d\chi/db|} \tag{3.2}$$

(where the absolute sign is necessary because $d\chi/db$ can be negative). Of course, the total cross section is just the *integral of the differential cross section*:

$$\sigma = \int d\sigma = \int I(\theta)\, d\omega = 2\pi \int_0^\pi I(\theta)\sin\theta\, d\theta \tag{3.3}$$

Given the deflection function $\chi(b)$ at the energy E_T of the experiment, we can easily calculate $I(\theta)$, the angular distribution of the scattering in the center-of-mass (c.m.) system (Fig. 3.5), except near $\theta = 0$, where Eq. (3.2) is divergent! This divergence of Eq. (3.2) is a defect of the *classical-*mechanical approach to what is really a *quantal* problem. There are, in fact, three troubles: (1) the divergence for large values of b, where $\theta \to 0$ (this is further discussed in Sec. 3.1.8), (2) the divergence at the glory impact parameter, where $\chi(b_g) = 0$ and thus $\sin\theta = 0$ (also discussed in Sec. 3.1.8), and (3) the divergence at the rainbow impact parameter, where $d\chi/db = 0$ (discussed in Sec. 3.1.9). In all three cases the correct quantal treatment

---

[†] There is an interesting complication when the deflection function is such that there may be more than one value of b corresponding to a given $\theta = |\chi|$ (see Fig. 3.3). This is discussed in Section 3.1.9.

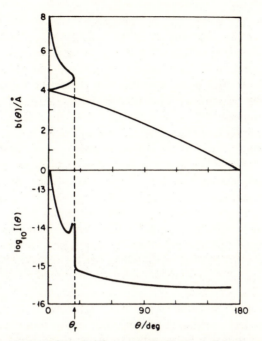

Fig. 3.5. Classically calculated [Eq. (3.2)] angular distribution of elastic scattering $I(\theta)$ for a realistic molecular system. Plotted here are $I(\theta)$ $(cm^2\,sr^{-1})$ and the corresponding $b(\theta)$ (Å), for the same collision energy. Note the classical *rainbow* divergence at θ_r.

yields a finite value. The physical manifestations of these classical singularities are, however, an important source of information on the intermolecular potential well.

3.1.4 Scattering as a probe of the potential

As an example, let us evaluate the angular distribution expected for the rigid-sphere potential. Using Eq. (3.1) for the deflection function, Eq. (3.2) yields

$$I(\theta) = \tfrac{1}{4}d^2 \tag{3.4}$$

independent of θ! This implies *isotropic* scattering, independent of collision energy, and from Eq. (3.3), $\sigma = \pi d^2$.

For a more realistic intermolecular potential function, however, the angular distribution of the scattered molecules will be both *anisotropic and energy dependent*. It is found to be concentrated in the low-angle region (see Fig. 3.5). Since the potential has a long "range," there is a large contribution to the scattering from ever-larger annular areas ($2\pi b\,db$) at large b and thus small $|\chi|$ (see Fig. 3.3). Thus the low-angle scattering is mainly due to (and thus tells us about) the long-range part of the potential. The wide-angle scattering comes from the repulsive collisions at small b

[note that $\chi(b \to 0) \to \pi$] and is approximately that of the hard-sphere model to the extent that the realistic deflection function (see Fig. 3.3) resembles that for hard spheres, Eq. (3.1), at small b.

Thus an experimental knowledge† of the angular distribution $I(\theta)$ at several energies enables us to "sample" the deflection function over a wide range of impact parameters and thus provides a sensitive probe of the intermolecular potential.

To actually implement this program we will need a *quantitative* relation between $\chi(b)$ and $V(R)$. We need *more* than the qualitative information that low-angle (forward) scattering corresponds to large impact parameter collisions which sample the long-range tail of the potential and that large-angle (backward) scattering is from low-impact parameter collisions which reach the inner "hard core" of the potential. We need to know the angular range over which to look for the effects of the potential well. We have already noted that the presence of the potential minimum implies that $d\chi/db = 0$ for $b = b_r$ (see Fig. 3.3), which occurs near R_e and hence leads to a "divergence" in the angular distribution. We need to know more about this so-called rainbow effect.

In the following Sections 3.1.5 to 3.1.10 we will develop a more *quantitative* classical mechanical description of *scattering as a probe of the dynamics* and of the quantal manifestations of the dynamics. We shall also note the information to be gained from the study of the velocity dependence of the collision cross section $\sigma(v)$.‡ Section 3.2 is the "payoff." What have we learned from collisional experiments about intermolecular forces?

3.1.5 From the potential to the deflection function

The computation of the differential cross section requires that we know the *deflection function* $\chi(b)$. This enables us to relate the *observable* angular distribution $I(\theta)$ to the desired *intermolecular potential*.

We evaluate the deflection function using an equation of motion, Eq. (2.24) for the angle of orientation ψ, of the interparticle vector **R** (see Fig. 2.7):

$$\dot{\psi} = bv/R^2 \tag{3.5}$$

To integrate Eq. (3.5) we change from a time derivative to a spatial derivative, that is, $d\psi/dR = \dot{\psi}/\dot{R}$. Using Eq. (2.31′), we have

$$\dot{R} = v\left(1 - \frac{V(R)}{E_T} - \frac{b^2}{R^2}\right)^{1/2} \tag{3.6}$$

---

† Appendix 3A discusses the problem of extracting the c.m. angular distribution from the observed scattering in the laboratory system.

‡ Contrary to our naive expectation, it is *not* the short-range repulsive part of the potential that is usually revealed in this fashion. In fact, at low collision energies (i.e., $E_T \approx kT$, the so-called thermal energy range), it is the *long-range, attractive* part of the potential that determines the overall magnitude of $\sigma(v)$ (see Sec. 3.1.8).

and so

$$\frac{d\psi}{dR} = -\frac{b}{R^2}\left(1 - \frac{V(R)}{E_T} - \frac{b^2}{R^2}\right)^{-1/2} \tag{3.7}$$

The minus sign arises because ψ increases, from an initial value of zero (at $R \to \infty$), as R decreases (see Fig. 2.8). We can now integrate Eq. (3.7) from the precollision condition up to the so-called turning point of the relative motion, $R = R_0$, $\psi = \psi_0$:

$$\psi_0 = \int_\infty^{R_0} \frac{d\psi}{dR}\, dR = b\int_{R_0}^\infty \frac{dR}{R^2}\left(1 - \frac{V(R)}{E_T} - \frac{b^2}{R^2}\right)^{-1/2} \tag{3.8}$$

The integrand is well defined, since R_0 is the positive root of the equation $1 - V(R_0)/E_T - b^2/R_0^2 = 0$. From Fig. 2.8, we have

$$\chi = \pi - 2\psi_0 = \pi - 2b\int_{R_0}^\infty \frac{dR}{R^2}\left(1 - \frac{V(R)}{E_T} - \frac{b^2}{R^2}\right)^{-1/2} \tag{3.9}$$

Equation (3.9) is quite general and exact within the classical framework.

Thus, *if we knew the potential $V(R)$, we could calculate the deflection function $\chi(b, E_T)$* via Eq. (3.9) at any desired collision energy E_T. We should then be able to predict the angular distribution of the scattering, using Eq. (3.2).

A useful approximate "high-energy" form of Eq. (3.9) is obtained in the limit of large impact parameters when $R_0 \approx b$. In this limit, $V(R)/E_T \ll 1$. When $V(R_0)/E_T \ll b^2/R_0^2$, it can be shown that the variable $E_T\chi$ is a function of b *only* (independent of E_T). We can then introduce a quantity $\tau(b)$, which we write as

$$\tau(b) = E_T\chi(b, E_T) \tag{3.10}$$

It is found that the b dependence of τ is essentially that of $V(b)$. Thus at large impact parameters the functional dependence of the deflection function on b and E_T is of the form

$$\chi(b, E_T) \propto V(b)/E_T \tag{3.11}$$

In terms of the variable τ, that is, using Eq. (3.10), we can rewrite Eq. (3.2) in the form

$$\theta \sin\theta\, I(\theta) = \tfrac{1}{2}\tau\left|\frac{db^2}{d\tau}\right| \tag{3.12}$$

where there is no explicit dependence upon E_T. Hence, in this high-energy approximation, a plot of $\theta \sin\theta\, I(\theta)$ versus τ is independent of the collision energy.

For a long-range potential of the form $V \propto R^{-s}$, $\theta \sin\theta\, I(\theta) \propto \tau^{-2/s}$. Thus for small angles ($\sin\theta \approx \theta$),

$$\theta \sin\theta (E_T\theta)^{2/s} I(\theta) \simeq E_T^{2/s}\theta^{(2s+2)/s} I(\theta) = \text{const.} \tag{3.13}$$

For realistic molecular systems, we shall see in Section 3.2 that the attractive long-range potential is dominated by an inverse sixth-power R dependence. We note that for $s = 6$, Eq. (3.13) implies that $I(\theta) \propto \theta^{-7/3} E_T^{-1/3}$, which has been found to be in accord with the experimental low-angle scattering data. For ion–molecule scattering $s = 4$ (as discussed in Sec. 2.4.7), so that $I(\theta) \propto \theta^{-5/2} E_T^{-1/2}$.

3.1.6 The differential cross section

Once we know the deflection function, we are in a position to calculate the angular distribution of the scattering in the c.m. system† and then predict the observable scattering pattern in the laboratory. We ask how many molecules $d\dot{N}(\theta, \phi)$ will be scattered per unit time into an element of solid angle $d\omega$ ($= \sin \theta \, d\theta \, d\phi$) in some *specified* direction θ, ϕ, where θ is the c.m. deflection with respect to the initial relative velocity direction and ϕ is the azimuthal angle (see Fig. 3.4). First let us recall that the *total* number of beam molecules scattered into *all angles* per unit time is given [see Eq. (2.9)] by

$$\dot{N} = Z n_A n_B \, \Delta V = n_A n_B v \sigma \, \Delta V = I_A N_B \sigma \qquad (3.14)$$

where ΔV is the volume of the scattering region, v the relative speed, n_A and n_B the respective number densities, $N_B = n_B \, \Delta V$ is the total number of target molecules, and $I_A = n_A v$ is the beam intensity (the incident flux of A molecules per unit area per unit time). We can then express the desired differential quantity as

$$d\dot{N}(\theta, \phi) = \dot{N} P(\theta, \phi) \, d\omega \qquad (3.15)$$

where the angular distribution $P(\theta, \phi)$ is a normalized probability density function (with dimensions of reciprocal solid angle), that is,

$$\iint P(\theta, \phi) \sin \theta \, d\theta \, d\phi = 1 \qquad (3.16)$$

Thus the number of A particles scattered per unit time per unit solid angle in the direction θ, ϕ is

$$\frac{d\dot{N}(\theta, \phi)}{d\omega} = n_A n_B v \sigma \, \Delta V P(\theta, \phi) = I_A N_B \sigma P(\theta, \phi) \qquad (3.17)$$

The product $\sigma P(\theta, \phi)$ is termed the *differential* (solid-angle) *cross section*, with dimensions of area per unit solid angle.

For a spherically symmetrical potential such as $V(R)$, there is no ϕ dependence of the scattering‡ and $P(\theta, \phi)$ is really only a function of θ (all

---

† This is equivalent to the angular distribution from a hypothetical experiment in which beam particles of mass μ are scattered by a fixed center of force (see Fig. 2.11).

‡ For a central potential, the collision trajectory is confined to a plane (see Sec. 2.2.5).

azimuthal planes are equally probable, so that the scattering is cylindrically symmetrical around v). Thus we can integrate over ϕ to obtain the intensity scattered into an annular cone θ, $\theta + d\theta$ (see Fig. 3.4):

$$d\dot{N}(\theta) = \int_0^{2\pi} d\phi \frac{d\dot{N}(\theta, \phi)}{d\phi} = 2\pi \sin\theta \frac{d\dot{N}(\theta, \phi)}{d\omega} = I_A N_B 2\pi \sin\theta \, I(\theta) \, d\theta$$

(3.18)

Here we used $I(\theta) = \sigma P(\theta, \phi)$ for the differential (solid-angle) cross section, [see Eq. (3.2)]. Thus

$$\sigma = 2\pi \int_0^\pi d\theta \sin\theta \, I(\theta)$$

(3.19)

In Sections 3.1.4 and 3.1.5 we completed our analysis of the elastic scattering experiment. We established a quantitative route from the potential $V(R)$ to the deflection function $\chi(b, E_T)$ to the differential cross section $I(\theta)$ to the observable angular distribution of the flux of scattered molecules $d\dot{N}(\theta, \phi)$ to the total cross section σ at a given E_T.

Let us now consider the opposite route, from the observed differential cross section to the potential. This is the famous "inversion problem" common to all aspects of scattering. We shall restrict ourselves to the simplest case, where the relation between the deflection function $\chi(b)$ and the impact parameter is one to one, that is, $\chi(b)$ is monotonic. What this means is either that the potential is purely repulsive or, if it is not, that we examine only such angles for which the relation is one to one. Such is the case for deflections at angles beyond the rainbow (see Fig. 3.5). Of course, in the latter case we shall only be able to determine the part of the potential responsible for the wide-angle scattering (which, from Fig. 3.3, is the repulsive part).

Consider the "partial" or "incomplete" total cross section $\sigma_>(\theta)$ defined by [see Eq. (3.3)]

$$\sigma_>(\theta) = 2\pi \int_\theta^\pi d\theta \sin\theta \, I(\theta)$$

(3.20)

Using Eq. (3.2) for $I(\theta)$, we have

$$\sigma_>(\theta) = 2\pi \int_\theta^\pi d\theta b \left| \frac{db}{d\theta} \right| = \pi \int_\theta^\pi d\theta \left| \frac{db^2}{d\theta} \right| = \pi [b(\theta)]^2$$

(3.21)

Hence, given the observed differential cross section $I(\theta)$, we compute $\sigma_>(\theta)$ by integration as in Eq. (3.20) and thence $b(\theta)$ via Eq. (3.21). But $b(\theta)$ and the potential are related by Eq. (3.9). To invert Eq. (3.9) we proceed as follows. One evaluates an integral $I(x)$, defined as

$$I(x) = \pi^{-1} \int_{b=x}^\infty db \frac{\theta(b)}{(b^2 - x^2)^{1/2}}$$

(3.22)

for a series of arbitrary values of the dummy parameter x. It can be shown that

$$V(R) = E\{1 - \exp[-2I(x)]\} \tag{3.23}$$

where

$$R = x \exp[I(x)] \tag{3.24}$$

Equations (3.23) and (3.24) together yield the following interpretation of x:

$$x = R[1 - V(R)/E]^{1/2} \tag{3.25}$$

but this is not required for the analysis.

For each chosen x one has a numerical value of $I(x)$ via Eq. (3.22) and knows R from Eq. (3.24). Equation (3.23) yields $V(R)$ for that value of R. The procedure is repeated for each x (and the corresponding R values), resulting in a unique set of points $V(R)$. This determines the repulsive part of the potential curve up to a potential energy equal to the collision energy E. The potential as a function of x (and hence of R) is given by Eq. (3.23).

The two inherent limitations of this method are clear from the derivation. In Eq. (3.21) we have assumed that there is only one value of b corresponding to a given value of θ and, also, Eq. (3.23) determines $V(R)$ only up to the innermost ($b = 0$) distance of closest approach $V(R_0) = E$ [see Eq. (2.34)].

The approach thus far has been classical mechanical and therefore all results must be regarded as approximations. Only the full quantum mechanical treatment can be considered correct. In more complex situations, using classical mechanics is simpler and hence we must ascertain the validity of the classical results at least for this simple case. This is our next task.

3.1.7 The quantum mechanical approach to elastic scattering

In many ways the quantum mechanical treatment of elastic scattering is similar to the classical approach. The essential difference is the "superposition" principle unique to quantum mechanics: When an event can be realized in several ways, its probability is computed by adding together (i.e., superposing) the probability amplitudes for each way in which it occurs and then taking the square of the sum. In writing out such a sum squared there will be many cross-terms, each a product of two different amplitudes. It is these cross-terms that give rise to the "interference structure" (to be discussed later) which is so typical of quantum mechanics. As will also be true in our problem, the classical limit obtains when these cross-terms are so numerous and their oscillation is so rapid that they tend to average out.

Rather than a swarm of trajectories of all impact parameters (see Fig. 3.2), in the time-independent quantum mechanical scattering approach we have a wave function consisting of a sum of partial waves:

$$\psi(\mathbf{R}) = \sum_{l=0}^{\infty} (2l + 1)i^l \psi_l(R) P_l(\cos \theta) \tag{3.26}$$

where θ is the polar angle with respect to the incident velocity and l is the angular momentum quantum number (see Sec. 2.2.5). The wave function $\psi(\mathbf{R})$ is itself a probability amplitude and it is written as a superposition of amplitudes for the different values of the angular momentum. The $P_l(\cos \theta)$ are the Legendre polynomials describing (as in many other problems) the angular part. At a given collision energy E, the Schrödinger equation $(E - H)\psi = 0$, after the angular part is factored out, leads to the radial equation

$$\left(k^2 + \frac{d^2}{dR^2} - \frac{l(l+1)}{R^2} - \frac{2\mu}{\hbar^2} V(R)\right) G_l(R) = 0 \tag{3.27}$$

Equation (3.27) is an effective Schrödinger equation for the relative motion with $\psi_l(R) = G_l(R)/kR$, k is the wave number, $E = \hbar^2 k^2/2\mu$, or $p = \hbar k$, so that $k = p/\hbar = 2\pi/\lambda$, where λ is the de Broglie wavelength.

The new feature of scattering problems as compared to bound state problems is that the boundary conditions, namely, the behavior of the wave function as $R \to \infty$, need to be explicitly discussed. Common to all elastic scattering problems is that the incident motion is in the direction of decreasing R and the scattered motion is in the direction of increasing R. Hence, when R is large enough such that $V(R) \to 0$, we expect the wave function $G_l(R)$ to have the form

$$G_l(R) \to \exp[-i(kR - l\pi/2)] - S_l \exp[i(kR - l\pi/2)] \tag{3.28}$$

where S_l is the scattering amplitude for the lth partial wave. The phase of the incident and scattered waves is determined by solving Eq. (3.27) first for the case where there is no potential. One then finds that the solution that precludes the motion reaching the origin is

$$G_l(R) \propto kRj_l(kR) \to \begin{cases} \sin(kR - \frac{1}{2}l\pi), & R \to \infty \\ 0, & R \to 0 \end{cases} \tag{3.29}$$

where $j_l(kR)$ is the regular spherical Bessel function of order l. It is an oscillatory function having essentially its asymptotic $(R \to \infty)$ form all the way back to the point where its first maximum occurs; it then declines to zero. Since $l + \frac{1}{2} \approx kb$, the innermost inflection of $G_l(R)$ occurs at about the classical turning point $R = b$.

The solution of the free motion is thus of the form of Eq. (3.28) with a unit scattering amplitude. In the presence of a potential the turning point for the relative motion is no longer at $R = b$. The corresponding wave function will thus have a phase somewhat different from that for the free motion. What matters to us is the change, or *shift*, in the phase. The general solution of Eq. (3.27) will thus have the form

$$G_l(R) \rightarrow \begin{cases} e^{i\delta_l}\sin(kR - \frac{1}{2}l\pi + \delta_l), & R \rightarrow \infty \\ 0, & R \rightarrow 0 \end{cases} \tag{3.30}$$

where δ_l is the *phase shift*. Comparing Eqs. (3.30) and (3.28), we find the scattering amplitude to be

$$S_l = \exp(2i\delta_l) \tag{3.31}$$

Note that $|S_l|^2 = 1$. All the incident flux is recovered. The only role played by the potential (in elastic scattering) is to shift the phase of the wave function. The magnitude of this shift is computed by solving Eq. (3.27) (in general, numerically) and determining δ_l from its definition by using the asymptotic form of Eq. (3.30).

The amplitude of the scattered wave at the angle θ from the direction of the incident wave, $f(\theta)$, is defined by comparing the full wave function $\psi(\mathbf{R})$ to $\phi(\mathbf{R})$, the wave function for free motion,

$$\psi(\mathbf{R}) = \phi(\mathbf{R}) + \frac{e^{ikR}}{R}f(\theta), \qquad R \rightarrow \infty \tag{3.32}$$

Using Eqs. (3.30)–(3.26), we finally obtain

$$f(\theta) = (2ik)^{-1}\sum_{l=0}^{\infty}(2l+1)(e^{2i\delta_l}-1)P_l(\cos\theta) \tag{3.33}$$

The scattered intensity is

$$I(\theta) = |f(\theta)|^2 \tag{3.34}$$

and $I(\theta)$ is identified with the differential cross section $d\sigma/d\omega$.

Two features of our result, Eq. (3.33), deserve immediate comment. One is that $S_l - 1 \equiv \exp(2i\delta_l) - 1$ rather than S_l itself appears in the scattering amplitude at the angle θ. For sufficiently large l, the wave function hardly samples the potential (being kept away by the centrifugal barrier). Hence as l increases, the phase shift must ultimately decline to zero, or $S_l \rightarrow 1$ when $l \gg A$, where $A \equiv kR_0$, with R_0 the "range" parameter of the potential (Fig. 3.6). Hence the sum of Eq. (3.33) is effectively a sum over a finite number of terms and thus convergent! The classical divergencies of the differential cross section have been eliminated.

Still, the sum of Eq. (3.33) will, in general, contain many terms. The reason is that at all but the very lowest relative velocities, atoms and molecules are heavy enough that their de Broglie wavelength $\lambda = h/\mu v$ is significantly shorter than the range R_0 of the potential. But if $\lambda \ll R_0$, then $A \gg 2\pi$. But we must retain terms up to $l \gg A$ in the sum of Eq. (3.33). Since the Legendre polynomials are oscillatory functions of θ, there will be very many highly oscillatory terms when we compute $|f(\theta)|^2$ (Fig. 3.7). At the higher relative velocities, as the de Broglie wavelength gets shorter (and

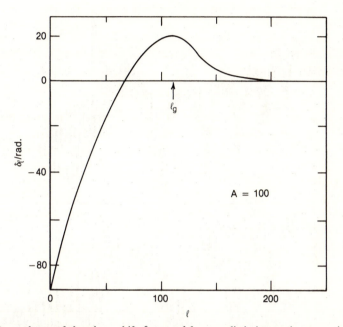

Fig. 3.6. Dependence of the phase shift δ_l upon l for a realistic interaction potential, with a "reduced collision wavenumber" $A = 100$. Note the large negative values of δ_l at small l, due to the predominance of repulsion, and the positive values at larger l, associated with the long-range attractive part of the potential. The initial slope $(d\delta_l/dl)_{l=0}$ is $\pi/2$. The maximum in $\delta_l(l)$ occurs at l_g corresponding to the glory impact parameter $l_g \simeq b_g k$.

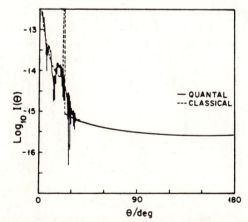

Fig. 3.7. Comparison of the quantal versus the classical calculation of $I(\theta)$ (see Fig. 3.6) for the same potential at the same collision energy. Whereas the classical curve diverges at $\theta = 0$ and at θ_r, the oscillatory quantal cross section is finite everywhere.

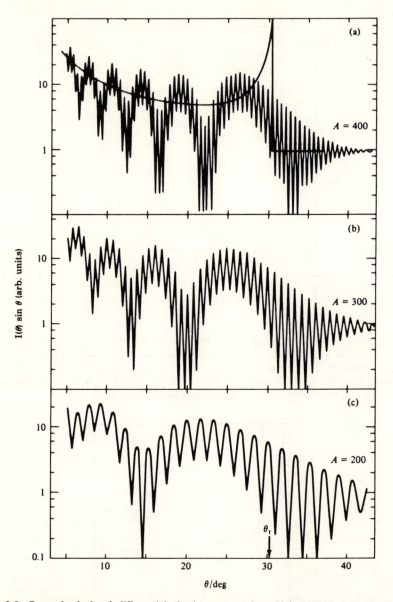

Fig. 3.8. Quantal-calculated differential elastic cross sections $I(\theta)$ multiplied by $\sin \theta$ for a realistic potential function corresponding to collisions at the same "reduced energy" (E_T in units of the potential well depth ϵ), but for different values of the reduced wave number A. Large A implies "more classical," since more partial waves contribute to the scattering amplitude [Eq. (3.33)]. The classical cross section is independent of A and is shown as the smooth curve in (a). [Adapted from H. Pauly, in Bernstein (1979).]

more terms need be included), the oscillations become more rapid and tend increasingly to average out to the classical result (Figs. 3.7 and 3.8).

Much as the deflection function serves as an intermediate construct between the potential and the observed differential cross section in classical mechanics, so does the set of phase shifts in quantum mechanics. Indeed, the relation between the two is even closer than this formal similarity might suggest. While our argument will be sketchy, the final conclusion will be correct. Consider the partial wave sum, Eq. (3.33), for the scattering amplitude. Although all l values contribute to the scattering at a given angle θ, since the Legendre polynomials at a given θ are a rapidly varying function of l, only a narrow range of l values contributes significantly. The range is determined by the observation that at a given θ, $P_l(\cos \theta)$ varies with l as $\sin(l\theta)$. The main contribution to the sum is where it is hardly oscillatory as a function of l. Regarding l as a continuous variable, this point of "stationary phase" is at

$$\theta = \frac{2d\delta_l}{dl} \tag{3.35}$$

The result of Eq. (3.35) is a "semiclassical" correspondence between the l dependence of the phase shift (see Fig. 3.6), and the deflection angle at the impact parameter b, $l \approx kb$.

Starting with a given potential $V(R)$, we have traveled both the classical and the quantal route to the differential cross section. We now continue to consider the "inverse problem": What can we learn about $V(R)$ from the observation of the angular distribution? This is the subject of the remainder of Section 3.1.

3.1.8 The total cross section and the glory effect

We recall from Section 2.2.7 that the total collision cross section can be written as the integral

$$\sigma = 2\pi \int_0^{b_c} b \, db = \pi b_c^2 \tag{3.36}$$

where b_c is the largest value of b that can lead to a "measurable" angular deflection.

For finite-ranged potentials, Eq. (3.36) is meaningful,† but for a realistic intermolecular potential, since $\chi(b)$ remains finite for very large b, Eq. (3.36) would predict infinite cross sections. However, *quantum mechanics* shows that the cross section is *finite*, since we are unable to resolve very small angles when dealing with the scattering of de Broglie "matter waves." Thus the largest value of b, say b_c, for which χ is just large enough to be observable, is given by $\chi(b_c) = \lambda/b_c$, where λ is the de Broglie wavelength

---

† For example, for the rigid-sphere potential, $\chi = 0$ for $b > b_c = d$ so $\sigma = \pi d^2$.

of the colliding particle system. A measurement of σ is therefore equivalent to a measurement of $\chi(b)$ at the *one* value $b = b_c$. But b_c is energy dependent and so a measurement of the velocity dependence of σ is tantamount to measuring $\chi(b)$ over a range of b (at large b). In particular, when b_c is large, one can deduce the *long-range* part of the potential function from $\sigma(v)$. For the long-range form $V(R) = -C_s R^{-s}$ we have, using the high-energy approximation of Eq. (3.11),

$$\chi(b_c) \propto \frac{V(b_c)}{E_T} \quad \text{or} \quad \frac{h}{\mu v b_c} \propto \frac{C_s}{b_c^s \mu v^2}$$

or

$$b_c \propto (C_s/hv)^{1/(s-1)} \tag{3.37}$$

Thus

$$\sigma = p_s (C_s/hv)^{2/(s-1)} \tag{3.38}$$

In a more exact treatment p_s is found to depend on s but is approximately equal to 2π. Since b_c *is* large at low velocities, $\sigma(v)$ at thermal energies is sensitive mainly to the long-range part of the potential.

Referring to Fig. 3.9, we note that a plot of $\log \sigma(v)$ versus $\log v$ is essentially linear (at low velocities), in accord with Eq. (3.38), and so can be used for the estimation of both s and C_s. The "glory undulations" superimposed on the linear dependence will be discussed shortly.

The quantal derivation proceeds as follows. First, by integrating $I(\theta)$ over solid angle and using the orthogonality of the Legendre polynomials, we obtain the exact expression

$$\sigma = \frac{4\pi}{k^2} \sum_{l=0}^{\infty} (2l + 1)\sin^2 \delta_l \tag{3.39}$$

As in the classical limit, the exact quantal form of Eq. (3.39) is a sum over

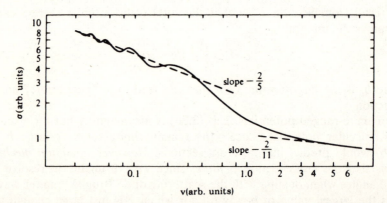

Fig. 3.9. Log–log plot of the total elastic scattering cross section versus the relative speed calculated for a Lennard–Jones (12,6) potential, showing glory extrema superimposed on the $-\frac{2}{5}$ slope (for an R^{-6} long-range attraction). At high speeds, the asymptotic slope is $-\frac{2}{11}$, corresponding to the R^{-12} repulsion. [Adapted from H. Pauly in Eyring (1974–1975).]

partial waves, with no cross-terms. To approximate the sum we note that when many partial waves contribute (i.e., $kR_0 \gg 1$), the phase shift is large and rapidly varying with l. As l increases from 0 on, δ_l varies over a range of many times π. Hence $\sin^2 \delta_l$ is oscillating rapidly between 0 and 1 and can therefore be approximated by its mean value of $\frac{1}{2}$. For a sufficiently high l the phase shift will, however, cease to vary and will start to monotonically decline to zero (see Fig. 3.6). This "random phase shift" approximation is thus valid up to the highest value of l, say, L_c, for which $\delta_l \approx \pi/2$ for the last time. Past this $l = L_c$ value, δ_l will remain below $\pi/2$ and will decline with increasing l. The contribution to the cross section from all partial waves with $l \leq L_c$ is therefore

$$\sigma = \frac{4\pi}{k^2} \sum_{l=0}^{L_c} \tfrac{1}{2}(2l + 1) = \frac{2\pi L_c^2}{k^2} = 2\pi b_c^2 \qquad (3.40)$$

where $b_c \approx L_c/k$ is the classical "cutoff" impact parameter corresponding to the angular momentum L_c. The actual cross section will be somewhat larger than Eq. (3.40) owing to an additional contribution from partial waves with $l > L_c$. The increment is small, some 20% for an R^{-6} long-range potential. What is more important is the factor 2 rather than 1 as given by the classical theory [Eq. (3.36)]. The origin of this factor 2 is purely quantal and is termed the shadow scattering effect. It arises from a forward peak in the angular distribution, the integrated contribution of which is about πb_c^2. It is present even for the rigid-sphere case.

We still need to relate L_c to the potential and the collision energy. It turns out that our choice of L_c as the largest value of l for which $\delta_l = \pi/2$ is the same as our choice of b_c as the largest value of b for which the deflection function equals λ/b. The quantal result is thus also Eq. (3.38). The quantal (i.e., correct) value of p_s is just above 2π.

Figure 3.9 shows the velocity dependence of the cross section for velocities at the thermal and epithermal range. The *undulatory velocity dependence* of $\sigma(v)$ about the mean line [which is given by Eq. (3.38)] is a manifestation of a quantum-mechanical interference phenomenon. This can be understood in terms of the classical trajectories shown in Fig. 3.2. Forward scattering ($\theta = 0$) arises *both* from trajectories at very large impact parameters ($b > b_c$) *and* from the so-called glory trajectory at $b = b_g$. It is a general feature of quantum mechanics that whenever two seemingly distinct classical trajectories lead to the same observable result, they produce an interference pattern.† In the present example, the interference is between the trajectory at $b = b_g$ and a trajectory at very high b. This interference can either diminish or enhance the forward scattering (and the total cross section σ), depending upon the magnitude of the (velocity-dependent) phase difference between the outgoing partial waves corresponding to the two trajectories.

---

† The often-quoted example is the "two-slit" experiment, where an electron can pass through two slits in a screen.

The velocity-dependent interference is manifested in the undulatory velocity dependence of $\sigma(v)$. At the *glory extrema* [i.e., the maxima and minima in $\sigma(v)$] the interference is either most constructive or most destructive. To emphasize these deviations it is customary to plot $\sigma v^{2/(s-1)}$ versus v (or $1/v$), since in the absence of the interference effect $\sigma \propto v^{-2/(s-1)}$ [Eq. (3.38)]. For closed-shell molecules, s is usually found to be 6. The glory impact parameter is identified in Fig. 3.6. At that angular momentum [see Eq. (3.35)], $d\delta_l/dl = 0$. Hence about that value of l the phase shift is only slowly varying with l and the "random phase approximation" is not valid. The required correction (which is known) yields the glory undulations, as well as the velocities v_N at which they occur. They can be indexed starting from the highest velocity maximum $(N = 1)$, with $(N - \frac{3}{8})\pi = \delta_{max}(v_N)$. At larger velocities, the extrema are spaced linearly in $1/v$, so a plot of $N - \frac{3}{8}$ versus v_N^{-1} is linear near the origin.

Examination of Figs. 3.2 and 3.3 suggests that for $\theta < \theta_r$, there will be interference among the *three* trajectories that lead to the same observed deflection. Let us now turn to this feature, the so-called rainbow effect.

3.1.9 Rainbow scattering as a probe of the potential well

To introduce the phenomenon of the rainbow we return to the classical mechanical description of scattering. As pointed out in Section 3.1.2 (see Fig. 3.3), there may be three (and sometimes more!) values of b corresponding to the same angle θ (i.e., $|\chi|$). *Each* of these impact parameters contributes to the intensity scattered at the angle θ, so that Eq. (3.2) becomes

$$I(\theta) = \sum_{i=1}^{3} \frac{b_i}{\sin\theta \, |d\chi/db|_i}, \qquad \theta < \theta_r \qquad (3.41)$$

where $\theta = |\chi(b_i)|$.

For $\theta > \theta_r$ there is only one contributing term, but for $\theta < \theta_r$ there are three. Thus there is a discontinuity in $I(\theta)$ at θ_r. But when θ is very near the extremum angle θ_r, we have a *range* of impact parameters giving rise to scattering at the *same* angle, that is, $|d\chi/db|_{b_r} = 0$. This induces a *divergence* in $I(\theta)$. This discontinuity and divergence at θ_r is an artifact of the classical approximation.

In the quantal scattering treatment the rainbow effect arises naturally upon consideration of the interference terms and there is no divergence. The angular distribution exhibits an oscillatory interference pattern with an envelope showing successive humps and valleys for $\theta < \theta_r$, leading to a large maximum near the so-called primary rainbow angle θ_r. The other bumps are termed supernumerary rainbows. This pattern is followed by a drop in the cross section on the "dark" side of the rainbow† (Fig. 3.7).

---

† The colorful terminology arises from an interesting optical analogy between the scattering of molecular rays and that of light rays (see the glory effect; Sec. 3.1.8). Basically, the rainbow effect arises from an interference among rays originating from different impact parameters scattered at the same θ.

Figure 3.8 shows a comparison of the quantal versus the classically calculated angular distribution at higher velocities (large A), where very many partial waves contribute.

That the rainbow [the minimum in $\chi(b)$] reflects the presence of a minimum in the potential at $R = R_e$ can also be seen from Eq. (3.11). Since the rainbow impact parameter approximately equals R_e, the rainbow angle is governed by the ratio of the potential well depth, $\epsilon = |V(R_e)|$, to the collision energy,

$$\theta_r \propto \epsilon/E_T \tag{3.42}$$

Experimental observations of the rainbow effect provided early evidence for the existence (and magnitude) of the shallow van der Waals potential well for many nonreactive systems.

There is a relation between the number of supernumerary rainbows and the index N of the glory extremum (see Fig. 3.10).

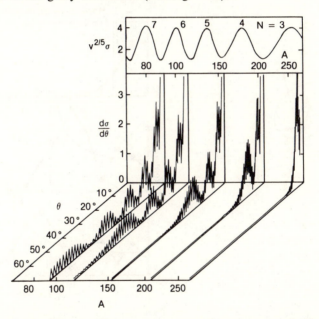

Fig. 3.10. Bottom: Differential (polar) elastic scattering cross section $d\sigma/d\theta$ versus θ as a function of the reduced velocity parameter $A = kR_e$, calculated for a specified Lennard–Jones (12,6) system, showing the rainbow pattern superimposed on the quantal oscillations. Top: The glory effect, that is, a plot of $v^{2/5}\sigma$ versus A showing the glory extrema (numbers are glory indices N). The exponent $\frac{2}{5}$ comes from the fact that $s = 6$ for the long-range part of the potential. [Adapted from U. Buck, in Lawley (1975).]

3.1.10 Rainbows in inelastic scattering

The inter*molecular* potential will, in general, depend not only on the relative separation but also on the relative orientation of the molecules. Even in the simplest situation, that of an atom–diatom collision, the potential will depend not only on R but also on the angle γ between **R** and the axis of the diatomic molecule. Of course, the potential will also depend on the separation of the two atoms in the diatomic but, as a limiting case, we can consider the diatomic molecules as a rigid rotor.

Even if the diatomic molecule is initially nonrotating, the noncentral force during the collision will set it in rotational motion. Let J be the (classical) angular momentum of the diatomic molecule after the collision. If the force were central, then to every initial orbital angular momentum L [or impact parameter b, $L = \mu v b$; see Eq. (2.29)] there corresponds a unique deflection angle θ. Because the reverse relation is not one to one (to the same θ there may be more than one L), we have the interference pattern in elastic scattering. Now, to each initial L *and* initial orientation γ there corresponds a unique deflection function χ *and* final rotational angular momentum J. As in the case of elastic scattering, $\chi(L, \gamma)$ and $J(L, \gamma)$ can be determined by

(a)

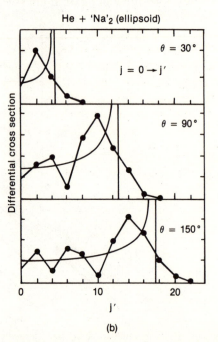

(b)

Fig. 3.11. Rotational rainbow effect for the scattering of a model atom–rigid ellipsoid system. (a) A projectile strikes the ellipsoid (at some arbitrary orientation angle γ) and is scattered at angle θ. The recoil momentum Δp imparts an angular momentum $\Delta \mathbf{J} = \mathbf{r} \times \Delta \mathbf{p}$ to the ellipsoid. For a *given* θ, as γ is varied, $\Delta J(\gamma)$ must pass through an extremum. Thus the classical orientation-averaged cross section for rotational excitation, say $d\sigma(\Delta J; \theta)/d\omega$, will show a singularity at this ΔJ. [Adapted from W. Schepper, U. Ross, and D. Beck, *Z. Phys.*, *A290*, 131 (1979).] (b) Calculations on the model system simulating He + Na$_2$ at $E_T = 0.1$ eV. Plotted are the differential cross sections $d\sigma$ $(j = 0 \rightarrow j'; \theta)/d\omega$ versus j' at the indicated scattering angles. The solid curves depict the classical results and the points represent calculations via the so-called "sudden" approximation, discussed in Chapters 5 and 6. [Adapted from H. J. Korsch and R. Schinke, *J. Chem. Phys.*, *75*, 3850 (1981).]

solving the classical equations of motion, usually numerically. The one simplifying aspect is that, typically, the rotational motion is slow on the time scale of the collision. Hence, unless the relative velocity is quite low, it is often reasonable to assume a *sudden* collision, that is, assume that the molecule does not rotate appreciably within the "duration" of the encounter. If this is the case, it is possible to express both $\chi(L, \gamma)$ and $J(L, \gamma)$ as integrals, just as was done for $\chi(L)$ in Section 3.1.5 [see Eq. (3.9)].

Both $\chi(L, \gamma)$ and $J(L, \gamma)$ are unique functions of the initial conditions. The converse is not necessarily true. There will, in general, be more than one set of initial conditions that leads to the same observed θ, J outcome, where $\theta = |\chi(L, \gamma)|$ and $J = |J(L, \gamma)|$. When many "adjacent" trajectories lead to the same observed J, there will be an extremum in the orientation averaged $J(L)$ and thus a so-called *rotational rainbow*, that is, a maximum in the intensity of scattering at a given θ as a function of J.

In contrast to the situation for elastic scattering, a rainbow is possible for inelastic scattering even when the potential is purely repulsive. Figure 3.11 shows the computational results for an extreme case where the diatomic molecule is taken to be a rigid ellipsoid. The "smoothing" of the classical singularity in the quantal computations is also evident.

The rainbow is also evident in the angular distributions. There is, classically, a sharp edge followed by a monotonic decline, with the rainbow angle shifting in the backward direction as the final J increases (Fig. 3.12).

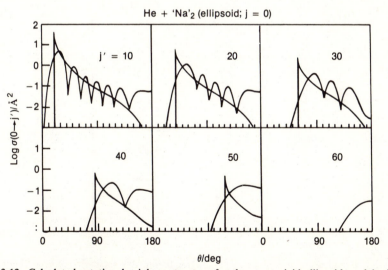

Fig. 3.12. Calculated rotational rainbow structure for the atom–rigid ellipsoid model, similar to that in Fig. 3.11, simulating $He + Na_2$. Plotted are differential cross sections for specified rotational transitions from $j = 0$ to j' (indicated). Here j is the quantum mechanical rotational quantum number, $J \sim \hbar j$. The classically calculated rainbow singularities are seen to move to larger angles with increasing Δj. The oscillatory curves are calculated by the quantum mechanical "sudden" approximation. [Adapted from D. Beck, U. Ross, and W. Schepper, *Z. Phys.*, *A293*, 107 (1979).]

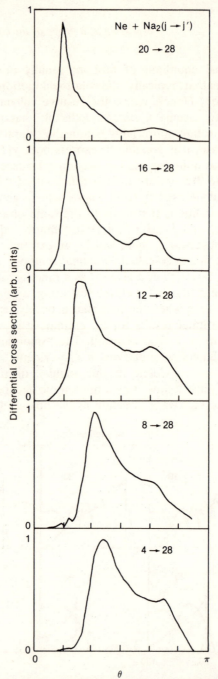

Fig. 3.13. Experimental rotational rainbow structure in the angularly resolved rotationally inelastic scattering of Na_2 by Ne at $E_T = 0.19\,eV$. Plotted are the differential cross sections $d\sigma(j \to j'; \theta)/d\omega$ for specified j values, all for $j' = 28$. Note the shift of the rotational rainbow peak to larger angles as Δj increases. The experimental arrangement is shown in Fig. 5.5. [Adapted from P. L. Jones, U. Hefter, A. Mattheus, J. Witt, K. Bergmann, W. Müller, W. Meyer, and R. Schinke, *Phys. Rev.*, *A26,* 1283 (1982).] Recent experiments for low initial j's have resolved the secondary rainbows for this system. See U. Hefter, P. L. Jones, A. Matthews, J. Witt, K. Bergmann, and R. Schinke, *Phys. Rev. Lett.*, *46*, 915 (1981).

Experiments where both the state and the angle are being resolved have been carried out; Fig. 3.13 is a recent example.

To sum up, scattering measurements serve as a probe of collision dynamics and can reveal detailed information on the intermolecular potential. In Section 3.2 we shall see some of the results.

Appendix 3A: Relation between scattering in the center-of-mass and laboratory systems

In a typical angular distribution experiment at thermal energies one crosses a collimated beam of A particles with one of type B, each with a fairly well-defined velocity, say, v_1 and v_2, respectively. The beams intersect in a small region of overlap (of volume ΔV) and the angular distribution of the deflected molecules is then measured. The intensity distribution of one of them, say, A, is observed as a function of the angle Θ of deflection from the original beam direction. If the B molecules are very heavy (in the limit of a "stationary target gas"), the observed laboratory (L) angular distribution $I_L(\Theta)$ is very nearly the true differential cross section in the c.m. system, since the laboratory deflection angles Θ are essentially equal to the c.m. angles θ.

However, in more typical cases the A and B velocities are more nearly comparable in magnitude. A velocity vector diagram is presented in Fig. 3A.1 showing the relation between velocities in the c.m. and laboratory systems.

In the laboratory system, the center of mass moves with a constant velocity v_{cm}, which is unchanged by the collision (see Sec. 2.2.2). In the c.m. system, the center of mass is at rest and the deflection θ is the deflection of the relative velocity vector v'. The final velocity of a particle in the laboratory system is obtained by a vector addition of its final relative velocity *with respect to the c.m.* and the *velocity of the c.m.* itself. Of course, since the collision is elastic, the *magnitude* of the relative velocity v' does not change. Thus the particle speeds with respect to the c.m., given in terms of the final relative speed v' by

$$w_1' = m_2 v'/(m_1 + m_2)$$
$$w_2' = m_1 v'/(m_1 + m_2)$$

(3A.1)

do not change either. The final *laboratory* velocities

$$v_1' = v_{\text{c.m.}} + w_1'$$
$$v_2' = v_{\text{c.m.}} + w_2'$$

(3A.2)

are, however, different from their initial values in both *magnitude* and *direction*. From Eq. (3A.2) and the corresponding relation before the collision, we can relate Θ and θ. The necessary trigonometric relationships can be deduced directly from the diagram. Figure 3A.2 illustrates two

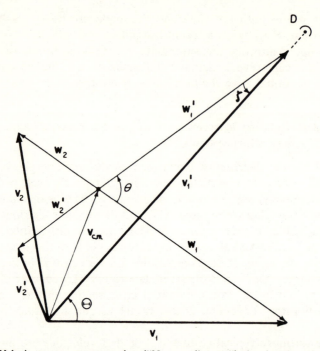

Fig. 3A.1. Velocity vector representation ("Newton diagram") showing the relation between laboratory and c.m. velocities. The velocities of the incident particles are v_1 and v_2, respectively (here shown intersecting at an angle greater than 90°); the final laboratory velocities are v_1' and v_2'. The detector D, set at lab angle Θ (measured from the direction of v_1), sees particles of type 1 that have been elastically scattered through the c.m. deflection angle θ. The angle ξ is that required in the c.m. → lab transformation. A given c.m. angle θ specifies both ξ and Θ.

particular possible outcomes of a collision as viewed in the lab system. From Fig. 3A.2a it is not immediately clear that the double-primed final velocities are due to forward $\theta \approx 30°$ scattering in the c.m. and that the primed final velocities are due to a rebound ($\theta \approx 180°$) collision. Figure 3A.2b showing the lab-to-c.m. transformation makes this conclusion evident, however. Figure 3A.2c shows the relative motion of the molecules with respect to the c.m.

The remaining problem is to convert the observed (laboratory) scattering intensity $I_L(\Theta)$ to the corresponding c.m. intensity $I(\theta)$. Because of the possibility of more than one c.m. angle of scattering leading to the same laboratory angle, the transformation is best carried out from the c.m. to the laboratory. An assumed c.m. intensity $I(\theta)$ is converted to the corresponding lab intensity $I_L(\Theta)$. For each θ, conservation of intensity yields

$$I_L(\Theta)\, d\Omega_L = I(\theta)\, d\omega \qquad (3A.3)$$

where $d\Omega_L$ is the element of solid angle in the lab system corresponding to

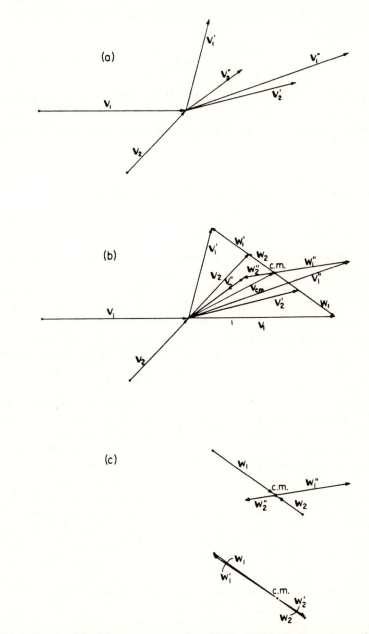

Fig. 3A.2. The need for lab-to-c.m. conversion. (a) The initial velocity vectors and two sets (primed and double primed) of final velocity vectors corresponding to elastically scattered molecules 1 and 2 (in the lab system). Without actually carrying out the lab-to-c.m. conversion, assisted by the light construction lines shown in (b), it is quite difficult to answer even such a simple question as which collision corresponds to "forward" and which to "backward" scattering. The relative motion per se for the two collisions is shown in (c).

$d\omega$ in the c.m. system. In terms of the solid-angle ratio (the "Jacobian" of the transformation),

$$I_{\mathrm{L}}(\Theta) = I(\theta)\frac{d\omega}{d\Omega_{\mathrm{L}}} \tag{3A.4}$$

From Fig. 3A.1, the required Jacobian (or ratio of area elements) is†
$v_1'^2/w_1'^2 |\cos \xi|$. The procedure must be repeated for each θ (leading to a given Θ) and a sum is required to yield the total (observable) lab intensity $I_{\mathrm{L}}(\Theta)$.

3.2 Intermolecular potentials from experiment and theory

3.2.1 Sources of interaction potentials

Accurate inter*atomic* potential functions for *stable* diatomics have been known for a long time, thanks to extensive spectroscopic studies. The potential in the vicinity of the equilibrium position of the (usually deep) well has been determined for both ground and excited molecular states. Such studies have also yielded information on repulsive parts of potential curves, especially when light absorption or emission leads to dissociation. The facile formation of cold "van der Waals" molecules during supersonic expansion has extended the spectroscopic approach to the shallow wells of more weakly bound pairs.

The spectroscopic techniques are limited to exploring the potential in a limited range of interparticle distances (the "Franck–Condon region") sampled by the transition. Newer techniques, such as emission from an excited dissociative state, are constantly extending the accessible range. The molecular beam scattering method remains the best general procedure to use over the entire range of the potential, particularly in the repulsive regime, where the inversion is unique. It is at its best in determining the "spherical" part of the potential, that is, the potential averaged over the orientation of the molecules. The noncentral (i.e., orientation-dependent) part of the potential requires a combination of scattering and internal state analysis (e.g., see Sec. 3.1.10) or of scattering with reagent state selection and preferably alignment. This will be even more so when we recognize, in Chapter 4, that the inter*molecular* potential depends also on the internal (vibrational) coordinates of the molecules.

Accurate *ab initio* quantum-mechanical computations of inter*atomic* potential curves for simple (few-electron) systems are feasible, even for excited states. Inter*molecular* potentials are also becoming available, and examples of dynamical computations for such potentials will be cited (e.g.,

---

† The factor $(v_1'/w_1')^2$ is due to the two different radii used to define an element of solid angle at D, and ξ is the angle between the directions of v_1' and w_1', respectively.

Fig. 3.27). In what follows we shall look at some results and see the interplay of theory with experiment.

3.2.2 *Potential curves from beam scattering*

The initial collisional experiments were measurements of the *velocity dependence* of the *collision cross section* at high energies (see Fig. 2.3), with the aim of determining the purely repulsive part of the potential. The results obtained, for example, for rare gases and relatively inert molecules, were that the potentials were essentially of the exponential form of Eq. (2.14) and the characteristic range parameters ρ were in accord with expectations (based upon crystal energy, compressibility data, etc., as well as the best available *ab initio* computations).

After a time, beam scattering cross section measurements at lower energies, with adequate energy and angular resolution, could be made and the results used to determine the long-range "tail" and the attractive "well" of the potential for a large variety of van der Waals molecules. Figure 2.3 showed the velocity dependence of the collision cross section $\sigma(v)$ for the essentially repulsive H–rare gas systems.

The best probe of the potential is, however, the *angular distribution* (the differential cross section). Figure 3.14 shows experimental high-resolution angular distributions for two rare gas systems. The potential functions deduced from these observations are shown in Fig. 3.15.

Extensive work has yielded accurate van der Waals potential wells and repulsive curves for all the rare-gas diatoms. Figure 3.16 shows the resulting potentials for the homonuclear rare-gas pairs. Note the increase in both the well depth and range of the potential with increasing polarizability of the atoms (polarizability is a measure of the deformability as well as the "volume" of the electron cloud of an atom).

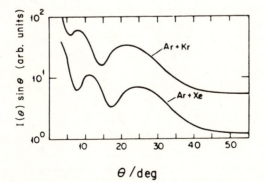

Fig. 3.14. Angular distribution of the scattering of Ar by Kr and Xe at a collision energy of about 10^{-20} J. Plotted is $I(\theta) \sin \theta$ versus θ. [Adapted from J. M. Parson, T. P. Schafer, P. E. Siska, F. P. Tully, Y. C. Wong, and Y. T. Lee, *J. Chem. Phys.*, *53*, 3755 (1970).]

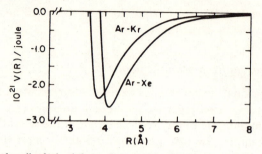

Fig. 3.15. Potential wells derived from the angular distribution results of Fig. 3.14 for the Ar–Kr and Ar–Xe interactions.

It has been found that within a family of similar diatoms (e.g., rare-gas pairs and alkali systems) the resulting potentials can be scaled (approximately) to a common *reduced* functional form, that is, their *shape* is often similar. We can define a reduced separation by $z \equiv R/R_e$, where R_e is the position of the minimum in $V(R)$, and a reduced potential energy by $V^* \equiv V/\epsilon$, where ϵ is the well depth. Then it is found that $V^*(z)$ is a nearly universal shape, at least for the family of related atoms or molecules. Figure 3.17 shows a reduced plot for the rare-gas homonuclear diatoms. This explains the success of the well-known *rule of corresponding states* in the "macro" world. Figure 3.18 shows a reduced plot of the second virial

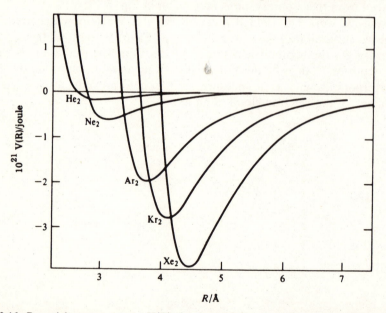

Fig. 3.16. Potential energy curves $V(R)$ for homonuclear rare-gas dimers derived from experiments on the elastic scattering of crossed beams of rare gases. [Adapted from J. M. Farrar, T. P. Schafer, and Y. T. Lee, in *Transport Phenomena*, A.I.P. Conference Proceedings, No. 11 (1973), J. Kestin, ed.]

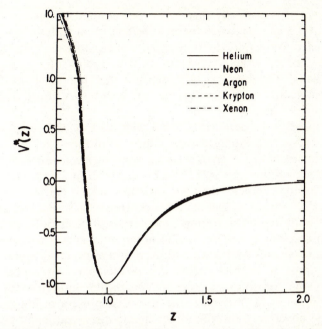

Fig. 3.17. "Reduced" potentials for homonuclear rare-gas pairs. Here the reduced potential $V^* = V/\epsilon$ is plotted versus the reduced separation $z \equiv R/R_e$, ϵ and R_e are the well depth and separation at the minimum of the potential. (Same source as in Fig. 3.16.) For a thorough review, see Maitland et al. (1981).

coefficient, $B^*(T^*) = B(T)/\frac{2}{3}\pi N_A R_e^3$ (where $T^* = kT/\epsilon$ is the reduced temperature), illustrating the "rule."

For closed-shell atom–atom systems the potential is, of course, spherically symmetrical, but for atom–molecule or molecule–molecule interactions it may depend also upon the relative orientations of the molecule(s) as well as on their mutual separation. The *orientation-averaged* or central part

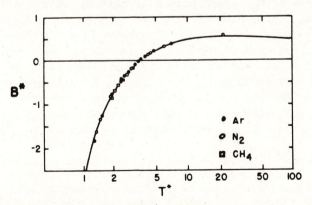

Fig. 3.18. The reduced second virial coefficient, $B^*(T^*)$, versus temperature $T^*(=kT/\epsilon)$ for Ar, N_2 and CH_4, illustrating the rule of corresponding states.

of the potential is normally that which is measured by ordinary scattering experiments.† Such results are found to be consistent with the potentials that had been deduced approximately from temperature-dependent transport coefficients, second virial coefficients and liquid- and solid-state properties, where available. For these ground electronic state potentials the well depths ranged from about 1 meV (for He–He) to about 0.6 eV (for the singlet state of Na_2).

Fewer studies have been carried out with *electronically excited* atoms or molecules.‡ One interesting feature is the role of the "size" of the electron cloud. Comparison of the cross sections $\sigma(v)$ for He scattered by He, He*, and Li has shown that the $1s2s$ He* and the $1s^2 2s$ Li atoms are similar and both significantly "larger" than the ground-state $1s^2$ He atom. (Note that the polarizabilities of He, He*, and Li are, respectively, 0.2, 46, and 22 Å^3).

More recently elastic scattering experiments of beams of *ions* have been carried out resulting in the evaluation of potential curves for *ion–atom* and *ion–molecule* systems over a wide range of internuclear separations.

In summary, it is now becoming possible to determine two-body potentials with good accuracy, even for complicated molecules for which the exact computation of intermolecular forces at present is intractable. It is noteworthy, however, that even rather simple theory can be of help to us in providing useful information on the *long-range* part of the potential. This is sometimes very valuable to us in the molecular collision field and will be discussed in Section 3.2.4.

3.2.3 The intermolecular potential

It is only in Chapter 4 that we shall discuss the nature of the potential that allows for a chemical reaction. Intermediate between reactive and elastic scattering is, however, inelastic scattering collisions which change not only the direction of the relative velocity but also its magnitude. Overall conservation of energy is ensured by a corresponding change in the internal energy of the collision partners. To discuss such inelastic collisions we must therefore explicitly recognize that molecules have internal degrees of freedom and we must couple these internal degrees of freedom to the relative motion. It is only because of such a coupling that a force can act on the internal coordinates during the relative motion. In other words, only such terms in the potential that depend on both the relative separation R between the molecules' center of mass *and* on the internal coordinates r of the molecules can lead to inelastic transitions. As R varies during the

---

† Certain sophisticated experiments have been carried out with aligned molecules, giving information on the orientation-dependent (*anisotropic*) forces of interaction.

‡ One of the problems is that they usually interact according to more than one potential energy curve. In other words, several different electronic states of the diatomic dissociate to the same electronic state of the separated atoms. This problem occurs also for open-shell, ground state atoms and makes the analysis much more difficult.

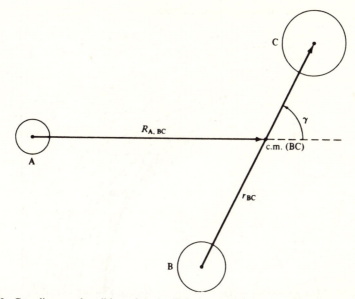

Fig. 3.19. Coordinates describing the A + BC interaction. The orientation angle is γ [$=\arccos(\hat{\mathbf{R}} \cdot \hat{\mathbf{r}})$].

collision these terms will lead to a varying force acting on the internal coordinates r. We consider here only the simplest process, that of a "rigid" diatomic molecule colliding with a spherically symmetrical (e.g., rare-gas) atom. The only internal coordinate is then the orientation γ of the diatomic molecule (Fig. 3.19).

The potential between the atom and the rigid diatomic molecule will depend on the distance R between the two centers of mass and the orientation γ of R relative to the axis of the diatom. By defining γ as in Fig. 3.19, we have ensured that it is the only relevant coordinate. Rotating the figure (at a fixed R) about the diatomic axis will clearly not alter the potential. Regarded as a function of two variables, R and γ, the potential is most easily plotted as contour lines of equal magnitudes of the potential in polar coordinates, as in Fig. 3.20. The radial coordinate is R and the polar angle is γ.

The Kr–HCl system shown in Fig. 3.20 is also a typical example of a van der Waals bound system. The minimum in the potential at $\gamma = 0$ is evident but, as is typical for such molecules, there is only a small rise in the potential as γ varies (at constant R) about the minimum. The barrier to rotation of HCl is small. The steep rise in the potential that allows us to approximate the potential (at higher collision energies) as that of a rigid ellipsoid is quite evident, particularly so for the He–LiH system shown in Fig. 3.21.

The anisotropy of the potential implies that in a general, "off-center" collision the molecule can be rotationally excited (or de-excited). For the

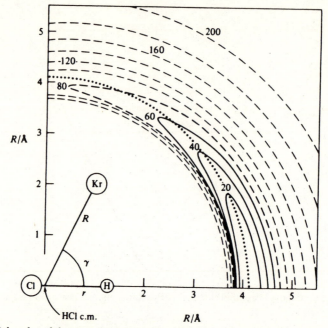

Fig. 3.20. Polar plot of the potential for the Kr·HCl molecule. Energies are in cm^{-1} relative to the minimum at $\gamma = 0$. [Adapted from J. M. Hutson, A. E. Barton, P. R. Langridge-Smith, and B. J. Howard, *Chem. Phys. Lett.*, *73*, 218 (1980).]

analysis of such inelastic collisions it is useful to adopt another representation of the potential, one that separates the spherical part. (The spherical or orientation-averaged potential can only cause a deflection of the relative motion but cannot induce an inelastic transition.) Mathematically, the separation is achieved by expressing the potential $V(R, \gamma)$ in terms of radial and angular parts:

$$V(R, \gamma) = \sum_{\lambda=0} V_\lambda(R) P_\lambda(\cos \gamma) \qquad (3.43)$$

The $P_\lambda(\cos \gamma)$ are the Legendre polynomials, and their orthogonality property [plus the fact that $P_0(x) = 1$] implies that

$$V_0(R) = 2\pi \int_0^\pi V(R, \gamma) \sin \gamma \, d\gamma \qquad (3.44)$$

is the orientation-averaged part of the potential.

For van der Waals pairs, $V_0(R)$ is indeed the dominant part of the potential (Fig. 3.22). Hence computing the collision trajectory as if $V_0(R)$ were the entire potential will not be an unreasonable approximation. Conversely, measuring the deflection will serve primarily to determine $V_0(R)$. To obtain information about the anisotropic parts of the potential we

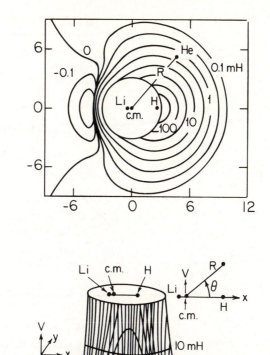

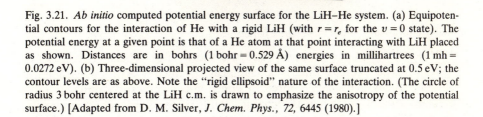

Fig. 3.21. *Ab initio* computed potential energy surface for the LiH–He system. (a) Equipotential contours for the interaction of He with a rigid LiH (with $r = r_e$ for the $v = 0$ state). The potential energy at a given point is that of a He atom at that point interacting with LiH placed as shown. Distances are in bohrs (1 bohr = 0.529 Å) energies in millihartrees (1 mh = 0.0272 eV). (b) Three-dimensional projected view of the same surface truncated at 0.5 eV; the contour levels are as above. Note the "rigid ellipsoid" nature of the interaction. (The circle of radius 3 bohr centered at the LiH c.m. is drawn to emphasize the anisotropy of the potential surface.) [Adapted from D. M. Silver, *J. Chem. Phys.*, **72**, 6445 (1980).]

must probe the rotational state of the molecule (see Sec. 3.1.10). The rotational excitation is due only to the higher order ($\lambda = 1, 2, \ldots$) terms in Eq. (3.43).

In the region of the well, the spectroscopy of the van der Waals bound species is an important tool. The study of the photodissociation of such molecules, particularly when combined with the analysis of both the angular and the rotational state distribution of the fragments, provides the advantages of both the spectroscopic and the beam-scattering approach. As in the more usual case of stable molecules, this photofragmentation process is often referred to as a "half-collision" since the system is "started off" inside

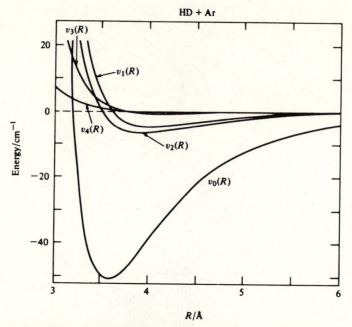

Fig. 3.22. Radial functions $V_\lambda(R)$ as employed in Eq. (3.43) in the Legendre representation of the anisotropic potential, here for HD·Ar. [Adapted from H. Kreek and R. J. LeRoy, *J. Chem. Phys.*, *63*, 338 (1975).]

the interaction region and the fragments are observed when the interaction is over.

3.2.4 *Long-range intermolecular forces*

The very long-range attraction (and the short-range repulsion) consists of "physical-type" forces that lack chemical specificity. Attractive chemical binding forces ("exchange forces") contribute significantly only at intermediate separations when the charge clouds of the approaching molecules begin to overlap. The *short-range* repulsion is primarily due to the energy increase if we bring too many nuclei and too many electrons into a small volume. The long-range attractive London or dispersion potential is due to quantum mechanical effects that lead to long-range attraction between any two polarizable systems.

An oversimplified classical theory of the long-range force is easily formulated if we associate a transition dipole moment with every allowed electronic transition of the atom or molecule. Consider two atoms where, for simplicity, each atom has only one excited state with transition dipole moment μ from the ground state. The field due to this dipole at a distance R away from the first atom is $E_1 = \mu_1/R^3$. When we place the second atom, with polarizability α_2, at a distance R from the first atom, we gain in energy

$-\alpha_2 E_1^2/2$. A similar gain is achieved owing to the polarizability of the first atom being in the field of the second atom. The resulting interaction is the sum

$$V(R) = -\tfrac{1}{2}(\alpha_2 E_1^2 + \alpha_1 E_2^2) = -\frac{\alpha_2\mu_1^2 + \alpha_1\mu_2^2}{2R^6} = -\frac{C_{\text{disp}}}{R^6} \qquad (3.45)$$

The energy gain due to the interaction of the polarizability of one system and the field of the transition dipoles of the other is the so-called *dispersion energy*. For atom–atom or atom–diatom interactions typical values of the dispersion constant C are in the range 10^{-77} to 10^{-79} J-m^6.

When we consider the long-range atom–*molecule* dispersion interaction, it is necessary to recognize that the polarizability of a molecule will depend on its orientation to the field. For the general case of an atom–diatomic molecule the long-range potential can be expressed as

$$V(R) = -(C_{\text{disp}}/R^6)[1 + a_1 P_1(\cos\gamma) + a_2 P_2(\cos\gamma) + \cdots] \qquad (3.46)$$

If the diatomic molecule is homonuclear, symmetry demands that only even terms be included. For the interaction of He with the nearly spherical H_2 molecule, $a_2 = 0.12$, but the anisotropy parameters a_2 can be much higher in general.

When one of the molecules is *charged* (or has a permanent dipole moment), the polarizability of the second molecule can interact with the field due to the *permanent* charge (or the dipole), giving rise to *induction energy*. At a distance R, the field due to an ion of charge q is $E = q/R^2$, so that the spherical part of the ion–molecule induction energy is

$$V(R) = -\tfrac{1}{2}\alpha E^2 = -\alpha q^2/2R^4 \qquad (3.47)$$

where α is the polarizability of the neutral atom or the orientation-averaged polarizability of the molecule. Similarly, the permanent dipole–atom induction energy is

$$V(R) = -\alpha\mu^2/2R^6 = -C_{\text{ind}}/R^6 \qquad (3.48)$$

where μ is the permanent dipole and $C_{\text{ind}} = \tfrac{1}{2}\alpha\mu^2$.

Other expressions for these so-called *asymptotic* intermolecular potentials are available for dipole–dipole cases, and so forth, including explicit *orientation-dependent* terms. The two most important cases are the electrostatic permanent dipole–permanent dipole potential

$$V(R) = -\frac{\boldsymbol{\mu}_1 \cdot \boldsymbol{\mu}_2 - 3(\mathbf{n}\cdot\boldsymbol{\mu}_1)(\mathbf{n}\cdot\boldsymbol{\mu}_2)}{R^3} \qquad (3.49)$$

and the exceedingly long-range ion–permanent dipole potential

$$V(R) = -\boldsymbol{\mu}\cdot\mathbf{n}/R^2 \qquad (3.50)$$

where $\boldsymbol{\mu}$ is the (vector) dipole moment and $\mathbf{n}$ is a unit vector in the direction of $\mathbf{R}$ so that $\boldsymbol{\mu}\cdot\mathbf{n} = \mu\cos\gamma$, where γ is the angle between the axis of the dipole and the relative orientation. The long-range nature of the dipole–dipole potential was already evident in the large cross section for the scattering of the ionic molecule CsCl by the polar molecule CH_2F_2 (see Fig. 2.1). Indeed, when operative, these electrostatic forces tend to dominate the scattering, at least at low collision energies.

The knowledge of the long-range part of the potential often suffices to account for most of the dynamical behavior of suitable systems, provided that the main observable effect comes from glancing collisions, that is, large b and thus large intermolecular separations. We have already seen two examples: the estimation of the reaction cross section for reactions without energy thresholds (in Sec. 2.4.7) and the velocity dependence of the total elastic scattering cross sections $\sigma(v)$ for atom–atom scattering [Eq. (3.38)].

Of course, it would be desirable to have available accurate *ab initio* quantum mechanically calculated van der Waals potentials, not only for large internuclear separations but also in the neighborhood of the potential well and the repulsive region as well. This is becoming possible for few-electron systems and in a few cases such as He–He, Li–He, and He–He* and for simple bound diatomics. The results are now achieving "spectroscopic accuracy." Surely in the future it will become feasible to compute such two-body potentials for many systems of practical interest.

Thus far we have acquired a bit of insight into the concepts and methods of the molecular collision technique and have been initiated into the relations between the *intermolecular forces* and the *dynamics of elastic collisions*; so we can begin looking into the more interesting question of the molecular dynamics of *reactive* collisions. We can now take up from where we left off in Section 2.4 and consider the *angular distribution* of the products from a binary reactive collision.

3.3 Angular distribution in direct reactive collisions

3.3.1 Concept of a direct reactive collision

The observation via infrared chemiluminescence of the very specific energy disposal in exoergic reactions (see Sec. 1.2) was preceded by the discovery that in crossed molecular beam scattering there is a strong *preferential angular disposition* of the *products* for different reactions. For example, in reactions such as

$$CH_3I + K \rightarrow KI + CH_3$$

the KI product molecules are observed in the "backward" hemisphere, with respect to the incident K atom. On the other hand, for reactions such as

$$K + I_2 \rightarrow KI + I,$$

the KI is scattered strongly "*forward*," that is, in the direction of the

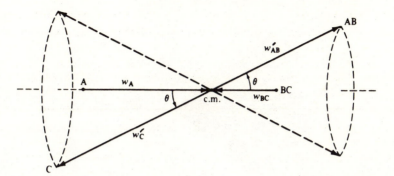

Fig. 3.23. Illustration of a collision A + BC (in the c.m. system) leading to the products AB and C ejected at the scattering angle θ with velocities $\mathbf{w}_{AB}$ and $\mathbf{w}_C$, respectively. Because of the axial symmetry (assuming unpolarized reagents), the products are emitted in cones with no ϕ dependence.

incident K atom. This is the typical *stripping mode* behavior (spectator limit) found earlier in certain nuclear reactions.

A view of such an $A + BC \rightarrow AB + C$ atom exchange reaction in the center-of-mass system is shown in Fig. 3.23. The pair of antiparallel [see Eq. (2.17)] incident relative velocity vectors $\mathbf{w}_A$ and $\mathbf{w}_{BC}$ define the initial relative velocity vector $\mathbf{v}$ [Eq. (2.18)]. In a reactive scattering event both the magnitude and direction of the final relative velocity vector $\mathbf{v}'$ can change. Shown in Fig. 3.23 is a particular pair of outgoing relative velocity vectors for the products $\mathbf{w}'_{AB}$ and $\mathbf{w}'_C$, scattered with a particular c.m. deflection angle θ. For unpolarized, nonoriented reagents the product flux pattern is cylindrically symmetrical around the incident relative velocity axis; that is, there is no azimuthal angle (ϕ) dependence of the scattered intensity per unit solid angle ($dw \equiv \sin \theta \, d\theta \, d\phi$) in the c.m. system.

In an actual experiment, where many impact parameters contribute, the outcome is more complex than that depicted in Fig. 3.23. The products can be scattered over an entire angular range *and* can emerge with a distribution of velocities. (The balance of energy is made up by a counterpart distribution of internal excitation of the AB product and, possibly, electronic excitation of C or even AB).

Figure 3.24 shows the predominant "cones" of the scattered KI product in the preceding reactions. Such *anisotropic* angular distributions, when first observed in the primitive molecular beam scattering studies of the early 1960s, were considered rather surprising. Such *preferential* scattering indicates that the process of reaction (i.e., the atomic rearrangement) must be over before the colliding pair of reactants have time to execute one or more rotations about one another. Obviously, if the reactants formed a "complex" that stuck together for more than a few classical rotation periods, the products would fly off in a random way, and not in such a

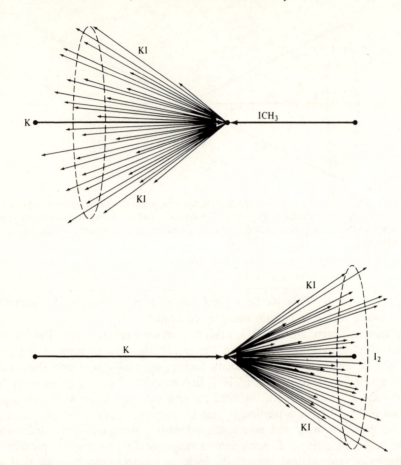

Fig. 3.24. The "cones" containing most of the products' intensity (in the c.m. system). For the $CH_3I + K$ reaction (top), the KI product appears primarily in the backward (rebound) direction; not so for the KI from $K + I_2$ (bottom), which shows typical stripping behavior. The cones shown contain most, but not all, of the angular distribution of the KI flux.

concentrated, "directed" cone disposed so specifically with respect to the initial relative velocity vector.†

Thus we have an experimental or "operational" definition of a *direct reaction*, that is, whenever the angular distribution is *not symmetric* with respect to the forward versus backward hemisphere. This establishes a ratio of times, such that if the "time of reaction" is less than, say, one rotational period of the combined system (roughly a picosecond), we can say the reaction is of the "direct" type.

Figure 3.24 contains more than just the preferential cone of angular scattering. It also shows the range of final velocities (and hence recoil energies). When these are examined in detail (in Chap. 5), it is found that

† In Chapter 7 we will discuss reactions that do proceed by the "long-lived" complex mode.

for both reactions, but especially for $K + I_2 \rightarrow KI + I$, the final translational distribution is quite narrow, thereby providing additional evidence for the direct nature of these reactions.

3.3.2 Angular distributions for direct reactions

We saw in Section 3.1 how the differential *elastic* cross section $I(\theta)$ serves as a sensitive probe of the dynamics of such collisions. In a similar fashion, one can introduce the *differential reactive* cross section $I_R(\theta)$, except that here we mean the number of *product* molecules scattered at a c.m. angle θ (per unit time and per unit solid angle) divided by the incident flux of *reactant* molecules. In other words, we modify Eq. (3.17) to read

$$\frac{d\dot{N}_R(\theta, \phi)}{d\omega} = (I_A N_B) I_R(\theta) \qquad (3.51)$$

where $I_R(\theta) = \sigma_R P_R(\theta, \phi)$. Here σ_R is the integrated *reaction cross section*

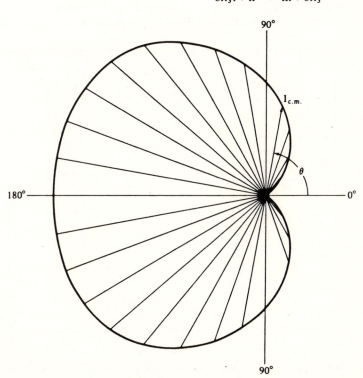

$$CH_3I + K \longrightarrow KI + CH_3$$

Fig. 3.25. Polar plot of the c.m. angular distribution of KI from the crossed-beam reaction of $CH_3I + K$ at $E_T \approx 0.2\,eV$. The length of the arrow for any given angle is the intensity of the KI scattering into that direction. [Adapted from R. B. Bernstein and A. M. Rulis, *Discuss. Faraday Soc.*, 55, 293 (1973).]

and $P_R(\theta, \phi)$ a normalized probability density function for the angular scattering. Thus, when $I_R(\theta)$ is integrated over all solid angles, it gives us the total reaction cross section σ_R (see Sec. 2.4). Figure 3.25 shows a polar plot of $I_R(\theta)$ for the angular distribution of KI from $K + CH_3I$, emphasizing once more the rebound nature of this reactive collision. As we will argue in Section 3.3.5, this reflects the primarily lower impact parameters contributing to the reactive events. But this should be typical for reactions with a barrier in their post-threshold regime. Indeed so: Fig. 3.26 shows results of quantum mechanical computations of the angular distribution of the H_2 product in the $H + H_2 \rightarrow H_2 + H$ exchange compared with molecular beam measurements for the isotope exchange $D + H_2 \rightarrow HD + D$. The strong backscattering implies that small impact parameter collisions are mainly responsible for reaction.

Anticipating the developments ahead, we can already expect that at higher energies, where the range of impact parameters that contribute to the reactive event is larger, the products' angular distribution will shift to more forward directions. Figure 3.27 is an illustration of this effect as calculated for the $D + H_2 \rightarrow HD + H$ exchange reaction.

The introduction of lasers has also opened up the possibility of directly observing the angular distribution in the center of mass system from a

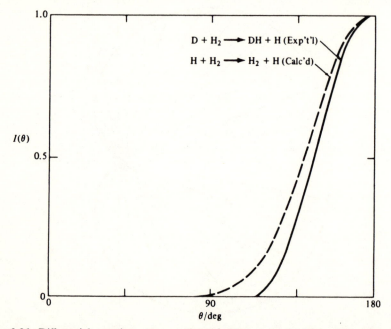

Fig. 3.26. Differential reaction cross sections (arbitrary units). The solid curve is the experimentally derived c.m. angular distribution of HD from the crossed-beam reaction of $D + H_2$ at $E_T \approx 0.5$ eV. [Data of J. Geddes, H. F. Krause, and W. L. Fite, *J. Chem. Phys.*, 56, 3298 (1972).] The dashed curve denotes quantal computations for the $H + H_2$ reaction at $E_T = 0.45$ eV. [Results of A. Kuppermann and G. C. Schatz, *J. Chem. Phys.*, 65, 4668 (1976).]

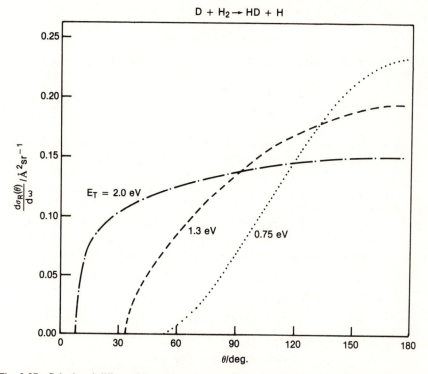

Fig. 3.27. Calculated differential reaction cross sections, via the classical trajectory method, for the $D + H_2 \rightarrow HD + H$ reaction at indicated values of E_T. [Results of H. R. Mayne and J. P. Toennies, *J. Chem. Phys.*, *75*, 1794 (1981).] Molecular beam scattering measurements of the differential cross section at $E_T \approx 1$ eV [R. Götting, H. R. Mayne, and J. P. Toennies, *J. Chem. Phys.*, *80*, 2230 (1984)] have confirmed the shape and magnitude of the calculated result.

Doppler shift. The experimental arrangement and principle of the method will be discussed in Section 5.2.

We now turn to a more systematic discussion of the concept of reaction probability in the analysis of angular distributions in direct reactions.

3.3.3 Optical model approach to direct reactions

The optical model provides a unified framework, based upon general principles, for a simplistic analysis of the collision dynamics. We begin with the form of the total collision cross section for reactants with an impact parameter in the range b to $b + db$,

$$d\sigma = 2\pi b \, db = d(\pi b^2) \tag{3.52}$$

As we emphasized in Section 2.4, this result is obtained in terms of the flux of the colliding reactants and is independent of whatever may happen as a result of the collision. If reaction is possible, a fraction $P(b)$ of collisions at the impact parameter b would lead to reaction; thus

$$d\sigma_R = 2\pi b P(b)\, db \qquad (3.53)$$

We have already invoked this first result of the model, or its integrated form,

$$\sigma_R = 2\pi \int_0^\infty b P(b)\, db \qquad (3.54)$$

to estimate the reaction cross section σ_R in terms of the opacity function $P(b)$. Such estimates were based upon specific model assumptions (see Sec. 2.4). However, neither the form of Eq. (3.54) nor the implication—that reactions at high b are more heavily weighted in their contribution to σ_R—is dependent on any specific assumptions.

Before we try to interpret the angular distribution of the *reactive* scattering in terms of the opacity function, let us examine the influence of $P(b)$ upon the concomitant *non*reactive scattering accompanying reaction.

3.3.4 Nonreactive scattering

The cross section for nonreactive collisions can be expressed in terms of the fraction $1 - P(b)$ of collisions that do not lead to reaction,

$$d\sigma_{NR} = 2\pi b [1 - P(b)]\, db \qquad (3.55)$$

The nonreactive cross section is necessarily smaller than it would have been had the reaction not taken place, that is, if $P(b) = 0$,

$$d\sigma_{NR}^0 = 2\pi b\, db \qquad (3.56)$$

where $d\sigma_{NR}^0$ is the cross section if $P(b)$ were zero for all b. The equality sign corresponds to the case where reaction is of negligible probability, so that $d\sigma_{NR}^0 = d\sigma$. In picturesque terms, we can say that the *nonreactive cross section is quenched* by the occurrence of the reaction. The closer the reaction probability $P(b)$ to unity, the more severe the quenching.

While we cannot measure the differential cross section $d\sigma_{NR}/db$, we *can* measure the angular distribution $I_{NR}(\theta)$ of the nonreactively scattered molecules. As in ordinary elastic scattering, we expect that $P(b)$ will contribute primarily at low b values and hence the quenching of $d\sigma_{NR}$ will correspond to the (near) absence of nonreactively scattered molecules in the backward direction. Figure 3.28 shows the computational results for $I_{NR}^0(\theta)$ and $I_{NR}(\theta)$ for an assumed form of $P(b)$, shown in the insert. The quenching, that is, the loss of nonreactive scattering in the backward direction, is evident. In quantitative terms we thus have

$$I_{NR}(\theta) = [1 - P(b)] I_{NR}^0(\theta) \qquad (3.57)$$

The implications of Eq. (3.57) are clear: $I_{NR}^0(\theta)$ is quenched in the presence

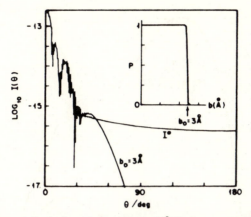

Fig. 3.28. Calculation of the backward quenching of $I^0(\theta)$ due to reaction. Here $I^0(\theta)$ is the angular distribution that *would* have obtained in the *absence* of reaction. Owing to reaction (opacity function shown in insert, with a cutoff impact parameter b_0), the non-reactive angular distribution is quenched in the backward direction. [Adapted from R. D. Levine and R. B. Bernstein, *Isr. J. Chem.*, 7, 315 (1969).] This "loss" of wide-angle scattering is concomitant with a "gain" of reactive scattering (not shown).

of reaction; the greater the range of impact parameters over which reaction is possible, the wider the angular range of which $I_{NR}^0(\theta)$ is quenched.

All our considerations thus far imply a unique relation between the deflection function χ and the impact parameter b. Of course, this is true for *elastic* scattering. It is found that such a condition is *approximately* the case also for *direct-mode* collision (i.e., one completed within a few vibrational periods). When the rearrangement time is longer, say, of the order of a rotational period or so, there is no longer a nearly one-to-one relation between b and χ. Only for a *direct* reaction is there a good "*correlation*" between b and χ; then we can take advantage of this to relate distributions over c.m. scattering angles θ with those over impact parameters b.

3.3.5 Angular distribution for rebound reactions

Figure 3.29 shows a comparison of the two angular distributions corresponding to the two KI-forming reactions. When we look at them "scaled properly," we note that the entire backscattered cross section for the KI from the rebound reaction fits under the backward "tail" of the angular distribution for the predominantly forward-scattered KI.

We interpret this in terms of the reaction probability as a function of impact parameter, the opacity function $P(b)$ (see Sec. 2.4). Collisions of low impact parameter (nearly head-on) usually lead to reaction and, because they are mainly head-on, the product diatomic rebounds backward. However, for K reacting with I_2, encounters at long range, that is, with very large impact parameters, are also effective in yielding reaction; in other

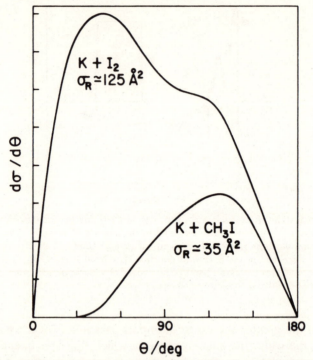

Fig. 3.29. The complete angular distribution of KI for the $K + I_2$ and $CH_3I + K$ reactions for $E_T \simeq 12$ kJ mol^{-1}. Plotted is the polar differential reaction cross section $2\pi \sin \theta I_R(\theta)$ versus θ. When integrated over θ, this gives directly the reaction cross section σ_R. [Adapted from the experimental results of K. T. Gillen, A. M. Rulis, and R. B. Bernstein, *J. Chem. Phys.*, *54*, 2831 (1971); A. M. Rulis and R. B. Bernstein, *J. Chem. Phys.*, *57*, 5497 (1972).]

words, $P(b)$ is near unity out to very large b (e.g., $b \approx 7$ Å). For these large-b encounters the K atom simply "picks up" the I atom and carries it forward with very little deflection, and the remaining I atom must sadly recoil in the opposite direction with a velocity determined by momentum conservation. This is the so-called *stripping mechanism*, which is characterized by the forward scattering of the product molecule.

The much larger differential reaction cross section for the stripping reaction is due to the much larger range of b values for which reaction is probable [see Eq. (3.54)]. Why this is so will be discussed in Chapter 4, Section 4.2, where we will begin to consider the forces during a reactive collision. We shall then find that "sideways scattering" is also possible in certain direct reactions.

It is possible to carry the general considerations even further by expressing the differential reactive cross section $I_R(\theta)$ in terms of a reference cross section [see Eq. (3.57)]

$$I_R(\theta) = P[b(\theta)]I_R^0(\theta) \qquad (3.58)$$

To make practical use of such an expression, one needs the "reference" angular distribution. For direct reactions it is often reasonable to assume that since the actual rearrangement is so rapid, the net angle of deflection is the sum of the angle of deflection of the reactants coming in and that of the products going out. For explicit results we need, however, to adopt specific models. One such model is that of *rebound reactions*. Here one assumes that reaction only occurs on close collisions when the reactants are subject to the short-range repulsive part of the intermolecular potential. The rearrangement thus takes place "at close quarters" and the newly formed products recede under the influence of the short-range repulsion. Hence the net deflection is that typical of hard-sphere scattering, for which the nonreactive scattering cross section, $I_{NR}^0(\theta)$, is independent of θ [see Eq. (3.4)]. The angular dependence of $I_{NR}(\theta)$ is thus entirely due to the b dependence of $P(b)$. Such a model therefore predicts *backward scattering* of the products, since $P(b)$ contributes only at low b values. Even in the stripping limit, there is always a rebound component, owing to the low-b reactive collisions.

As we increase the collision energy, the rebound character at low energy changes to more forward scattering at higher energy. There are two factors that govern this behavior. One is quite general. Since $E_T\chi(b) = \tau(b)$ [Eq. (3.10)], collisions at a given impact parameter [and hence given $\tau(b)$] will lead to decreasing deflections as the collision energy is increased. Moreover, for reactions with an energy threshold, the range of b values that lead to reaction increases with increasing E_T (see Sec. 2.4). Both factors in Eq. (3.58) therefore tend to favor forward scattering as E_T increases.

We can summarize as follows: *Direct reactions* that occur primarily at low impact parameter are characterized by a *small reaction cross section*; there is predominantly *backward scattering* of the products and the nonreactive angular distribution is quenched at wide angles. Reactions that occur with a high probability over a *wide range of impact parameters* will have a *large reaction cross section*, the products will appear mostly in the *forward direction*, and the nonreactive cross section will be severely quenched except in the very forward direction. As the energy is increased, the tendency toward forward scattering of the products is favored (see Fig. 3.27).

Throughout this chapter and Chapter 2 we have been discussing elementary chemical reactions using concepts borrowed from *"two-body"* collision theory. It is now time to break fresh ground and consider the implications of the *poly*atomic nature of a chemically reactive system.

Suggested reading

Section 3.1

Books

Bowman, J. M. (ed.), *Molecular Collision Dynamics,* Springer-Verlag, Berlin, 1983.
Brandsen, B. H., *Atomic Collision Theory,* 2nd ed., Benjamin, New York, 1983.
Geltman, S., *Topics in Atomic Collision Theory,* Academic, New York, 1969.

Review Articles

Beck, D., "Elastic Scattering of Nonreactive Atomic Systems," in Schlier (1970).
Bernstein, R. B., "Quantum Effects in Elastic Molecular Scattering," *Adv. Chem. Phys., 10,* 75 (1966).
Child, M. S., "Semiclassical Methods in Molecular Collision Theory," in Miller (1976).
Child, M. S., "Low Energy Atom–Atom Collisions," in Gianturco (1982).
Gianturco, F. A., and Palma, A., "Rotational Rainbows in Collisions Involving van der Waals Molecules," in Jortner and Pullman (1982).
Miller, W. H., "The Semiclassical Nature of Atomic and Molecular Collisions," *Acc. Chem. Res., 4,* 161 (1971).
Pauly, H., and Toennies, J. P., "The Study of Intermolecular Potentials with Molecular Beams at Thermal Energies," *Adv. At. Mol. Phys., 1,* 195 (1965).
Reuss, J., "Scattering from Oriented Molecules," *Adv. Chem. Phys., 30,* 389 (1975).
Schinke, R., and Bowman, J. M., "Rotational Rainbows in Atom–Diatom Scattering," in Bowman (1983).
Stolte, S., and Reuss, J., "Elastic Scattering Cross Sections: Noncentral Potentials," in Bernstein (1979).
Thomas, L. D., "Rainbow Scattering in Inelastic Molecular Collisions," in Truhlar (1981).

Section 3.2

Books

Arrighini, P., *Intermolecular Forces and Their Evaluation by Perturbation Theory,* Springer-Verlag, Heidelberg, 1981.
Goodisman, J., *Diatomic Interaction Potential Theory,* Academic, New York, 1973.
Hirschfelder, J. O. (ed.), *Intermolecular Forces, Adv. Chem. Phys., 12* (1967).
Israelachvili, J. N., *Intermolecular and Surface Forces,* Wiley, New York, 1985.
Maitland, A., Rigby, M., Smith, E. B., and Wakeham, W. A., *Intermolecular Forces: Their Origin and Determination,* Clarendon, Oxford, 1981.
Margenau, H., and Kestner, N. R., *Theory of Intermolecular Forces,* Pergamon, Oxford, 1969.
Pullman, B. (ed.), *Intermolecular Interactions from Diatomics to Biopolymers,* Wiley, New York, 1978.
Pullman, B. (ed.), *Intermolecular Forces,* Reidel, Boston, 1981.

Review Articles

Amdur, I., and Jordan, J. E., "Elastic Scattering of High-Energy Beams: Repulsive Forces", *Adv. Chem. Phys., 10,* 29 (1966).
Andersen, T., and Neitzke, H. P., "Shapes of Atoms Excited in Hard or Soft Collisions," in Eichler et al. (1984).
Baudon, J., "Low Energy Collisions with Excited Atoms," in Eichler et al. (1984).
Bernstein, R. B., and Muckerman, J. T., "Determination of Intermolecular Forces

via Low-Energy Molecular Beam Scattering," *Adv. Chem. Phys.*, *12*, 389 (1967).

Buckingham, A. D., "Permanent and Induced Molecular Moments and Long-Range Intermolecular Forces," *Adv. Chem. Phys.*, *12*, 107 (1967).

Certain, P. R., and Bruch, L. W., "Intermolecular Forces," *Int. Rev. Sci. Phys. Chem.*, *1*, 113 (1972).

Cross, R. J., Jr., "Determination of Intermolecular Potentials Using High-Energy Molecular Beams," *Acc. Chem. Res.*, *8*, 225 (1975).

Dalgarno, A., "New Methods for Calculating Long-Range Intermolecular Forces," *Adv. Chem. Phys.*, *12*, 143 (1967).

Este, G. O., Knight, D. G., Scoles, G., Valbusa, U., and Grein, F., "Interaction of Hydrogen Atoms with Polyatomic Molecules Studied by Means of Scattering Experiments and Hybrid Hartree–Fock Plus Damped Dispersion Calculations," *J. Phys. Chem.*, *87*, 2772 (1983).

Ewing, G. E., "Vibrational Predissociation of van der Waals Molecules and Intermolecular Potential Energy Surfaces," in Truhlar (1981).

Gerber, R. B., "The Extraction of Intermolecular Potentials from Molecular Scattering Data: Direct Inversion Methods," in Pullman (1981).

Haberland, H., Lee, Y. T., and Siska, P. E., "Scattering of Noble-Gas Metastable Atoms in Molecular Beams," *Adv. Chem. Phys.*, *45*, 487 (1981).

Hirschfelder, J. O., and Meath, W. J., "The Nature of Intermolecular Forces," *Adv. Chem. Phys.*, *12*, 3 (1967).

Klemperer, W., "The Rotational Spectroscopy of van der Waals Molecules," *Faraday Discuss. Chem. Soc.*, *62*, 179 (1977).

Kutzelnigg, W., "Quantum Chemical Calculation of Intermolecular Interaction Potentials, Mainly of van der Waals Type," *Faraday Discuss. Chem. Soc.*, *62*, 185 (1977).

LeRoy, R. J., "Vibrational Predissociation of Small van der Waals Molecules," in Truhlar (1984).

LeRoy, R. J., and Carley, J. S., "Spectroscopy and Potential Energy Surfaces of van der Waals Molecules," *Adv. Chem. Phys.*, *42*, 353 (1980).

Lindinger, W., "State Selected Ion-Neutral Interactions at Low-Energies," in Eichler et al. (1984).

Luck, W. A. P., "Studies of Intermolecular Forces by Vibrational Spectroscopy," in Pullman (1981).

McCourt, F. R. W., and Lui, W.-K., "Anisotropic Intermolecular Potentials and Transport Properties in Polyatomic Gases," *Faraday Discuss. Chem. Soc.*, *23*, 241 (1982).

Mason, E. A., and Monchick, L., "Methods for the Determination of Intermolecular Forces," *Adv. Chem. Phys.*, *12*, 329 (1967).

Mueller, C. R., Smith, B., McGuire, P., Williams, W., Chakraborti, P., and Penta, J., "Intermolecular Potentials and Macroscopic Properties of Argon and Neon from Differential Cross Sections," *Adv. Chem. Phys.*, *21*, 369 (1971).

Pople, J. A., "Intermolecular Bonding," *Faraday Discuss. Chem. Soc.*, *73*, 7 (1982).

Schaefer, H. F., III, "Interaction Potentials: Atom–Molecular Potentials," in Bernstein (1979).

Schlier, C., "Intermolecular Forces," *Annu. Rev. Phys. Chem.*, *20*, 191 (1969).

Scoles, G., "Two-Body, Spherical, Atom–Atom, and Atom–Molecular Interaction Energies," *Annu. Rev. Phys. Chem.*, *31*, 81 (1980).

Stwalley, W. C., "Observations on State-Resolved Cross Sections for Long-Range Molecules," in Brooks and Hayes (1977).

Toennies, J. P., "Elastic Scattering," *Faraday Discuss. Chem. Soc., 55,* 129 (1973).

Van der Avoird, A., "Intermolecular Forces: What Can Be Learned from Ab Initio Calculations," in Pullman (1981).

Wahl, A. C., "The Calculation of Energy Quantities for Diatomic Molecules," *Int. Rev. Sci. Phys. Chem., 1,* 41 (1972).

Winn, J. S., "A Systematic Look at Weakly Bound Diatomics," *Acc. Chem. Res., 14,* 341 (1981).

Section 3.3

Book

Brooks, P. R., and Hayes, E. F. (eds.), *State-to-State Chemistry,* American Chemical Society, Washington, D.C., 1977.

Review Articles

Bernstein, R. B., "Reactive Scattering: Recent Advances in Theory and Experiment," *Adv. At. Mol. Phys., 15,* 167 (1979).

Farrar, J. M., and Lee, Y. T., "Chemical Dynamics," *Annu. Rev. Phys. Chem., 25,* 357 (1974).

Greene, E. F., Moursund, A., and Ross, J., "Elastic Scattering in Chemically Reactive Systems," *Adv. Chem. Phys., 10,* 135 (1966).

Grice, R., "Reactive Scattering," *Adv. Chem. Phys., 30,* 247 (1975).

Herschbach, D. R., "Reactive Scattering in Molecular Beams," *Adv. Chem. Phys., 10,* 319 (1966).

Herschbach, D. R., "Reactive Scattering," *Faraday Discuss. Chem. Soc., 55,* 233 (1973).

Herschbach, D. R., "Molecular Dynamics of Chemical Reactions," *Pure Appl. Chem., 47,* 61 (1976).

Kinsey, J. L., "Molecular Beam Reactions," *Int. Rev. Sci. Phys. Chem., 9,* 173 (1972).

4
The polyatomic approach
to chemical dynamics

Any chemical reaction worthy of the name necessarily involves the interaction of at least three atoms. We have thus far largely side-stepped this point, adopting approximate "two-body" approaches with the hope of simulating some of the gross features of the dynamical behavior of the *three-body system*. However, if we wish to consider such detailed questions as the role of internal energy or the origin of steric factors, we must explicitly take cognizance of the *molecular* character of the reactants and the products. To do so we must somehow generalize the concept of an interatomic potential and then learn how to calculate collision trajectories representing the transformation of reactants into products. After that, we need to examine theoretical approximations that can help us estimate the overall reaction rate. We begin with the experimental evidence on the *role of energy in chemical reactivity*.

4.1 Energy and chemical change

4.1.1 Energy partitioning

We have seen that energy is a factor of central importance in the dynamics of chemical change. Energy is needed to drive the reaction forward; energy determines the magnitude of the cross section (and thus the reaction rate); energy governs the details of the collision trajectory, and energy is what we can get out of the products.

Let us consider first *energy disposal*. Exoergic reactions can release their energy in a number of ways. One possibility is to form *electronically excited* products that can then emit visible (or ultraviolet) light, for example,

$$Cl + K_2 \longrightarrow K^* + KCl$$
$$\downarrow h\nu$$
$$K$$

or†

$$NO_2 + Ba \longrightarrow BaO(A^1\Sigma) + NO$$

$$\downarrow hv$$

$$BaO(X^1\Sigma)$$

Another mode of energy disposal is the production of vibrotationally excited products, as we saw in Section 1.2. One of the many examples is the infrared chemiluminescent reaction

$$O + CS \rightarrow S + CO^\dagger.$$

In all these cases the source of the exoergicity is the replacement of a weaker bond by a stronger bond (see Fig. 1.1); however, another route is via *chemical* activation, where the energy is provided by the formation of a new bond (see Sec. 4.2.8). Thus the reaction sequence

$$CH_3 + CF_3 \rightarrow CH_3CF_3{}^\dagger$$

$$CH_3CF_3{}^\dagger \rightarrow CH_2CF_2 + HF^\dagger$$

releases HF vibrationally excited up to $v = 4$. Even electronically excited products can be formed *via* chemical activation. An example is the thermolysis of aromatic endoperoxides,

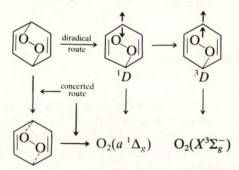

which can lead to high (up to over 90%) yields of singlet (excited) oxygen molecules.

The energy released can also be supplied by *physical* activation. Photo-elimination reactions, for example, are often found to yield vibrationally excited products.

---

† Such chemiluminescence reactions of Ba atoms have already been used in the study of the upper atmosphere, through the release of Ba vapor from rocket-borne containers.

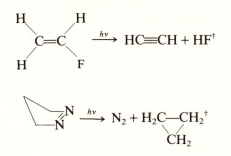

These reactions that *selectively* populate product vibrotational states can be used to provide the population *inversion* needed for chemical laser action.

Finally, the reaction exoergicity can, of course, be released in the most degraded form, namely, relative translation of the products. Figure 4.1 shows the translational energy release in the exoergic reaction

$$CH_3I + K \rightarrow KI + CH_3$$

Just as in the other cases of energy release, the original energy can here too be supplied via physical activation. An example is the photodissociation reaction

$$HI \xrightarrow{hv} \begin{cases} H + I(^2P_{3/2}) \\ H + I(^2P_{1/2}) \end{cases}$$

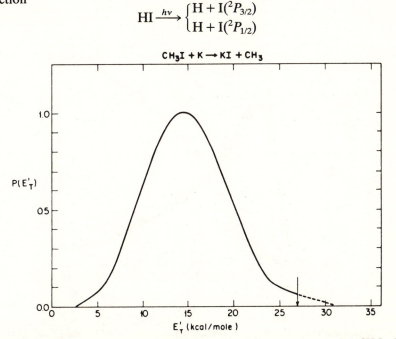

Fig. 4.1. Distribution of the final translational energy $P(E'_T)$ for the reaction $CH_3I + K \rightarrow KI + CH_3$. The initial relative translational energy $E_T = 1.8 \text{ kcal mol}^{-1}$. The total available energy is indicated by the arrow. The dashed "tail" of the curve is spurious. [Adapted from A. M. Rulis and R. B. Bernstein, *J. Chem. Phys.*, *57*, 5497 (1972).]

which has been employed as a practical source of H atoms with high translational energies. These so-called *hot atoms* are used to study chemical reactivity at above-thermal translational energies. For example, at $\lambda = 266$ nm the energy of the H atom is such that the $H + D_2$ reaction occurs with a center-of-mass collision energy $E_T = 1.3$ and 0.55 eV, respectively. By varying λ, one can change the collision energy and probe the translational energy dependence of both the reaction cross section (see Fig. 2.14) and the energy disposal. These fast H atoms can also be used in energy transfer collisions.

4.1.2 Energy consumption

Energy consumption in reactions with a potential energy barrier is highly selective. Especially important in overcoming the barrier to reaction is the reactants' vibrational energy (see Sec. 4.3.4). Translational energy also can be used to overcome even large reaction endoergicities, for example, by the use of hot atoms. Bond-breaking energies can be readily achieved using seeded supersonic beams (see Appendix 5B), and it has thereby become possible to study a variety of *collision-induced dissociation* processes, for example,

$$Xe + CsCl \rightarrow Xe + Cs^+ + Cl^-$$

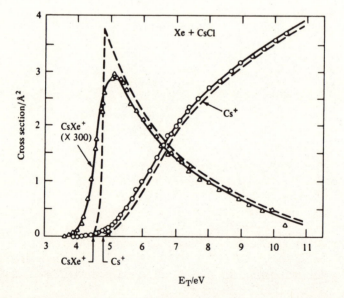

Fig. 4.2. Experimental measurements of absolute cross sections for Cs^+ and $CsXe^+$ formation from collisions of $CsCl + Xe$ as a function of the average relative translational energy (points and curves). Dashed curves are deconvoluted results from which the two indicated thresholds have been deduced. [Adapted from S. H. Sheen, G. Dimoplon, E. K. Parks, and S. Wexler, *J. Chem. Phys.*, *68*, 4950 (1978).]

which has a threshold† of 4.85 eV. The increase of the dissociation cross section with collision energy E_T (Fig. 4.2), is stronger than linear. For such dissociation processes as well, it is found that the cross section shows a nearly exponential increase with increasing *vibrational* energy of the reagent molecule.

A long-established route to "drive" endothermic reactions has been electronic excitation via *light absorption. Organic photochemistry* provides many examples where a particular reaction outcome can be realized by using light of the proper wavelength. Very often when we overcome the barrier with photon energy, a *completely different* (often unexpected) type of reaction product is formed, for example,

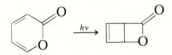

Similarly, the use of *electronically excited reactants* also often produces surprises—and new synthetic routes! The use of electronically excited mercury atoms can initiate new chemical reactions, for example,

$$CH_3Cl + Hg^* \rightarrow HgCl + CH_3$$

Excited oxygen molecules may add directly to double bonds,‡ for example, the stereospecific reaction

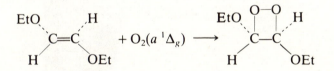

We are about to discuss potential energy functions and their main features. Novel modes of reactivity are possible when the reaction is taking place on a different (i.e., excited) electronic potential energy surface. The topology of the excited surface can be completely different from the ground-state one. A striking example is the reaction of an electronically excited $H(2s)$ atom with H_2. The supermolecule, an electronically excited H_3 molecule, is metastable, so the potential energy surface contains a well.

An important goal of photoselective chemistry is to achieve product specificity on a *given* potential energy surface. This goal is not easily achieved. An understanding of the inter*molecular* potential and of the dynamics that it can give rise to is then essential.

---

† The threshold energy here is essentially the endoergicity. The latter is equal, of course, to the dissociation energy of the salt molecule into ions.
‡ Hence the use of electronically excited oxygen molecules in the disinfection of wastewater.

4.2 Three-body potential energy functions and chemical reactions

4.2.1 Potential energy surfaces

How can we generalize the concept of an inter*atomic* potential? We need to know the interaction energy as a function of the *configuration* of the system *throughout* the *rearrangement* from reactants to products (Fig. 4.3). Even in the simplest case of three interacting atoms, the potential energy is a function of three coordinates, for example, the three coordinates that specify the three interatomic distances.† However, even then it is difficult to

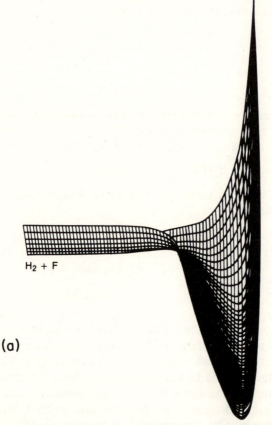

$H_2 + F$

(a)

$H + HF$

Fig. 4.3. (a) Perspective plot of *ab initio* computed potential energy surface for the collinear FH_2 system, appropriate to describe the $F + H_2 \rightarrow HF + H$, viewed from the exit valley. [Adapted from C. F. Bender, P. K. Pearson, S. V. O'Neil, and H. F. Schaefer III, *J. Chem. Phys.*, *56*, 4626 (1972).]

---

† The interaction potential is a function of the positions of the three nuclei and hence, in principle, a function of nine coordinates. Imagine now the three atoms as the vertices of a triangle in an otherwise empty space. The interparticle interaction energy will not be changed if the triangle is rotated or translated as a rigid body. Hence the potential is a function of only the three coordinates required to specify the atomic triangle.

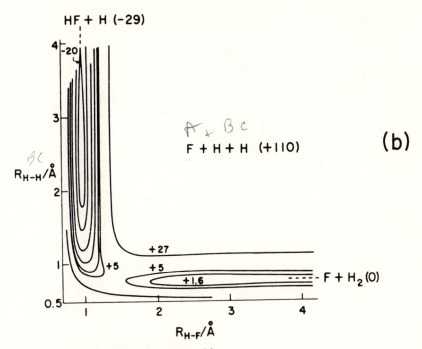

HF + H (−29)

4 −20

F + H + H (+110)

A + B c

(b)

R_{H-H}/Å

BC

3

2

1

+5   +27

+5

+1.6   F + H$_2$ (0)

0.5

1   2   3   4

R_{H-F}/Å

Fig. 4.3. (b) Contour map corresponding to (a).

visualize. The problem can be simplified by restricting consideration to limited geometries, for example, by examining the potential for the "collinear" (ABC) configuration, for which there are only two independent interatomic distances, R_{AB} and R_{BC}. In this case one can plot the potential as a function of the two coordinates as a *potential contour map*, showing equipotential lines in a manner made familiar by topographical maps. Figure 4.3a shows a potential energy function drawn in a perspective three-dimensional form. The same function is displayed as a contour map in Fig. 4.3b.

An understanding of the main topological features of the potential energy surface is the first step toward a qualitative understanding of the reaction dynamics. The "mountain pass" en route from the reactants' "valley" to that of the products, termed the "minimum energy path" or the "reaction coordinate," is a dominant feature of many potential surfaces, explaining why the energy threshold for reaction is often much smaller than a bond dissociation energy. As pointed out by Eyring, Polanyi, and Evans in the 1930s, an elementary bimolecular reaction, say,

$$H + H—H \rightarrow H—H + H$$

proceeds *not* in sequential steps, that is, by breaking the "old" bond and subsequent formation of the "new" bond, but via the *concerted* motion of the nuclei in a *continuous* transformation from the reactants' valley to the

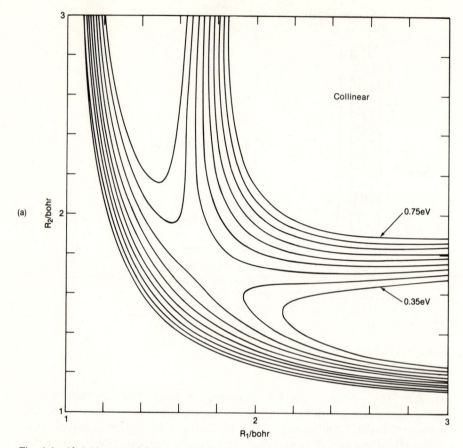

Fig. 4.4. *Ab initio* potential energy contour maps for H_3 in the collinear ($\gamma = 0°$) and bent ($\gamma = 50°$) configurations. Here R_1 and R_2 are in bohrs and the contour energies in eV. The contour lines are 0.05 eV apart. Note the (required) symmetry about the diagonal. The potential energy surface is from the parametrization [by D. G. Truhlar and C. J. Horowitz, *J. Chem. Phys.*, *68*, 2466 (1978); *71*, 1514 (1979)] of converged *ab initio* computations [by P. Siegbahn and B. Liu, *J. Chem. Phys.*, *68*, 2457 (1978).] It is designated the LSTH surface.

products' valley. A potential energy surface thus serves to mediate between the reactants' and the products' configuration.†

† The concept of a potential energy determining the motion of the nuclei is an approximate one, depending for its validity on the Born–Oppenheimer separation, familiar from diatomic molecules. When the Born–Oppenheimer approximation is valid, the potential surface is invariant to isotopic substitution. In essence, this approximation requires that only one electronic state be occupied at any configuration of the nuclei. This is a good approximation when dealing with "slow" collisions (i.e., if the speed of the nuclear motion is very much less than that of the electrons). However, the Born–Oppenheimer approximation is sometimes inapplicable in certain molecular collisions, as we have seen (e.g., the K + IBr reaction) and will see in Section 6.6. In general, the formation of an electronically excited state can occur as soon as the energy exceeds the necessary "threshold." Then we must consider simultaneously the ground- and excited-state surfaces.

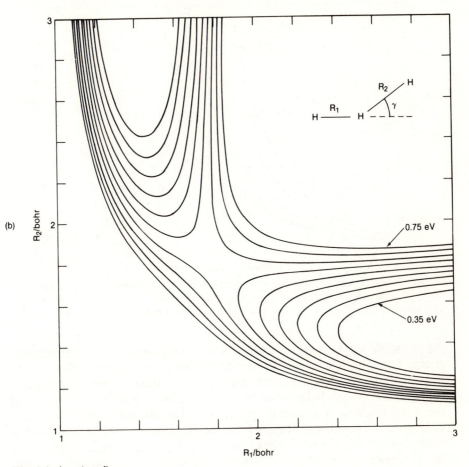

(b)

Fig. 4.4. (*continued*)

One of the most significant recent trends in applied quantum chemistry is the development of computationally tractable schemes for the evaluation of three-atom potential surfaces with accuracies on the order of kilojoules per mole. Such calculations must be better than Hartree–Fock (i.e., they must take into account electron correlation) in order to be of sufficient accuracy to be usable in dynamical problems. Semiempirical potential surfaces designed to capture the essential features of the dynamics are discussed further in Section 4.2.3.

The best known *ab initio* potential surface is that for the $H + H_2$ system, shown in Fig. 4.4 for two different approach configurations.

4.2.2 The reaction path

To better understand the origin of a threshold energy for reaction, let us consider the "cost" in energy required to cross over from the reactants' to

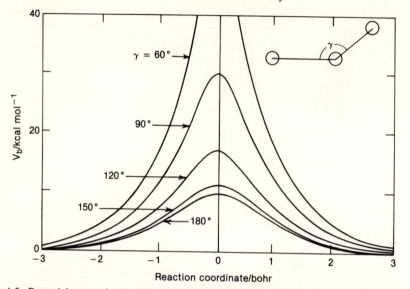

Fig. 4.5. Potential energy barriers for the reaction of $H + H_2$ at several fixed bond angles as a function of the reaction coordinate (along the minimum-energy path) for the LSTH surface. The global minimum of the barrier is for the collinear approach (180°), with a value of 9.8 kcal mol^{-1}.

the products' regions. We can define a *reaction path* as the line of minimal energy from the reactant to the product valley. From Figure 4.5 it is seen that the minimum energy path for the H_3 system is for the collinear configuration (which happens to be the one of lowest energy). We note that the barrier along this energy profile is relatively small, that is, $E_b \approx$ 40 kJ mol^{-1}. The energy required for an H-atom transfer in a concerted motion of the three atoms is thus less than 10% of the dissociation energy of H_2. It is the presence of a low-energy mountain pass between the reactants' and the products' valleys that favors the concerted (i.e., bimolecular) mechanism of atom (or group) transfer.

Figure 4.6 shows a semiempirical potential energy profile for a highly exoergic system, $F + H_2 \rightarrow HF + H$. Here the barrier is much smaller than for H_3 and, furthermore, is in the "entrance" valley. Over most of the reaction path the potential energy decreases monotonically. For such a reaction, the exoergicity is released early along the reaction path and is available to be pumped into the vibration of the emerging new bond. A surface with an early release of the exoergicity is often referred to as an *attractive* surface. A surface of the opposite type, that is, a late release, is termed *repulsive*. (Of course, if a surface is attractive for the forward reaction, it is repulsive for the reverse reaction!)

A useful (but not invariable) correlation† is that exoergic reactions will

---

† Organic chemists will recognize this as a special case of Hammond's postulate. This states that in a series of reactions, the more exoergic the reaction, the more the transition state (see Sec. 4.4.6) will resemble the reagents.

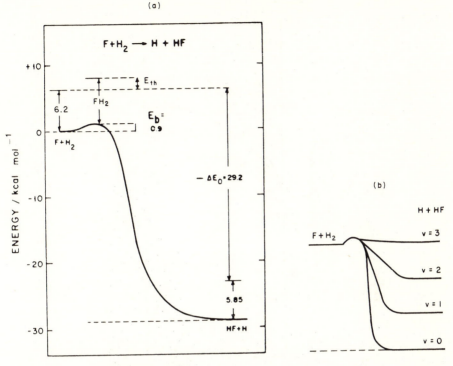

Fig. 4.6. (a) Potential energy profile along the reaction coordinate for collinear FH_2, based on a semiempirical potential surface. The energy barrier here is $E_b = 0.9\ \text{kcal mol}^{-1}$; the zero-point energies of H_2 and HF are 6.2 and 5.85 kcal mol^{-1}, respectively, while that of the transition state is 6.9 kcal mol^{-1}, so that $E_{th} = 1.6\ \text{kcal mol}^{-1}$. $\Delta E_0 = -29\ \text{kcal mol}^{-1}$. [Adapted from J. T. Muckerman, *J. Chem. Phys.*, *54*, 1155 (1971).] (b) Energetically accessible vibrational levels of HF product from the $F + H_2$ reaction, at threshold.

have an early barrier. In a series of atomic halogen (X) reactions, for example, $H_2 + X \rightarrow H + HX$, where there is a common exchanged atom, we thus expect the barrier to shift into the products' valley as the reaction becomes more endoergic (Fig. 4.7). Both *ab initio* and semiempirical potential-energy surfaces do indeed show such correlations.

Since the reaction path passes through the local minima of the surface, the potential energy increases upon sideways deviations from the path. Hence near the barrier the potential surface has the form of a saddle. The location of the barrier is thus referred to as the *saddle point* (sometimes as the *col*) of the surface. As we move along the reaction path, the curvature of the potential (for a displacement perpendicular to the path) changes from the force constant of the AB reactant to the force constant of the BC product. At the saddle point it is the force constant for the symmetrical stretch of ABC. The asymmetrical stretch at that point is the reaction path itself.

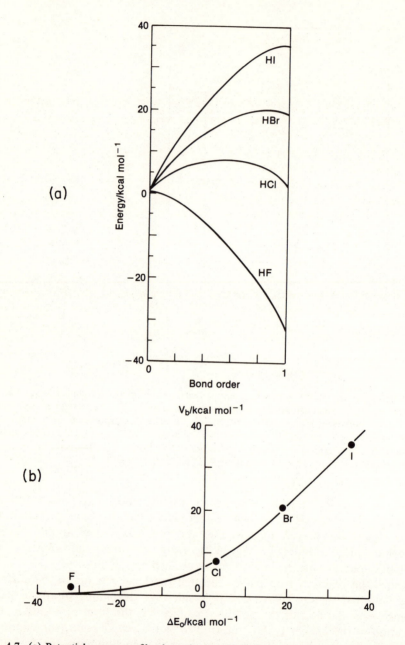

Fig. 4.7. (a) Potential energy profile along the reaction path for the $H_2 + X \rightarrow H + HX$ series of reactions. A common plot is achieved by measuring distance along the reaction path in terms of the order of the HX bond. In this semiempirical computation, the location of the barrier clearly shifts in the direction of the products as the endoergicity of the reaction increases. A related correlation is shown in (b), where the barrier height (points, *ab initio* computed values; line, semiempirical result) increases with increasing endoergicity. [Adapted from N. Agmon and R. D. Levine, *Isr. J. Chem.*, **19**, 330 (1980).]

Later we shall see the relation of the potential-energy barrier to the energy threshold for the reaction (see Sec. 4.3.4).

4.2.3 Determining the potential energy surface

Steady progress is being made towards the accurate computation of potential surfaces for systems of chemical interest. (The extension beyond the Hartree–Fock level so as to include the correlation energy presents computational difficulty.) However, certain quantal approximation schemes have provided surfaces with good qualitative "topographical" features. One such case is that of the $Li + F_2$ system, shown in Fig. 4.8. Here we see no barrier at all—the reaction path is "downhill all the way" and there should be no threshold energy in the reaction. This "early release" of the exoergicity can be understood qualitatively in terms of a simple model, which will be discussed in Section 4.2.5.

For stable triatomic molecules, the potential energy in the vicinity of the three-body well can be directly determined from spectroscopic data. An illustration of this is given for the HCN molecule in Fig. 4.9. Such a potential energy surface would be appropriate for the $H + CN$ collision in a collinear geometry. For other triatomics, the most stable configuration would be

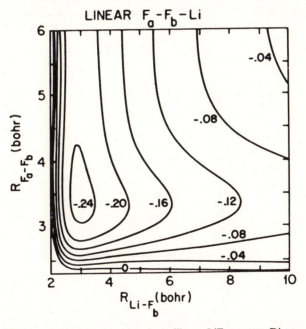

Fig. 4.8. Potential energy contour map for the collinear LiF_2 system. Distances are in bohrs and contour energies in hartrees ($1 h = 627.5 \text{ kcal mol}^{-1}$). [The surface is the result of an "atoms-in-molecules" calculation of G. G. Balint-Kurti, *Mol. Phys.*, **25**, 393 (1973). See the *ab initio* calculations on the LiF_2 ground-state molecule with C_{2v} symmetry by B. Maessen and P. E. Cade, *J. Chem. Phys.*, **80**, 2618 (1984).]

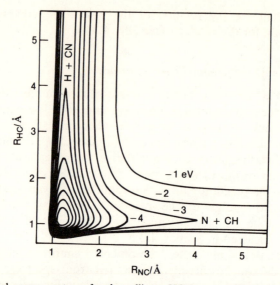

Fig. 4.9. Potential energy contours for the collinear HCN system. Distances are in angstroms and energies in electron-volts. The potential was obtained using observed overtones of the stretching vibrations to construct an algebraic Hamiltonian. [Adapted from I. Benjamin and R. D. Levine, *Chem. Phys. Lett.*, *117*, 314 (1985). For a compendium of methods and results for fitting "flexible" potential-energy functions to spectroscopic observations, see J. N. Murrell et al. (1984).]

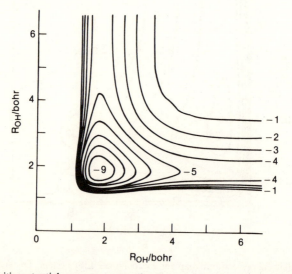

Fig. 4.10. *Ab initio* potential energy contours for the water molecule at an HOH bond angle of 104.5°. Note the symmetry about the diagonal. Bond distances are in bohrs and energies in eV. [Adapted from J. S. Wright, D. J. Donaldson, and R. J. Williams, *J. Chem. Phys.*, *81*, 397 (1984).]

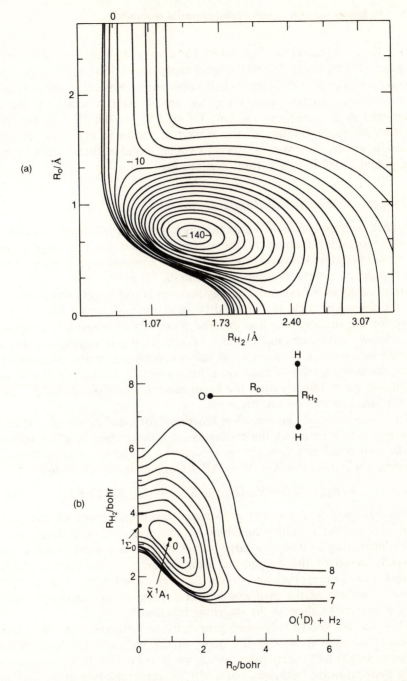

Fig. 4.11. (a) Semiempirical (diatomics in molecules) potential energy contours for H_2O, maintaining C_{2v} symmetry. The abscissa R_{H_2} is the H–H internuclear separation, the ordinate R_O is the distance of the O atom from the midpoint of the H–H bond. Distances are in angstroms and energies in $kcal\,mol^{-1}$, taking zero as the floor of the reactant valley for $R_O \rightarrow \infty$. [Adapted from P. A. Whitlock, J. T. Muckerman, and E. R. Fisher, *J. Chem. Phys.*, **76**, 4468 (1982).] (b) *Ab initio* calculated potential contours for H_2O analogous to (a). Distances are in bohrs and energies in eV, measured above a zero defined by the ground-state (X^1A_1) minimum of H_2O. The absolute accuracy of the computations, based on comparison with experimental energetics, is within $0.35\,eV$; the relative precision is about $5\,kcal\,mol^{-1}$. [Adapted from G. Durand and X. Chapuisat, *Chem. Phys.*, **96**, 381 (1985).]

bent, for example, H_2O (Fig. 4.10). Here we can, as in Fig. 4.4, draw a contour plot by fixing the HOH bond angle not at $180°$ but at some other value, for example, the equilibrium bond angle. Note that the surface is drawn in coordinates appropriate for an $H + OH$ reaction. If we are interested in the insertion reaction† $O(^1D) + H_2 \rightarrow OH(^2\Pi) + H$, the more appropriate representation of the surface is shown in Fig. 4.11. In many cases, however, the reaction proceeds via an *unstable* intermediate, that is, the potential will have a barrier (e.g. see Fig. 4.6).

4.2.4 Semiempirical potential surfaces

We have seen that *ab initio* computations can sometimes provide us with the main topographic features of the surface, at least for the simple, "few-electron" systems. However, there is a real *"computational* barrier" to the generation of purely theoretical potentials. Thus the most widely used method of obtaining potential energy surfaces is still largely empirical. One adopts a semiempirical (or, if need be, a completely empirical) functional form for the surface, which allows the possibility of *interpolating* between the "known" reactant and product valleys. Dynamical computations are carried out and the parameters in the potential are varied systematically from the initial guesses (based on a knowledge of the barrier E_b, in turn estimated approximately from the experimental activation energy E_a) so as to best simulate the known dynamics.

One procedure with an empirical functional form makes use of "switching functions". It is based on the realization that the surface in effect switches off the old bond and switches on the new one. For the simple case of a collinear surface for the $A + BC \rightarrow AB + C$ reaction, we would use

$$V(R_{BC}, R_{AB}) = V_{BC}(R_{BC})f_1(R_{AB}) + V_{AB}(R_{AB})f_2(R_{BC}) \qquad (4.1)$$

where V_{BC} and V_{AB} are the *presumed known* potentials of the stable diatomics and the switching functions f_1 and f_2 have the range from zero to unity, increasing to unity for large values of their argument. Such a form correctly describes the asymptotic form of the potential (and hence the correct overall energetics) and brings out very explicitly the concept of the potential surface as the *mediator* between *reactants* and *products*. But of course we have to guess at the switching functions!

The use of known diatomic potentials to estimate the three-atom potential function is also at the heart of the so-called London–Eyring–Polanyi–Sato (LEPS) semiempirical scheme based on the London equation. In its most primitive form, one starts with the representation of the diatomic

---

† The oxygen atom in its ground electronic state is $O(^3P)$. The $O(^3P) + H_2$ collision occurs on a surface that is not correlated (in a Born–Oppenheimer, adiabatic sense) with the ground electronic state of H_2O. To reach that surface one needs either to start with an excited, $O(^1D)$, oxygen atom or to make an electronically nonadiabatic transition. In the latter case the system "hops" from one surface to another.

potential function in terms of a Coulomb (Q) contribution and an exchange (J) contribution,

$$V(R_{AB}) = Q_{AB} \pm J_{AB} \qquad (4.2)$$

where the plus sign refers to the singlet electronic state and the minus sign to the triplet.† Both Q and J are functions of R, reducing to zero for very large R. When the three atoms are in close proximity, their potential is then written as

$$V_{ABC} = Q_{AB} + Q_{BC} + Q_{CA}$$
$$\pm \left[\tfrac{1}{2}(J_{AB} - J_{BC})^2 + \tfrac{1}{2}(J_{BC} - J_{CA})^2 + \tfrac{1}{2}(J_{CA} - J_{AB})^2 \right]^{1/2} \qquad (4.3)$$

When one atom is removed to infinity, the Coulomb and exchange terms for the two broken bonds vanish and Eq. (4.3) reduces to Eq. (4.2). Figure 4.12 is an example of a so-called optimized LEPS surface "tailored" for the $O(^3P) + H_2$ system.

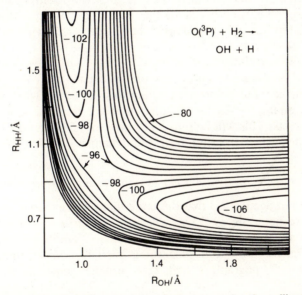

Fig. 4.12. Semiempirical (LEPS) potential energy contours for the collinear OH_2 system, appropriate to the reaction $O(^3P) + H_2 \rightarrow OH + H$. Distances are in Å, energies in kcal mol^{-1}. [LEPS surface by B. R. Johnson and N. W. Winter, *J. Chem. Phys.*, *66*, 4116 (1977).]

Strictly speaking, the London equation, Eq. (4.3), is valid only when all three atoms have but one valence electron (i.e., are in the ^2S state, as in $H + H_2$); often, however, this is not the case [e.g., $F(^2P) + H_2$ or $O(^3P) + H_2$]. The considerable advantage of having the potential as an explicit function of the three interatomic coordinates often leads to the London

† Actually, J is negative and $|J| > Q$ near the equilibrium separation of the diatomic.

equation being employed outside its range of validity. A more proper procedure is the method of diatomics in molecules (DIM). For an ABC system, the method consists in considering the possible products of electronic states of AB with electronic states of C, and similarly of BC with those of A, and of AC with those of B. The resulting set of states is used as a basis in which the electronic Hamiltonian is diagonalized.†

The input to the DIM method is similar to that of LEPS: the potential energy curves (bound and repulsive) for all "relevant" AB, BC, and AC states. For each nuclear configuration, the matrix representation of the electronic Hamiltonian in the DIM basis needs to be diagonalized. Although such matrices need not be large [e.g., 4×4 for $X(^2P) + H_2$], they must be diagonalized a large number of times if the entire potential energy surface is to be generated.‡

A considerable amount of work has been done on the properties of semiempirical surfaces. There is evidence to suggest that in a family of related reactions of decreasing barrier heights, the location of the barrier occurs correspondingly earlier along the reaction coordinate. Moreover, in a family of related reactions, the barrier height decreases with an increase in the reaction exoergicity (see Fig. 4.7).

To be able to correlate the main features of the surface with the observed dynamics, we first need to know how one calculates the collision dynamics on a *given* surface, but before we attempt this, let us see what we can learn from certain qualitative surface features that are responsible for important dynamical effects. This will help us learn how to infer the major qualitative features of the surface from the observed dynamics.

4.2.5 Qualitative interpretations: The example of the "harpoon mechanism"

We have seen in Section 2.3 that there are many fast exoergic reactions that have no "threshold energy." One example is the $Li + F_2$ reaction, whose potential surface was discussed in Section 4.2.3. We noted the early release of the exoergicity and the absence of a barrier. Can we understand the origin of this "topography" and then predict some of its dynamical implications?

Long before any computed or approximate potential-energy surfaces were available for such reactions (e.g., alkali + halogen compounds), a simple model was proposed to explain the main body of observations, namely, the extremely rapid reaction rates that implied reaction cross sections up to 200 Å^2! The picture envisaged by Polanyi, the "harpoon mechanism," led to the Magee theory, which put it on a semiquantitative basis.

---

† Like the London equation (which is a special case), the diatomics-in-molecules method reduces to a proper diatomic limit when one atom is removed to infinity.

‡ Trajectory computations (see Sec. 4.3), of course, require the potential at very many points (and anew for each trajectory). This makes it necessary to construct interpolation representations of the surface, just as for the *ab initio* generated case.

The first stage of the reaction is envisaged as the transfer of the valence electron of the alkali atom to the halogen molecule. Such a transfer (of a "light" electron) is possible even when reactants are far apart (e.g., 5–10 Å). Once the transfer takes place and an ion pair (e.g., $Li^+F_2^-$) is formed, the strongly attractive Coulombic force brings the two oppositely charged ions together, followed by formation of the stable LiF molecule and ejection of the F atom. The metal atom has, in effect, used its valence electron as a "harpoon" in order to pull in the halogen molecule, employing as a "rope" the Coulombic interionic attraction.

A simple estimate of the range of the harpoon can be obtained easily by considering the energetics of the charge transfer. First we realize that the ionization potential of the metal (which may be an alkali or alkaline-earth atom) exceeds the electron affinity of the halogen molecule, so that at large separations charge transfer is endoergic and cannot take place at "thermal" collision energies. However, as the reactants approach one another, the formation of the ion pair *can* occur owing to the gain in energy from the *Coulombic attraction itself* of the newly formed ion pair.

Figure 4.13 is a schematic drawing of the potential curves appropriate for a discussion of the $K + Br_2$ reaction. From Fig. 4.13 one can see that the largest separation R_x at which charge transfer can take place on energetic grounds is the solution of the equation (see Fig. 4.13)

$$-e^2/R_x + \Delta E_0 = -C/R_x^6 < 0 \qquad (4.4)$$

where ΔE_0 is the endoergicity, given by the difference between the ionization potential of the alkali and the so-called vertical electron affinity of

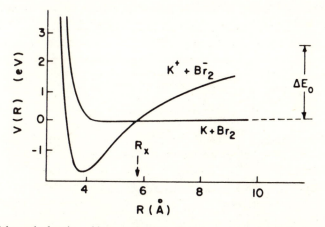

Fig. 4.13. Schematic drawing of intersecting potential curves to model the harpoon reaction of $K + Br_2$. The nearly flat curve represents the covalent $K + Br_2$ interaction at large R, the upper curve to the ionic $K^+ + Br_2^-$ potential. As R is decreased to the crossing point R_x, the KBr_2 system switches from the covalent to the ionic form and follows the lower curve until K^+Br^- forms, with the ejection of a Br atom. The asymptotic separation ΔE_0 is the difference between the ionization potential of K and the (vertical) electron affinity of Br_2.

the halogen molecule and $-C/R^6$ is the long-range dispersion interaction, negligible compared to e^2/R at R_x.

Thus Eq. (4.4) can be simplified to yield an explicit formula for R_x:

$$R_x/\text{Å} \approx \frac{e^2}{\Delta E_0} = \frac{14.4}{\Delta E_0 \,(\text{eV})} \tag{4.5}$$

Owing to the large Coulombic attraction, reaction follows "immediately" and efficiently after the electron transfer. Thus we expect that $P(b) = 1$ up to $b_{max} \approx R_x$. (Note that for $b > R_x$ the reactants will never get "close enough," i.e., to within R_x.) Hence we expect $\sigma_R \approx \pi R_x^2$.

Equation (4.5) implies that R_x will increase for a decrease in the ionization potential of the metal; thus σ_R should increase for the sequence Li, Na, . . . , Cs, reacting with any given halogen.

This simple mechanism thus offers a qualitative explanation for both the large magnitude of the alkali–halogen reaction cross sections and the trends with the metal and halogen involved. Even larger cross sections can be expected for electronically excited alkali atoms [whose ionization potential is smaller; see Eq. (4.5)]. The harpoon model also provides an understanding of the mechanism of the collisional ionization reactions, for example, $K + Br_2 \rightarrow K^+ + Br_2^-$, and explains why the threshold energy for such reactions equals the endoergicity, without any additional barrier (see Fig. 2.12).

The essential feature of the harpoon mechanism, the crossing over, at a finite separation of an ionic potential energy curve (with an asymptotically higher energy) and a covalent one is found in other systems as well. One example already mentioned (see Sec. 1.3.1) is the lasing species in excimer lasers. Figure 4.14 shows schematic potential energy curves for the rare-gas

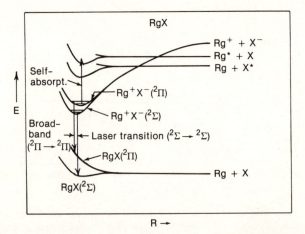

Fig. 4.14. Generalized potential energy diagram for rare-gas halides (RgX). The strongly bound (Rg$^+X^-$) ionic states can be formed *via* collision of a metastable rare-gas atom and a halogen; the halides then emit (e.g., *via* the lasing transition, $^2\Sigma \rightarrow {}^2\Sigma$) to the covalent Rg$X(^2\Sigma)$ state (very weakly bound, thermally unstable). [Adapted from C. K. Rhodes (ed.), *Excimer Lasers*, Springer-Verlag, Berlin, 1979).]

monohalides, RgX. The electronically excited rare-gas atom Rg* behaves like an alkali atom.†

4.2.6 The steric factor: Qualitative and quantitative considerations

We have already discussed the origin of the *steric effect* in chemical reactions as due to preferential orientation (see Sec. 2.4.9). In Fig. 2.21 and again in Fig. 4.5, we show two views of the barrier for the $H + H_2$ reaction, revealing that it is lowest for the collinear configuration. Thus, at collision energies only slightly higher than the barrier, the preferred orientation for reaction is expected to be essentially collinear. As the collision energy increases, the range of "acceptance angles" increases, that is, the "cone of acceptance" grows. The line-of-centers model introduced in Section 2.4.9 implies that a greater range of impact parameters can then contribute to reaction.

The results for the angle-dependent barrier to reaction shown in Figs. 2.21 and 4.5 are based on accurate *ab initio* computations. (Similar results for systems with more than three electrons are becoming available.) Can one, however, understand the origin of the orientation dependence from simple ideas of chemical bonding? This is important not only for its own sake but also for demonstrating that electronic structure theory provides a unified approach encompassing chemical reactivity.

A simple interpretation of the bonding "during" a reactive collision can be obtained via the standard molecular-orbital approach. The molecular orbitals (MOs) are constructed as a linear combination of atomic orbitals (AOs). The number of independent MOs is equal to the number of AOs used. To arrange these MOs in order of increasing energy, we note that when an MO acquires another node, its energy is higher. Once the order of the orbitals is established, the electronic configuration is determined by assigning two electrons (of opposite spin) to each orbital, starting with the one lowest in energy, until all electrons are assigned.

To discuss the $H + H_2$ system, consider first the H_2 reagent. Using a $1s$ AO on each H atom (denoted by a sphere), we have two independent linear combinations:

$$\sigma^*(1s)$$

$$\sigma(1s)$$

The dots represent the nuclei and the shading defines the phase of the atomic orbitals. In the σ orbital, the two $1s$ functions have the same phase. The MO has no node and hence it is a low-energy bonding obital. In the σ^* (antibonding) MO, the two atomic functions have opposite phases. There is a node midway between the two nuclei. The ground state configuration of H_2 is $\sigma(1s)^2$. (Thus far everything is at the level of first-year chemistry.)

---

† The ionization potentials of $Ar^*({}^3P_2)$ and $Kr^*({}^3P_2)$ are 4.21 and 4.08 eV, respectively, compared to 4.34 and 4.18 eV for K and Rb, respectively.

Now let an H atom (with its 1s orbital) collinearly approach H_2. The 1s orbital can add in phase or out of phase to either σ or σ^*:

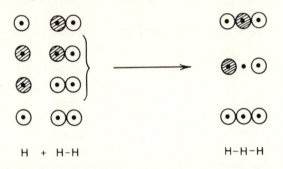

The in-phase addition to σ gives rise to a lowest (node-free) H_3 MO. The out-of-phase addition to σ^* (producing a node at each bond) gives rise to a strongly antibonding H_3 MO. The two middle "alternatives" shown in the left column are really the same: a node at one H—H bond, no node at the other. An alternative description is to add them together so as to have the orbital picture on the right, which shows that this orbital is somewhat antibonding: It has a node midway between the two end nuclei.

Three electrons are to be assigned: Two go to the lowest, bonding orbital; the third is assigned to the higher-energy, somewhat antibonding orbital. Thus, in a collinear configuration, there is expected to be a small energy barrier for the approach of $H + H_2$.

Now let the H atom approach at an angle ($<180°$) with respect to the H_2 bond. The corresponding orbitals for H_3 are pictured as follows:

The energy of the bonding orgital and the strongly antibonding orbital are going to be nearly unaffected by the change in the H_3 bond angle. But what was a somewhat antibonding orbital is now more strongly antibonding, because its node is between two nuclei that are now nearer to one another. The third electron, which is assigned to that orbital, will thus have a higher energy. therefore the barrier to a noncollinear (sideways) approach of H to H_2 is higher and should increase monotonically as the bond is bent further.

The detailed computations (e.g., see Fig. 4.5) verify this qualitative conclusion.

A computation-free example is the case of the H_3^+ ion, where only two electrons need be assigned, and they go into the lowest bonding orbital. We thus expect that H_3^+ will be stable and, moreover, that H^+ can approach H_2 essentially freely from all directions. Such is indeed the case.

Now let us get further into chemistry. Consider the approach of a hydrogen atom to a halogen molecule, beginning with a homonuclear one, say, Cl_2. Again, let the initial approach be collinear, say, along the z axis. The $1s$, $2s$, $2p$, and $3s$ Cl orbitals are far too low-lying in energy to interact with the $1s$ H orbital. Only the $3p$ orbitals are relevant. Of the $3p$ orbitals of each Cl atom, two are orthogonal to the Cl_2 bond and hence by symmetry do not interact with the spherically symmetrical $1s$ H orbital, which approaches along the z direction. Only the $\sigma(3p_z)$ and $\sigma^*(3p_z)$ can be combined with the $1s$ H. To construct the HClCl orbitals, consider the available Cl_2 orbitals:

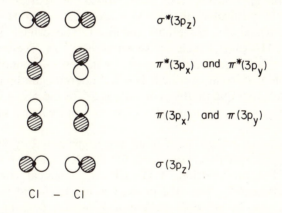

$$\sigma^*(3p_z)$$

$$\pi^*(3p_x) \text{ and } \pi^*(3p_y)$$

$$\pi(3p_x) \text{ and } \pi(3p_y)$$

$$\sigma(3p_z)$$

Cl — Cl

and then add the $1s$ H orbital in and out of phase:

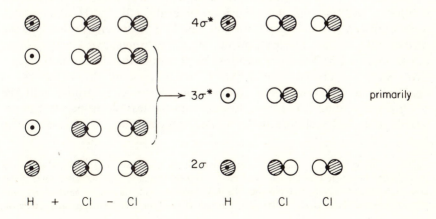

$4\sigma^*$

$3\sigma^*$ primarily

2σ

H + Cl — Cl   H   Cl   Cl

As in H_3, we have three independent MOs made out of three atomic ones, $1s$ on H and $3p_z$ on each of the Cl atoms. In contrast to the H_3 case, here it is found that the middle orbital is primarily bonding on H—Cl and antibonding on Cl—Cl, rather than "equal mixture" as in H_3. The rationale is, of course, that the H—Cl bond is stronger than the Cl—Cl bond.

Each Cl atom has 5 $3p$ electrons and H brings 1 for a total of 11: Two are in the bonding HClCl orbital; Four are in the bonding $\pi(3p_x)$ and $\pi(3p_y)$ Cl_2 orbitals, and another four in the antibonding ones. Thus 10 electrons have been assigned. As in H_3, the remaining "odd" electron is assigned to the middle orbital of HClCl, that is, that orbital which is antibonding on Cl—Cl but bonding on H—Cl.

Thus we conclude that there should not be a larger barrier for collinear approach and that the energy release upon reaction should be "repulsive," the exoergicity released as a *repulsion* between the two Cl atoms.†

Say now that the H atom approaches Cl_2 sideways, in the x–z plane. The $\pi(3p_y)$ MO remains orthogonal to the plane and hence, by symmetry, is indifferent to the approaching $1s$ H atom. Not so for the $\pi(3p_x)$ and $\pi^*(3p_x)$ MOs of Cl_2. In particular, the $\pi^*(3p_x)$ acquires strong H—Cl bonding character and hence is considerably stabilized. The orbital energies as a function of the H—Cl—Cl angle are shown in Fig. 4.15. The stabilization of the $\pi^*(3p_x)$ orbital is countered, however, by the less efficient overlap in the 2σ orbital. Will a collinear or a bent H + XY intermediate be more stable? Here we require real quantitative computations. In the absence of precise, *ab initio* results, we often make use of more approximate semiempirical results.

Barrier heights as a function of bond angle, computed by the diatomics-in-molecules approximation, are shown for H + XY and X + HY reactions in Fig. 4.16. For the reaction of H + Cl_2, the collinear configuration is favored. Experimentally, the HCl product is indeed found to be strongly backscattered, that is, the HCl "rebounding" with a peak in the reaction cross section at $\theta = 180°$.

One aspect of the general H + XY case that still remains to be discussed is which "end" of the heteronuclear molecule will the H atom attack preferentially? To answer that, we note that in a heteronuclear diatomic the binding MO is more localized on the more electronegative atom; whereas the antibonding MO is more localized on the atom that is less electronegative. Now the crucial orbital $(3\sigma^*)$, which governs the dynamics, is essentially an in-phase linear combination of the $1s$ H orbital with the $\sigma^*(3p)$ antibonding XY orbital, localized on the less electronegative atom. The H atom will therefore preferentially approach the less electronegative

---

† The HClCl complex will thus dissociate along the Cl_2 bond axis. The products' scattering angle is then the angle that the Cl_2 bond makes with the initial relative velocity. Since the H atom is light and usually fast moving, the Cl_2 molecule will not rotate much during the interaction time. The angle of scattering is thus indicative of the preferred approach geometry.

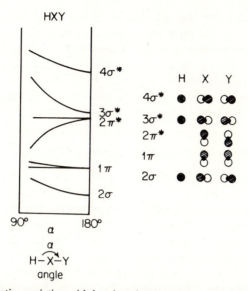

Fig. 4.15. The qualitative variation with bond angle of the energy of the higher MOs of HXY (X and Y are halogen atoms, with Y the more electronegative). Such Walsh diagrams are widely used for understanding the geometry of *bound* molecules [A. D. Walsh, *J. Chem. Soc.*, *2266* (1953)]. They can also be used to advantage in understanding the origin of steric effects in reaction dynamics. [See D. R. Herschbach, *Faraday Discuss. Chem. Soc.*, *55*, 233 (1973), and C. F. Carter, M. R. Levy, and R. Grice, *Faraday Discuss. Chem. Soc.*, *55*, 357 (1973).] Such diagrams also remind us that the same forces operate in bound and unbound molecular systems.

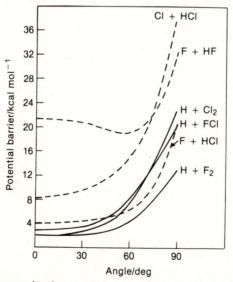

Fig. 4.16. Potential energy barrier versus bond angle for various atom–molecule reactions (solid curves, $H + XY \rightarrow HX + Y$; dashed curves $X + HY \rightarrow HX + Y$), calculated from semi-empirical (DIM) potential energy surfaces. [Adapted from I. Last and M. Baer, *J. Chem. Phys.*, *80*, 3246 (1984).]

halogen atom. Thus, for the reaction of H with ICl, HI will therefore be preferred over HCl as the principal product. (This is *despite* the obvious fact that the reaction to form HCl is more exoergic!)

Qualitative considerations cannot replace detailed computations of the height of the barrier versus the approach angle; however, such arguments do provide valuable insight and also must serve temporarily for more complex systems (for which actual computations are not currently feasible).

4.2.7   *The steric effect: Polar map representation*

Thus far we have characterized the steric effect in terms of the dependence of the activation barrier upon the orientation angle (i.e., the bond angle; see Fig. 4.5). For a more detailed understanding, we examine potential-energy contour maps at several different (fixed) angles of approach (e.g., see Fig. 4.4) From such plots we can trace out the dependence of the saddle-point geometry, as well as the barrier energy, upon the orientation angle. A less detailed but instructive alternative representation of the steric effect is by a polar map. This consists of equipotential contours plotted as a function of the coordinates R and γ locating the approaching atom with respect to the center-of-mass and symmetry axis of the target molecule, keeping a fixed (usually equilibrium) geometry for the latter.

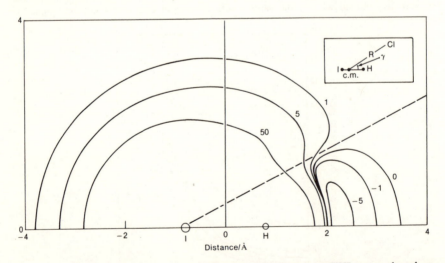

Fig. 4.17. A polar representation of a potential contour map for the ClHI system, based on a semiempirical LEPS computation. The potential energy contours (energy in kcal mol^{-1}) show the dependence of the potential on the approach angle γ of the Cl and on R, the distance to the HI center of mass (see insert) at a fixed H–I separation. The dashed line shows the cone of approach within which the chlorine can abstract the H atom. [Adapted from C. A. Parr, J. C. Polanyi, and W. H. Wong, *J. Chem. Phys., 58*, 5 (1973).]

Figure 4.17 shows such a plot for the approach of Cl to HI (based on a semiempirical surface for the $Cl + HI \rightarrow I + HCl$ reaction). The bond distance in HI is held constant, at its equilibrium value,† as we vary the distance and orientation. The contours show clearly the steric hindrance to the H-atom abstraction by the bulky iodine atom, leaving only a narrow *cone of approach* along which the chlorine can attack the small hydrogen atom.

The I–Cl repulsion evident in Fig. 4.17 is a typical "bulky group" steric effect often invoked in organic chemistry. By assigning so-called van der Waals radii to atoms and thence to functional groups, one can assess the steric accessibility (to an approaching reagent) of a particular reaction site in a given isomer. An even more elaborate approach is to determine an entire "force field." This is nothing but our potential energy surface, albeit for a system with many degrees of freedom. The "molecular mechanics" approach in physical organic chemistry is concerned with the design of such surfaces on a basis of largely empirical foundations. Having a very wide data base, these force fields can be tuned to a high degree of perfection.

In the polar plot the bond distance in the reagent diatomic molecule is held constant. Let us, however, draw a set of polar plots, shown in Fig. 4.18 for $H + H_2$, at increasing H—H bond distances. As is evident from the drawings, the cone of acceptance "opens up" upon stretching of the reagent bond. This provides an interpretation for the enhancement of reactivity in (nearly) thermoneutral reactions upon vibrational excitation of the reagents.

4.2.8 Unimolecular reaction and collision-induced dissociation

The potential-energy surface for a stable polyatomic molecule has a *well* rather than a *saddle* somewhere along the reaction path (see Sec. 4.2.3). At low vibrational energies the molecule is confined to the well. As the energy is increased, the molecule can ultimately cross the barrier separating the well from one of the exit valleys and dissociate into fragments.

A *unimolecular dissociation* can be considered to be a *half-collision* on such a potential surface. To achieve the necessary energy for dissociation the molecule must first be activated. The only two processes that are confined to the well region itself are (infrared) multiphoton activation (see Sec. 7.3) and direct overtone excitation.‡ Figure 4.19 shows an energy-level diagram for the unimolecular dissociation of tetramethyldioxetane to a

---

† In the course of an actual reaction, the HI distance gradually *increases* as the Cl approaches, so such a polar map is not quite a faithful indication of the steric situation.

‡ For a strictly harmonic molecule and for light of ordinary intensity the selection rule for vibrational transitions is $\Delta v = \pm 1$. For a realistic anharmonic molecule there is a small probability for $\Delta v > 1$, that is, a direct access of an "overtone" mode ($v > 1$) from the ground state ($v = 0$). When the photon flux is high, even small absorption cross sections are sufficient for a large number of molecules to absorb per unit time.

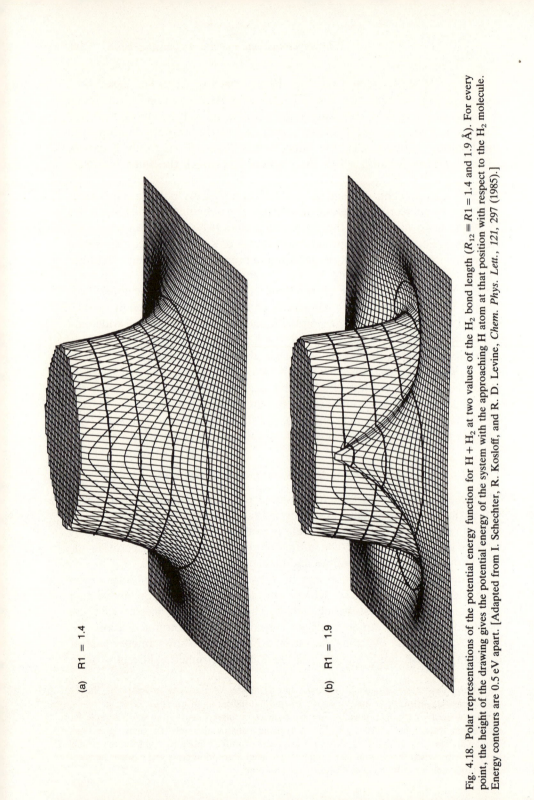

(a) R1 = 1.4

(b) R1 = 1.9

Fig. 4.18. Polar representations of the potential energy function for H + H$_2$ at two values of the H$_2$ bond length ($R_{12} \equiv R1 = 1.4$ and 1.9 Å). For every point, the height of the drawing gives the potential energy of the system with the approaching H atom at that position with respect to the H$_2$ molecule. Energy contours are 0.5 eV apart. [Adapted from I. Schechter, R. Kosloff, and R. D. Levine, *Chem. Phys. Lett.*, 121, 297 (1985).]

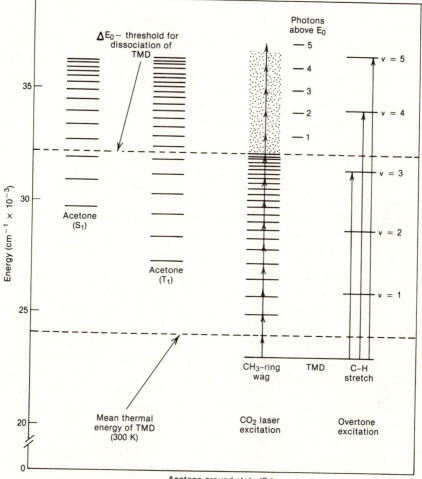

Fig. 4.19. Energy–level diagram showing vibrational states of tetramethyldioxetane (TMD), which can be accessed by CO_2 laser excitation *via* the CH_3 ring wag or by visible laser excitation *via* the CH stretch vibration overtones. When the threshold ΔE_0 for dissociation of TMD is exceeded, acetone in vibrationally excited levels of one or both of the indicated electronic states (T_1, S_1) can be formed as a product, as well as its ground state (S_0). [Adapted from T. R. Rizzo, B. D. Cannon, E. S. McGinley, and F. F. Crim, *Laser Chem.*, **2**, 321 (1983); see also Y. Haas, S. Ruhman, G. D. Greenblatt, and O. Anner, *J. Am. Chem. Soc.*, **107**, 5068 (1985).]

ground and an electronically excited acetone molecule:

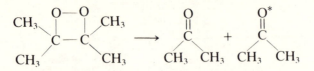

It takes direct excitation of the $v = 4$ overtone of the CH stretch before enough energy is provided to overcome the barrier to this four-center dissociation. Alternatively, it requires at least nine photons of the CH_3 wag frequency to provide enough energy to reach the barrier.

Energy-rich molecules in the well region can also be prepared by chemical means. This is particularly the case if there is more than one entry (or exit) valley. An example of such a "chemical activation" is

$$CH_3 + CF_3 \rightarrow [CH_3CF_3]^\dagger \rightarrow CH_2 = CF_2 + HF$$

A potential energy profile for HF elimination is shown in Fig. 4.20. The energy of the newly formed C—C bond (over $400\ kJ\ mol^{-1}$) is available as excitation energy of the newly formed CH_3CF_3 molecule. The barrier for (the four-center) HF elimination is estimated to be about $300\ kJ\ mol^{-1}$, so the chemically activated molecule has sufficient energy to overcome that barrier.

The energy-rich CH_3CF_3 molecule will, however, spend some time in the well region before it dissociates. The reason, as first pointed out by Lindemann and Hinshelwood, is that there are very many vibrational degrees of freedom ($3s - 6 = 18$) of CH_3CF_3, only one of which becomes the reaction path. It will therefore take time before over $300\ kJ\ mol^{-1}$ (of the total excitation of about $400\ kJ\ mol^{-1}$) is localized in just one of these 18 modes. This argument is equally applicable for a nonlinear triatomic molecule ($3s - 6 = 3$). The excitation energy is partitioned between the bend motion and the symmetric and the antisymmetric stretches. Only the latter is "along the reaction path," and it will take some time before the anharmonicities of the surface channel enough energy into the anti-symmetric stretch.† It follows that the existence of a hollow along the reaction path will give rise to a "sticky" collision, with the colliding molecules spending a "long" time (e.g., longer than a picosecond) in the region of the well. This implies the existence of a long-lived "complex" or intermediate, an old concept in reaction kinetics.

At higher translational energies it becomes possible for the actual reaction trajectories to deviate more and more from the *minimum-energy path,* and eventually it is energetically unnecessary that the system exit along the product valley at all. Provided that the available energy exceeds

---

† Note also that this *intra*molecular vibrational redistribution of energy prior to dissociation is even more evident in the case of Fig. 4.19, where the locus of initial deposition of the excitation is far from the reaction site.

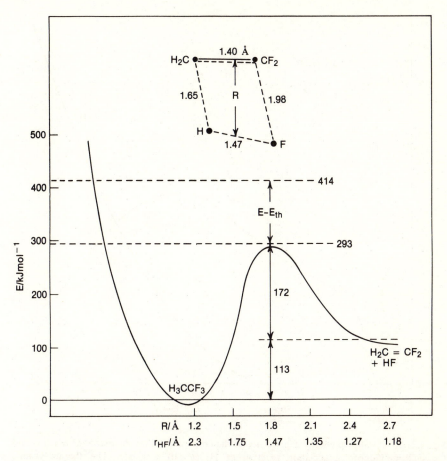

Fig. 4.20. Energy profile for HF elimination from CH_3CF_3 prepared by chemical activation with an available energy $E = 414\ kJ\ mol^{-1}$. The barrier height of $E_b = 293\ kJ\ mol^{-1}$ is close to that inferred from experimental data. The endoergicity of HF elimination from ground-state HF is $113\ kJ\ mol^{-1}$. Shown also is the (shrinking) HF distance along the reaction path and the geometry of the four-membered ring at the barrier. [Adapted from R. M. Benito and J. Santamaria, *Chem. Phys. Lett., 109,* 478 (1984).]

the thermochemical requirement (the dissociation energy of the reactant molecule), the atoms can separate, that is, the system can exit to the *plateau* corresponding to the free atoms (see Figs. 4.3 and 4.4).

One well-studied example is the reaction of H_2^+ with He, (see Fig. 4.21), which at low translational energies proceeds to $HeH^+ + H$ but, as E_T increases past the threshold $[E_T > \Delta E_0 = D_0(H_2^+) = 2.65\ eV]$, goes almost exclusively into dissociation:

$$H_2^+ + He \rightarrow H^+ + H + He$$

This collision-induced dissociation process is an illustration of our "rough correlation" of energetic effects (see Sec. 2.4): As the translational energy is

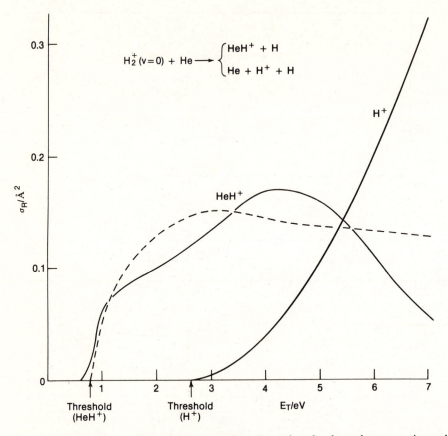

Fig. 4.21. Translational energy dependence of the cross sections for the exchange reaction and the collision-induced dissociation reaction of $H_2^+(v = 0)$ with He. The thermodynamic threshold energies are 0.8 and 2.65 eV, respectively, indicated by the arrows [see Eq. (2.61)]. The solid curves represent the first experimental results (adapted from W. A. Chupka, J. Berkowitz, and M. E. Russell, *Proc. VI ICPEAC,* MIT Press, 1969, p. 71); the dashed curve represents recent experiments [adapted from T. Turner, O. Dutuit, and Y. T. Lee, *J. Chem. Phys., 81,* 3475 (1984).]

increased, newly "allowed" reaction paths take over at the expense of the old ones. (Youth conquers all!)

4.3 The classical trajectory approach to reaction dynamics

4.3.1 From the potential surface to the dynamics

Since the 1930s it has been recognized that there is a direct route from the potential energy surface to the dynamics of the collision, namely, the (numerical) solution of the classical mechanical equations of motion for the atoms. This procedure has already been discussed for two-body elastic collisions (see Sec. 2.2). Such studies can simulate all the essential *nonquantal* features† of the reaction dynamics. These computational studies are carried out for two purposes: first, as a "diagnostic" of general trends, that is, features of the dynamics arising from different topographical features of the surface and from changes in reactants' energies, masses, and so on, and, second, to investigate particular reactions in the hope of accounting for observed chemical dynamical behavior.

The basic procedure is to choose a set of initial conditions and solve the classical equations of motion for the three (or more) particles under the influence of the assumed potential. For each initial condition one computes the time development of the coordinates of each particle (in the center-of-mass system). The results of such a *trajectory* of the system can be displayed graphically in several ways, one of the most instructive being plots of $R_{ij}(t)$ for each pair of atoms i, j. For a three-particle system such curves are easy to understand (Fig. 4.22). But what can we *do* with the results?

4.3.2 The need for averaging trajectory results

Each trajectory shown in Fig. 4.22 corresponds to a *particular* choice of initial values for the coordinates and momenta of the participating atoms. To obtain quantities of physical interest (such as cross sections, rates, etc.) it is necessary to *average* over some or all of these initial conditions. For example, even to determine such intermediate quantities as the opacity function $P(b)$ at fixed energy E, it is necessary to average over all possible initial *orientations* of the reactants at the given b. (We have seen in Sec. 2.4.9 how this averaging can lead to the concept of a steric factor.) In principle (and sometimes in practice), one can measure the reaction cross section for oriented reagents. The corresponding trajectory computation (as in Fig. 2.22) should, then, not average over orientations but present the cross section as a function of the orientation angle. There is, however, another and deeper reason for an *inevitable* need to average over initial conditions, a need that is present even for the most perfect experiment.

An essential difference between classical and quantal mechanics is the number of initial conditions that may be specified for a given trajectory. In classical mechanics one must specify both the position and the momentum for each degree of freedom. In quantum mechanics the uncertainty principle implies that if, say, the momentum is well specified, the value of the position can be anywhere within its possible range. Since molecules are inherently quantal, a complete specification of initial conditions for a collision in a system of *n degrees of freedom consists of n* quantum numbers.

---

† There are often important differences between classical- and quantal-dynamical results (see Sec. 3.1). Even so, classical mechanics is very useful conceptually, in that it provides an easily visualized *picture* of the dynamics. It also provides the correct overall trends, even when it fails in the details.

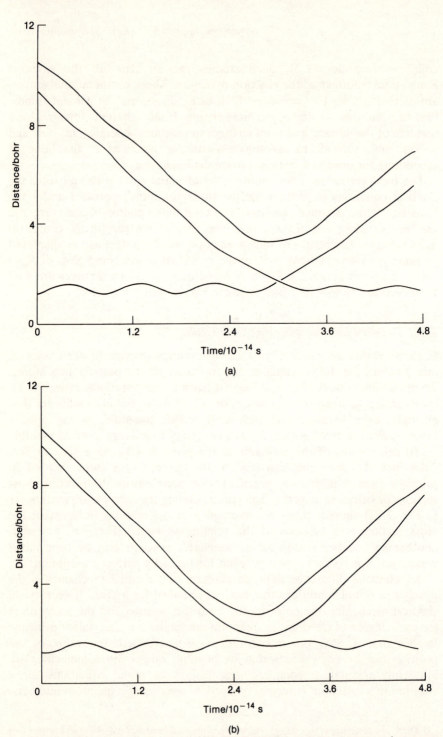

Fig. 4.22. Classical trajectories calculated for $H + D_2$ collisions at $E_T = 79\ kJ\ mol^{-1}$ for D_2 in its ground vibrational state. Only (a) is a reactive trajectory leading to $HD + D$. (Computations by I. Schechter, unpublished.)

In contrast, a classical trajectory for the system requires $2n$ initial conditions. The method of classical trajectories *mimics* this quantal aspect by running many classical trajectories where, of the $2n$ initial conditions, n are held constant (the same n as in the quantal case) while the other n are varied. The final outcome is determined by averaging over the variable initial conditions.†

Consider the trajectories shown in Fig. 4.22. They differ only in the n initial conditions that are to be averaged over. Now each trajectory has a definite outcome, for example, a reaction either did or did not occur. When we perform an *average* over initial conditions, the outcome is no longer necessarily definite: Some trajectories have led to reaction, others have not. Hence we have as the probability for reaction the fraction of trajectories that (at a common set of values for the n initial conditions that are held fixed) have led to reaction. By averaging over initial conditions, we *mimic* the probabilistic aspects of quantum mechanics.

The sampling of initial conditions for the purpose of executing an average over them is often done by a so-called Monte Carlo (i.e., random selection) procedure. Thus the computation of observable dynamical quantities by the use of averages over classical trajectories is often referred to as the *Monte Carlo method*. Rudimentary aspects of the method as applied to our problem are discussed in Section 4.3.7. There are also certain *systematic* (i.e., nonrandom) procedures for the selection of initial conditions that can be more efficient than Monte Carlo in producing averages of given reliability.‡

4.3.3 Energy disposal in exoergic reactions

More than half a century ago the "atomic flame" experiments of M. Polanyi showed that in such reactions as

$$K + Br_2 \rightarrow KBr^\dagger + Br$$

the newly formed salt molecule is *highly vibrationally excited* (as denoted by the dagger), so much so that in a subsequent collision with an alkali it can cause *electronic* excitation of the alkali:

$$KBr^\dagger + K \rightarrow KBr + K^*$$

The excited alkali decays by fluorescence (light emission),

$$K^* \rightarrow K + h\nu$$

leading to a visible "flame," with a color characteristic of the alkali metal used.

---

† For example, in the simplest case of the $A + BC$ reaction, one must average over the phases of the vibrational and rotational motions of BC.

‡ Depending upon the degree of detail characterizing the observable quantity, a larger or smaller number of trajectories will be required to achieve any desired degree of precision in the result. Typically about 10^3 trajectories leading to reaction suffice for establishing the coarser features of the dynamics of the reaction.

At the same time Eyring, Polanyi, Evans, and others used a form of the London equation to generate the first "realistic" potential energy surfaces. Then they sought to interpret the dynamics of such reactive collisions (including the vibrational excitation of the products) using these surfaces. Ever since, energy disposal in exoergic reactions has been interpreted and discussed in terms of the analysis of classical trajectories on potential energy surfaces.

One of the correlations that has emerged is as follows: For given reactant masses (and at not too high energies) the efficiency of the conversion of the reaction exoergicity into the product vibration will be greater for a more attractive potential energy surface.

Figure 4.23 shows a simple view of this correlation. An early release of the exoergicity occurs while the attacking atom A is still approaching the reactant BC. The repulsion between the reactants at close separation "pushes" this exoergicity into product vibration.

Alternative representations are also shown in Fig. 4.23, where a typical trajectory is superimposed on the potential energy surface. In Fig. 4.23a, the early release of the exoergicity appears initially as kinetic energy in the AB direction. This leads to the so-called bobsled effect. As the potential contours bend round (because of AB and BC repulsions) the trajectory cannot follow the minimum-energy path but tends to continue in a straighter path (due to the large momentum in the AB direction), leading to a bobsled motion. Ultimately, the influence of the short-range AB repulsion converts the exoergicity into vibration of the AB product. This is in accord with our expectations of an early release of exoergicity in alkali–halogen reactions based on the harpoon model (see Fig. 4.8; see also Sec. 4.2.5).

All of our discussion thus far on the relations between the topography of the potential surface and the dynamics of the system has ignored the important question of the influence of the *masses* of the particles upon the dynamical behavior of the system. It was recognized early in the game that there would be strong *mass effects* on any given potential surface (since, after all, Newton's laws involve the masses rather directly!). In order to see these effects most clearly, a *mass-weighted* coordinate system was devised upon which to plot the potential energy surface.

In this representation the system can be characterized by a point-mass particle moving (without friction) on a *physical surface* under the influence of gravity (see Appendix 4A). One can then more easily understand the influence of the initial translational energy, or of the initial vibrational energy, of the reactants upon the trajectories and the exoergicity disposal.

Because of the mass effect, it is possible to have high product vibrational excitation even on a repulsive, late downhill surface, if only the important mass factor $\cos^2 \beta$ [that of Eq. (1.5)] is close to unity.† In this limit the

---

† The significance of this ubiquitous mass factor is further considered in Appendix 4A, which describes the *mass-weighted* coordinate system.

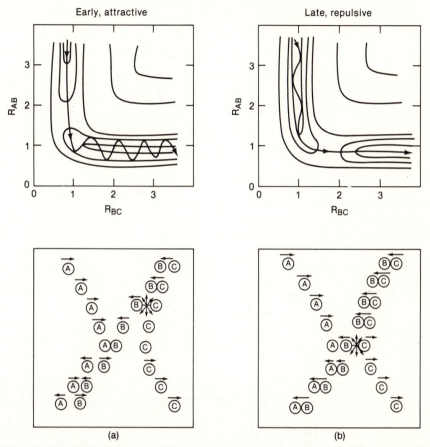

Fig. 4.23. Schematic portrayal of (a) attractive energy release on a potential energy surface with an early barrier and (b) repulsive energy on a surface with a late barrier. In (a) the reaction exoergicity is released while the A atom and the BC molecule are approaching each other, leading to vibrational excitation of the AB product. In (b) the energy is released when the atom A has approached BC and is almost at rest, so the AB product "rebounds" and most of the energy is released as product translation. [Adapted from J. C. Polanyi, *Acc. Chem. Res.,* **5,** 161 (1972).]

exchanged atom is very light, and so even on a repulsive surface, the exoergicity is released while the incident atom is still approaching.

Figure 4.24 compares the results of classical trajectory computations, using semiempirical potential energy surfaces with the measured vibrational energy distribution for the $H + F_2 \rightarrow HF + F$ and $H + Cl_2 \rightarrow HCl + Cl$ reactions.

Note that the same problem with initial conditions exists with the final ones. The problem is even more serious with the latter, in that we *do* have control over initial conditions. We can specify, for example, that for all trajectories the initial vibrational energy of F_2 will be that of the F_2 molecule

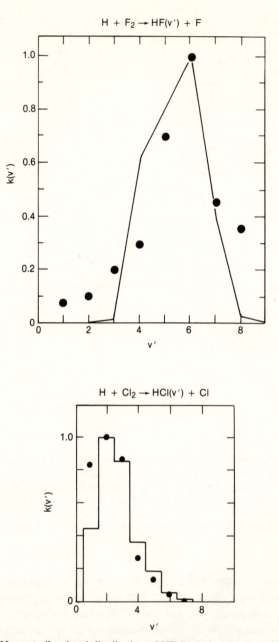

Fig. 4.24. (Top) Nascent vibrational distribution of HF from the reaction of $H + F_2$ expressed in terms of the relative rate coefficients $k(v')$, comparing trajectory calculations on a semiempirical potential surface (denoted by dots connected by line segments) with experimental data (circles). [Adapted from J. C. Polanyi, J. L. Schreiber, and J. J. Sloan, *Chem. Phys.*, **9**, 403 (1975).] (Bottom) Same as above for the analogous reaction of $H + Cl_2$ (histogram presentation of trajectory results). [Adapted from M. D. Pattengill, J. C. Polanyi, and J. L. Schreiber, *J. Chem. Soc. Faraday Trans. 2*, **72**, 897 (1976).]

in the $v = 0$ state.† The trajectories emerge from the interaction region with a continuous range of HF vibrational energy so that some sort of "quantization" is needed before the results can be compared with experiment. The pragmatic solution is to divide the continuous classical vibrational energy axis into "bins." Each bin is centered about a quantal value for the vibrational energy. All trajectories whose final HF vibrational energies are in the same bin are assigned the same final vibrational quantum number. In this fashion the continuous classical distribution is converted into a discrete histogram. There are several more "correct" procedures, but the convenience of the binning or histogrammic approach (termed the quasiclassical method) makes it resistable to improvement.

Classical trajectory computations have been successful in simulating the energy disposal (and many other observations) in elementary chemical reactions. That is not to say that there are no quantal effects (over and above the quantization of initial and of *final* conditions). The gross average dynamical features of the reactions are, however, well reproduced, and experimental groups often carry out trajectory computations in conjunction with their laboratory measurements.

4.3.4 Energy requirements for reactions with a barrier

For exoergic reactions with a *barrier* along the reaction path and for *all* endoergic reactions, we seek to obtain information on the energy requirements for reaction. Obviously we require a collision energy E_T (more properly, a total energy E consisting of E_T plus the reagent internal energy E_I) in excess of the threshold‡ E_{th} to achieve even one "reactive" trajectory. For $E < E_{th}$ it is clear that any trajectory must return through the entrance valley; however, for $E > E_{th}$ it is not obvious whether E_T or E_I is more effective in overcoming the barrier.

As already implied in Section 4.3.3, there is a correlation between the location of the barrier along the reaction path and the form of initial energy most conducive to reaction. *Translational* energy is most effective for passage across an *early* barrier (i.e., the col in the entrance valley), whereas *vibrational* energy of the reactant molecule is more efficient for surmounting a *late* barrier (i.e., in the "exit valley").

These requirements can be interpreted (Fig. 4.25) in terms of the availability of kinetic energy in the proper coordinate. For an early barrier, one requires momentum along R_{AB}, whereas to overcome a late barrier momentum is needed in the R_{BC} coordinate. The energy requirements of reactions will obviously become less restrictive at higher collision energies (or for $\cos^2 \beta \to 1$), where an efficient interchange of vibrational and

---

† We typically do that by specifying the initial amplitude and phase of the vibration. The amplitude is kept fixed, the phase is varied.

‡ The relation between the threshold energy and the barrier height is illustrated in Fig. 4.6.

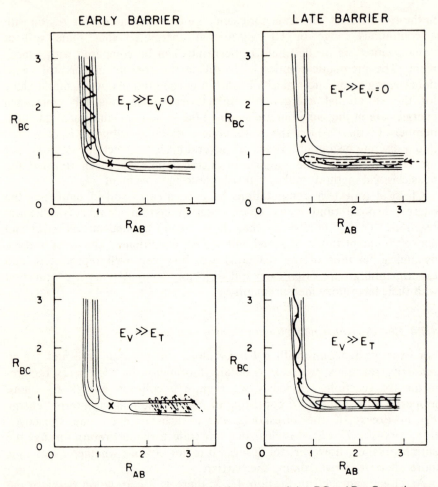

Fig. 4.25. Influence of reactant energy for the thermoneutral $A + BC \rightarrow AB + C$ reaction on a LEPS surface with a barrier ($\times$). In all four cases $E > E_b$, so that reaction is allowed on energetic grounds. When the relative translational energy E_T is high (top row), there is ample kinetic energy for motion along R_{AB}. Reaction thus occurs for a barrier in the entrance valley (i.e., along R_{BC}, left column) but fails if the barrier is in the exit valley (i.e., along R_{BC}, right column). The opposite is found when the reactant diatomic has high vibrational excitation E_v, but E_T is low ($E_T < E_b$, bottom row). The vibrational energy (i.e., the kinetic energy along R_{BC}) *helps* in surmounting a late barrier. [Adapted from J. C. Polanyi and W. H. Wong, *J. Chem. Phys.*, *51*, 1439 (1969).]

translational energy along the reaction path becomes more likely. Figure 4.26 shows computational results for the reaction

$$O(^3P) + H_2(v) \rightarrow OH + H$$

It depicts the cross section versus the initial translational energy for various H_2 vibrational states.

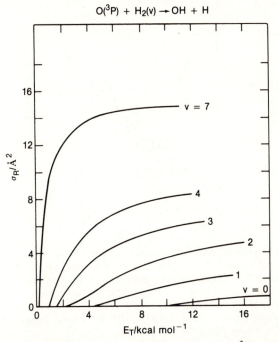

Fig. 4.26. Reaction cross section versus E_T for the reaction of $O(^3P)$ with H_2 in specified vibrational states v, calculated *via* classical trajectories using a semiempirical potential energy surface (see Fig. 4.12). [Adapted from M. Broida and A. Persky, *J. Chem. Phys.*, *80*, 3687 (1984).]

It is clear that all forms of energy are not equally effective in promoting reaction. The relative importance of translational versus internal energy in overcoming a potential barrier is currrently a subject of active investigation, both theoretically and experimentally, to be discussed further in Section 5.5.

4.3.5 Dynamics by classical trajectories

The method of classical trajectories takes us by way of a given potential-energy surface to the observed outcome of the collision. It gives us a qualitative understanding of the role of energy. However, it provides much more. Foremost is the angular distribution of the products, that is, the correlation between the directions of final and initial relative velocities. Other correlations can also be explored, and Fig. 4.27 is a example for the

$$F + C_2H_4 \rightarrow [C_2H_4F] \rightarrow C_2H_3F + H$$

reaction.† Generating a detailed distribution for products' attributes requires that very many trajectories be run in order to generate a reasonably

† The reaction proceeds *via* a very short-lived energy-rich C_2H_4F intermediate.

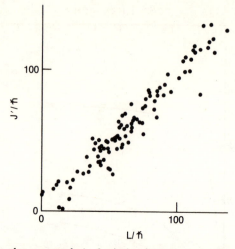

Fig. 4.27. Correlation plot generated *via* classical trajectory computations for the reaction of F with C_2H_4 at $E_T = 20$ kcal mol^{-1}. Shown is the final rotational angular momentum J' of the C_2H_3F product versus L, the orbital angular momentum of the reagents, both in units of $\hbar$. In Appendix 4B we interpret the strong $L \rightarrow J'$ correlation as due to the low reduced mass of the products. A total of 6674 trajectories were run, of which 2465 formed the C_2H_4F short-lived intermediate, of which 106 dissociated to the H + C_2H_3F products. [Adapted from W. L. Hase and K. C. Bhalla, *J. Chem. Phys.*, 75, 2807 (1981).]

noise-free† final distribution. Since we are solving a set of $2n$ (coupled) classical equations of motion, the computational labor is significant. It is therefore often the case that highly detailed final state distributions are generated by making simplifying assumptions about the dynamics.

An aspect of collision dynamics much explored by classical trajectory computations is the role of reagent rotation (and the complementary aspect—that of rotational energy disposal). For low rotational energies, the results are quite sensitive to the details of the potential energy surface, in large part because of the dependence of the barrier to reaction on the mutual orientation (see Sec. 2.4.9). This sensitivity to the assumed interaction is found not only for atom–molecule exchange reactions but also for more complex processes, for example, dissociative adsorption of molecules on metal surfaces (see Sec. 7.5). It is likely, therefore, to emerge

---

† Recall that the initial conditions are selected in a Monte Carlo fashion. Since each initial classical condition yields a *distinct* classical final state, increasing the number of trajectories changes the final distribution until the number of trajectories is large enough to achieve statistical significance. When we classify the final states into bins and N trajectories fall in a given bin, the fractional uncertainty in the answer is about $N^{1/2}/N = N^{-1/2}$. If there are just two bins (e.g., reaction did or did not occur), a small number of trajectories suffices. If we desire a high-resolution analysis of the product, we need many bins, with many (say, $N = 100$) trajectories terminating in each one. The required total number of trajectories can be overwhelming.

as an important diagnostic check of empirical and semiempirical potential energy surfaces. With increasing rotational excitation, the molecule rotates faster. At every initial mutual orientation there is then about the same probability for reaction. In general, the primary effect is the additional energy that is made available. There can, however, be specific effects, and a celebrated example is the dependence of the HF/DF branching ratio on the HD rotational state in the F + HD reaction (Fig. 4.28). A possible

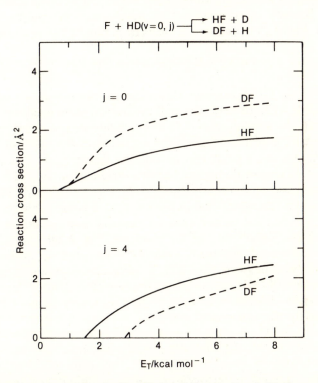

Fig. 4.28. Reaction cross section for F + HD ($v = 0, j$) versus translational energy determined by classical trajectory computations. Formation of DF is more probable for $j = 0$, but less so at higher j values. [Adapted from J. T. Muckerman in Henderson (1981).]

interpretation of the computational result is as follows. At $j = 0$ and thermal velocities, there is a slight preference toward DF formation. At higher j values this is reversed, since the center of mass of HD is nearer to D and so when the HD molecule rotates, the H atom sweeps out a larger sphere than that of the D atom. The approaching F atom is thus preferentially intercepted by the H atom before it gets close enough to the D atom.

4.3.6 Direct versus compound collisions

For a given potential surface (and masses), classical mechanics can be used to illustrate the progress of the reaction through plotting the several interatomic distances as a function of time, as we saw in Fig. 4.22 for the H-atom transfer reaction $H + H_2 \rightarrow H_2 + H$. This is a *direct reaction* in that the H–H bond is continually extending as the H atom approaches and the switchover between the two bonds occurs within a very short time interval, corresponding to about one vibrational period. This is typical for all direct reactions. By way of contrast, let us have a look at another trajectory calculation, Fig. 4.29, a similar representation but for the reaction

$$O(^1D) + H_2 \rightarrow [HOH] \rightarrow OH + H$$

In contrast to the $O(^3P)$ reaction, this is the one occurring on a surface with the H_2O stable well. To reach the well the oxygen atom needs to "insert" in the H_2 bond. As a result, the H_2O has a highly excited bend mode, as it "inverts" a couple of times before the energy is transferred from the bend to the asymmetrical stretch and onward to dissociation. The reaction is clearly not direct, since during the bending–inversion motion of H_2O the H_2 bond distance oscillates. The lifetime of the energy-rich H_2O molecule is, however, only a few vibrational periods and so is comparable to or shorter (depending on the total energy) than the classical rotation time of H_2O. The resulting angular distribution of the products (Fig. 4.29b) has almost, but not quite, a forward–backward symmetry characteristic of longer-living complex intermediates (see Sec. 7.2). Trajectory computations can also be used to simulate the OH rotational distribution (Fig. 4.30).

A qualitative difference is evident in the results (Fig. 4.31) of the computation for the (gas-phase) reaction

$$KCl + NaBr \rightarrow KBr + NaCl$$

we note the "snarled trajectory" or "sticky collision" feature suggesting very long-lived *complex formation*. There is also experimental evidence that this reactive collision involves a long-lived intermediate, or "complex." This is then a so-called *compound reaction* (i.e., nondirect), which can arise from a deep well along the reaction path. Thus we see the two extreme modes of reactive collisions: the *direct* versus the compound or *complex* mode.

What we have learned in this section is that certain dominant features of the potential energy surface seem to govern the chemical reactivity. Can we use this knowledge to develop some simpler (albeit more approximate) theoretical approach and thus obviate the need for extensive trajectory calculations? Reactions for which the saddle-point region of the potential surface plays the major role would be the most likely to yield to such an attack. We shall explore this possibility in Section 4.4.

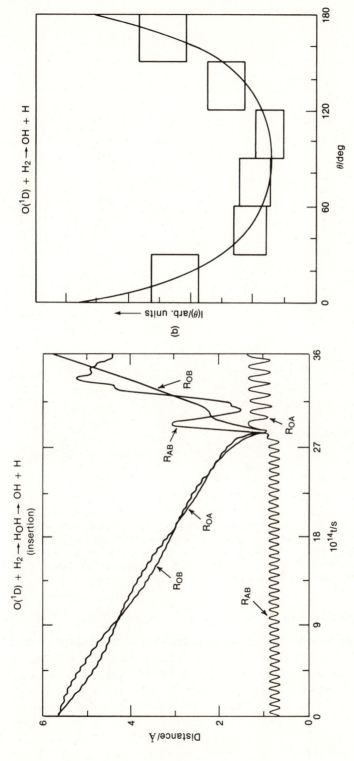

Fig. 4.29. (a) Internuclear distances (in angstroms) versus time (in units of 10^{-14} s) for a selected trajectory on a semiempirical potential at $E_T = 0.5\,\text{kcal mol}^{-1}$, showing the insertion of the O(^1D) into the H–H bond. Here A and B denote the individual hydrogen atoms. (b) Differential reaction cross section $I(\theta)$ for the O(^1D) + H$_2$ → OH + H reaction, from quasiclassical trajectory calculations on a semiempirical potential energy surface, at $E_T = 5\,\text{kcal mol}^{-1}$. Computational results are shown as rectangular boxes indicative of the range of angles (i.e., the bin size) and statistical uncertainty of the cross sections. The solid curve represents the prediction of a short-lived complex model. [Adapted from P. A. Whitlock, J. T. Muckerman, and E. R. Fisher, *J. Chem. Phys.*, **76**, 4468 (1982).]

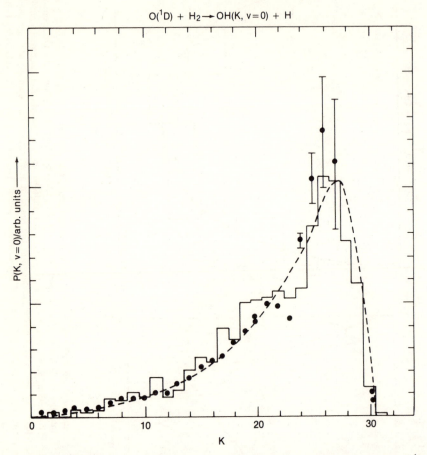

Fig. 4.30. Rotational state distribution of the nascent OH ($v = 0$) from the reaction $O(^1D)$ with H_2 at $E_T = 0.5$ kcal mol^{-1}. Here K is the OH rotational quantum number. Points represent experimental data. The histogram was derived from quasi-classical trajectory calculations using a semiempirical potential surface. The dashed curve is a so-called linear surprisal fit (see Sec. 5.5). [Adapted from A. C. Luntz, W. A. Lester, Jr., and H. H. Gunthard, *J. Chem. Phys.*, *70*, 5908 (1979). See also J. E. Butler, G. M. Jursich, I. A. Watson, and J. R. Wiesenfeld, *J. Chem. Phys.*, *84*, 5365 (1986).]

4.3.7 Monte Carlo sampling

The application of Monte Carlo sampling of initial conditions in the method of classical trajectories is illustrated here for the particular problem of computing the reaction cross section σ_R given in terms of the opacity function $P(b)$ by [see Eq. (2.45)]

$$\sigma_R = 2\pi \int_0^\infty b P(b)\, db \tag{4.6}$$

Now $P(b)$ is the probability of reaction at impact parameter b. Let us

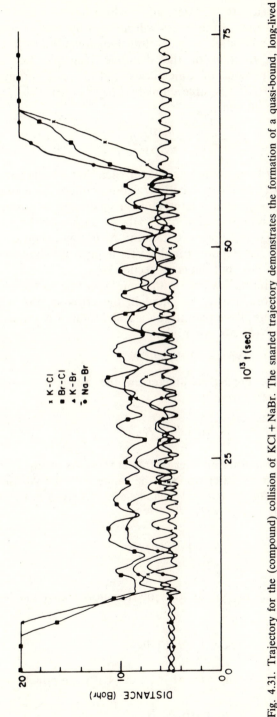

Fig. 4.31. Trajectory for the (compound) collision of KCl + NaBr. The snarled trajectory demonstrates the formation of a quasi-bound, long-lived complex, which happens here to dissociate into the products KBr + NaCl. By varying the initial conditions, it has been found that the dissociation of the complex to the *products* or back to *reactants* is essentially independent of the initial conditions. Over the long lifetime of the collision complex, there is ample time for energy exchange between the different vibrational modes and hence all "memory" of the initial conditions is erased. This is not the case for "direct" collisions. [Adapted from P. Brumer and M. Karplus; P. Brumer, Ph.D. dissertation (1972).]

therefore run $N(b)$ classical trajectories, all of which have the same initial parameter b. Not all of these trajectories will necessarily lead to reaction. The reason is that they will differ in some other initial conditions (e.g., the phase of vibration or rotation of the diatomic reagent). Let $N_r(b)$, $N_r(b) \leq N(b)$, be the number of trajectories that do exit through the products valley. Then if $N(b)$ is large enough, $P(b) = N_r(b)/N(b)$, and so we need to evaluate the integral

$$\sigma_R = 2\pi \int_0^B b \frac{N_r(b)}{N(b)} \, db \qquad (4.7)$$

where B is a sufficiently large impact parameter such that no reaction takes place for $b \geq B$ [i.e., $N_r(b) \approx 0$ beyond B]. We could replace the continuous integration over b by a sum over a large number of discrete b values [i.e., generate a histogramic representation of $P(b)$]. At each such b value we would need to run $N(b)$ trajectories, determine $N_r(b)$, and then compute the sum. This is feasible but requires running very many b values. However, what we are after is not $P(b)$ but the cross section σ_R.

Hence let us sample initial b values with some attention to the physics. Higher b values are more heavily weighted in the cross section, via the annulus $2\pi b \, db$. Therefore, if we plan to run a grand total of N trajectories, let us allot the number $N(b) \, \Delta b$, where

$$N(b) = (2\pi b/\pi B^2)N \qquad (4.8)$$

to initial values of the impact parameter in the range b to $b + \Delta b$. Integrating Eq. (4.8) over b from 0 to B,

$$\int_0^B N(b) \, db = N \qquad (4.9)$$

shows that N is indeed the total number of trajectories. As in Eq. (4.7), let $N_r(b) \, \Delta b$ be the number of trajectories in the range b to $b + \Delta b$ that did lead to reaction. Then, for the *particular choice* of $N(b)$ given in Eq. (4.8), we have, by substituting in Eq. (4.7),

$$\sigma_R = \pi B^2 \frac{\int_0^B N_r(b) \, db}{N} \qquad (4.10)$$

We now introduce N_r as the grand total of trajectories that did lead to reaction,

$$N_r \equiv \int_0^B N_r(b) \, db \qquad (4.11)$$

Since $N_r(b) \leq N(b)$, $N_r \leq N$. In terms of N_r, Eq. (4.10) can now be written as

$$\sigma_R = \pi B^2 N_r/N \qquad (4.12)$$

We reiterate that while Eq. (4.7) is general, Eq. (4.12) is a special case, valid only with the choice of Eq. (4.8) for $N(b)$.

The operational procedure, based on Eq. (4.12), is thus as follows. Randomly select initial impact parameters, but do so from a "biased" distribution $f(b) = 2b/B^2$ [$f(b)$ is uniform, but in b^2 rather than in b]. Thereby higher b values are given a preference of the quantitative form dictated by Eq. (4.8). Do so for a total of N trajectories and count the number N_r that were reactive. The reaction cross section is given by Eq. (4.12).

We have clearly gained in accuracy for a given amount of computer time. We need far fewer trajectories to determine σ_R than would be necessary to evaluate it by explicit integration. What's the trade-off? It is simply that the opacity function as a function of b is only poorly determined. The integral is known to a far higher accuracy than the integrand.

The Monte Carlo sampling is equally valid and useful for other averages. Say we require the thermal reaction rate constant $k(T) = \langle v\sigma_R \rangle$. The thermal averaging can be done in this way using about the same number of trajectories that were previously used to determine the reaction cross section at one particular velocity. The computation will be statistically accurate for $k(T)$ but not for the velocity dependence of the reaction cross section. If we do want both $\sigma_R(v)$ and $k(T)$, it is necessary to use far more trajectories so as to represent σ_R well at each desired value of the velocity. This is the strength and the weakness of the Monte Carlo method. The number of trajectories required for a given accuracy of one value of some quantity is roughly independent of how highly resolved or, conversely, how averaged is the quantity sought.

Appendix 4A: Mass-weighted coordinate systems

The solution of the classical equations of motion to yield trajectories usually requires numerical integration by computer. However, we can obtain considerable physical insight into the dynamics simply by examining the potential energy contour map using a *mass-weighted* coordinate system. We shall illustrate this approach for the case of a collinear collision where one can exactly simulate the actual trajectory by the process of "rolling a ball" on the potential-energy surface. The Monte Carlo averaging is then simply performed by varying the initial conditions for the motion of the ball. We examine first the mathematical transformation to the mass-weighted coordinates and then the physical implications.

Consider the collision $A + BC \rightarrow AB + C$. Then $V(R_{AB}, R_{BC})$ is a function of only two of the three internuclear separations, $R_{AB} = R_B - R_A$ and $R_{BC} = R_C - R_B$. We now introduce two new coordinates, Q_1 and Q_2, by the transformation

$$Q_1 = aR_{AB} + bR_{BC} \cos \beta \tag{4A.1}$$

$$Q_2 = bR_{BC} \sin \beta \tag{4A.2}$$

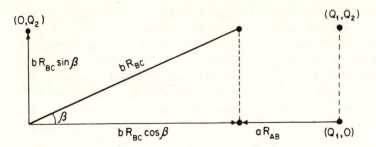

Fig. 4A.1. The construction of the skewed coordinates Q_1 and Q_2 in terms of the "physical" internuclear distances R_{AB} and R_{BC} [Eq. (4A.1) and (4A.2)]. [Adapted from J. O. Hirschfelder, *Int. J. Quant. Chem. IIIS*, 17 (1969).]

Here a, b, and $\cos \beta$ depend on the masses only,

$$a = [m_A(m_B + m_C)/M]^{1/2} \qquad (4A.3)$$
$$b = [m_C(m_B + m_A)/M]^{1/2}$$

$$\cos^2 \beta = m_A m_C/(m_B + m_C)(m_A + m_B) \qquad (4A.4)$$

($M = m_A + m_B + m_C$, the total mass).

The geometrical significance of this change of variable is illustrated in Fig. 4A.1. If we use Q_1 and Q_2 as two cartesian axes, the effects of the transformation is to *skew* the two bond distances at an angle β to one another. Hence, if we regard the potential energy as a function of Q_1 and Q_2 and draw it in the Q_1–Q_2 plane, the entrance and exit valleys will asymptotically be at an angle β to one another (Fig. 4A.2). Such a contour map is known as a (mass-weighted) *skewed axis* representation.

The change of variables in Eqs. (4.A1) and (4.A2) expresses the coordinates Q in terms of the bond distances, which are the natural variables of the potential energy surface. Collision theorists will prefer to work with the distances R_{A-BC} of A to the center of mass of BC and R_{BC}. In terms of these, the definition of Q_2 is unchanged, but an equivalent

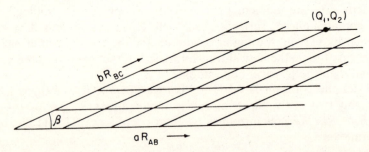

Fig. 4A.2. The grid of *mass-weighted* coordinates. The point (Q_1, Q_2) on the grid is the same point as in Fig. 4A.1. [Same source as for Fig. 4A.1].

definition of Q_1 is

$$Q_1 = aR_{A-BC} \qquad (4A.5)$$

Similarly, for the products of the reaction the collision theorist will use R_{C-AB} and R_{AB}. In terms of these, we can use

$$Q_1' = bR_{C-AB} \qquad (4A.6)$$

$$Q_2' = aR_{AB} \sin \beta \qquad (4A.7)$$

The primed coordinates are rotated by an angle $\beta + \pi$ with respect to the unprimed ones (Fig. 4A.3).

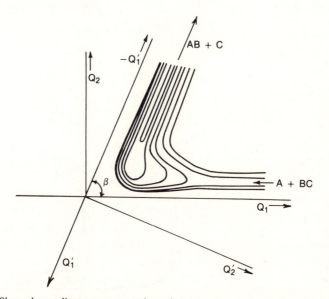

Fig. 4A.3. Skewed coordinate representation of a potential energy surface for the collinear reaction $A + BC \rightarrow AB + C$. The Q' coordinates have been rotated by $\pi + \beta$ with respect to the Q coordinates. [Adapted from F. T. Smith, *J. Chem. Phys.*, **31**, 1352 (1959).]

For such a coordinate system it is found that the kinetic energy for the collision (in the center-of-mass system) has the so-called diagonalized form

$$T = \tfrac{1}{2}(\dot{Q}_1^2 + \dot{Q}_2^2) = \tfrac{1}{2}(\dot{Q}_1'^2 + \dot{Q}_2'^2) \qquad (4A.8)$$

What have we gained by replacing the (physical) coordinates R_{AB} and R_{BC} by Q_1 and Q_2? Simply, that Eq. (4A.8) has the interpretation of a kinetic energy of a point particle (often refered to as the system point) of unit mass, whose position is specified by the two coordinates Q_1 and Q_2. When we express the potential surface as a function of Q_1 and Q_2, the solutions of the classical equations of motion for Q_1 and Q_2, that is, $Q_1(t)$ and $Q_2(t)$, are identical to the solution for the motion of the point particle on the potential surface $V(Q_1, Q_2)$. Rather than numerically solve the

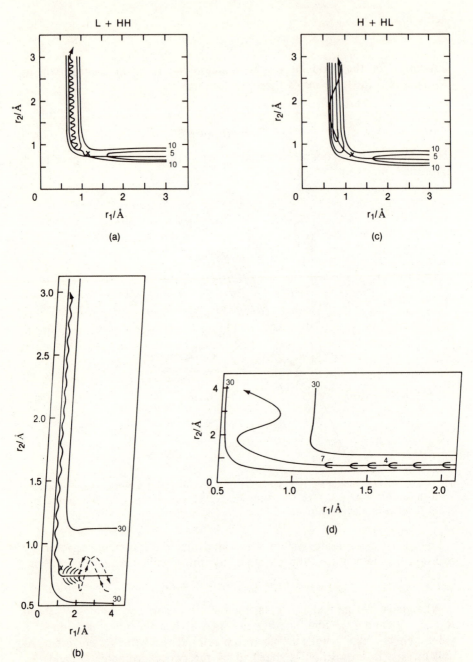

Fig. 4A.4. (a) Plot of a LEPS potential energy surface, showing a collinear L + HH trajectory at $E_T = 9\,\text{kcal mol}^{-1}$, where the barrier ($E_b = 7\,\text{kcal mol}^{-1}$) is indicated by the cross, and (b) skewed and scaled representations. (c) Rectilinear plot, as in (a), showing a collinear H + HL trajectory at the same E_T and (d) skewed and scaled representations. [Adapted from B. A. Hodgson and J. C. Polanyi, *J. Chem. Phys.*, *55*, 4745 (1971).]

classical equation of motion, we can simulate the solution by letting a ball (of unit mass) roll along the surface.†

Figure 4A.4 illustrates the role of the masses in going over from an ordinary to a skewed and a scaled representation of the same potential energy surface. The symbols L and H represent light and heavy atoms (with a mass ratio of 1:80), and the two cases shown are the $L + HH \rightarrow LH + H$ and $H + HL \rightarrow HH + L$ reactions.

Such effects as the efficient conversion of the exoergicity into product vibration on an attractive surface can be readily visualized in the skewed coordinate system. On an attractive surface, the ball is rolling "downhill" as Q_1 decreases and hence it enters the "bend" along the reaction path with a high speed. Rather than make its exit along the products' valley, it will climb the shoulder of the potential (the bobsled effect) and thereby convert much of the exoergicity to product vibration. As $\cos^2 \beta$ increases towards unity, the bend becomes sharper and the efficiency of this "conversion" increases. The sampling of initial conditions is easily carried out by varying the starting conditions for the motion of the system point on the surface.

Instructive as the picture of rolling a ball on the surface may be, we should not forget its critical limitation: It is for a collinear collision. It does not describe, for example, the role of reagent orientation or the centrifugal barrier. It also tends to make us overlook the very many broad-side approach reactions or even the observation that it is not unknown for an atom A to approach BC preferentially from the B side but to "switch partners" and depart with C.

Appendix 4B: Role of angular momentum

The possible states of the products may be limited not just by the conservation of energy but also by the conservation of total angular momentum. The total angular momentum $\mathbf{J}$ is conserved,‡ where $\mathbf{J}$ is the (vector) sum of the orbital angular momentum $\mathbf{L}$ (see Sec. 2.2.5) and the rotor's angular momentum $\mathbf{j}$, in an atom-diatomic system

$$\mathbf{J} = \mathbf{L} + \mathbf{j} \qquad (4B.1)$$

which implies the "triangular" inequality

$$|L - j| \leq J \leq L + j \qquad (4B.2)$$

The reduced mass of, say, the $F + H_2$ (or D_2) system is atypically low [and $L = \mu v b$, Eq. (2.29)]. Also, since the reaction cross section in the

---

† The technique of solving equations of motion by constructing a model subject to the same equations is known as *analog* (as opposed to digital) computation.

‡ In Section 5.6.2 we shall consider the implications of angular momentum conservation for the expression of the cross section in terms of the reaction probabilities.

post-threshold regime is small (≈ 2 Å^2), the range of impact parameters that contribute to the reactive collisions is small. Hence the range of L is unusually narrow. The H_2 (or D_2) molecules have large rotational spacings. Hence, at thermal equilibrium, the most probable j values for H_2 (D_2) are also atypically small. It follows that the range of values of J that are possible for reactive collisions is quite limited. Estimates suggest that $J \leq 30\,\hbar$, as compared to a range of hundreds or thousands for usual heavy-particle collisions with large cross sections (say, $K + Br_2$, where $J \leq 700\,\hbar$).

The total angular momentum can also be written in terms of the products' angular momenta,

$$\mathbf{J} = \mathbf{L}' + \mathbf{j}' \tag{4B.3}$$

or

$$|L' - j'| \leq J \leq L' + j' \tag{4B.4}$$

where $L' = \mu' v' b'$, and since μ' is even smaller than μ and the ranges of b' and b are roughly the same, the range of L' is also limited. This is particularly the case when E_I' is large and E_T' small.

The final conclusion is that the range of j',

$$0 \leq j' \leq L' + J \tag{4B.5}$$

is also quite narrow. When we arrange the final states in order of increasing internal energy, we expect, therefore, that the distribution would show "bands" corresponding to the narrow range of the final rotational energies possible for each vibrational state (as found experimentally).

In contrast, in a more typical situation, say, for the reaction $Cl + HI \rightarrow HCl + I$, the different bands overlap, owing to the wider range of j'. Figure 4B.1 shows the relative distribution of internal energy, as determined via chemiluminescence. The individual discrete levels have been "smeared" to simulate the shape of $d\sigma/dE_T'$ that would result from a molecular beam velocity analysis.† Note the contrast between the distributions for $Cl + HI$ and for $F + D_2$. The manifolds of rotational states of different vibrational levels do not overlap for the $F + D_2$ reaction, as can also be seen in the angular distribution of Fig. 7.1.

Other examples where the conservation of angular momentum influences the product state distribution include the reaction

$$K + HBr \rightarrow H + KBr$$

where the L can be quite large and so $j \ll L$ and $J \approx L$. The reaction is not very exoergic and the most probable E_T' is nearly the same as E_T, but

---

† The plot in Fig. 4B.1 is obtained from the stick diagram, Fig. 7.4, by converting the discrete distribution into a continuum one. At a given v', $d\sigma/dE_T = -d\sigma/dE_R'$. However, $d\sigma/dE_R'$ is not the envelope of the stick diagram. Rather, the envelope is $d\sigma/dj' \propto P(v', j')$, and $d\sigma/dE_R' = (d\sigma/dj')/(dE_R'/dj')$. The reduced variable f_T is the fraction of the total energy E in translation, i.e., $f_T \equiv E_T/E$.

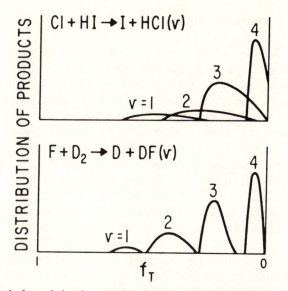

Fig. 4B.1. Disposal of translational energy for given product vibrational states in two exoergic reactions. Shown is $P(f_T \mid v)$ versus f_T for the Cl + HI and F + D$_2$ reactions. [Adapted from K. G. Anlauf, P. E. Charters, D. S. Horne, R. G. McDonald, D. H. Maylotte, J. C. Polanyi, W. J. Skrlac, D. C. Tardy, and K. B. Woodall, *J. Chem. Phys.*, *53*, 4091 (1970).]

$\mu' \ll \mu$ and hence $L' \ll L$. It thus follows that $j' \approx L$; thus the rotational state distribution of KBr is a reflection of the distribution of impact parameters that contribute to the reactive collisions. The j' distribution should thus mimic the opacity function $2\pi b P(b)$, which is likely to peak toward high b values. Experiment shows that a significant fraction of the reaction exoergicity is released in the form of rotational excitation of KBr.

The correlation $j' \approx L$ just discussed for the K + HBr reaction and shown in Fig. 4.27 for the F + C$_2$H$_4$ reaction is valid irrespective of the precise details of the forces that operate during the collision. It is an example of a *kinematic* correlation. The very description of the motion imposes the result. When the departing atom is not light, the correlation is more complicated. A convenient summary (cf. Appendix 4C) is

$$\mathbf{j}' = \mathbf{L}\sin^2\beta + \mathbf{j}\cos^2\beta + \mathbf{d}\cos^2\beta$$
$$\mathbf{L}' = \mathbf{L}\cos^2\beta + \mathbf{j}\sin^2\beta - \mathbf{d}\cos^2\beta$$

$$(4B.6)$$

where $\mathbf{d}$ is a vector that does depend on the dynamics and β is our ubiquitous angle† ($\cos\beta \to 0$ for a light ejected atom). For every type of collision, kinematic considerations govern, at least in part, the detailed outcome. Unraveling the respective contributions of kinematic and dynamic correlations is an active area of current research.

---

† It is a straightforward problem in mechanics to derive Eq. (4B.6) from the definition of β shown in Fig. 4A.3.

Appendix 4C: Kinematic models

Models seek to bypass the need for solving the classical equations of motion. By making plausible assumptions (as in Secs. 2.4.6–2.4.9) we are then able to make predictions which can be readily tested and which provide useful guidelines. The spectator model (Sec 1.4) is the simplest model which correlates the states of the products with those of the reactants. We now seek to examine it in some further detail and to incorporate in the model the possibility of forces operating during the collision. To retain the simplicity of the model we shall assume that the switching of partners is very rapid, i.e., that the forces are impulsive. As an example, consider the $H + Cl_2$ reaction (Sec. 4.2.6). In the c.m. system, the light H atom approaches Cl_2 very rapidly. As soon as it is within the range of chemical forces, the HCl bond is formed, while the two Cl atoms repel one another. It is this 'instant' product repulsion [cf. also Fig. 4.23(b)] that one needs to incorporate in the model. The resulting, so called DIPR (*direct interaction product repulsion*, see also Sec. 7.6.4) model is just one of several that can be developed on the basis of our derivation below.

The starting point is the set of equations (4A.2)–(4A.7) which we write (in vector notation) as (cf. Fig. 4A.3)

$$\mathbf{Q}_1' = -\cos \beta \; \mathbf{Q}_1 + \sin \beta \; \mathbf{Q}_2$$
$$\mathbf{Q}_2' = \sin \beta \; \mathbf{Q}_2 + \cos \beta \; \mathbf{Q}_1. \tag{4C.1}$$

With the definition $\mathbf{L} = \mu \mathbf{R} \times \dot{\mathbf{R}}$ where (cf. 4A.5), $\mathbf{Q}_1 = a\mathbf{R}$ so that $\mathbf{L} = \mathbf{Q}_1 \times \dot{\mathbf{Q}}_1$ and similarly for $\mathbf{j}$, $\mathbf{L}'$ and $\mathbf{j}'$ one readily derives (4B.6) with $\mathbf{d} = \tan \beta \; (\mathbf{Q}_1 \times \dot{\mathbf{Q}}_2 + \mathbf{Q}_2 \times \dot{\mathbf{Q}}_1)$. Of course, to evaluate $\mathbf{d}$ we need some assumption since, so far, no simplifications have been made.

In the spectator model $\dot{\mathbf{Q}}_2 = 0$ and since $E_T = \dot{Q}_1^2/2$ (and similarly for E_T') we recover Eq. (1.5) from (4C.1): $E_T' = \cos^2 \beta \, E_T$. In the DIPR model, the sudden repulsion imparts a large "velocity," say $\mathbf{q}$, along the old bond $\mathbf{Q}_2$, so that

$$\dot{\mathbf{Q}}_1' = -\cos \beta \; \dot{\mathbf{Q}}_1 + \sin \beta \; \mathbf{q}. \tag{4C.2}$$

Hence the products' velocity (i.e., $\dot{\mathbf{Q}}_1'$) has a component along the direction of the old bond. For a light atom attack ($\sin \beta \to 1$), that component can dominate so that the products' angular distribution reflects the distribution of the orientation of the old bond (with respect to the initial velocity) at the instant of reaction (cf. Fig. 5.27). For reactions where there is no preferred cone of acceptance so that $\mathbf{q}$ is randomly oriented with respect to $\dot{\mathbf{Q}}_1$,

$$E_T' = \cos^2 \beta \, E_T + \sin^2 \beta \, E_{PR}, \tag{4C.3}$$

where $E_{PR} = q^2/2$ is the energy repulsively released along $\mathbf{Q}_2$.

When the reaction proceeds preferentially via a bent three-atom transition state $\hat{\mathbf{Q}}_1 \times \hat{\mathbf{Q}}_2$ is finite and hence the repulsive release (where $\dot{\mathbf{Q}}_2$ is along $\mathbf{Q}_2$) makes a significant contribution to $\mathbf{d}$ and hence to $\mathbf{j}'$. The only case where such rotational excitation would not be evident is if the incident (or departing) atom is very light so that $\cos \beta \to 0$ whence [from (4B.6)] $\mathbf{j'} \sim \mathbf{L}$. Sections 7.6.3 and 7.6.4 discuss the implications of such kinematic considerations for the alignment of reaction products.

4.4 From microscopic dynamics to macroscopic kinetics

Dynamics is the description of the motion under the influence of a force (or a potential). When we begin with an assumed potential energy surface, compute classical trajectories, and thereby evaluate a thermal, bimolecular rate constant, we are following a dynamical procedure. Why do we intuitively associate *dynamics* with detailed molecular level-type of experiments and *kinetics* with bulk-type experiments leading to reaction rate or transport coefficients? Part of the reason, at least, is that macroscopic observations are necessarily averages over the detailed processes taking place. Such averaging typically "washes out" details. We have often seen that loss of detail. An obvious example is the poor leverage provided by macroscopic transport coefficients for the determination of intermolecular forces. A less obvious example was encountered in Section 3.1.8. The integral collision cross section is not sensitive to the precise value of the phase shift of each partial wave. The summation of many l values (or the classical integration over the impact parameter) tends to erase any "fine structure."

The purpose of this section is threefold: (1) to exhibit explicitly the averaging required in going from a detailed cross section to a thermal rate constant, (2) to explore the use of symmetry and specifically the symmetry under time reversal to derive relations that are valid for any potential, and (3) to explore the possibility that more averaged rates can be computed or approximated more readily than more detailed ones. What can we say when details of the potential energy surface are not available? Our results under conditions (1) or (2) will be exact, but not so when we seek approximations useful for averaged rates.

4.4.1 State-to-state rate constants

Computations and, increasingly so, experiments can both select the state of the reagents and probe the distribution of states of the products. The ultimate experiment would therefore measure the differential cross section for the reaction of a single quantum state of the reagents to form a single quantum state of the products at a sharp value of the initial relative velocity. Let us first integrate over the final scattering angle to obtain the corresponding state-to-state (integral) reaction cross section $\sigma_R(i \to f)$. Next suppose that the reagents are in definite (internal) quantum states but have a spread in the relative velocity. We then need to *average* over that distribution.† Take the simplest case when the translation of the reagents has a (Maxwellian) thermal distribution at the temperature T. Then, in the center-of-mass system the normalized distribution function for the relative velocity is

$$f(v)\, d\mathbf{v} = (\mu/2\pi \mathbf{k}T)^{3/2} \exp(-\mu v^2/2\mathbf{k}T)\, d\mathbf{v} \tag{4.13}$$

where μ is the reduced mass and $\mathbf{k}$ is Boltzmann's constant. The state-to-state rate constant at the temperature T is then given by

$$k(i \to f; T) \equiv \langle v\sigma_R(i \to f) \rangle \tag{4.14}$$

or explicitly by

$$k(i \to f; T) = \int d\mathbf{v} f(v) v \sigma_R(i \to f) \tag{4.15}$$

Putting $d\mathbf{v} = v^2\, d\hat{v}\, dv$, where $d\hat{v}$ is the element of the solid angle in the direction of the velocity vector $\mathbf{v}$ that integrates out to 4π, and changing the variable of integration from v to the relative kinetic energy $E_T = \mu v^2/2$, with $dE_T = \mu v\, dv$, we obtain

$$k(i \to f; T) = \left(\frac{8\mathbf{k}T}{\pi\mu}\right)^{1/2} \int_0^\infty d\left(\frac{E_T}{\mathbf{k}T}\right)\left(\frac{E_T}{\mathbf{k}T}\right) \exp\left(-\frac{E_T}{\mathbf{k}T}\right)\sigma_R(i \to f) \tag{4.16}$$

We have intentionally written the integral in terms of the dimensionless variable $E_T/\mathbf{k}T$ so as to exhibit its dimensions, namely, that of the average relative velocity, $(8\mathbf{k}T/\pi\mu)^{1/2}$ times an area (i.e., an overall dimension of volume/time). We must, however, caution against interpreting Eq. (4.16), which is $k = \langle v\sigma \rangle$, as $\langle v \rangle \langle \sigma \rangle$. It is true that the first factor in Eq. (4.16) is precisely $\langle v \rangle$, but the integral is not $\langle \sigma \rangle$.

Having determined the state-to-state rate constant, let us for a moment throw all caution to the winds and assume, as in Chapter 2, that the reaction cross section is independent of the initial and final states and depends only on the translational energy. The simplest is the "Arrhenius" or line-of-centers functional form Eq. (2.61),

$$\sigma_R = \pi d^2(1 - E_0/E_T) \tag{4.17}$$

for $E_T \geq E_0$, and zero otherwise. Then Eq. (4.16) becomes

$$k(T) = \pi d^2(8\mathbf{k}T/\pi\mu)^{1/2} \exp(-E_0/\mathbf{k}T) \tag{4.18}$$

that is, the standard "collision theory" rate constant. The temperature dependence of $k(T)$ as given by Eq. (4.18) is known as Arrhenius-like, $k(T) = A \exp(-E_a/\mathbf{k}T)$, with the pre-exponential factor A having a $T^{1/2}$ temperature dependence. But measured values of A often differ con-

---

† Note that our procedure is not symmetrical with respect to reagents and products. We *sum* over all unresolved (e.g., scattering angle) states of the *products*, while we *average* over the *reagents*. Bear with us—we shall shortly show that this is the correct way.

siderably from the naive prediction $A = \pi d^2 \langle v \rangle$. Of course, we already know what went wrong. To begin with, the assumed cross section of Eq. (4.17) ignores the existence of a (possibly E_T dependent; see Sec. 2.4.9) steric factor. Furthermore, the assumption that the reaction cross section depends only on the translational energy of the reagents is clearly at variance with the results of both computational (see Fig. 4.26) and experimental (see Fig. 1.3) studies. It is therefore necessary that we find out how to compound properly the detailed rate constants to yield the overall rate coefficient.

4.4.2 Sum over final states, average over initial states

The title of this subsection is the rule for compounding detailed rate constants. The reasoning behind this rule is so obvious that we shall go through its derivation.

Consider an elementary bimolecular reaction in the gas phase, say,

$$F + H_2(i) \rightarrow HF(f) + H \qquad (4.19)$$

where we have completely specified the internal states of reagents and products (and, for simplicity of notation, taken the F and H atoms to be in their ground state, which is a good assumption for hydrogen but not so for fluorine). Using square brackets to denote concentrations, we have a second-order rate law for the formation of $HF(f)$:

$$\frac{d[HF(f)]}{dt} = k(i \rightarrow f; T)[F][H_2(i)] \qquad (4.20)$$

where $k(i \rightarrow f; T)$ is the state-to-state rate constant at the temperature T. But the reaction between F and $H_2(i)$ can produce HF in states other than f. The total rate of HF production is necessarily the sum of the rates to produce HF in specific states:

$$\frac{d[HF]}{dt} = \sum_f \frac{d[HF(f)]}{dt} = \sum_f k(i \rightarrow f; T)[F][H_2(i)] \qquad (4.21)$$

But only the detailed rate constant depends on f; hence we can sum

$$\frac{d[HF]}{dt} = k(i; T)[F][H_2(i)] \qquad (4.22)$$

where

$$k(i; T) = \sum_f k(i \rightarrow f; T) \qquad (4.23)$$

is the reaction rate constant for production of all final states of HF from $F + H_2(i)$. So far we have the first part: the sum over final states, or Eq. (4.23).

The mixture of reagents may contain H_2 molecules in different quantum states. Let p_i be the mole fraction of H_2 molecules in the quantum state i,

$$p_i = [H_2(i)] \Big/ \sum_i [H_2(i)] = \frac{[H_2(i)]}{[H_2]} \tag{4.24}$$

The overall rate of HF production is again the *sum* of the rate of its production from the different states of H_2,

$$\frac{d[HF]}{dt} = \sum_i k(i; T)[F][H_2(i)] \tag{4.25}$$

Note that the reasoning in Eqs. (4.21) and (4.25) is identical: We are summing over all the independent processes that produce HF. The origin of our second part is revealed when we substitute $[H_2(i)] = p_i[H_2]$ from Eq. (4.24):

$$\frac{d[HF]}{dt} = \sum_i p_i k(i; T)[F][H_2] \tag{4.26}$$

Again we can factor out the part containing the summation,

$$\frac{d[HF]}{dt} = k(T)[F][H_2] \tag{4.27}$$

where

$$k(T) = \sum_i p_i k(i; T)$$

$$= \sum_i p_i \sum_f k(i \rightarrow f; T) \tag{4.28}$$

Equation (4.27) will be recognized as the operational definition of the reaction rate constant $k(T)$. The averaging over initial states, evident in Eq. (4.26) or (4.28), is simply an implication of the definition of a reaction rate.

In deriving Eq. (4.28) we were careful not to specify the distribution p_i of internal states of the reagents. As is obvious, the magnitude of $k(T)$ will depend on what the distribution is. In particlular, if the distribution is varying with time (owing to, say, depletion of reagent states by reaction), $k(T)$ will not be independent of time and thus not be a rate "constant." It is for this reason that in dealing with fast chemical reactions we need the detailed, state-to-state rate constants. Ordinarily, energy transfer collisions are rapid enough so that on the time scale of the chemical reaction the internal states of the reagents have had time to settle down to a steady (most commonly, thermal) distribution. But when the reaction cross sections begin to be comparable with those for energy transfer, the utility of the notion of an overall $k(T)$ begins to break down.†

† In the laboratory we know what to do: dilute with an inert buffer gas to promote relaxation. But for a technological process (e.g., an internal combustion engine) or in nature (e.g., the atmosphere), that may not be feasible. Chemical kineticists are becoming increasingly aware that the implicit assumption that the internal states of the reagents are in an equilibrium distribution is not necessarily valid.

Finally, a word about reagent orientation. Unless a special effort is made, all reagent orientations will be equally probable. Since all such orientations have the same energy, it is customary to average over orientation early in the game, already at the level of the cross section. The "degeneracy-averaged" cross section for a diatomic molecule in a given vibrotational state $\sigma_R(vj)$ is defined [see Eq. (4.28)] by

$$\sigma_R(vj) = (2j + 1)^{-1} \sum_{m_j=-j}^{j} \sigma_R(vjm_j) \qquad (4.29)$$

where $g_j \equiv 2j + 1$ is the degeneracy (i.e., number of equal energy quantum states) of the rotational state j and m_j is the orientation quantum number.

4.4.3 Detailed balance

Detailed balance is a relation between detailed rate constants that is valid irrespective of the nature of the interaction potential between the molecules. The considerable utility of the relation makes it worth deriving.

There are two routes that one can follow. One is to start from the dynamics. The equations of motion of classical mechanics are invariant under time reversal. What this means is that, given a set of initial conditions for a classical trajectory, one can run the trajectory until, say, the colliding molecules have separated with or without reaction, as the case may be. Now stop the integration of the equation of motion, reverse the direction of all momenta, and start integrating again. The trajectory will return the system to the original initial conditions.† For each and every trajectory going over from the reagents' to the products' valley there is an exact "reversed" counterpart going over from the products' to the reactants' valley. Since we have already argued that the reaction cross section is determined by the number of reactive trajectories, there must be a relation between the cross section for the forward and reverse reactions. Both are computed, after all, from the very same set of reactive trajectories. Where they do differ involves only the total number of trial trajectories. A proper counting leads to the result that the (degeneracy-averaged) state-to-state cross sections are related by

$$g_j k^2 \sigma_R(vj \rightarrow v'j') = g_{j'} k'^2 \sigma_R(v'j' \rightarrow vj) \qquad (4.30)$$

The only care we must take is to remember the origin of the result. It is derived by examining the very same set of reactive trajectories, that is, when both sides of Eq. (4.30) are evaluated at the same total energy E. From conservation of energy,

$$E = E_T + E_{vj} = E'_T + E_{v'j'} + \Delta E_0 \qquad (4.31)$$

---

† The computer code must satisfy this criterion. If it does not, it suffers from excessive round-off errors.

where $-\Delta E_0$ is the exoergicity of the reaction. The wave vectors k and k' are related to the translational energy as usual by $E_T = \hbar^2 k^2/2\mu$ and $E_T' = \hbar^2 k'^2/2\mu'$, where μ and μ' are the reduced masses of the reagents and products, respectively.

When the result in Eq. (4.30) is derived from the dynamics, it is known as "microscopic reversibility." Using it in Eq. (4.16) leads to the relation between rate constants known as detailed balance. Since we did not derive Eq. (4.30), we shall derive detailed balance and leave it to the dedicated reader to reproduce our final result from microscopic reversibility.

Detailed balance is the statement that for a system at equilibrium the rate of each process, however detailed, is exactly balanced by the rate of the reversed process.† Consider, therefore, a mixture of fluorine and hydrogen that has come to both thermal and chemical equilibrium. One of the detailed equilibria in this mixture is

$$F + H_2(v, j) \rightleftarrows HF(v', j') + H \tag{4.32}$$

By the principle of the detailed balance, the *rates* of the two opposing reactions in Eq. (4.32) must be equal:

$$k(vj \rightarrow v'j'; T)[F][H_2(v, j)] = k(v'j' \rightarrow vj; T)[H][HF(v', j')] \tag{4.33}$$

Note that Eq. (4.33) is the essence of the quantitative argument. It is the actual reaction rates, and not the rate constants, that are equal at equilibrium. The rate constants are well defined,‡ whether or not the system is in chemical equilibrium.

We now rearrange Eq. (4.33) so as to explicitly exhibit its considerable practical merit. To do so, recall that the system is also in chemical equilibrium and hence has an equilibrium constant

$$K(T) = \{[H][HF]/[F][H_2]\}_{eq} \tag{4.34}$$

Then recall that the system is also in thermal (i.e., Boltzmann) equilibrium. Hence the mole fraction of, say, H_2 molecules in a given vibrotational state,

$$p(v, j) = [H_2(v, j)]/[H_2] \tag{4.35}$$

is also known, being given by the Boltzmann factor

$$p(v, j) = g_j \exp[-E(v, j)/kT]/Q_{H_2}$$

$$Q_{H_2} = \sum_v \sum_j g_j \exp\left(\frac{-E(v, j)}{kT}\right) \tag{4.36}$$

---

† One of the earliest applications was by Einstein to a molecular system in equilibrium with photons. By equating the rates of absorption and emission, he argued that there must be a process of stimulated emission, proportional to the number density of photons. This process is the basis for laser action (see App. 5A).

‡ The *only* condition is that the *translational* motion of both reagents and products have a thermal distribution at the same temperature T.

Hence, in terms of known factors, Eq. (4.33) becomes

$$\frac{k(vj \rightarrow v'j'; T)}{k(v'j' \rightarrow vj; T)} = \frac{[\mathrm{H}][\mathrm{HF}(v', j')]}{[\mathrm{F}][\mathrm{H}_2(v, j)]} = K(T)\frac{p(v', j')}{p(v, j)} \tag{4.37}$$

Using the standard statistical mechanics expression for the equilibrium constant in terms of partition functions, one can after quite a few manipulations derive Eq. (4.37) from microscopic reversibility (4.30).

The only condition on the validity of Eq. (4.37) is that translational motion be in thermal equilibrium. As we shall discuss in some detail in Chapter 6, translational equilibration is a very fast process. Somewhat slower, but still quite rapid, is rotational relaxation. Say, therefore, that both translation and rotation have reached thermal equilibrium. One can then derive from Eq. (4.37) the relation

$$\frac{k(\text{all} \rightarrow v'; T)}{k(v' \rightarrow \text{all}; T)} = K(T)p(v') \tag{4.38}$$

relating the vibrational energy disposal in the forward reaction to the role of reagent vibrational excitation in the reversed reaction. Here $p(v')$ is the fraction of molecules in the vibrational state v' at thermal equilibrium.

4.4.4 The thermal reaction rate constant

In this section we carry out the grand summing over final states and averaging over initial states. In the final result, Eq. (4.41), all the dynamical details, at a given energy, have been collected together into one function, $N_{\ddagger}(E)$.

Let us start with the general explicit expression for the thermal rate constant in an atom–diatom exchange reaction [e.g., Eq. (4.28)]

$$k(T) = \sum_{v} \sum_{j} \left\{ \left[g_j \exp\left(\frac{-E_{vj}}{\mathbf{k}T}\right) \Big/ Q_{BC} \right] \sum_{v'} \sum_{j'} k(vj \rightarrow v'j'; T) \right\} \tag{4.39}$$

where the first factor is the (Boltzmann) probability of the v, j state of the diatomic at thermal equilibrium [see Eq. (4.36)]. The detailed rate constant is itself given as an average over the thermal distribution of the initial relative velocity, Eq. (4.16). In terms of the wave number k, with $E_T = \hbar^2 k^2 / 2\mu$, we rewrite Eq. (4.16) as

$$k(vj \rightarrow v'j'; T) = \left(\frac{h^2}{2\pi\mu\mathbf{k}T}\right)^{3/2} \int_0^\infty dE_T h^{-1}\left(\frac{k_j^2}{\pi}\right)\sigma(vj \rightarrow v'j') \exp\left(\frac{-E_T}{\mathbf{k}T}\right) \tag{4.40}$$

The factor in front of the integral in Eq. (4.40) will be recognized as the reciprocal of Q_T, the translational partition function, for the relative motion, per unit volume.

Now let us substitute Eq. (4.40) in Eq. (4.39) and change the order of summations (over internal quantum states) and integration (over the translational energy). By performing the summations first, we can write the result as

$$k(T) = Q^{-1} \int_0^\infty dE h^{-1} N_{\ddagger}(E) \exp\left(\frac{-E}{kT}\right)$$ (4.41)

All the dynamics are now in $N_{\ddagger}(E)$, defined by

$$N_{\ddagger}(E) = \sum_v \sum_j g_j\left(\frac{k_j^2}{\pi}\right) \sum_{v'j'} \sigma(vj \to v'j')$$ (4.42)

where, as always, $\hbar^2 k_j^2/2\mu = E - E_{vj}$ is the translational energy of the $A + BC(v, j)$ collision when the total energy is E. The internal partition function Q_{BC} of BC has been multiplied by the partition function per unit volume $(2\pi\mu kT/h^2)^{3/2}$ for the A–BC relative motion to give Q, the overall partition function. The translational and internal Boltzmann factors $\exp(-E_{vj}/kT) \exp(-E_T/kT)$ have been multiplied, using Eq. (4.31) to yield an overall energy Boltzmann factor $\exp(-E/kT)$. One Planck constant is "left over." To make sure it really belongs there, let's check the dimensions: $N_{\ddagger}(E)$ is dimensionless (since the wave vector k has the dimension of an inverse length). Hence the entire integral has the dimension of $(time)^{-1}$ [(energy-time)$^{-1}$ from h times energy from the variable of integration] and Q is the partition function per unit volume. Hence the dimensions of $k(T)$ (namely, volume/time) are just right for a bimolecular rate constant. The h in Eq. (4.41) really belongs in there. What our considerations imply is that we can write the reaction rate when the energy is in the range E to $E + dE$ as $k(E) \, dE$, where, from Eq. (4.42).

$$k(E) = N_{\ddagger}(E)/h$$

$$= \frac{1}{h} \sum_i g_i\left(\frac{k_i^2}{\pi}\right) \sigma(i)$$ (4.43)

Not only have we compounded all the dynamics into one dimensionless function $N_{\ddagger}(E)$, but that function has the same value for the forward and reversed reaction! [Use Eq. (4.30) in Eq. (4.42)†.] Surely $N_{\ddagger}(E)$ must have a clear physical interpretation! Indeed, in the following we shall provide it. In the meantime let us complete all our integrations in this mathematical section. At $E = 0$ no reaction is possible; hence $N_{\ddagger}(E) \to 0$ as $E \to 0$. On the other hand, there are no molecules at very high energies, so $\exp(-E/kT) \to 0$ as $E \to \infty$. One can therefore integrate Eq. (4.41) by parts:

$$k(T) = Q^{-1} \int_0^\infty h^{-1} N_{\ddagger}(E) \exp\left(\frac{-E}{kT}\right) dE$$

$$= -(kT/h)Q^{-1} \int_0^\infty N_{\ddagger}(E)\left[\frac{d}{dE} \exp\left(-\frac{E}{kT}\right)\right] dE$$

$$= (kT/h)Q^{-1} \int_0^\infty \frac{dN_{\ddagger}(E)}{dE} \exp\left(-\frac{E}{kT}\right) dE$$ (4.44)

---

† Indeed, the rate constant for the reversed reaction is given by Eq. (4.41), except that Q is now the overall partition function for AB + C.

The result, Eq. (4.44), is still exact, provided that $N_{\ddagger}(E)$ is obtained by an exact route. But the only one so far discussed is via an exact dynamical computation using Eq. (4.42).

4.4.5 Activation energy and the reactive reactants

The Arrhenius activation energy,

$$E_a = -\mathbf{k}\frac{d \ln k(T)}{d(1/T)} \tag{4.45}$$

serves as a convenient and a familiar measure of the temperature dependence of the reaction rate constant. Our intention here is threefold: to obtain an exact interpretation, following Tolman, of E_a; to argue that it is equally instructive to consider the activation energy of detailed rate constants; and to begin our identification of $N_{\ddagger}(E)$.

Substituting Eq. (4.41) in Eq. (4.45) and writing $k = P/Q$ so that $\ln k = \ln P - \ln Q$ and $d \ln P = P^{-1} dP$, we find that the activation energy can be written as a difference of two special averages:

$$E_a = \langle E^* \rangle - \langle E \rangle \tag{4.46}$$

The second term comes from differentiating the partition function,

$$\langle E \rangle = -\mathbf{k}\frac{d \ln Q}{d(1/T)} \tag{4.47}$$

and is just the average energy of all the reactants. The first term,

$$\langle E^* \rangle = \int_0^\infty dE\, E N_{\ddagger}(E) \exp\left(\frac{-E}{\mathbf{k}T}\right) \Big/ \int_0^\infty dE\, N_{\ddagger}(E) \exp\left(\frac{-E}{\mathbf{k}T}\right) \tag{4.48}$$

arises from differentiating the integral in Eq. (4.41). It also looks like some kind of average. In fact, it is the average energy of the distribution $f(E) = N_{\ddagger}(E) \exp(-E/\mathbf{k}T)$. This is the distribution of total energy in those A–BC pairs that *do* react, that is, $\langle E^* \rangle$ is average energy of those collisions which lead to reaction. It is the mean energy of these reactive reactants, which is the first term in Eq. (4.46). (See Fig. 4.36 below.)

To have a more intuitive picture, recall the procedure for a Monte Carlo trajectory computation. At each energy E we select $N(E)$ trajectories. These will differ, *inter alia*, in the partition of energy among the internal excitation of BC and the A–BC relative motion. Only $N_r(E)$ of these trajectories will lead to reaction. These are the "reactive reactants," that is, those collisions (at the total energy E) that do lead to reaction.

Similar considerations apply to the detailed rate constants [see Eq. (4.40)], for example,

$$k(vj \rightarrow v'j') = Q_T^{-1} h^{-1} \int_0^\infty dE_T \frac{k_j^2}{\pi} g_j \sigma(vj - v'j') \exp\left(\frac{-E_T}{\mathbf{k}T}\right) \tag{4.40}$$

Hence

$$-\mathbf{k}\, d \ln k(vj \to v'j')/d(1/T) = \langle E^* \rangle_{vj \to v'j'} - \langle E_T \rangle \tag{4.49}$$

where $\langle E_T \rangle$ is the mean relative translational energy at the temperature T,

$$\langle E_T \rangle = -\mathbf{k}\, d \ln Q_T/d(1/T) = \tfrac{3}{2}\mathbf{k}T \tag{4.50}$$

and the first term is the mean translational energy of those collisions that convert reagents in the state vj to products in the state $v'j'$.

4.4.6 The configuration of no return

We begin our attempt to find useful approximation methods for the overall reaction rate. Common to many of those is the concept of a configuration of *no return* such that when the system has reached this critical configuration, it will necessarily proceed to reaction and not "turn back" to the reagents' region. Our simple reaction models of Section 2.4 have, in fact, already invoked this concept. There the critical configuration was a relative separation that, if the system reached it, reaction followed. Computing the reaction rate becomes thereby quite simple. No longer is it necessary to follow the trajectories all the way from the reagents' to the products' valley; it is sufficient to monitor at the critical configuration and count the rate at which molecules pass through.

The existence of such a unique configuration is an approximation. Strictly speaking, only when the nascent products have well separated can we be *sure* that they will indeed proceed to form the final products. When there is a barrier along the reaction path, the location of the barrier, that is, the saddle point, is a useful choice for a point of no return.† The reason is as follows. The theory is meant to compute the *thermal* reaction rate constant. Unless the barrier is small compared to $\mathbf{k}T$, most trajectories that do make it to the barrier have very little excess (i.e., translational) energy. Once the barrier is crossed, the motion is rapidly downhill, with the trajectory gaining in kinetic energy. Such a trajectory is unlikely to wind its way back to the narrow passage at the saddle point. To do so requires a very concerted motion, with as much energy as possible along the reaction path. It is much more likely that such a trajectory will exit into the products' valley.

Figure 4.32 shows the trajectory-generated results for the mean number of crossings at a saddle point for the higher-barrier reaction $(H + H_2)$ and the low-barrier case $(F + H_2)$, both confined to collinear collisions. In both cases we see that trajectories that just make it past the barrier indeed do not recross, at least in the post-threshold regime. At higher energies, however, the "no neturn" assumption does break down. This is not unexpected, since for higher-energy trajectories the saddle-point region is no longer a narrow

---

† For reactions without an energy threshold, as we have seen in Section 2.4.7, location of the maximum of the effective potential provides a reasonable choice.

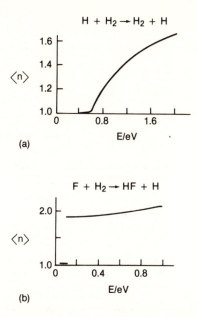

(a)

(b)

Fig. 4.32. Mean number of crossings at the "configuration of no return" versus total energy calculated for (a) the collinear $H + H_2$ exchange reaction and (b) for the collinear $F + H_2$ reaction. [Adapted from E. Pollak and R. D. Levine, *J. Chem. Phys.*, 72, 2990 (1980); 86, 4931 (1982). See also Ch. Schlier, *Chem. Phys.*, 77, 267 (1983).]

pass. It is no longer necessary to channel most of the energy into translational energy along the reaction path in order to clear the barrier.

The key assumption of the theory is thus that there exists a single configuration of no return, called the *transition state*. Once the system has reached this configuration, it is assumed to proceed further and form the products. That path (on the surface) along which the reactants reach the critical configuration and along which the supermolecule in the transition state "deforms" to yield the products is the *reaction coordinate*. The theory we are setting up is clearly a theory of direct reactions, with a single barrier separating reactants and products. But this is not invariably the case. It follows that we shall have to modify our theory to accommodate reactions whose potential energy surfaces have a well. Bear with us: We shall do so shortly.

4.4.7 Evaluating the reaction rate

Contrast parts (a) and (b) of Fig. 4.33. Both show a set of trajectories at a given total energy E. Figure 4.33a depicts the simple situation where the transition state is a strict configuration of no return. Trajectories that originate in the reactants' side either fail to reach the transition state or cross it and proceed to form products. All the trajectories that cross the

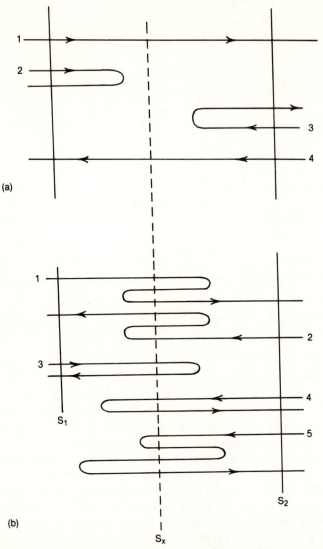

Fig. 4.33. (a) Schematic drawing of different types of trajectories for the case of a transition state that is a real configuration of no return (see text). Trajectories 1 and 4 are "reactive" and 2 and 3 "nonreactive." (b) Same as in (a) but for the case in which trajectories can recross the transition state. Trajectory 1 is reactive (with three crossings) and trajectory 2 its inverse. Trajectories 3, 4, and 5 are nonreactive (with two, two, and four crossings, respectively). Here S_1 and S_2 denote initial and final states, and S_x the transition state.

transition state from left to right started as the separated reagents and terminate as the separated products. The rate of reaction is just the rate of passage from left to right. In Fig. 4.33b trajectories recross the transition state. Those that do so an odd number of times proceed to form products, and those that recross an even number of times return to the valley from

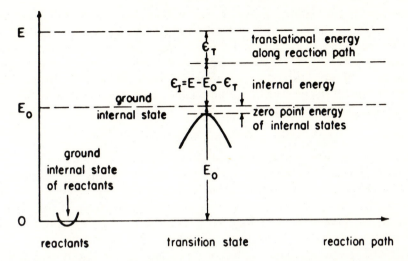

Fig. 4.34. Energetics of the transition state: Schematic energy profile along the reaction coordinate for a potential with a barrier. Here E_0 is the zero-point energy of the *internal* vibrations of the supermolecule at the transition state (measured from the zero-point internal state level of the reactants), E is the total energy, and $\epsilon_I = E - E_0 - \epsilon_T$ is the energy available for the internal states of the supermolecule at the transition state. [Adapted from R. A. Marcus, *J. Chem. Phys.*, 43, 2658 (1965).]

which they came. Life is simpler when there is a strict configuration of no return.†

What we are after is the thermal reaction rate constant $k(T)$. Let us begin by computing the reaction rate when the total energy of the reactants is in the range E to $E + dE$. Given that the transition state is a configuration of no return, all the possible trajectories through the transition state are reactive. Thus the computation is really just proper bookkeeping; we just need to learn how to count.

For proper counting, the use of quantum mechanics is unavoidable.‡ In particular, we require the result that the number of trajectories with momentum (along the coordinate q) in the range p to $p + dp$ be $dp\, dq/h$, where dq is an element of length. The number per unit length along q is thus dp/h. We are now ready.

First, we must consider the energy balance at the transition state (Fig. 4.34). Let E_0 be the minimal energy measured from ground-state reagents. At a given total energy E (measured also above the zero-level of the

† But even otherwise, the rate of passage at the transition state remains an upper bound to the reaction rate.

‡ We require a quantal result because of a flaw in the classical mechanical counting of translational states that was first realized in connection with black body radiation density and which led to the introduction of Planck's constant, h. The same failing was subsequently realized in connection with the contribution of the translational motion to the entropy (the Sackur–Tetrode formula) and hence to the other thermodynamic properties of gases.

reagents) the translational energy ϵ_T for motion along the reaction coordinate, can be at most $E - E_0$. In general, this available energy, $E - E_0$, will be distributed between ϵ_T and the energy in the other degrees of freedom of the transition state. These other degrees of freedom correspond to bound motions as in an ordinary stable molecule. We shall refer to them as the internal degrees of freedom of the supermolecule at the transition state. The energy balance thus reads

$$E = E_0 + \epsilon_I + \epsilon_T \qquad (4.51)$$

where ϵ_I is the energy of the internal modes.

Our first task is to calculate the *rate of passage* of the supermolecules across the point of no return along the reaction coordinate q. Let us consider a *subset of the supermolecules*, namely, an ensemble of them, each with its *translational energy* (along q) *in the range* ϵ_T to $\epsilon_T + d\epsilon_T$. (Later we will integrate over all ϵ_T's to take into account all systems.)

For the given ϵ_T there is a corresponding linear speed and linear momentum along q; let us denote them by $v^{\ddagger}(=\dot{q})$ and $p^{\ddagger} = m^{\ddagger}v^{\ddagger} = (2m^{\ddagger}\epsilon_T)^{1/2}$. Here we use the superscript $\ddagger$ to denote properties at the transition state; $m^{\ddagger}$ is a mass appropriate to the one-dimensional transition of the system, along q. (However, it disappears from the final result!) The range of ϵ_T can be written in terms of $p^{\ddagger}$: $d\epsilon_T = p^{\ddagger} dp^{\ddagger}/m^{\ddagger}$.

Thus the *rate* (the number of supermolecules per unit time) *of passage of systems* in the *specified* ϵ_T range is $dr^{\ddagger} = v^{\ddagger} dN^{\ddagger}$, where $dN^{\ddagger}$ is the number per unit length along q (i.e., a one-dimensional number density), that is, $dN^{\ddagger} = dp^{\ddagger}/h$. This gives for the rate of passage

$$dr^{\ddagger} = \frac{v^{\ddagger} dp^{\ddagger}}{h} = \frac{p^{\ddagger} dp^{\ddagger}}{hm^{\ddagger}} = \frac{d\epsilon_T}{h} \qquad (4.52)$$

The *rate of passage through* the *point of no return* along the reaction coordinate is $d\epsilon_T/h$, *irrespective* of the value of ϵ_T itself. The only assumption made so far is that of a point of no return; all the systems reaching it proceed further and ultimately reach the products' valley.

In the second step we realize that a given total energy E may be *partitioned in* many ways between ϵ_T and ϵ_I, that is, between the motion along the reaction coordinate and the internal modes of the transition state. We now invoke the assumption that there is no reagent state selection so that *all* trajectories with energy in the range E to $E + dE$ that cross the configuration of no return are to be counted.

Each internal state of the supermolecule (at the transition state) contributes $d\epsilon_T/h$ to the rate of passage, so the *total rate* $k(E) dE$ when the total energy is in the range E to $E + dE$ is the *sum of all the contributions*:

$$k(E) dE = \sum_{n_{\ddagger}} dr^{\ddagger} = \sum_{n_{\ddagger}} \frac{d\epsilon_T}{h} = \sum_{n_{\ddagger}} \frac{dE}{h} \qquad (4.53)$$

The sum is over all the $n_{\ddagger}$ internal states of the supermolecule whose ϵ_I is in the allowed range 0 to $E - E_0$; for a given ϵ_I we have $d\epsilon_T = dE$. The last equality defines our "differential reaction rate," that is, $k(E) \, dE$ is the *rate of the reaction when the energy E is in the range E to $E + dE$*. Hence the dimensions of $k(E)$, the reaction rate per energy interval, are $(energy \times time)^{-1}$. The central result is thus

$$k(E) = \sum_{n_\ddagger} \frac{1}{h} = \frac{N_\ddagger(E - E_0)}{h} \tag{4.54}$$

where $N_\ddagger(E - E_0)$ is the number of internal states of the supermolecule at the transition state for which $\epsilon_I \leq E - E_0$.

If we had a good knowledge of the potential energy surface near the saddle point, we could *calculate* the *energies* of the allowed *internal states* and simply *count up* how many had $\epsilon_I \leq E - E_0$. As a short-cut we could use approximate formulas for the *density of internal states* $\rho_\ddagger(\epsilon_I) = dN_\ddagger/d\epsilon_I$ and then replace the sum by an integral:

$$N_\ddagger(E - E_0) = \int_{\epsilon_I=0}^{E-E_0} \rho_\ddagger(\epsilon_I) \, d\epsilon_I \tag{4.55}$$

With Eq. (4.55) we have provided not only an interpretation for $N_\ddagger(E)$ first introduced on purely mathematical grounds in Eq. (4.41) but also an alternative (and feasible) route for its determination. No dynamical computation whatever is called for. All that is required is the knowledge of the potential energy function at the configuration of no return.

4.4.8 Activated complex theory for k(T)

A particularly useful form† for $k(T)$ arises out of the so-called *activated-complex theory* of rate processes. We shall derive it here in a way that delineates the assumptions involved.

The first assumption is that the *reactants* are in *thermal equilibrium*. This is simply a condition on the experimental conditions for which the theory is applicable.‡ When the reactants are in thermal equilibrium, Eq. (4.45)

---

† First proposed by H. Eyring in the 1930s; it had a profound influence on the development of chemical kinetics.

‡ This is not a trivial condition and can be violated for a number of reasons. Thus endoergic reactions will often proceed *preferentially* from reactants that are vibrationally excited. The reaction thus serves to *deplete* such states; only if the repopulation of these states by energy-transferring collisions is faster than the reaction rate will the reagents remain in thermal equilibrium. Similarly, the specific *release of* reaction *exoergicity* tends to displace the system from thermal equilibrium. Hence the latter condition obtains only when the collisional mechanism that restores equilibrium occurs on a time scale shorter than that of reaction. Since this condition may not be met (particularly for fast reactions) or can be made to "fail" (by keeping the system in thermal disequilibrium), molecular dynamics is also relevant to reactions in bulk, in particular in those cases (e.g., fast combustion reactions) where thermal equilibrium is not maintained.

applies. We can then use Eq. (4.55) to conclude that $dN_{\ddagger}/d(E-E_0) = \rho_{\ddagger}(E-E_0)$. Changing the variable of integration in Eq. (4.44) from E to $E_I = E - E_0$ and factoring $\exp(-E_0/kT)$ out, we have

$$k(T) = \frac{kT}{hQ}\exp\left(\frac{-E_0}{kT}\right)\int_0^{\infty} dE_I\rho_{\ddagger}(E_I)\exp\left(\frac{-E_I}{kT}\right)$$

$$= \frac{kT}{h}\frac{Q_I^{\ddagger}}{Q_TQ_I}\exp(-E_0/kT) \tag{4.56}$$

Here we identified the integral as the partition function $Q_I^{\ddagger}$ for the internal states of the supermolecule at the transition state configuration. We have also factored out the partition function Q of the reagents as a product of the internal partition function and the partition function for the relative translation.

With Eq. (4.56) we have completed the line of reasoning that was begun in Section 4.4.4. The computation of the thermal rate constant has been reduced to the computation of partition functions, requiring no dynamical knowledge whatsoever. It does require, however, a knowledge of the potential energy surface in the saddle-point region; both E_0 and the energies of the internal states at the transition state configuration are needed. The latter information is rarely available; one usually assumes a structure for the supermolecule in the transition state and calculates a value of $Q_I^{\ddagger}$. However, when E_0 and $Q_I^{\ddagger}$ can actually be computed from a well-defined potential surface, one finds that Eq. (4.56) can often provide results of useful accuracy. Figure 4.35 shows the activated-complex theory prediction of $k(T)$ for the hydrogen ortho–para exchange reaction compared with experimental results.

The origin of the activation energy and its relation to the threshold energy becomes clearer when we consider the integrand in Eq. (4.41). For reaction with an energy threshold, $k(E)$ is zero below the threshold and is an increasing function of E beyond the threshold. In the transition state approach

$$k(E) = \begin{cases} 0, & E \le E_0 \\ N_{\ddagger}(E-E_0)/h, & E > E_0 \end{cases} \tag{4.57}$$

Fig. 4.35. Arrhenius plot of the rate constant for the reaction $H + p\text{-}H_2 \rightarrow o\text{-}H_2 + H$, that is, the temperature dependence of $k_H \equiv k + k'$ (where the prime denotes the reverse reaction) over the temperature range 299–549 K. [Adapted from B. C. Garrett and D. G. Truhlar, *Proc. Nat. Acad. Sci. U.S.A.*, **76**, 4755 (1979). The experimental points are from measurements of K. A. Quickert and D. J. LeRoy, *J. Chem. Phys.*, **53**, 1325 (1970); **54**, 5444 (1971), and A. A. Westenberg and N. de Haas, *J. Chem. Phys.*, **47**, 1393 (1967).] Calculation (a) is *via* transition state theory using the LSTH *ab initio* potential-energy surface of Fig. 4.4; calculation (b) employs a "scaled" version of this surface believed to be closer to a fully converged *ab initio* potential.

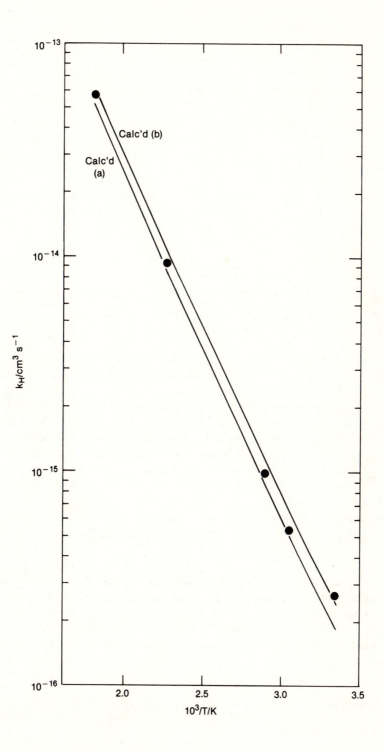

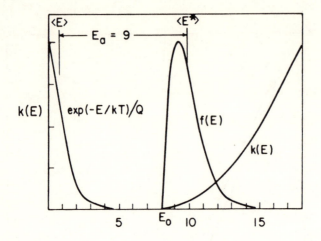

Fig. 4.36. The Tolman interpretation of the activation energy. Shown are $k(E)$, the Boltzmann factor $\exp(-E/kT)/Q$, and the product $f(E) = k(E)\exp(-E/kT)$. The average values $\langle E^* \rangle$ and $\langle E \rangle$ and the activation energy $E_a = \langle E^* \rangle - \langle E \rangle$ are shown. (Note that $E_a \neq E_0$.) [Adapted from R. Wolfgang, *Acc. Chem. Res.*, 2, 248 (1969).]

where $N_\ddagger$ represents the number of internal states, and hence rapidly increases with E above threshold.

The integrand in Eq. (4.41) is thus a *product* of *two functions*, one (the Boltzmann factor) *rapidly decreasing* with E (but increasing with T) and one [$k(E)$ itself] *rapidly increasing* with E beyond threshold, as illustrated in Fig. 4.36. It is often the case that $E_0 > \langle E \rangle$, so that only the values of $k(E)$ in the immediate energy region following the threshold determine the magnitude of $k(T)$. As the *temperature* is *increased*, so does the integrand and *hence so does* $k(T)$.

The *dominant contribution* to the *activation energy* E_a can be shown to be the magnitude of the *threshold energy* E_0. Thus, using Eq. (4.56), we obtain

$$E_a = E_0 - \tfrac{1}{2}kT + (\langle E_I^\ddagger \rangle - \langle E_I \rangle) \qquad (4.58)$$

where $\langle E_I^\ddagger \rangle$ is the average *internal* energy of the supermolecule in the transition state.† Since both the $\tfrac{1}{2}kT$ and the difference term in brackets are often small compared to E_0, we have a rough identification (but not an equality) of E_a with E_0. Moreover, we should remember that E_0 is the threshold energy (see Fig. 4.34), and not the potential energy barrier height.

---

† The origin of the $-\tfrac{1}{2}kT$ term is the one-dimensional motion along the reaction coordinate. This is counted in the mean energy of all reactants but not in the transition state (which is the supermolecule excluding the motion along the reaction coordinate).

4.4.9 The statistical approximation

The differential reaction rate $k(E)\,dE$ for reagents in the energy range E to $E + dE$ is an essential intermediate stage in the computation of the thermal reaction rate constant Typically, however, we do not measure $k(E)$. What we are really after are the state-to-state or "state-to-all" reaction rates. We could use the approach of Section 4.4.7, but a problem is immediately apparent: What we did there is to stand at the point of no return and count the crossing trajectories; we did not have to know which states of the reagents these came from, since all states at the total energy E were allowed—not so now. It is no longer enough to stand only at the point of no return. We need to know what the probability, say, $P(i)$, is of reaching the point of no return given that the reagent molecule was initially in the internal state i (at, of course, the total energy E).

Once the trajectories cross the point of no return, they proceed to form the products. The reaction cross section for reagents in the internal state i will then be given by the fraction $P(i)$ of the cross section to reach (or cross) the configuration of no return:

$$\sigma_R(i) = P(i)\sigma_R = (\pi/k_i^2)P(i)N_{\ddagger}(E) \tag{4.59}$$

Since the $P(i)$'s are normalized, $\sum_i P(i) = 1$, the cross sections add up to the total, $(\pi/k_i^2)N_{\ddagger}(E)$, for reagents in the energy range E to $E + dE$.

We are ultimately after the state-to-state reaction cross section. Do we therefore need a second type of probability, namely, that of the passage from the configuration of no return to products in the internal state j'? At this point detailed balance comes to the rescue with a resounding no. At a given total energy, the probability $P(i)$ to reach the configuration of no return from $A + BC(i)$ is the very same as the probability $P'(i)$ to form the products $A + BC(i)$ from the configuration of no return of the $AB + C$ reaction. We shall prove this in a moment, but let us pause to remember that we are assuming (from Sec. 4.4.6 on) that there is a unique configuration of no return en route from reagents to products. Also, we are working at a sharp value of the total energy.† The proof uses Eq. (4.30) after summing both sides over v, j. This gives

$$k_i^2\sigma_R(i) = \sum_f k_f^2\sigma_R(f \to i) \tag{4.60}$$

The right-hand side of Eq. (4.60) is the reaction rate for $AB + C$ in the energy range E to $E + dE$ forming products $A + BC$ in the internal state i. By assumption, there is a unique configuration of no return. Hence the rate

---

† Strictly speaking, we should be working at a sharp value of *all* conserved quantities. Besides the total energy, there is another important conserved quantity: the total angular momentum. If the reaction is such that there is a large change in the reduced mass, as, for example, in $Cs + HBr \to CsBr + H$, then the conservation of total angular momentum is an important restriction, (see App. 4B); otherwise it can often be overlooked.

on the right is $N_{\ddagger}(E)$ times the probability $P'(i)$ to reach A + BC(i) from the configuration of no return. The left-hand side of Eq. (4.60) follows on writing Eq. (4.59) as $k_i^2\, \sigma_R(i) = \pi N_{\ddagger}(E)P(i)$:

$$N_{\ddagger}(E)P(i) = N_{\ddagger}(E)P'(i) \tag{4.61}$$

It follows from Eq. (4.61) that the probability $P'(i)$ to reach A + BC(i) from the transition state equals the probability $P(i)$ to reach the transition state from A + BC(i). Hence the state-to-state cross sections have the separable form

$$\sigma_R(i \to f) = (\pi/k_i^2)N_{\ddagger}(E)P(i)P(f) \tag{4.62}$$

The statistical assumption made in Eq. (4.59) has led us inevitably to the conclusion that the distribution of AB(f) + C products, in the A + BC(i) reaction at the total energy E,

$$P(f) = \sigma_R(i \to f) \Big/ \sum_f \sigma_N(i \to f) = \frac{\sigma_R(i \to f)}{\sigma_R(i)} \tag{4.63}$$

is *independent* of the internal state of the reagents.

It is sometimes more convenient to speak not of the probability $P(i)$ to reach the configuration of no return to A + BC(i) (or *vice versa*) but of the *transmission* $T(i)$ of the configuration of no return, $T(i) = N_{\ddagger}(E)P(i)$. Then

$$\sigma_N(i \to f) = \frac{\pi}{k_i^2} \frac{T(i)T(f)}{N_{\ddagger}(E)} \tag{4.64}$$

Either way, we are left with a major problem: how to calculate the distribution $P(f)$ of products, Eq. (4.63). Section 5.5 takes up that issue.

4.4.10 Compound collisions: Unimolecular versus bimolecular rates

The discussion in Sections 4.4.6–4.4.9 applies when it is reasonable to assume that there is only one "bottleneck" en route from reagents to products. We then take up our position at the narrow pass and compute the rate of crossing and thus the rate constant. Even for direct reactions with a simple barrier the assumption may fail, for there may be entropic as well as energetic barriers.† It will certainly fail when the potential energy surface has a well in its midst, flanked by two barriers, as in Fig. 4.37. For such a surface we expect that, at least for lower-energy collisions, the trajectories will "spend some time" in the well (see Secs. 4.2.8 and 4.3.6).

Under such circumstances we can again invoke the concept of a point of

---

† For that reason a more proper procedure is to determine the configuration of no return by a variational procedure. At a given E, one can compute $N_{\ddagger}(E)$ for different positions along the reaction coordinate. At each energy the configuration of no return is the position where the density of states $dN_{\ddagger}/dE$ is minimal. Of course, it may (and does) happen that this optimal position will shift with energy.

no return. We alert the reader, however, to please note the distinct sense in which we are about to employ this approximation. Referring to Fig. 4.37, we place *two* configurations of no return, one on each side of the well. The one on the reagents' side is for entry to the well. All trajectories that cross it reach the well. The same for the products' side. To get from the reagents to the products (or *vice versa*), one needs to cross *both* configurations of no return.

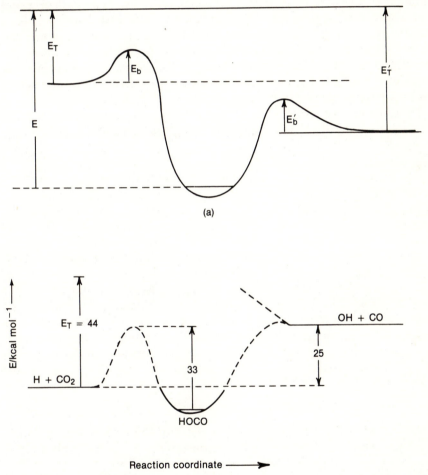

(a)

(b)

Fig. 4.37. (a) Schematic illustration of the minimum-energy path for a potential energy surface with a well flanked by two barriers (characteristic of certain complex-mode reaction systems). (b) Realistic potential curve for the system $H(^2S) + CO_2(^1\Sigma_g^+) \leftrightarrow HOCO \leftrightarrow OH(^2\Pi) + CO(^1\Sigma^+)$. The overall reaction is endoergic by some 25 kcal mol^{-1}, but experiments with translationally hot H atoms (e.g., for $E_T = 44$ kcal mol^{-1}, as shown) yield OH + CO product *via* a long-lived complex believed to be HOCO. The two barriers are associated with bent transition state structures of the intermediate. [Adapted from J. Wolfrum, *J. Phys. Chem.*, *89*, 2525 (1985).]

Trajectories that originate in the well region (i.e., unimolecular dissociation starting from a bound ABC molecule) and cross the configuration of no return on the left are assumed to cross only once and then proceed to form "reactants" A + BC; those that reach the configuration of no return on the right also cross once and proceed to form "products" AB + C.

For *direct* bimolecular reactions, those trajectories that cross the first (and only) bottleneck are assumed to proceed directly to products. For *compound* reactions those trajectories that cross the first bottleneck are assumed to form a complex. The complex can then dissociate either back into the reagents or into the products, but because the complex is assumed to be long lived, the formation and dissociation are taken to be independent. The probability of a trajectory being in the complex region is given in terms of counting at the two critical configurations, irrespective of whether the trajectory originated in the reagents' or products' valleys.

To compute the cross section, we employ our assumption that the formation and decay of the complex are independent. Hence at a given total energy E

$$\sigma(i \to f) = \sigma_c(i) P_c(f) \tag{4.65}$$

where $\sigma_c(i)$ is the cross section for the formation of the complex from state i of the reagents and $P_c(f)$ is the probability of the decay of the complex into state f of its dissociation products. It is at this point that we differ from the corresponding discussion for direct reactions in Section 4.4.9. The "dissociation products" of the complex include both the reagents and the products of the overall reaction. Through summing over states of both reagents and products, $P_c(f)$ is normalized to unity.

If we sum $P_c(f)$ over the states of the products only, we obtain the *branching fraction F*, that is, the fraction of complexes (at the energy E) that dissociate into products. Since the dissociation of the complex is unimolecular, the branching ratio Γ, where $\Gamma = F/(1 - F)$, is the ratio of the two unimolecular rates for dissociation into reagents and products, respectively. In the configuration of no return, the two rates are simply $N_{\ddagger}^{(r)}(E - E_0^{(r)})/h$ and $N_{\ddagger}^{(p)}(E - E_0^{(p)})/h$,

$$\Gamma = N_{\ddagger}^{(p)}(E - E_0^{(p)})/N_{\ddagger}^{(r)}(E - E_0^{(r)}) \tag{4.66}$$

where $N_{\ddagger}^{(\alpha)}(E - E_0^{(\alpha)})$ is the number of states of the transition state (apart from the reaction coordinate) on the products' $(\alpha = p)$ and reagents' $(\alpha = r)$ sides, and $E_0^{(\alpha)}$ are the heights of the respective barriers.

The total reaction cross section from reagents in the state i is therefore given by

$$\sigma_R(i) = \sigma_c(i) \frac{N_{\ddagger}^{(p)}(E - E_0^{(p)})}{N_{\ddagger}(E)} \tag{4.67}$$

The second factor in Eq. (4.67) is the branching fraction, where

$$N_{\ddagger}(E) = \sum_{\alpha} N_{\ddagger}^{(\alpha)}(E - E_0^{(\alpha)}) \tag{4.68}$$

Of course, the reaction cross section is *smaller* than that for complex formation, since the complex can dissociate back into the reagents.

The cross section for complex formation $\sigma_c(i)$ can be computed as in Section 4.4.9, since $\sigma_c(i)$ is the cross section for originating in state i and crossing once a configuration of no return. Using (4.59) and defining $T_c(i) = N_\ddagger(E)P_c(i)$,

$$\sigma(i \to f) = (\pi/k_i^2)N_\ddagger(E)P_c(i)P_c(f)$$

$$= \frac{\pi}{k_i^2} \frac{T_c(i)T_c(f)}{N_\ddagger(E)}. \tag{4.69}$$

But isn't that just Eq. (4.64)? No. Here $N_\ddagger(E)$, defined in Eq. (4.68), is the *sum* of the rates of passage for the *two* dissociation modes of the complex.

The formal similarity between Eqs. (4.64) and (4.69) should not hide the difference in the underlying chemistry. The total reaction cross section for compound processes is not given by $\sigma_c(i)$, which is the cross section for clearing the first configuration of no return.†

Explicit results, devoid of the dynamics, are now available for $k(E)$ in two limiting situations: (1) direct reactions with a single configuration of no return, where

$$hk(E) = N_\ddagger(E - E_0), \qquad E \geq E_0 \tag{4.57}$$

and (b) compound reactions with two configurations of no return separated by a deep well so that they act independently of one another:‡

$$hk(E) = \frac{N_\ddagger^{(r)}(E - E_0^{(r)})N_\ddagger^{(p)}(E - E_0^{(p)})}{N_\ddagger(E)}, \qquad E \geq E_0^{(p)}, E_0^{(r)} \tag{4.70}$$

with $N_\ddagger(E)$ given by Eq. (4.68).

4.4.11 Photoselective chemistry

With the advent of lasers considerable interest in photoselective chemistry has developed. If, however, the dissociation of a photoexcited molecule (which is our "complex") were independent of its mode of formation, then one could not control its fragmentation by pumping it in different ways as in Fig. 4.19. In other words, the rapid energy transfer among vibrational modes that is characteristic of compound dynamics precludes the possibility of observing selective chemistry following the deposition of initial excitation selectively in some mode (or region) of the molecule, e.g., Fig. 4.38.

---

† The branching fraction in Eq. (4.67) therefore represents the fraction of trajectories that cross the first configuration of no return on the reagents' side and do make it to the products. It can therefore be regarded as a correction factor, when the assumption of a strict no return is invalid. Note that this is a correction factor for the theory of *direct* reactions. The correction replaced Eq. (4.57) by $k(E) = \kappa(E)N_\ddagger(E - E_0)$, where $\kappa(E)$ is the *transmission factor* (just our branching fraction).

‡ An interesting problem is the intermediate region between Eqs. (4.57) and (4.70).

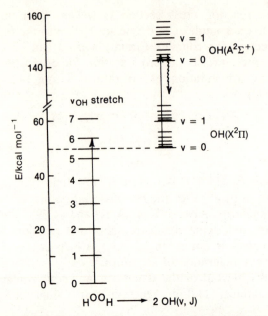

Fig. 4.38. Energy level diagram for the $6\nu_{OH}$ overtone vibration-induced decomposition of hydrogen peroxide and laser-induced fluorescence detection of the OH product. [Adapted from T. R. Rizzo, C. C. Hayden, and F. F. Crim, *J. Chem. Phys., 81*, 4501 (1984).]

One of the earliest experiments aimed at probing this aspect preceded lasers and used chemical activation instead. Energy-rich methylcyclopropane was formed both by the addition of triplet methylene to propene,

$$CH_2(^3\Sigma_g^-) + CH_3CH=CH_2 \longrightarrow CH_3-CH-CH_2$$
$$\underset{CH_2}{\diagdown\diagup}$$

and by the insertion reaction of singlet methylene into cyclopropane,

$$CH_2(^1A_1) + CH_2-CH_2 \longrightarrow CH_3-CH-CH_2$$
$$\qquad\quad\underset{CH_2}{\diagdown\diagup} \qquad\qquad\quad \underset{CH_2}{\diagdown\diagup}$$

The "energy-rich" methylcyclopropane is about equally energy rich by both routes, yet in the first one the energy is initially primarily localized in the ring and in the second one the "hot spot" is the side bond. The methylcyclopropane mostly† isomerizes to various butene isomers. The

---

† Why does it not revert back to the reagents? Because the threshold energy for isomerization is much lower and hence the excess energy at the appropriate transition state is higher. The number of internal states of the transition states for isomerization is therefore very much higher, since $N_{\ddagger}(x)$ increases with the excess energy x as x^{s-1}, where s is the number of vibrations in the complex. Isomerization has by far the higher rate.

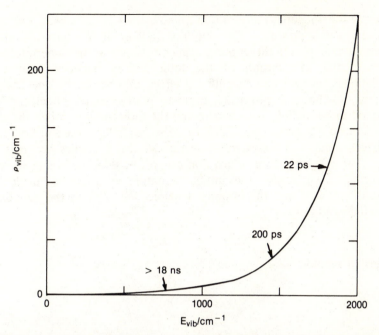

Fig. 4.39. The density of vibrational states of anthracene versus vibrational energy and characteristic times of intramolecular vibrational energy redistribution as determined from time-resolved fluorescence (on the picosecond scale). [Adapted from P. M. Felker and A. H. Zewail, *J. Chem. Phys.*, *82*, 2975 (1985), see also J. B. Hopkins, D. E. Powers, and R. E. Smalley, *J. Chem. Phys.*, *72*, 5039 (1980); *73*, 683 (1983).] Note that the excess vibrational energy in this experiment is far below that which is made available in a typical chemical activation experiment.

branching fractions for the different isomers are found to be about equal [differing only within the range allowed by the total energy not being quite the same in both reactions].

Subsequent experiments using chemical activation have verified and extended these results. Once an energy-rich molecule has survived about 10–100 vibrational periods, the initial excitation will be "scrambled" over many vibrational modes; that is, intramolecular vibrational energy redistribution is rapid. The rate of intramolecular vibrational energy distribution depends on both structural and energetic factors (see also Secs. 7.2 and 7.3). A clear-cut example of the structural aspect† is the study of the fluorescence of alkyl benzenes, $C_6H_5 \cdot (CH_2)_n CH_3$. Following (electronic and vibrational) excitation of the benzene ring, vibrational energy can move to the alkyl tail. The rate of this redistribution is found to depend on the length of the side chain. It is least for ethyl benzene ($n = 1$), and by $n = 4$ it is already in the picosecond time scale. The rate also increases with excess vibrational energy (Fig. 4.39).

† Which we can understand in terms of the density of states; see also Fig. 6.5 below.

Should we conclude, therefore, that one cannot excite a molecule so as to achieve selective products? Certainly not, in general. We have already seen numerous examples in direct-mode reactions where at the same total energy the particular distribution of the total energy between internal and translational energies had a significant effect. Of course, the time scales are different, so that one promising route to photoselective chemistry (to be discussed in Sec. 6.5.6) is to laser-pump the molecule just when the energy is required (i.e., during the collision). Also, as will be seen in Sec. 7.2, as the excess energy is considerably increased, the dynamics becomes more direct. There is, therefore, much that can yet be done, although so far it is clear that just because the initial excitation *is* selective, it does not necessarily follow that the ensuing dynamics will "retain the memory" of the initial conditions.

Suggested reading

Section 4.1

Books

Ben-Shaul, A., Haas, Y., Kompa, K. L., and Levine, R. D., *Lasers and Chemical Change*, Springer-Verlag, New York, 1981.

Birks, J. B. (ed.), *Organic Molecular Photophysics*, Vols. 1 and 2, Wiley, New York, 1973, 1975.

Cowan, D. O., and Drisko, R. L., *Elements of Organic Photochemistry*, Plenum, New York, 1984.

Duley, W. W., and Williams, D. A., *Interstellar Chemistry*, Academic, New York, 1984.

Faraday Discussions of the Chemical Society, Kinetics of State Selected Species, *67*, (1979).

Fontijn, A. (ed.), *Gas-Phase Chemiluminescence and Chemi-Ionization*, North-Holland, Amsterdam, 1985.

Fontijn, A., and Clyne, M. A. A. (eds.), *Reactions of Small Transient Species*, Wiley, New York, 1983.

Gardiner, W. C., Jr. (ed.), *Combustion Chemistry*, Springer-Verlag, New York, 1984.

Jortner, J., Levine, R. D., and Rice, S. A. (eds.), *Photoselective Chemistry, Adv. Chem. Phys.*, *47* (1981), Parts 1 and 2.

Lahmani, F. (ed.), *Photophysics and Photochemistry Above 6 eV*, Elsevier, Amsterdam, 1985.

McGowan, J. W. (ed.), *The Excited State in Chemical Physics*, Parts 1 and 2, Wiley, New York, 1975, 1981.

Polanyi, M., *Atomic Reactions*, Williams and Norgate, London, 1932.

Smith, I. W. M. (ed.), *Physical Chemistry of Fast Reactions: Reaction Dynamics*, Plenum, New York, 1980.

Turro, N. J., *Modern Molecular Photochemistry*, Benjamin, Menlo Park, 1978.

Wasserman, H. H., and Murray, R. W., *Singlet Oxygen,* Academic, New York, 1979.
Zewail, A. H. (ed.), *Advances in Laser Chemistry,* Springer-Verlag, Berlin, 1978.

Review Articles

Agrawalla, B. S., and Setser, D. W., " Hydrogen Abstraction Reactions of F, Cl and O Atoms Studied by Infrared Chemiluminescence and Laser-Induced Fluorescence in a Flowing Afterglow Reaction," in Fontijn (1985).
Basov, N. G., Oraevsky, A. N., and Pankratov, A. V., "Stimulation of Chemical Reactions with Laser Radiation," in Moore (1974).
Bauer, S. H., "How Energy Accumulation and Disposal Affect the Rates of Reactions." *Chem. Rev., 78*, 147 (1978).
Bauer, S. H., "Four Center Metathesis Reactions." *Annu. Rev. Phys. Chem., 30*, 271 (1979).
Bernstein, R. B., "State-to-State Chemistry via Molecular Beams and Lasers," in Glorieux *et al.* (1979).
Berry, M. J., "Chemical Laser Studies of Energy Partitioning into Chemical Reaction Products," in Levine and Jortner (1976).
Bersohn, R., "Final State Distributions in the Photodissociation of Triatomic Molecules," *J. Phys. Chem., 88*, 5145 (1984).
Campbell, I. M., and Baulch, D. L., "Chemiluminescence in the Gas Phase." *Spec. Per. Rep., 3*, 42 (1978).
Carrington, T., and Polanyi, J. C., "Chemiluminescent Reactions." *Int. Rev. Sci. Phys. Chem., 9*, 135 (1972).
Clary, D. C., and Henshaw, J. P., "Light-Heavy-Light Chemical Reactions," in Clary (1986).
Dalgarno, A., "Collisions in the Ionosphere." *Adv. At. Mol. Phys., 4*, 381 (1968).
Ferguson, E. E., Fehsenfeld, F. C., and Albritton, D. L., *"Ion Chemistry of the Earth's Atmosphere,"* in Bowers (1979).
Franklin, J. L., "Energy Distribution in the Unimolecular Decomposition of Ions," in Bowers (1979).
Frimer, A. A., "The Reaction of Singlet Oxygen with Olefins: The Question of Mechanism." *Chem. Rev., 79*, 359 (1979).
Golde, M. F., "Reactions of Electronically Excited Noble Gas Atoms." *Spec. Per. Rep., 2*, 13 (1976).
Green, S., "Interstellar Chemistry: Exotic Molecules in Space." *Annu. Rev. Phys. Chem., 32*, 103 (1981).
Grice, R., "Effect of Translational Energy on Reaction Dynamics." *Faraday Discuss. Chem. Soc., 67*, 16 (1979).
Hamilton, C. E., and Leone, S. R., "Nascent Product Vibrational State Distributions of Thermal Ion–Molecule Reactions Determined by Infrared Chemiluminescence," in Fontijn (1985).
Holmes, B. E., and Setser, D. W., "Energy Disposal by Chemical Reactions," in Smith (1980).
Houston, P. L., "Initiation of Atom–Molecule Reactions by Infrared Multiphoton Dissociation." *Adv. Chem. Phys., 47*, 625 (1981).
Kempter, V., "Electronic Excitation in Collisions Between Neutrals." *Adv. Chem. Phys., 30*, 417 (1975).

King, D. L., and Setser, D. W., "Reactions of Electronically Excited-State Atoms." *Annu. Rev. Phys. Chem., 27,* 407 (1976).

Kneba, M., and Wolfrum, J., "Bimolecular Reactions of Vibrationally Excited Molecules." *Annu. Rev. Phys. Chem., 31,* 47 (1980).

Knudtson, J. T., and Eyring, E. M., "Laser-Induced Chemical Reactions." *Annu. Rev. Phys. Chem., 25,* 255 (1974).

Lee, E. K. C., "Laser Photochemistry of Selected Vibronic and Rotational States." *Acc. Chem. Res., 10,* 319 (1977).

Leone, S. R., "Infrared Fluorescence: A Versatile Probe of State-Selected Chemical Dynamics." *Acc. Chem. Res., 16,* 88 (1983).

Leventhal, J. J., "The Emission of Light from Excited Products of Charge Exchange Reactions," in Bowers (1984).

McDonald, J. D., "Creation and Disposal of Vibrational Energy in Polyatomic Molecules." *Annu. Rev. Phys. Chem., 30,* 29 (1979).

McElroy, M. B., "Chemical Processes in the Solar System: A Kinetic Perspective." *Int. Rev. Sci. Phys. Chem., 9,* 127 (1975).

McGowan, J. W., Kummler, R. H., and Gilmore, F. R., "Excitation and De-excitation Processes Relevant to the Upper Atmosphere." *Adv. Chem. Phys., 28,* 379 (1975).

Marx, R., "Charge Transfers at Thermal Energies: Energy Disposal and Reaction Mechanisms," in Ausloos (1979).

Menzinger, M., "Electronic Chemiluminescence in Gases." *Adv. Chem. Phys., 42,* 1 (1980).

Moore, C. B., and Smith, I. W. M., "Chemical Reaction of Vibrationally Excited Molecules." *Faraday Discuss. Chem. Soc., 67,* 146 (1979).

Nicolet, M., "Atmospheric Chemistry." *Adv. Chem. Phys., 55,* 63 (1985).

Ogryzlo, E. A., "Chemiluminescent Association Reactions in the Upper Atmosphere," in Fontijn (1985).

Oldershaw, G. A., "Reactions of Photochemically Generated Hot Hydrogen Atoms." *Spec. Per. Rep., 2,* 96 (1976).

Ottinger, C., "Electronically Chemiluminescent Ion–Molecule Exchange Reactions," in Bowers (1984).

Ottinger, C., "Electronic Chemiluminescence from Ion-Molecule Reactions," in Fontijn (1985).

Porter, R. N., "State-to-State Considerations in Reactions in Interstellar Clouds," in Brooks and Hayes (1977).

Rowland, F. S., "Experimental Studies of Hot Atom Reactions." *Int. Rev. Sci. Phys. Chem., 9,* 109 (1972).

Russek, A., "Chemistry frcm Atom–Molecule Collisions," in Eichler *et al.* (1984).

Smalley, R. E., "Dynamics of Electronically Excited States." *Annu. Rev. Phys. Chem., 34,* 129 (1983).

Smith, I. W. M., "The Production of Excited Species in Simple Chemical Reactions." *Adv. Chem. Phys., 28,* 1 (1975).

Smith, I. W. M., "Reactive and Inelastic Collisions Involving Molecules in Selected Vibrational States." *Spec. Per. Rep., 2,* 1 (1976).

Smith, I. W. M., "Chemical Reactions of Selectively Energized Species," in Smith (1980).

Ureña, A. G., "Influence of Translational Energy upon Reactive Scattering Cross Section: Neutral–Neutral Collisions." *Adv. Chem. Phys. 66,* 214 (1987).

Watson, W. D., "Interstellar Chemistry." *Acc. Chem. Res., 10,* 221 (1977).

Watson, W. D., "Gas Phase Reactions in Astrophysics." *Annu. Rev. Astron. Astrophys., 16,* 585 (1978).

Whitehead, J. C., "The Distribution of Energy in the Products of Simple Reactions," in Bamford and Tipper (1983).

Wiesenfeld, J. R., "Atmospheric Chemistry Involving Electronically Excited Oxygen Atoms." *Acc. Chem. Res., 15,* 110 (1982).

Wilson, T., "Chemiluminescence in the Liquid Phase: Thermal Cleavage of Dioxetanes." *Int. Rev. Sci. Phys. Chem. 9,* 265 (1975).

Winn, J. S., "Chemiluminescent Chemi-ionization," in Fontijn (1985).

Wolfrum, J., "Atom Reactions," in Eyring et al. (1975).

Woodin, R. L., and Kaldor, A., "Enhancement of Chemical Reactions by Infrared Lasers." *Adv. Chem. Phys. 47,* 3 (1981).

Section 4.2

Books

Burkert, U., and Allinger, N. L., *Molecular Mechanics. Am. Chem. Soc.,* Monograph No. 177 (1982).

Davidovits, P., and McFadden, D. L.(eds.), *The Alkali Halide Vapors,* Academic, New York, 1979.

Faraday Discussions of the Chemical Society, 62, Potential Energy Surfaces, (1977).

Fukui, K., *Theory of Orientation and Stereoselection* (Springer Verlag, Berlin, 1970).

Herzberg, G., *Molecular Spectra and Molecular Structure,* Vol. II, Van Nostrand, New York, 1945.

Herzberg, G., *Molecular Spectra and Molecular Structure,* Vol. III, Van Nostrand, New York, 1966.

Hirst, D. M., *Potential Energy Surfaces: Molecular Structure and Reaction Dynamics,* Taylor & Francis, London, 1985.

Laidler, K. J., *Theories of Chemical Reaction Rates,* McGraw-Hill, New York, 1969.

Lester, W. A., Jr. (ed.), *Potential Energy Surfaces in Chemistry,* IBM, San Jose, Calif., 1970.

Murrell, J. N., Carter, S., Farantos, S. C., Huxley, P., and Varadas, A. J. C., *Molecular Potential Energy Functions,* Wiley, New York, 1984.

Pearson, R. G., *Symmetry Rules for Chemical Reactions,* Wiley, New York, 1976.

Salem, L., *Electrons in Chemical Reactions: First Principles,* Wiley, New York, 1982.

Schaefer, III, H. F., *Electronic Structure of Atoms and Molecules,* Addison-Wesley, New York, 1972.

Schaefer III, H. F., *Quantum Chemistry,* Clarendon Press, Oxford, 1984.

Truhlar, D. G. (ed.), *Potential Energy Surfaces and Dynamics Calculations,* Plenum New York, 1981.

Woodward, R. B., and Hoffmann, R., *The Conservation of Orbital Symmetry,* Verlag Chemie, Weinheim, 1971.

Review Articles

Balint-Kurti, G. G., "Potential Energy Surfaces for Chemical Reaction." *Adv. Chem. Phys., 30,* 137 (1975).

Bender, C. F., Meadows, J. H., and Schaefer III, H. F., "Potential Energy Surfaces for Ion–Molecule Reactions." *Faraday Discuss. Chem. Soc., 62*, 59 (1977).

Bersohn, R., "Reactions of Electronically Excited Atoms: Superalkalis in Super-halogens," in Levine and Jortner (1976).

Bottcher, C., "Excited-State Potential Surfaces and Their Applications." *Adv. Chem. Phys., 42*, 169 (1980).

Dunning, T. H., and Harding, H. B., "*Ab Initio* Determination of Potential Energy Surfaces for Chemical Reactions," in Baer (1985).

Dunning, T. H., Harding, L. B., Bair, R. A., Eades, R. A., and Shepard, R. L., "Theoretical Studies of the Energetics and Mechanisms of Chemical Reactions," in El-Sayed (1986).

Dykstra, C. E., "Potential Energy Barriers in Unimolecular Rearrangements." *Annu. Rev. Phys. Chem., 32*, 25 (1981).

Gordon, M. S., "Potential Energy Surfaces in Excited States of Saturated Molecules," in Truhlar (1981).

Hirst, D. M., "The Calculation of Potential Energy Surfaces for Excited States," *Adv. Chem. Phys., 50*, 517 (1982).

Jasinski, J. M., Frisoli, J. K., and Moore, C. B., "Unimolecular Reactions Induced by Vibrational Overtone Excitation." *Faraday Discuss. Chem. Soc., 75*, 289 (1983).

Kaufman, J. J., "Potential Energy Surface Considerations for Excited State Reactions." *Adv. Chem. Phys., 28*, 113 (1975).

Kuntz, P. J., "Features of Potential Energy Surfaces and Their Effect on Collisions," in Miller (1976).

Kuntz, P. J., "Interaction Potentials: Semiempirical Atom–Molecule Potentials for Collision Theory," in Bernstein (1979).

Kuntz, P. J., "Semiempirical Potential Energy Surfaces," in Baer (1985).

Mahan, B. H., "Electronic Structure and Chemical Dynamics," *Acc. Chem. Res., 8*, 55 (1975).

Malrieu, J. P., "Elementary Aspects of the Electronic Control of Photoreactions," in Glorieux et al. (1979).

Morokuma, K., and Kato, S., "Potental Energy Characteristics for Chemical Reactions," in Truhlar (1981).

Moylan, C. R., and Brauman, J. L., "Gas Phase Acid–Base Chemistry." *Annu. Rev. Phys. Chem., 34*, 187 (1983).

Scrocco, E., and Tomasi, J., "Electronic Molecular Structure, Reactivity and Intermolecular Forces: An Heuristic Interpretation by Means of Electrostatic Molecular Potentials." *Adv. Quant. Chem., 11*, 116 (1978).

Setser, D. W., Dreiling, T. D., Brashears, H. C., Jr., and Kolts, J. H., "Analogy Between Electronically Excited State Atoms and Alkali Metal Atoms." *Faraday Discuss Chem. Soc., 67*, 255 (1979).

Sidis, V., and Dowek, D., "Extension of the MO Promotion Model to Atom–Molecule Collisions Using Cubic and Cylindric Correlation Diagrams," in Eichler et al. (1984).

Simons, J., "Roles Played by Metastable States in Chemistry," in Truhlar (1984).

Simons, J., Jorgensen, P., Taylor, H., and Ozment, J., "Walking on Potential Energy Surfaces." *J. Phys. Chem., 87*, 2745 (1983).

Truhlar, D. G., Brown, F. B., Steckler, R., and Isaacson, A. D., "The Representation and Use of Potential Energy Surfaces in the Wide Vicinity of a Reaction Path for Dynamics Calculations on Polyatomic Reactions," in Clary (1986).

Tully, J. C., "Semiempirical Diatomics-in-Molecules Potential Energy Surfaces." *Adv. Chem. Phys., 42*, 63 (1980).

Zimmerman, H. E., "Some Theoretical Aspects of Organic Photochemistry." *Acc. Chem. Res., 15*, 312 (1982).

Section 4.3

Review Articles

Bunker, D. L., "Classical Trajectory Methods." *Methods Comput. Phys., 10*, 287 (1971).

Kuntz, P. J., "Collision-Induced Dissociation: Trajectories and Models," in Bernstein (1979).

Mahan, B. H., "Collinear Collision Chemistry. I. A Simple Model for Inelastic and Reactive Collision Dynamics." *J. Chem. Educ., 51*, 308 (1974).

Muckerman, J. T., "Applications of Classical Trajectory Techniques to Reactive Scattering," in Henderson (1981).

Pattengill, M. D., "Rotational Excitation: Classical Trajectory Methods," in Bernstein (1979).

Polanyi, J. C., "Some Concepts in Reaction Dynamics." *Acc. Chem. Res., 5*, 161 (1972).

Porter, R. N., "Molecular Trajectory Calculations." *Annu. Rev. Phys. Chem., 25*, 317 (1974).

Porter, R. N., and Raff, L. M., "Classical Trajectory Methods in Molecular Collisions," in Miller (1976).

Raff, L. M., and Thompson, D. L., "Classical Trajectory Approach to Reactive Scattering," in Baer (1985).

Sathyamurthy, N., "Effect of Reagent Rotation on Elementary Bimolecular Exchange Reactions." *Chem. Rev., 83*, 601 (1983).

Schatz, G. C., "Overview of Reactive Scattering," in Truhlar (1981).

Schatz, G. C., "Quasiclassical Trajectory Studies of State to State Collisional Energy Transfer in Polyatomic Molecules," in Bowman (1983).

Schlier, Ch., "Trajectory Calculations and Complex Collisions," in Hinze (1983).

Smith, I. W. M., "The Collision Dynamics of Vibrationally Excited Molecules." *Chem. Soc. Rev., 14*, 141 (1985).

Thompson, D. L., "Quasiclassical Trajectory Studies of Reactive Energy Transfer." *Acc. Chem. Res., 9*, 338 (1976).

Truhlar, D. G., and Dixon, D. D., "Direct-Mode Chemical Reactions II: Classical Theories," in Bernstein (1979).

Truhlar, D. G., and Muckerman, J. T., "Reactive Scattering Cross Sections: Quasiclassical and Semiclassical Methods," in Bernstein (1979).

Section 4.4

Baer, M. (ed.) *The Theory of Chemical Reaction Dynamics*, CRC Press, Boca Raton, Fla., 1985.

Benson, S. W., *Thermochemical Kinetics*, 2nd ed., Wiley, New York, 1976.

Christov, S. G., *Collision Theory and Statistical Theory of Chemical Reactions*, Springer-Verlag, Berlin, 1980.

Clary, D. C. (ed.), *The Theory of Chemical Reaction Dynamics*, Reidel, Boston, 1986.

Faraday Discussions of the Chemical Society, 75, Intramolecular Kinetics (1983).

Forst, W., *Theory of Unimolecular Reactions*, Academic, New York, 1973.

Glasstone, S., Laidler, K. J., and Eyring, H., *The Theory of Rate Processes*, McGraw-Hill, New York, 1941.

Robinson, P. J., and Holbrook, K. A., *Unimolecular Reactions*, Wiley, New York, 1972.

Review Articles

Aquilanti, V., "Resonances in Reactions: A Semiclassical View," in Clary (1986).

Baer, M., and Kouri, D. J., "The Sudden Approximation for Reactions," in Clary (1986).

Baer, T., "The Dissociation Dynamics of Energy Selected Ions." *Adv. Chem. Phys., 64*, 111 (1986).

Basilevsky, M. V., "Transition State Stabilization Energy as a Measure of Chemical Reactivity." *Adv. Chem. Phys., 33*, 345 (1975).

Bigeleisen, J., and Wolfsberg, M., "Theoretical and Experimental Aspects of Isotope Effects in Chemical Kinetics." *Adv. Chem. Phys., 1*, 15 (1958).

Bloembergen, N., and Zewail, A. H., "Energy Redistribution in Isolated Molecules and the Question of Mode-Selective Laser Chemistry Revisited." *J. Phys. Chem. 88*, 5459 (1984).

Callear, A. B., "Basic RRKM Theory," in Bamford and Tipper (1983).

Chesnavich, W. J., and Bowers, M. T., "Statistical Methods in Reaction Dynamics," in Bowers (1979).

Connor, J. N. L., "The Distorted Wave Theory of Chemical Reactions," in Clary (1986).

Crim, F. F., "Selective Excitation Studies of Unimolecular Reaction Dynamics." *Annu. Rev. Phys. Chem., 35*, 657 (1984).

Gardiner, W. C., Jr., and Olson, D. B., "Chemical Kinetics of High Temperature Combustion." *Annu. Rev. Phys. Chem., 31*, 377 (1980).

Gardiner, W. C., Jr., and Troe, J., "Rate Coefficients of Thermal Dissociation, Isomerization and Recombination Reactions," in Gardiner (1984).

Hase, W. L., "Variational Unimolecular Rate Theory." *Acc. Chem. Res., 16*, 258 (1983).

Jortner, J., and Levine, R. D., "Photoselective Chemistry." *Adv. Chem. Phys., 47*, 1 (1981).

Kaufman, F., "Kinetics of Elementary Radical Reactions in the Gas Phase." *J. Phys. Chem., 88*, 4909 (1984).

Kaufman, F., "Rates of Elementary Reactions: Measurements and Applications." *Science, 230*, 393 (1985).

Keck, J. C., "Variational Theory of Reaction Rates," *Adv. Chem. Phys., 13*, 85 (1967).

Kouri, D. J. and Baer, M., "Arrangement Channel Quantum Mechanical Approach to Reactive Scattering," in Clary (1986).

Laidler, K. J., and King, M. C., "The Development of Transition-State Theory." *J. Phys. Chem., 87*, 2657 (1983).

Light, J. C., "Complex-Mode Chemical Reactions: Statistical Theories of Bimolecular Reactions," in Bernstein (1979).

Light, J. C. "The R-Matrix Method," in Clary (1986)

Marcus, R. A., "On the Theory of Intramolecular Energy Transfer." *Faraday Discuss. Chem. Soc.*, *75*, 103 (1983).

Marcus, R. A., "Statistical Theory of Unimolecular Reactions and Intramolecular Dynamics." *Laser Chem.*, *2*, 203 (1983).

McIver, J. W., Jr., "The Structure of Transition States: Are They Symmetric?" *Acc. Chem. Res.*, *7*, 721 (1974).

Menzinger, M., and Wolfgang, R., "The Meaning and the Use of the Arrhenius Activation Energy." *Angew. Chem.*, *8*, 438 (1969).

Miller, W. H., "Importance of Nonseparability in Quantum Mechanical Transition-State Theory," *Acc. Chem. Res.*, *9*, 306 (1976).

Miller, W. H., "Symmetry-Adapted Transition-State Theory: Nonzero Total Angular Momentum," *J. Phys. Chem.*, *87*, 2731 (1983).

Miller, W. H., "On the Question of Mode-Specificity in Unimolecular Reaction Dynamics." *Laser Chem.*, *2*, 243 (1983).

Miller, W. H., "Reaction Path Models for Polyatomic Reaction Dynamics—From Transition State Theory to Path Integrals," in Clary (1986).

Noid, D. W., Koszykowski, M. L., and Marcus, R. A., "Quasiperiodic and Stochastic Behavior in Molecules." *Annu. Rev. Phys. Chem.*, *32*, 267 (1981).

Oref, I., and Rabinovitch, B. S., "Do Highly Excited Reactive Polyatomic Molecules Behave Ergodically?" *Acc. Chem. Res.*, *12*, 166 (1979).

Parmenter, C. S., "Vibrational Energy Flow Within Excited Electronic States of Large Molecules." *J. Phys. Chem.*, *86*, 1735 (1982).

Pechukas, P., "Statistical Approximations in Collision Theory," in Miller (1976).

Pechukas, P., "Transition State Theory." *Annu. Rev. Phys. Chem.*, *32*, 159 (1981).

Pollak, E., "Periodic Orbits and the Theory of Reactive Scattering," in Baer (1985); in Clary (1986)

Quack, M., and Troe, J., "Unimolecular Reactions and Energy Transfer of Highly Excited Molecules." *Spec. Per. Rep.*, *2*, 175 (1976).

Quack, M., and Troe, J., "Statistical Methods in Scattering," in Henderson (1981).

Rice, S. A., "An Overview of the Dynamics of Intramolecular Transfer of Vibrational Energy." *Adv. Chem. Phys.*, *47*, 117 (1981).

Rizzo, T. R., Cannon, B. D., McGinley, E. S., and Crim, F. F., "Local Mode Excitation for Time Resolved Unimolecular Decay Studies." *Laser Chem.*, *2*, 321 (1983).

Römelt, J., "Calculations on Collinear Reactions using Hyperspherical Coordinates," in Clary (1986).

Ross, J., Light, J. C., and Schuler, K. E., "Rate Coefficients, Reaction Cross Sections, and Microscopic Reversibility," in Hochstim (1969).

Schatz, G. C., "Recent Quantum Scattering Calculations on the $H + H_2$ Reaction and its Isotopic Counterparts," in Clary (1986).

Smalley, R. E., "Vibrational Randomization Measurements with Supersonic Beams." *J. Phys. Chem.*, *86*, 3504 (1982).

Troe, J., "Unimolecular Reactions." *Int. Rev. Sci. Phys. Chem.*, *9*, 1 (1975).

Troe, J., "Atom and Radical Recombination Reactions." *Annu. Rev. Phys. Chem.*, *29*, 223 (1978).

Troe, J., "Inter- and Intramolecular Dynamics of Vibrationally Highly Excited Polyatomic Molecules," in Jortner and Pullman (1982).

Truhlar, D. G., and Garrett, B. C., "Variational Transition State Theory." *Annu. Rev. Phys. Chem., 35*, 159 (1984).

Truhlar, D. G., and Garrett, B. C., "Variational Transition-State Theory." *Acc. Chem. Res. 13*, 440 (1980).

Truhlar, D. G., and Wyatt, R. E., "History of H_3 Kinetics," *Annu. Rev. Phys. Chem., 27*, 1 (1976).

Truhlar, D. G., Isaacson, A. D., Skodje, R. T., and Garrett, B. C., "Incorporation of Quantum Effects in Generalized Transition-State Theory." *J. Phys. Chem., 86*, 2252 (1982).

Truhlar, D. G., Hase, W. L., and Hynes, J. T., "Current Status of Transition-State Theory." *J. Phys. Chem., 87*, 2664 (1983).

Truhlar, D. G., Isaacson, A. D., and Garrett, B. C., "Generalized Transition State Theory." in Baer (1985).

Walker, R. B., and Hayes, E. F., "Reactive Scattering in the Bending-Corrected Rotating Linear Model," in Clary (1986).

Zellner, R., "Bimolecular Reaction Rate Coefficients," in Gardiner (1984).

5
The practice of molecular reaction dynamics

We have thus far touched upon many of the experimental findings of molecular reaction dynamics and have briefly outlined some of the theoretical ideas used in their interpretation. It is time to dig deeper. What is it that we would like to measure, what are the experiments that we can actually do, what information are we likely to obtain, and what theoretical techniques are available for the interpretation of such experiments and for the design of new ones? In the course of this chapter we hope to convey the essence, if not the details, of the current methodologies employed by the practitioners of molecular reaction dynamics.

5.1†. הרצוי לעומת המצוי

There is still something of a "technology gap" between the very detailed experiments we would *like* to do and the experiments we are *able* to do. In this section we first consider the main elements in an idealized experiment and then examine the currently available methods to see how closely they approach the realization of the ideal.

5.1.1 Desiderata

To keep the discussion simple, we confine our attention to an elementary reaction involving only three atoms, say,

$$H + ICl \rightarrow HI + Cl$$

Even in this simple case there is an alternative energetically possible reaction path,

$$H + ClI \rightarrow HCl + I$$

There will also be concomitant nonreactive (both elastic and inelastic) collisions. For each of the two reaction paths, the halogen atom may be

† The wish and the realization.

formed in its ground ($^2P_{3/2}$) or spin-orbit excited ($^2P_{1/2}$) electronic state, and the product diatomic, even though in the ground ($X^1\Sigma$) electronic state, may be in any of the large number of internal (vibrotational) states allowed by energy conservation.

A complete specification of the collision thus requires not only product identification (HCl versus HI), but also a knowledge of the reactants' and products' internal states, for example,

$$H(^2S) + ICl(vj) \rightarrow HI(v'j') + Cl(^2P_{1/2})$$

at some specified collision energy. In an ultimate experiment with aligned molecules, we would require specification of the quantum numbers m_j (and $m_{j'}$) that determine the orientation of the reactant (and product) molecule with respect to some fixed direction.† Usually, the more refined experiments employ molecular beam-laser methods, as will be seen in subsequent examples.

In the ultimate experiment one would determine the cross section and angular distribution of the products for a completely specified collision. In practice, we aim for the most detailed observations suitable for comparison with theory. This means (roughly in order of increasing detail) total reaction cross section at specified collision energies, angular distribution of products (if possible, accompanied by velocity distribution of products‡), internal-state analysis of the products, internal state selection of reactants, selection of orientation or of alignment of reactants, and analysis of the alignment (i.e., the polarization of angular momentum) of the products. Furthermore, the different attributes are by no means independent and we need to know their correlation!

As yet no single experiment has implemented all of these desiderata simultaneously, but *all* the above requirements (separately or in combination of several) have been realized at the present time. It is obvious that one of the serious limitations is detection sensitivity: The more stringent the specifications, the fewer molecules to be detected and the weaker the signal at the detector relative to "noise."§ The considerable progress in resolution has been accompanied by progress in another direction, that of increasing reactant complexity. Lasers, which have played an important role in providing better resolution in experiments with simple reactants, were instrumental in opening up this second front.

Of course, for larger polyatomics one does not require quite the same resolution of detail. With complex reagents or products (with many internal degrees of freedom), the more extensive summations (over final states) and

---

† Usually, the relative velocity vector. The orientation of the electronic orbitals for non-S states can also be determined. It is even not completely futuristic to specify the nuclear spin states.

‡ Given the final relative translational energy distribution, the internal energy of the products is determined via the conservation of energy.

§ We have often noted the corresponding limitations on computational methods.

averaging (over initial states) lead to results that are often not very deviant from prior (statistical) expectations. Moreover, collisions of larger molecules occur more often by a complex rather than by a direct mode (see Sec. 7.2). We have more modest goals for larger systems, simply to know "what the primary products *are*." An example is the crossed molecular beam experiment of O atoms with hydrocarbons,

$$O(^3P) + C_6H_5CH_3 \rightarrow \begin{cases} C_6H_5O + CH_3 \\ C_7H_7O + H \end{cases}$$

$$O(^3P) + C_6H_6 \rightarrow \begin{cases} C_6H_5O + H \\ C_6H_5OH \end{cases}$$

which provided definitive nascent products identification (unexpectedly, neither CO nor OH elimination products are detected as the primary products), as well as angular and recoil energy distributions (Fig. 5.1).

Let us now look at some of the specific experimental methods and typical results.

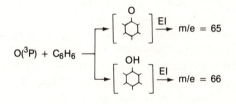

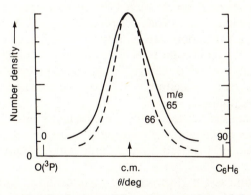

Fig. 5.1. Laboratory angular distributions of the primary products of the reaction of $O(^3P)$ with C_6H_6 ($E_T = 6.5\,\text{kcal mol}^{-1}$). One of these is C_6H_5OH, and the other C_6H_5O (plus H, undetected). Because of fragmentations of the parent ions in the electron-impact ionization process used for detection, the ions actually observed are the principal fragments of each of the above products. The angular distributions are peaked near the centroid angle, implying small center-of-mass product recoil velocities (zero in the case of the C_6H_5OH adduct). [Adapted from S. J. Sibener, R. J. Buss, P. Casavecchia, T. Hirooka, and Y. T. Lee, *J. Chem. Phys.*, *72*, 4341 (1980).]

5.2 Molecules, radiation, and laser interactions

The use of photons as "reagents", the detection of photons as "products," and, more recently, photons as "catalysts" are central to the current practice.

5.2.1 Chemiluminescence

This technique involves spectroscopic analysis of the radiation emitted from excited, nascent products of an elementary chemical reaction, with the goal of measuring the relative populations of the excited internal states of these product molecules. Some of the most detailed work has come from systems in which hydrogen halides are formed, emitting a resolvable vibrotational spectrum in the infrared. Other experiments have yielded vibrational state distributions, via the visible and ultraviolet emission from *excited* (*) electronic state products, for example,

$$Ca + F_2 \rightarrow CaF^*(v', j') + F$$

A typical infrared chemiluminescence experiment is carried out in a fast-flow system, under steady-state conditions. Because of the problem of collisional relaxation (mainly of the rotational distribution), it is necessary to work under very low pressure conditions. Typically two uncollimated jets (or, better yet, crossed beams) of reactants intersect and are pumped away rapidly. The radiation emitted from the reaction zone is collected and focused into an infrared spectrometer. From measurements of the relative intensities of the lines for various v,j transitions, and with a knowledge of the relevant Franck–Condon factors, relative populations of the parent state can be inferred. It is important to arrest the normally rapid relaxation of the nascent rotational excitation pattern within a given vibrational manifold. By extrapolation to "zero relaxation," it is possible to deduce reliable nascent internal state distributions.

The chemiluminescence technique yields *relative* rates for the formation of product states of known v',j'; however, absolute values of overall rate coefficients can be carried over from other (more conventional) experiments.

When the products are electronically excited, the radiative lifetime is far shorter, leading to higher emission intensity. It is then possible not only to monitor state populations but also even more detailed aspects, such as the polarization of the emission that provides information on the alignment of the products (see Sec. 7.6). Figure 5.2 shows a chemiluminescence spectrum with assignment of the vibrational transitions and the polarization distribution for CaF* from the Ca + F$_2$ reaction.

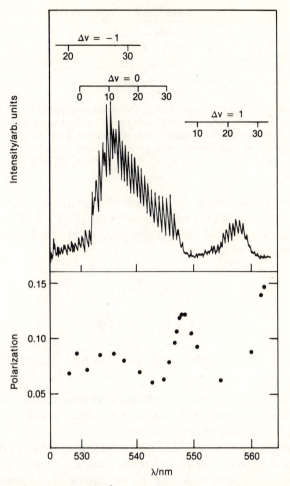

Fig. 5.2. Chemiluminescence of $CaF(B^2\Sigma^+)$ from the reaction of $Ca + F_2$. (Top) Chemiluminescence spectrum, showing the $\Delta v = 0$ main sequence (as well as the $\Delta v = \pm 1$ sequences). (Bottom) Polarization of the chemiluminescence, from which the alignment of the nascent CaF can be deduced (see Sec. 7.6) using DIPR model (Appendix 4C) computations. [Adapted from M. G. Prisant, C. T. Rettner, and R. N. Zare, *Chem. Phys. Lett.*, *88*, 271 (1982); *J. Chem. Phys. 81*, 2699 (1984).]

5.2.2 Chemical lasers

The chemical laser represents a marriage of the method of chemiluminescence and that of time-resolved (or kinetic) spectroscopy (see Sec. 6.1). The light emitted from the excited product molecules is monitored as a function of time following the initiation of the reaction. To achieve "lasing", a reaction that provides population inversion is carried out in an optical cavity, leading to stimulated emission of photons (see Appendix 6A). There

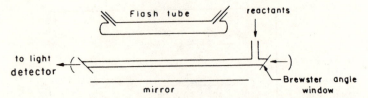

Fig. 5.3. Schematic diagram of a flash-initiated chemical laser. After filling the cavity with the proper mixture of reactants (in excess inert buffer gas), the flash lamp is fired and an intense light pulse photolyzes one of the precursor molecules, yielding the reactive species (atom or radical). This initiates the population-inverting reaction that leads to lasing. [Adapted from K. L. Kompa, J. H. Parker, and G. C. Pimentel, *J. Chem. Phys.*, *49*, 4257 (1968).]

are several ways of initiating the reaction, the most common being by a flash of light that is absorbed by a reactant molecule (Fig. 5.3). For example, one source of F atoms is CF_3I. In the presence of H_2 (and a buffer gas), the following processes ensue:

$$CF_3I \xrightarrow[\text{flash}]{h\nu(\text{UV})} F + CF_2I$$

$$F + H_2 \rightarrow H + HF^\dagger$$

$$HF^\dagger \xrightarrow[\text{stimulated}]{h\nu} HF + h\nu$$

The excited HF emits an infrared photon that then stimulates the emission of another nascent excited HF molecule to produce still another photon, and so on, thereby producing the cascade.

To achieve lasing, however, there must be enough *gain* in the cascade to overcome losses (due to window reflections, etc.). This gain depends on the extent of population inversion, that is, the excess of the population of the different vibrational levels over the equilibrium distribution of these levels. By varying the operating conditions and measuring the gain for the various transitions, it is possible to estimate the relative vibrational population distribution† of the nascent products. Figure 5.4 shows the time development of the laser emission for the $F + H_2$ reaction.

Many reactions have produced chemical lasers, all relying upon population inversion in the products for their operation. Table 5.1 lists the major types of processes that have been employed, mainly in the infrared. Indeed, even the burning of CS_2 in air has been used to make a CO chemical laser, providing an example of an "all-chemical" laser with no external pumping power required.

---

† Because of the relatively high pressure employed in many such experiments, rotational relaxation (within a given vibrational manifold) is essentially complete on the microsecond time scale of the experiment.

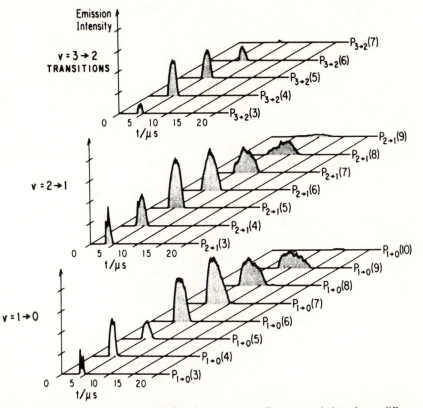

Fig. 5.4. Temporal evolution of laser pulses corresponding to emission from different vibrotational states of HF formed in the flash-initiated reaction [adapted from M. J. Berry, *J. Chem. Phys.*, *59*, 6229 (1973)]. Here $P_{v' \to v'-1}(j)$ denotes the $(v, j-1) \to (v-1, j)$ transition. The complete solution of the time dependence would require consideration of the whole "ladder" of coupled vibrational levels of the nascent HF molecules. The following processes need to be considered for each level: one gain term due to the formation of the state by the reaction and another for its formation via radiative decay from an upper state, a radiative decay term for the state, and a bimolecular $R \to R'$ or $R \to T$ term. The novel feature here is the radiative cascade among the rotational levels. (Because of the short time span, the vibrational population is essentially unrelaxed.)

To reduce the need for electrical pumping, chain reactions can also be used. In the H_2–F_2 mixture, the so-called HF chain is very effective:

$$F + H_2 \to HF\,(v \leq 3) + H$$

$$H + F_2 \to HF\,(v \leq 9) + F$$

The analogous reactions for Cl and Br are energetically unfavorable.†

---

† The $Cl + H_2$ reaction is just endothermic ($\Delta H^0 \approx 4\,\text{kJ mol}^{-1}$) and the $Br + H_2$ reaction is quite endothermic ($\Delta H^0 \approx 70\,\text{kJ mol}^{-1}$), so that they are too slow for the required fast chain propagation.

Table 5.1. Chemical lasers[a]

Classification	Reactions
Photodissociation	$IBr \rightarrow Br^* + I$
	$RI \rightarrow I^* + R$
	$H_3CNC \rightarrow CN^* + CH_3$
	$Cs_2 \rightarrow Cs^* + Cs$
	$OCS \rightarrow CO^\dagger + S$
	$H_2CCO \rightarrow CO^\dagger + CH_2$
	$ClNO \rightarrow NO^\dagger + Cl$
	$(CN)_2 \rightarrow CN^\dagger + CN$
	$HFCO \rightarrow HF^\dagger + CO$
Photoelimination	$H_2C{=}CHF \rightarrow HF^\dagger + C_2H_2$
	$H_2C{=}CHCl \rightarrow HCl^\dagger + C_2H_2$
	$H_3C{-}CCl_3 \rightarrow HCl^\dagger + H_2C{=}CCl_2$
Atom–diatom exchange	$F + H_2 \rightarrow HF^\dagger + H$
	$H + F_2 \rightarrow HF^\dagger + F$
	$Cl + HI \rightarrow HCl^\dagger + I$
	$Cl + HBr \rightarrow HCl^\dagger + Br$
	$CN + H_2 \rightarrow HCN^\dagger + H$
	$O + CH \rightarrow CO^\dagger + H$
	$O + CS \rightarrow CO^\dagger + S$
	$O + CN \rightarrow CO^\dagger + N$
	$O^* + CN \rightarrow CO^\dagger + N^*$
	$Xe^* + F_2 \rightarrow XeF^* + F$
Diatomic exchange	$O_2 + CH \rightarrow CO^\dagger + OH$
	$NO + CH \rightarrow CO^\dagger + NH$
Abstraction	$F + CH_4 \rightarrow HF^\dagger + CH_3$
	$H + ClN_3 \rightarrow HCl^\dagger + N_3$
	$H + O_3 \rightarrow OH^\dagger + O_2$
	$Ar + F_3N \rightarrow ArF^* + NF_2$
	$Xe_2^* + N_2O \rightarrow XeO^* + N_2 + Xe$
Insertion–elimination	$O + CHF \rightarrow HF^\dagger + CO$
	$O^* + CHF_3 \rightarrow HF^\dagger + F_2CO$
Addition–elimination	$O + H_2CCHF \rightarrow HF^\dagger + H_2CCO$
	$NF^* + C_2H_4 \rightarrow HF^\dagger + H_3CCN$
	$CH_3 + CF_3 \rightarrow HF^\dagger + C_2H_2F_2$
Chain reaction	$\begin{cases} F + H_2 \rightarrow HF^\dagger + H \\ H + F_2 \rightarrow HF^\dagger + F \end{cases}$
Energy transfer	$N_2^\dagger + CO_2 \rightarrow CO_2^\dagger + N_2$
	$DF^\dagger + CO_2 \rightarrow CO_2^\dagger + DF$
	$O_2^* + I \rightarrow I^* + O_2$
Radiative association	$O^* + Ar \rightarrow ArO^*$

[a] Adapted from Ben Shaul et al. (1981), p. 292.

5.2.3 Laser pump and laser probe experiments

The well-defined frequency, high power, and high degree of collimation of a laser beam make it an ideal source of photons, both for the preparation of the reagents and the "interrogation" of the products. A typical arrangement for a laser-crossed molecular beam experiment is shown in Fig. 5.5. Two crossed molecular beams are employed. The pump laser intersects one beam just before the collision region. The high flux of laser photons pumps a significant fraction of molecules to some desired quantum state. The sharply selected frequency (and tunability) ensures that the upper state is well defined (and, since the laser is tunable, different excited states can be reached). If the laser light is linearly polarized, the electric field **E** is in a direction perpendicular to the wave vector of the light. Thus, molecules that have absorbed† are aligned in the laboratory framework.

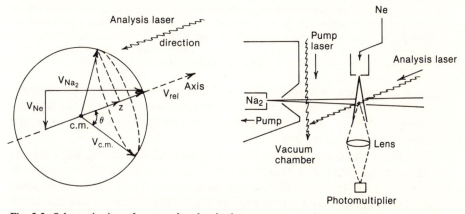

Fig. 5.5. Schematic view of a crossed molecular beam apparatus with a pump laser and a probe laser for state-to-state cross-section measurements. (The pump laser and optic axis are perpendicular to the plane of the figure.) [Adapted from J. A. Serri, C. H. Becker, M. B. Elbel, J. L. Kinsey, W. P. Moskowitz, and D. E. Pritchard, *J. Chem. Phys.*, *74*, 5116 (1981); see also K. Bergmann, U. Hefter, and J. Witt, *J. Chem. Phys.*, *72*, 4777 (1980).] For typical results, see Fig. 3.13.

The probe laser interrogates the collision products. Most collisions lead to products in their ground electronic state but with varying amounts of vibrotational excitation. A very common technique is that of *laser-induced fluorescence* (see Appendix 5A). In this procedure, a tunable laser is scanned over the appropriate wavelength region in the visible or near-ultraviolet, electronically exciting the product molecules whose vibrotational distribution is sought. Their fluorescence is monitored as they radiate back to the ground electronic state. Since the radiative lifetimes of electronically

---

† The absorption is proportional to $|\mu \cdot \mathbf{E}|^2$, where μ is the molecular transition dipole, which is fixed in the molecular frame. Note that the alignment is with respect to **E** and hence is defined in the laboratory frame.

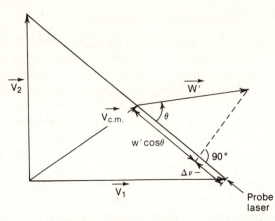

Fig. 5.6. Arrangement of a Doppler shift experiment to measure the velocity of (state-selected) products from a crossed molecular beam reaction. The probe laser is directed along the relative velocity vector and the Doppler shift of the laser-induced fluorescence from the nascent product (in a specific internal state) is measured. This $\Delta \nu$ is proportional to $(\cos \theta - 1)$ and thus determines $w' \cos \theta$, where w' is the c.m. recoil velocity of the product and θ its scattering angle. [Adapted from J. L. Kinsey, *J. Chem. Phys.*, **66**, 2560 (1977).]

excited species are short (e.g., 10^{-8} s), the molecules being probed will have hardly moved during the observation. Fluorescence lifetimes in the infrared are long ($>10^{-3}$ s), so that the molecule has ample opportunity to escape from the observation region.

The probe laser can also be linearly polarized. Observing the Doppler shift of the transition can then yield the (vector) velocity distribution of the product flux. With orientation of the wave vector of the probe laser along the initial relative velocity, the Doppler shift is due to the component of the final relative velocity parallel to the initial relative velocity (Fig. 5.6). In other words, the Doppler shift can determine directly the angular distribution in the center-of-mass (c.m.) system. Figure 5.7 shows the results of such a study for an atom–molecule reaction.

In a photodissociation experiment there is only one molecular beam. The pump laser is used to dissociate the molecule. Sometimes the dissociation is achieved using a single photon. However, at the very high powers currently available, it is possible to achieve multiple photon absorption† (see Sec. 7.3). The probe in a photodissociation experiment can be another laser‡ or an electron-impact mass spectrometer.

---

† Even probing reaction products using multiphon processes is feasible. This is particularly important when laser-induced fluorescence is ruled out by the absence of a suitable upper electronic state.

‡ An important practical consideration in the use of lasers is the magnitude of the beam intensity at a given wavelength. The energy content of a photon (with λ in nm) is

$$\epsilon = hc/\lambda = 1.987 \times 10^{-16}/\lambda \ \text{J} = 1.240 \times 10^3/\lambda \ \text{eV}$$

$NO_2 + H \rightarrow OH$ (v = 0, K = 17, J = 17.5) + NO

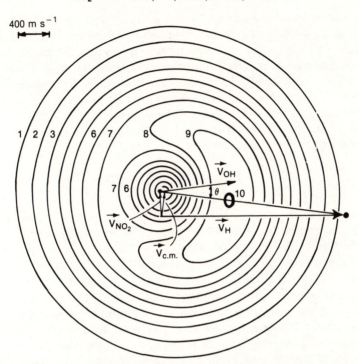

Fig. 5.7. Experimental results on the (vector) velocity distribution of the flux of OH in the specified state from the reaction of $H + NO_2$, using the Doppler shift laser-induced fluorescence technique. The initial laboratory velocity vectors of the H atoms and NO_2 molecules are shown, scattered at a c.m. angle θ (from the relative velocity vector). The maximal flux distribution is in the forward direction. Contours at successively smaller intensities are also indicated. [Adapted from E. J. Murphy, J. H. Brophy, G. S. Arnold, W. L. Dimpfl and J. L. Kinsey, *J. Chem. Phys. 70*, 5910 (1979).]

Thus, the number of photons per laser pulse (of j J) is

$$N = j/\epsilon = 5.034 \times 10^{15} j \lambda$$

For a cw laser the intensity (number of photons s^{-1} for a power of P W at the wavelength λ) is therefore

$$I = 5.034 \times 10^{15} P \lambda$$

Consider as an example a pulsed visible laser (e.g., a frequency doubled Nd:YAG) at $\lambda = 532$ nm delivering 100-mJ pulses at a repetition rate of 10 Hz. Then $N = 2.68 \times 10^{17}$ photons (of energy 2.33 eV) per pulse. For a pulse width of 5 ns the average power (during the pulse) will be 2×10^7 W, so $I = 5.36 \times 10^{25}$ photons s^{-1} (over the pulse duration). Note that for cw operation at $P = 1$ W, $I = 2.68 \times 10^{18}$ photons s^{-1}.

5.2.4 Photofragmentation spectroscopy

This is a technique for studying the dynamics of photodissociation. What is the fate of a molecule after it has absorbed a photon whose energy exceeds the threshold for dissociation of one or more bonds? In the simplest case the molecule undergoes an electronic transition from its ground state to an antibonding, "repulsive" electronic state that leads to dissociation. The fragments recoil in opposite directions (in the c.m. system), sharing the total available energy $E = h\nu - D$ between them. Some of this energy can go into internal excitation of the fragments, for example, vibrotational or even electronic excitation. An example is

$$CH_3I \xrightarrow[\text{266 nm}]{h\nu} CH_3 + I(^2P)$$

where a large fraction of the I atoms are in the excited ($^2P_{1/2}$) spin-orbit state and the CH_3 radical has its "umbrella mode" vibrationally excited (emitting infrared chemiluminescence). The balance of the available energy goes into translational recoil. An important goal is to deduce this partitioning of the available energy. This can be carried out by measuring the translational energy distribution of one or the other of the fragments. A higher resolution is provided if the fragments' internal state distribution is measured, usually by laser-induced fluorescence. Another goal is to determine the direction of the transition dipole moment with respect to the molecular axis.

In a typical photofragmentation experiment a molecular beam is crossed with a pulsed, intense beam of laser light, polarized in a specified plane (Fig. 5.8). The plane of polarization can be rotated so that the angle between the electric vector of the light and the direction of the emission of the fragments (detected, for example, by a mass spectrometer) can be varied. The resulting angular distribution of the products gives a direct indication of the angle between the bond axis and the transition dipole.† To determine the translational energy disposal the pulse of laser light triggers the detector, which records the number of particles arriving per unit time interval following the pulse. The distribution in "arrival time" is then converted to a distribution of final translational energy E'_T. One can thus determine the distribution of internal energy E'_I in the fragments from the known available energy E,

$$P_I(E'_I) = P_T(E - E'_I) = P_T(E'_T) \tag{5.1}$$

---

† The electric vector of the light excites the transition dipole of the molecule, which has a specified orientation with respect to the molecule frame (e.g., in a diatomic molecule, the transition dipole can be either parallel or perpendicular to the bond). The fact that the angular distribution of the products does often depend on the orientation of the plane of polarization indicates that the photodissociation occurs by a direct-mode (Sec. 3.3). If the energy-rich molecule stayed together for more than a classical rotational period, such a correlation would be lost. See also Section 7.2.

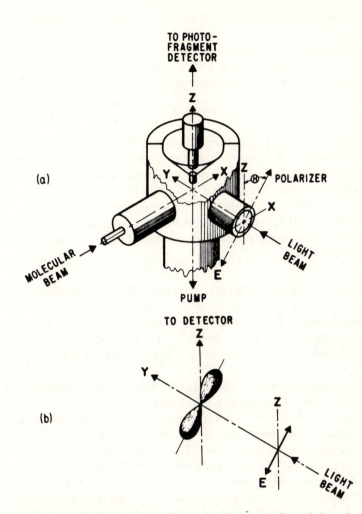

Fig. 5.8. (a) Photofragment spectrometer. The molecular beam is directed along the X axis, the light beam along Y, and those photodissociation fragments traveling in the Z direction are detected mass spectrometrically. The light is pulsed so that the time-of-flight of the photofragments can be measured, and thus their translational energy. The polarizer allows study of the dependence of the photodissociation yield upon the angle Θ between the electric vector of the light and the direction of the transition dipole of the absorbing molecule that may undergo photofragmentation, as shown in panel (b). [Adapted from G. E. Busch and K. R. Wilson, *J. Chem. Phys. 56*, 3626 (1972).] (b) Angular distribution of recoiling photofragments. When the rotation of the diatomic molecule is very slow, only those molecules that dissociate along the Z axis will yield fragments reaching the detector. The absorption probability of the molecule is proportional to $|\boldsymbol{\mu} \cdot \mathbf{E}|^2$, where $\boldsymbol{\mu}$ is the molecular transition dipole. Consider a diatomic molecule for which $\boldsymbol{\mu}$ is directed along the internuclear axis. The angular distribution of the photofragments as seen by the detector is from those molecules for which $\boldsymbol{\mu}$ is along the Z axis, $|\boldsymbol{\mu} \cdot \mathbf{E}|^2 \propto \cos^2 \Theta$. This distribution, which has a cylindrical symmetry about the $\mathbf{E}$ axis, is as shown.

where $P_T(E'_T) dE'_T$ is the fraction of products with final translational energy in the range E'_T to $E'_T + dE'_T$. Using both polarized light pumping and polarized light probing, one can determine not only the energy disposal, but also the relative alignment of the recoiling fragments.

For example, in the photodissociation of I_2 in the visible, $\lambda \simeq 500$ nm,

$$I_2(X^1\Sigma) \xrightarrow{h\nu} I_2(B^3\Pi_u) \rightarrow \begin{cases} I(^2P_{3/2}) + I^*(^2P_{1/2}) \\ 2I(^2P_{3/2}) \end{cases}$$

it was found from the translational spectrum that fragmentation into two ground-state atoms predominates and that the direction of the ejected I atoms (i.e., the direction of the bond) is *perpendicular* to the electric vector of the light absorbed, as expected for I_2 excited to the $B^3\Pi_u$ state. For triatomic (or polyatomic) molecules, the fragments can also carry vibrational excitation, and hence a higher resolution is required for state analysis.

An example of final state resolved recoil velocity analysis is the photodissociation of O_3:

$$O_3 \xrightarrow{h\nu} O_2^\dagger + O$$

The distribution of internal energy in O_2 following the photodissociation of O_3 (Fig. 5.9) shows it to have an average internal energy far in excess of thermal equilibrium. The specific energy disposal also shows that photodissociation reactions are often "direct." The population inversion in the products of such reactions have indeed been exploited for chemical laser action.

The photofragmentation of the linear OCS molecule is interesting in that it can be thought of as almost a collinear "half-collision". Using a F_2 excimer laser at $\lambda = 157$ nm (photon energy, 7.90 eV), the photolysis yields a sulfur atom in the electronically excited $(2.75 \text{ eV})^1S_0$ state. Since the S–CO bond-dissociation energy is 3.12 eV, the energy available for partitioning between relative translation E'_T and CO internal excitation E'_I is 2.03 eV. The time-of-flight distribution of the recoiling S atom (Fig. 5.10) reveals considerable internal (here, essentially only vibrational) excitation of the CO, comparable to that of $CO^\dagger$ produced in the (full) collision $O + CS \rightarrow CO + S$.

Laser-induced fluorescence (Appendix 5A) of the dissociation products has been extensively used in conjunction with photofragmentation. Figure 5.11 shows the state-resolved internal energy distributions for NO from the photodissociation of NO_2:

$$NO_2 \xrightarrow[337 \text{ nm}]{h\nu} NO + O$$

For larger molecules, both products need be probed. The simplest case is a tetratomic molecule producing two diatomic fragments (Fig. 5.12), for example,

$$NCNO \xrightarrow{h\nu} NC + NO$$

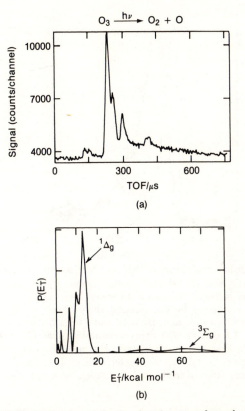

Fig. 5.9. (a) Time-of-flight distribution of the O_2 photofragment from the photodissociation of O_3 at 266 nm. The O_2 is detected at 30° with respect to the laser beam. (b) Data of (a) converted to c.m. translational energy distribution function, $P(E'_T)$. The most intense peak corresponds to the formation of $O(^1D)$ with $O_2(^1\Delta_g)$ in the $v' = 0$ state. Peaks indicating $O_2(^1\Delta_g)$ in states $v' = 1$, 2 and 3 are also observed. The weak, high-energy "tail" corresponds to the process yielding $O(^3P) + O_2(^3\Sigma_g^-)$. A detailed analysis of the results shows that the relative yields of the vibrational states are: 57% $v' = 0$, 24% $v' = 1$, 12% $v' = 2$, 7% $v' = 3$, and that some 17% of the remaining energy is disposed into O_2 rotational excitation. [Adapted from R. K. Sparks, L. R. Carlson, K. Shobotake, M. L. Kowalczyk and Y. T. Lee, *J. Chem. Phys.* **72**, 1401 (1980).]

A summary of energy disposal in many photodissociation experiments is provided in Fig. 5.13.

So far we have discussed the dissociation of strong chemical bonds using lasers in the visible/ultraviolet region. The photodissociation of weakly bound van der Waals-bound molecules using infrared lasers provides an important insight on vibrational translational coupling in (half) collisions on potential-energy surfaces which contain a well.† Figure 5.14 shows the

---

† To be sure, the well is shallow, but the excess energy can be made low. In Section 7.2 we shall see that an important parameter is the ratio of the available energy to the well depth.

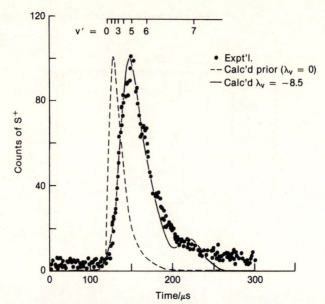

Fig. 5.10. Time-of-flight distribution of the number density of sulfur atoms from the photodissociation of OCS at 157 nm (flight path, 78 mm). Upper abscissa scale shows flight times calculated from individual v' states (assuming $j' = 0$, expected since OCS is a linear molecule), the highest allowed by the available energy being $v' = 7$. The dashed curve is simulated for the "prior" distribution, the solid curve for a linear vibrational surprisal (with $\lambda_v = -8.5$, see Sec. 5.5 below). The experiment reveals that $\langle E_{vib} \rangle$ is some 73% of the available energy, i.e., that the vibrational distribution of the CO photofragment is highly inverted. [Adapted from G. Ondrey, S. Kanfer, and R. Bersohn, *J. Chem. Phys.* **79**, 179 (1983).] The value of λ_v is about the same as that found for the $O + CS \rightarrow CO(v) + S$ reactive collisions.

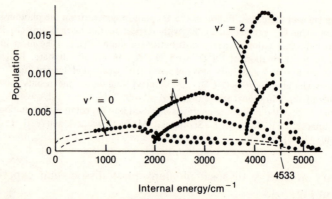

Fig. 5.11. Results from laser-induced fluorescence analysis of the population distribution of $NO(X^2\Pi_{1/2}, {}^2\Pi_{3/2}; v' = 0, 1, 2)$ from the photofragmentation of NO_2 at 337.1 nm (3.677 eV). Since the O–NO dissociation energy is 3.115 eV, the available energy is 0.562 eV or 4533 cm^{-1} (shown on the abscissa with a dashed vertical line denoting the maximum internal energy, for ground state NO_2). The mean initial internal energy of NO_2 is *ca.* 310 cm^{-1}, accounting for the "tails" of the distributions. Upper circles denote $^2\Pi_{1/2}$; Lower circles, $^2\Pi_{3/2}$ NO. Note the strong vibrational population inversion; approximately 70% of the available energy goes into internal excitation of the NO, mainly into vibration. The O atoms formed are essentially all in the $^3P_{0,1,2}$ state. [Adapted from H. Zacharias, M. Geilhaupt, K. Meier and K. H. Welge, *J. Chem. Phys.* **74**, 218 (1981).]

$$NCNO \xrightarrow{h\nu} CN + NO$$

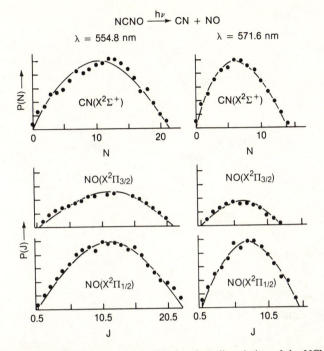

Fig. 5.12. Laser-induced fluorescence results for the photodissociation of the NCNO molecule at the two photolysis wavelengths specified: Plotted are the rotational state population distributions from the $CN(X^2\Sigma^+, v = 0, N)$ photofragments (upper) and the $NO(X^2\Pi_{3/2}, v = 0, j + 1/2)$ $NO(X^2\Pi_{1/2}, v = 0, j + 1/2)$ fragments (lower). The experimental points (open circles) are compared with solid curves calculated by a statistical (phase space) theory. Note that at $\lambda = 554.8$ nm, the available energy is 939 cm^{-1}, while at $\lambda = 571.6$ nm, it is only 411 cm^{-1}. [Adapted from S. Buelow, M. Noble, G. Radhakrishnan, H. Reisler, and C. Wittig, *J. Phys. Chem.* **90**, 1015 (1986); see also I. Nadler, J. Pfab, H. Reisler, and C. Wittig, *J. Chem. Phys.* **81**, 653 (1984).]

results for the infrared (CO$_2$ laser) photodissociation of the ethylene dimer:

$$(C_2H_4)_2 \xrightarrow{h\nu} 2C_2H_4.$$

Not only weakly bound species can be photodissociated using infrared photons. In Section 7.3 we discuss *multi*photon absorption and dissociation.

The ultimate probe in a photodissociation experiment is the molecule itself. Since by the very nature of the experiment it has been promoted to an upper electronic state, it will fluoresce. If the photodissociation is direct, there is only a very short time available to the molecule, as it moves on its way out to products. Detecting photons is, however, so very efficient that such emission *can* be monitored. It serves as a very detailed probe of a molecule in the very process of dissociation, as will be seen in Section 6.5.

It is also possible to work in the time (as opposed to the frequency) domain and determine directly the *lifetime* of the photoexcited state with

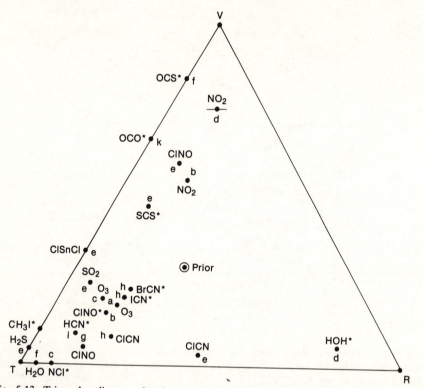

Fig. 5.13. Triangular diagram showing representative results of energy partitioning in the photofragmentation of triatomic molecules. Vertices *T, R,* and *V* correspond to distributing the available energy into relative translation of the atom–diatomic products, and rotation and vibration, respectively, of the diatomic product. Each point denotes the average energy partitioning of the specified molecule at a wavelength (λ, in nm) coded as follows: a, 532; b, 337; c, 266; d, 248; e, 193; f, 157; g, 150; h, 147; i, 130; j, 122; k, 90.1 nm. An asterisk means that the designated product is formed in an excited electronic state: S*, I*, CO*, NO*, CN*, OH*. Methyl iodide is included since a "triatomic approximation" has been used to describe its photofragmentation. The point labeled "prior" is for the simplest case of a statistical partitioning, for which $\langle f_T \rangle$, $\langle f_v \rangle$, and $\langle f_R \rangle$ are 3/7, 2/7, and 2/7, respectively [assuming for the diatomic the so-called rigid-rotor harmonic oscillator (RRHO) model; see Appendix 5C]. To be noted are the molecules whose points lie along the *VT* axis, for which $\langle f_R \rangle \simeq 0$. These are molecules whose dissociated excited states must be collinear (i.e., the half-collision is of zero impact parameter), resulting in no rotational angular momentum of the diatomic product. Note also those molecules for which $\langle f_v \rangle$ exceeds that of the prior; these are usually characterized by a significant difference in the appropriate bond length relative to that of the product diatomic. [Adapted from R. Bersohn, *J. Phys. Chem.* **88**, 5145 (1984).]

the aid of fast laser pulse techniques. Figure 5.15 shows an application to methyl iodide. A picosecond laser pulse is used to excite the molecule in a beam, and a delayed laser pulse probes the appearance of the I atoms, here by multiphoton ionization.

Prior to the development of such direct lifetime (τ) measurement techniques, one estimated τ from the so-called anisotropy parameter β of

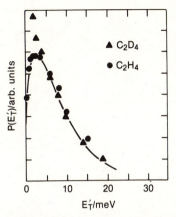

Fig. 5.14. Relative translational energy distributions for C_2H_4 and C_2D_4 from the photodissociation of the dimer molecules $(C_2H_4)_2$ and $(C_2D_4)_2$ using CO_2 laser irradiation at the appropriate wavelengths; $\bar{v} = 947.8\,cm^{-1}$ and $1073.3\,cm^{-1}$, respectively, for $(C_2H_4)_2$ and $(C_2D_4)_2$. The abscissa is E'_T in meV. The points are experimental; the solid curve is a Boltzmannized functionality $P(E'_T) \propto E_T'^{1/2} \exp(-cE'_T/\bar{E})$, where $\bar{E} = \langle E_{T'} \rangle = 8.8\,meV$ and $c = 1.5$. [Adapted from M. A. Hoffbauer, K. Liu, C. F. Giese, and W. R. Gentry, *J. Chem. Phys.* **78**, 5567 (1983).]

the photofragment distribution. The route from β to τ, however, is not rigorous, since a model is required. Direct, real-time (picosecond or less) measurements are clearly more reliable.

5.2.5 Lasers as effectors and detectors of chemical change

The last two subsections provided a clear illustration of a theme that runs throughout this volume: Lasers are being used in an essential way to promote the chemical change and to probe the products. Table 5.2 shows the primary laser sources available in the energy range of interest. What are the key properties of the lasers we use in the field of microlevel chemistry? One important aspect is tunability, i.e., the ability to select the wavelength at which lasing occurs. Dye lasers are truly tunable, as are lasers based on nonlinear optical effects. Molecular transition lasers have a number of discrete well-defined lasing wavelengths spanning much of the chemically important spectral region, from the infrared to the far-ultraviolet. Efficient Raman shifters and nonlinear harmonic-generating crystals for doubling and frequency mixing are available to extend the usable wavelength range. Recently, rare-gas jets have been used as nonlinear harmonic generation media to produce usable pulse intensities of radiation in the vacuum ultraviolet ($\lambda \lesssim 100\,nm$), making it possible to detect H and D atoms via resonant ionization. Last, but not least, the free-electron laser is beginning to take its place as the most tunable laser source, from the microwave all the way to the vacuum ultraviolet.

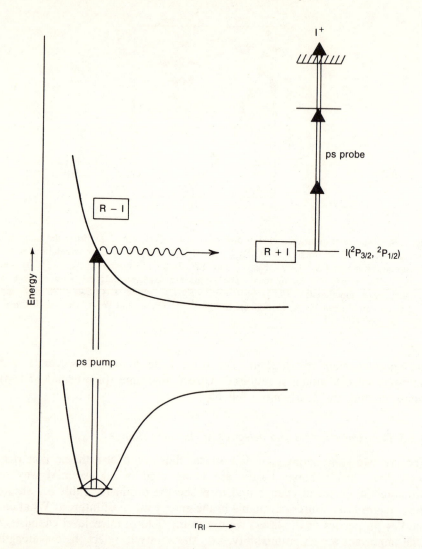

Fig. 5.15. Picosecond laser experiment to determine the lifetime of the $\bar{A}$ state of the CH₃I prior to photofragmentation to yield CH₃ + I, I*. (a) Schematic diagram of experiment, using a ps-pulse pump laser at 280 nm to excited the X-state CH₃I in a molecular beam to the repulsive $\bar{A}$ state. The I atoms are detected by a ps probe laser pulse at 304.0 nm to ionize the I* ($^2P_{1/2}$) state or at 304.7 to ionize the ground state I($^2P_{3/2}$) photofragments (via 2-photon resonance-enhanced ionization, to be discussed in Sec. 7.3). (b) Trace of the transient signal of I⁺ (identified via time-of-flight mass spectrometry) from the I* photofragment. (Similar results were found for the I photoproduct.) Points: experimental; solid curve: convoluted transient calculated from a knowledge of the pulse durations and assuming a lifetime $\tau \leq 0.5$ ps, setting an upper bound in accord with less direct photofragmentation data. [Adapted from J. L. Knee, L. R. Khundar, and A. H. Zewail, *J. Chem. Phys. 83,* 1966 (1985).] More recently (lower panel), it has been possible to carry out similar femtosecond photofragmentation experiments for the ICN photodissociation at 306 nm. (Response function also shown.) [N. F. Scherer, J. L. Knee, D. D. Smith, and A. H. Zewail, *J. Phys. Chem. 89,* 5141 (1985).]

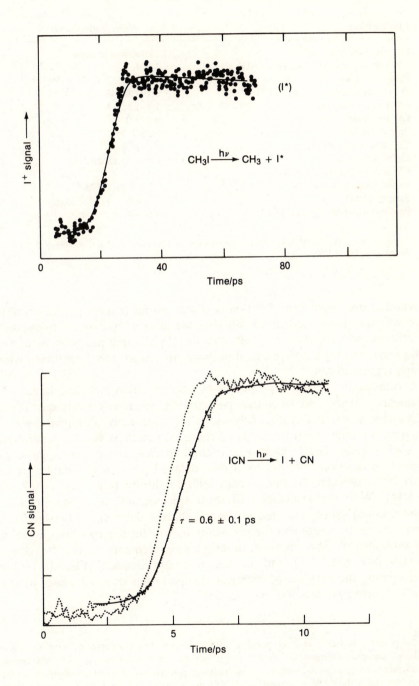

Table 5.2. Types of lasers[a]

Classification	Wavelengths (μm)
CW ion laser (e.g., Ar^+)	Lines, visible region
CW solid-state laser (e.g., Nd:YAG)	1.064
Pulsed solid-state laser (e.g., Nd:YAG)	1.064, 0.532, 0.355, 0.266
Pulsed excimer laser (e.g., ArF*)	0.193, 0.248, 0.308, 0.351
CW color center laser	1.4–1.6, 2.3–3.3
CW dye laser	0.4–1.0
Pulsed dye laser	0.26–0.95
CO_2 laser (discharge)	9.6–10.4
Molecular gas laser	Lines, in range 2–11
CW semiconductor diode laser	0.3–0.8
Pulsed metal vapor laser	0.51, 0.58, 0.628
Ring dye laser	0.4–0.8
Free-electron laser (pulsed FEL)	Vacuum uv to far ir

[a] Adapted from G. C. Pimentel (ed.), *Opportunities in Chemistry*, National Academy Press, Washington, D.C., 1985, p. 93.

Another important consideration is power output (energy per unit time). This will very much depend on whether the laser is operated continuously (continuous wave, cw) or in a pulsed mode. High output per pulse is readily achievable, but for many practical purposes this needs to be combined with a high repetition rate.†

Additional and desirable properties of laser radiation include collimation, monochromaticity, and, of course, polarization. We have already seen some such applications, and in what follows we shall see many additional ones.

Lasers are convenient initiators of chemical reactions via the generation of reactive (e.g., fast H atoms) or excited species. Using a laser pulse to trigger the reaction followed by a time-delayed probe laser enables us to study intermolecular dynamics under bulk conditions (e.g., Fig. 1.4, 2.14, or 5.16). With the availability of short (picosecond) or even ultrashort (femtosecond) pulses one can use much shorter delay times between the pump and probe lasers and thereby study *intra*molecular dynamics, even in the bulk phase. The progress in using measurements in the frequency domain (see Sec. 6.5) and in the real-time domain (Fig. 5.15) for interrogating the very act of chemical change brings us much closer to the ideals of molecular reaction dynamics.

† In general, higher power is possible using electron beam excitation, with electrical discharge pumping being next. 100 J/pulse is possible in e-beam rare-gas excimer lasers. 1–10 J/pulse can be achieved with electrical discharge pumped rare gas or CO_2 lasers. Since the pulses are quite short, peak powers of 10 MW are possible. Excimer lasers with kilohertz repetition rates are already available.

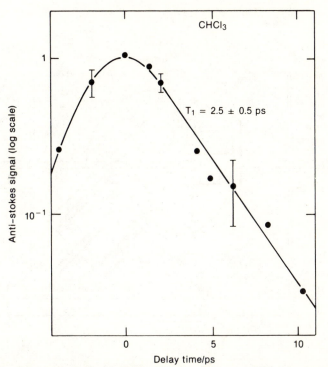

Fig. 5.16. The CH-stretch mode of $CHCl_3$ (0.1 M in CCl_4 at 295 K) is excited by a short (~4 ps) infrared laser pulse. The excess population of the vibrationally excited state is monitored by anti-Stokes scattering induced by a subsequent short probe pulse. Plotted is the anti-Stokes signal versus the delay time between the pump and probe lasers. At zero delay the peak intensities of both lasers coincide. The measured scattered signal is directly proportional to the momentary population of the excited state. [Adapted from A. Laubereau, S. F. Fischer, K. Spanner, and W. Kaiser, *Chem. Phys. 31*, 335 (1978).]

Appendix 5A: Laser-induced fluorescence

The method of laser-induced fluorescence (LIF) is to irradiate the molecules to be probed with a tunable laser and observe the total (undispersed) fluorescence emitted as the laser-excited states return to the ground state. The lifetimes of these states must be short, however (e.g., ≤100 ns). Figure 5A.1 shows a schematic apparatus diagram. Figure 5A.2 shows the LIF excitation spectra of BaBr produced via the reaction

$$HBr + Ba \rightarrow BaBr + H$$

that is, the fluorescence intensity versus laser wavelength. The BaBr product formed in the various v'', j'' states of the ground electronic state ($X^2\Sigma^+$) are excited to the $C^2\Pi_{3/2}$ state and fluoresce back to the ground state. The $C \rightarrow X$, $\Delta v = 0$ and -1 sequences are observed for BaBr.

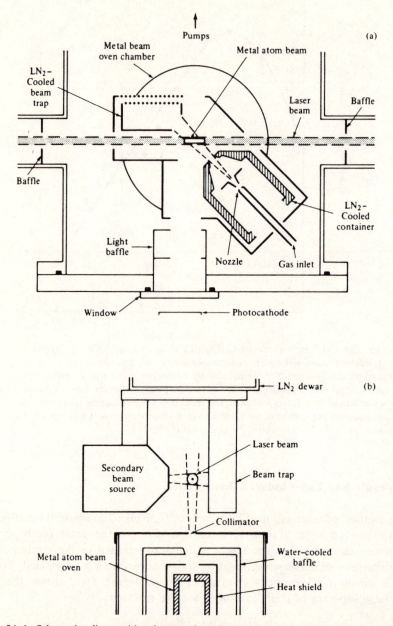

Fig. 5A.1. Schematic diagram(s) of crossed molecular beam laser-induced fluorescence apparatus for the detection of nascent products of reactions of alkaline earth atoms with hydrogen halides. [Adapted from H. W. Cruse, P. J. Dagdigian, and R. N. Zare, *Faraday Discuss. Chem. Soc. 55*, 277 (1973); also R. N. Zare, *ibid., 67*, 7 (1979).]

HBr + Ba ⟶ BaBr (X) + H

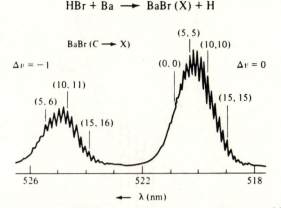

Fig. 5A.2. Laser-induced fluorescence from the BaBr product of the crossed-beam reaction of HBr with Ba. [Adapted from the references of Fig. 5A.1.]

Because of the small rotational (B_v) constants, individual product rotational states could not be resolved.

Figure 5A.3 is an oversimplified schematic drawing illustrating the essential features of the LIF process for BaX molecules. The potential-energy curve for the upper electronic state (C) lies more or less directly over that for the ground (X) state; the vibrational spacings of the upper state, ω', are greater than those for the ground state, ω''. For simplicity, we have assumed harmonic oscillator levels so that excitation frequencies are given by

$$v_{v'v''} = v_{00} + v'\omega' - v''\omega'' \qquad (5A.1)$$

Shown in Fig. 5A.2 are $\Delta v = 0$ excitation transitions, i.e., those for which $v'' = v'$, so that

$$v_{v'v''} = v_{00} + v'(\omega' - \omega'') \qquad (5A.2)$$

Also observed (at least for BaBr) are $\Delta v = -1$ transitions, for which $v' = v'' - 1$; thus,

$$v_{v'v''} = v_{00} + v'(\omega' - \omega'') - \omega'' \qquad (5A.3)$$

Shown at the bottom of Fig. 5A.3 is a schematic drawing of the LIF spectrum expected for a "high-temperature" vapor of BaBr, showing the band origin v_{00}, labeled $(0,0)$, and the remainder of the $\Delta v = 0$ series at higher frequencies, with constant spacings $\omega' - \omega''$ [from Eq. (5A.2)] and decreasing relative intensities. The $\Delta v = -1$ series is also shown, at longer wavelengths, with the $(0,1)$ line shifted from the $(0,0)$ position by the ground-state vibrational frequency ω'' [Eq. (5A.3)]. It should be noted that the fluorescence intensity plotted is the total undispersed fluorescence, i.e., the sum of the intensities of the downward transitions out of the given v' state to the various final v'' states.

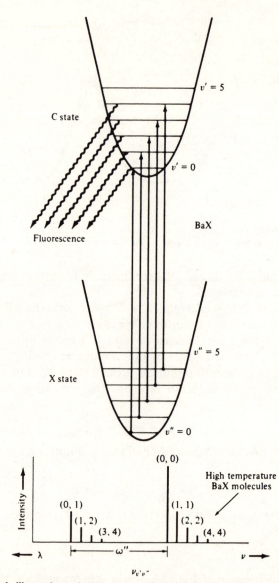

Fig. 5A.3. Sketch illustrating principle of laser-induced fluorescence. The example is for hot BaX molecules (e.g., BaBr)—see text.

The observation of the relative intensities $I_{v'v''}$ of these vibrational bands leads to a set of nascent vibrational populations $N_{v''}$ for the X-state BaX molecules.

One can use as an approximation

$$N_{v''} = I_{v'v''} \bigg/ \left(q_{v'v''} \rho(v_{v'v''}) \sum_v v_{v'v}^4 q_{v'v} \right) \tag{5A.4}$$

where $\rho(v)$ is the laser power density, $q_{v'v''}$ the Franck–Condon factor for the v'', v' transition, and the sum Σ_v is overall vibrational states v of the ground electronic states to which the particular $v'(C)$ state fluoresces. The limitation of Eq. (5A.4) is that it requires a knowledge of the Franck–Condon factors for each of the important transitions, implying good spectroscopic information on the excited (and ground) electronic states. For the BaX molecules, the $C \rightarrow X$, $\Delta v = 0$ system is the strongest, since $B'_v \approx B''_v$. Also, the q values vary by less than a factor of 2 over the range of states excited. The fluorescence sum consists mainly of the term for which $\Delta v = 0$, i.e., $q_{v'v''} \approx \delta_{v'v''}$. Thus, under favorable conditions, Eq. (5A.4) reduces simply to a linear proportionality:

$$N_{v'v''} \simeq I_{v'v''} \tag{5A.5}$$

Thus, the relative intensities of the observed LIF spectrum can, under favorable circumstances, display directly the relative vibrational populations of the X-state molecules (here, the nascent BaBr product of the HBr + Ba reaction).

Since its inception, many further developments of the LIF state analysis and detection technique have followed, and its sensitivity has enhanced sufficiently to make possible population measurements as a function of scattering angle. The Doppler shift of the LIF signal has been used to determine the velocity-angle distributions of the specific rovibrational states of scattered molecules (see Sec. 5.2.3).

5.3 Molecular and ion beam scattering

The experimental techniques used for scattering experiments with typical applications are discussed in this section. Appendix 5B describes some relevant properties of supersonic beams ("jets") that are being extensively employed in both collisional and spectroscopic studies.

5.3.1 Crossed molecular beams

The crossed-beam method is ideally suited to the study of detailed dynamics of reactive collisions. In common with the chemiluminescence technique it yields information on the energy disposal by means of product state analysis, but its unique additional feature is that it provides a measurement of the angular distribution of the scattered products. As we have already seen (Sec. 3.3), significant dynamical information can be derived from a consideration of the "shape" of such an angular distribution, not the least of which is that one can immediately classify the reaction into the categories, direct or compound, depending on whether the angular distribution of scattering in the c.m. system is anisotropic or symmetric.

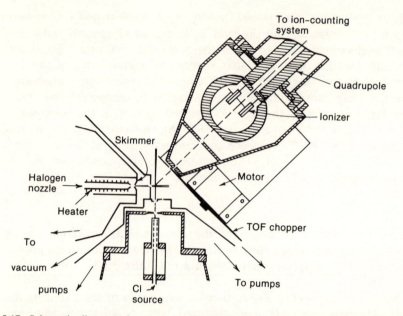

Fig. 5.17. Schematic diagram of a crossed molecular beam machine for the study of reactive scattering of Cl atoms with halogen molecules. The apparatus employs two separately pumped supersonic nozzle beam systems, a time-of-flight chopper, a rotatable ultrahigh vacuum detector chamber with a high-efficiency electron-impact ionizer–quadrupole mass filter and an ion counter. [Adapted from J. J. Valentini, Y. T. Lee, and D. J. Auerbach, *J. Chem. Phys.* 67, 4866 (1977).]

Figure 5.17 illustrates the design of a typical crossed molecular beam experiment. The reactant molecules (or atoms, or ions) are formed into collimated beams (preferably with narrow velocity distributions and typically from a supersonic nozzle beam source—see Appendix 5B), which are then allowed to intersect (usually at right angles) in a small region, in which the scattering occurs. The ambient pressure is kept low ($<10^{-6}$ Torr) to minimize secondary collisions. A discriminating detector (e.g., an electron-bombardment mass spectrometer) can be rotated at various angles with respect to the beam system and thus measure the angular distribution of the various scattered species (reactants as well as products) in the laboratory coordinate system. At the same time their velocity distribution can be measured (e.g., by the time-of-flight method). From the laboratory angular distribution and the velocity distribution of one or more of the scattered products, one can deduce the detailed differential reaction cross section in the c.m. system (to be discussed in Sec. 5.4).

Additional refinements are introduced for more detailed studies, including the use of seeded, supersonic "accelerated" beams (see Appendix 5B) allowing high translational energies E_T, and thus extending the range of investigation to reactions with an energy threshold, and devices for internal state selection of reactant molecules and product state analysis as, for example, in Fig. 5.5.

An example of the angular and velocity distributions of products is shown in Fig. 5.18 for the ICl products of the endoergic reaction $Cl + IC_2H_5 \rightarrow ICl + C_2H_5$. Many additional examples will be found in Sections 5.4 and Chap. 7. State- and angle-resolved products are shown for inelastic collisions in Sec. 6.3.2. For reactive collisions, angular resolution is often

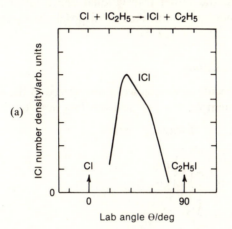

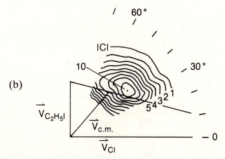

Fig. 5.18. (a) Laboratory angular distribution of ICl from the crossed-beam reaction $Cl + IC_2H_5 \rightarrow ICl + C_2H_5$. The Cl atom beam was formed from a high-pressure microwave discharge source; the mean speed of the Cl atom beam could be varied from 1 to 2 km s^{-1} by varying the ratio of Cl_2 to Ne or He carrier gas. The thermal C_2H_5I beam was at 90° to the Cl direction and the number density of ICl was measured by a rotatable mass spectrometer. The average relative translational energy E_T was 56 kJ mol^{-1} (cf. the endoergicity $\Delta E_0 = 10$ kJ mol^{-1}). The translational energy threshold for the reaction was measured to be *ca.* 15 kJ mol^{-1}. (b) Experimental flux-velocity contour map of the ICl product corresponding to (a) above. The average velocity of the centroid, $\mathbf{v}_{c.m.}$, is shown. Relative intensities for each contour are indicated. The results are interpreted in terms of a long-lived complex mechanism (with a small, so-called "stripping" component in the forward c.m. direction), as discussed in Section 7.2. [Adapted from S. M. Hoffmann, D. J. Smith, A. G. Ureña, and R. Grice, *Chem. Phys. Lett.* **107**, 99 (1984).]

combined with a time-of-flight analysis as discussed in Section 5.4. The use of laser-induced fluorescence (Appendix 5A) in a crossed-beam arrangement for the exoergic reaction

$$Ba + HF \rightarrow BaF + H, \qquad \Delta E_0 \approx -18 \text{ kJ mol}^{-1}$$

is illustrated in Fig. 5.19. Here the energy is provided in part by fast HF molecules produced by the supersonic expansion of HF in H_2 or He. A considerable range in the initial translational energy E_T could thereby be achieved. The dependence of the energy disposal on E_T is evident in the figure.

5.3.2 Ion–molecule reactions

Historically, ion–molecule reactions were discovered in the ionizers of mass spectrometers. For example, the H_3^+ ion was first detected in 1913 from the

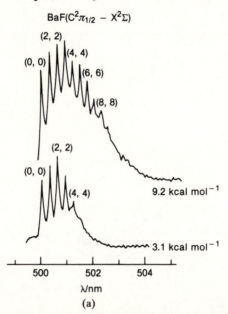

$$BaF(C^2\pi_{1/2} - X^2\Sigma)$$

(a)

Fig. 5.19. (a) Example of laser-induced fluorescence data in product-state analysis for BaF(v', j') formed in the collision of HF (i.e., a seeded supersonic beam of HF in a He carrier) with a thermal Ba beam. Excitation spectra of the BaF from reaction at $E_T = 9.2$ (top) and 3.1 kcal mol^{-1} (bottom). Note the increase in the population of the higher vibrational states of BaF with increased relative translational energy of the HF + Ba reagents. (b) Energy disposal (triangular) plots for the reaction of Ba with HF at three values of E_T: 1.6, 6.5, and 13.2 kcal mol^{-1}. Contours of equal reaction cross section into BaF product internal states (v', j'). Coordinates expressed in terms of fraction of total energy E released into product vibration (f'_v), rotation (f'_R), and relative translation (f'_T) of the recoiling BaF + H. The exoergicity of the reaction is $-\Delta E^0 = 4.4$ kcal mol^{-1}. There is a small activation barrier, $E_a \simeq 2$ kcal mol^{-1}. [Adapted from A. Gupta, D. S. Perry, and R. N. Zare, *J. Chem. Phys.*, *72*, 6237 (1980). Comparable results for the related reactions of Ba with HCl and HBr are by A. Siegel and A. Schultz, *J. Chem. Phys.*, *72*, 6227 (1980).]

(*in situ*) reaction

$$H_2^+ + H_2 \rightarrow H_3^+ + H$$

The reaction proceeds rapidly even at hydrogen pressures below 10^{-5} Torr. The rate constants of a large number of similar exoergic ion–molecule reactions were measured in such experiments, and their common independence of temperature taken as evidence for the Langevin capture model (Sec. 2.4.7). However, it was not until the development of the tandem mass spectrometer, flowing afterglow and the ion cyclotron resonance (ICR) techniques that the full gamut of ionic reactivity was explored. In the ICR method, a selected ion (of a given mass-to-charge ratio, m/z) in cyclotron orbit is allowed to react with neutral molecules at low pressures (*ca.* 10^{-6} Torr), with residence times ranging from μs to ms. From the time

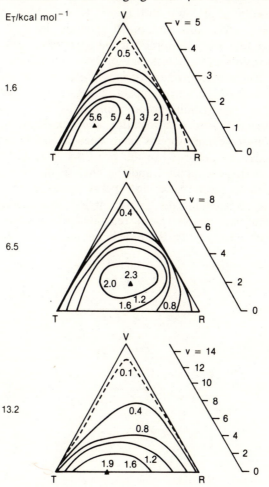

(b)

decay of the number density of the reagent ions and the (rate of) appearance of product ions, both qualitative chemical reactivity and quantitative kinetic data were obtained.

An example of new chemistry thus discovered is that from the reaction of CH_3F^+ with CH_3F, which proceeds rapidly through the following sequence of exoergic reactions:

$$CH_3F^+ + CH_3F \rightarrow CH_3FH^+ + CH_2F \text{ (protonation)}$$

$$CH_3FH^+ + CH_3F \rightarrow [\text{adduct}]^+ \rightarrow (CH_3)_2F^+ + HF$$

(formation of a fluoronium ion)

The fluoronium ion is sufficiently energized that it undergoes extensive intramolecular rearrangement and unimolecular dissociation as follows:

$$(CH_3)_2F^+ \rightarrow CH_2F^+ + CH_4(!)$$

Although flowing afterglows and the ICR technique provide accurate bimolecular rate constants for most exoergic ion–molecule reactions, alternative techniques are required for the study of the endoergic reactions. Here tandem mass spectrometers and ion beam techniques are widely used. Since beams of a vast assortment of ions (both closed- and open-shell) are readily prepared [from electron-impact, charge transfer, surface ionization, photoionization, laser multiphoton ionization, etc., followed by mass separation (magnetic or electrostatic) and acceleration], an enormous number of ion–molecule systems have been available for study. Elastic, inelastic, reactive, and charge transfer processes have all been studied using ion beam scattering techniques, and the analysis of the results is much the same as for neutral beam scattering experiments.

An example of a chemically significant ion–molecule reaction involves transition metal ion beams. Such reactions often proceed via complex formation. Not infrequently there will be more than one complex, for example,

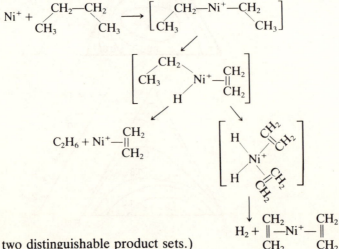

(Note the two distinguishable product sets.)

Product state analysis provides a mechanistically diagnostic tool. Figure 5.20 shows the products' relative translational energy distributions for the dissociation of two different collision complexes. Anticipating Section 7.2, strong coupling as the fragments recede (so-called final state or exit channel interactions) masks the expected (near) statistical product-state distribution when the collision proceeds via a long-lived complex.

Let us return to the experimental considerations.

In some ways the study of ion–neutral collision dynamics is technically simpler than similar neutral–neutral collision experiments, since ion beams can be more readily controlled than molecular beams. The crossed-beam technique is still the overall method of choice, but there is an important variant for ion–neutral experiments, namely, that of "merging beams," which provides for a very wide variation *and* excellent energy resolution even at low values of the initial relative kinetic energy. The merging-beams technique has been widely used for threshold determinations and studies of the post-threshold energy dependence of ion–neutral reactions.

A schematic of a typical apparatus is shown in Fig. 5.21. The neutral beam is generated by charge exchange of a fast beam of ions. The second beam of ions is deflected by a field so as to merge with the neutral beam. When the two beams are well collimated, the relative kinetic energy for an ion–molecule collision is determined by the difference in the velocity of the two beams. Fairly high laboratory kinetic energies of both beams can be used, yet low *relative* velocities can be obtained. Despite the energy spread in each beam, the distribution of *relative* translational energy is much narrower than the energy spread of either beam. This can be seen as follows: The relative translational energy is given in terms of the relative velocity of the reagent particles moving coaxially by

$$E_T = \tfrac{1}{2}\mu(v_2 - v_1)^2 = \mu[(E_2/m_2)^{1/2} - (E_1/m_1)^{1/2}]^2 \tag{5.2}$$

For simplicity, assume $m_1 = m_2$. Then,

$$E_T = \tfrac{1}{2}(E_2^{1/2} - E_1^{1/2}) \simeq \Delta E^2/8\bar{E} \tag{5.3}$$

where $\Delta E = E_2 - E_1$, $\bar{E} = (E_1 + E_2)/2$, and the approximation is valid for $\Delta E/\bar{E} \ll 1$. Thus, the fractional energy spread in E_T, say $\delta E_T/E_T$ can be estimated:

$$\frac{\delta E_T}{E_T} \simeq 2\frac{\delta \Delta E}{\Delta E} \tag{5.4}$$

Consider the example case of $E_1 = 980 \pm 1\,\text{eV}$, $E_2 = 1020 \pm 1\,\text{eV}$, so $\Delta E = 40 \pm 1.4\,\text{eV}$. Via Eqs. (5.3) and (5.4), $E_T = 0.200\,\text{eV}$ and $\delta E_T = \pm 0.014\,\text{eV}$!

The guided-beam technique (which was used to determine the energy dependence of the reaction cross section shown in Fig. 2.15) is similar to a beam-scattering cell arrangement. The difference is that multipole electrical fields are used to "trap" and guide the incident beam of ions and to ensure that the totality of scattered ions reaches the detector. There is some

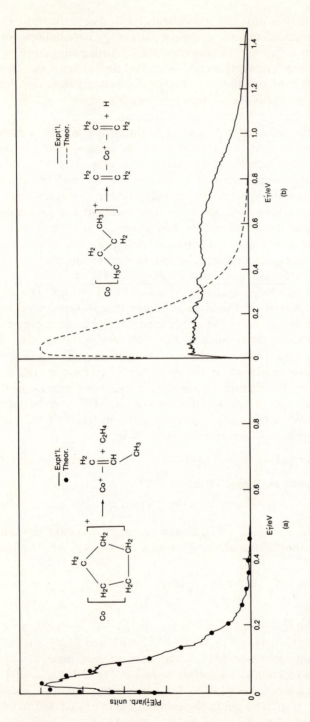

Fig. 5.20. Products' relative translational-energy distribution in the reactions of Co⁺ ions with organic molecules. (a) For the indicated reaction which proceeds without an "exit barrier." Good agreement obtains with a statistical distribution (dots). (b) For the indicated reaction with a barrier. [Adapted from A. M. Hanratty, J. L. Beauchamp, A. J. Illies, and M. T. Bowers, *J. Am. Chem. Soc.*, *107*, 1788 (1985). For details about the chemistry, see P. B. Armentrout and J. L. Beauchamp, *J. Am. Chem. Soc.*, *103*, 784 (1981).]

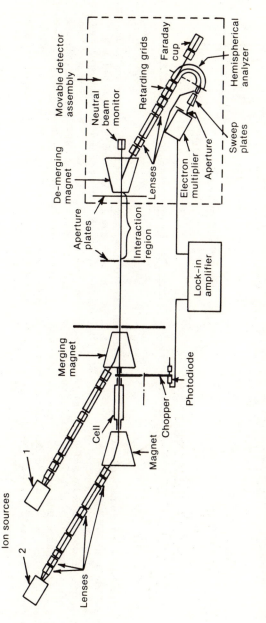

Fig. 5.21. Schematic drawing of merged-beam apparatus for study of ion–neutral reactions. The neutral beam is formed from a precursor ion beam (mass selected), which passes through a suitable low-pressure gas and undergoes charge transfer, e.g., $H_2^+ + M \rightarrow H_2 + M^+$, where the H_2 has the (fast) velocity of the original H_2^+ ions; the unwanted H_2^+ ions are swept out of the neutral beam, which is then merged with the co-reagent–ion beam. After traversing the interaction region, the desired product ion is separated from the reagent ions and the neutral by magnetic deflection. [Adapted from S. M. Trujillo, R. H. Neynaber, and E. W. Rothe, *Rev. Sci. Instrum.*, 37, 1655 (1966); it is also possible to provide energy analysis for the product ions: W. R. Gentry, D. J. McClure, and C. H. Douglass, *Rev. Sci. Instrum.*, 46, 367 (1975).]

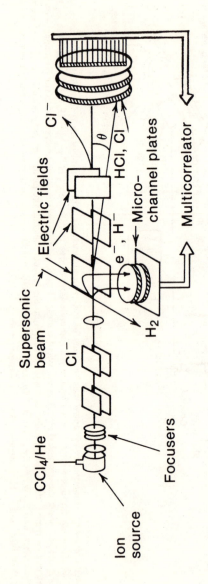

Fig. 5.22. Schematic diagram of a multicoincidence detection ion–neutral beam scattering apparatus. The key components are the microchannel plates (MCP), one of which serves as a position-sensitive detector (PSD). One is for the fast neutrals, the other for the charged particle (e^-, H^-) detection; both position and time encoders are employed with a multicorrelator that delivers a 4-bit position plus a 6-bit time word (corresponding to the difference in the arrival time of two correlated particles) to a multichannel analyzer and on-line computer. The Cl^- ion beam is formed from CCl_4 in a He discharge; the supersonic H_2 beam at right angles has an extremely narrow velocity distribution (<1%), $T \leq 0.01$ K, and is well collimated. The negative species formed in the collision zone are extracted by a transverse electric field and accelerated to a MCP. The fast neutral products are collected on a PSD with an angular resolution of $\leq 0.3°$ in the scattering plane. The undesired fast Cl^- ions (unreacted) are swept out of the beam. The data yield coincidence time-of-flight spectra at various laboratory scattering angles for collisions at E_T in the range from 5.6 to 12 eV (c.m.). [Adapted from M. Barat, J. C. Brenot, J. A. Fayeton, J. C. Houver, J. B. Ozenne, R. S. Berry, and M. Durup-Ferguson, *Chem. Phys.*, *97*, 165 (1985).]

sacrifice in kinetic energy resolution as compared to the merging-beams technique, but the internal states of the reagents are better defined (due to thermalization in the ion trap).

Negative ion beam reactions can also be studied, for example, the halide ion reactions with molecular hydrogen. A multicoincidence detection ion-beam scattering apparatus (Fig. 5.22) has been used to measure the differential cross sections for a variety of competing processes over a broad energy range:

		Products	$\Delta E_0/\text{eV}$
$Cl^- + H_2$	$\xrightarrow{a}$	$HCl + H^-$	2.91
	$\xrightarrow{b}$	$Cl + H_2 + e^-$	3.60
	$\xrightarrow{c}$	$HCl + H + e^-$	3.66
	$\xrightarrow{d}$	$Cl + H + H^-$	7.34
	$\xrightarrow{e}$	$Cl + H + H + e^-$	8.09

The vibrational excitation of the HCl product of reaction a is found to be small (≤ 1 eV), whereas all energetically allowed states are populated by reaction c (involving electron detachment from the H^-). A wealth of detailed data is available from such experiments, and their interpretation in terms of potential energy hypersurfaces is a subject of considerable attention.

Appendix 5B: Supersonic beams

When a gas at low pressure effuses out of a chamber through a small orifice, its velocity distribution is found to be exactly Maxwellian, being governed by the temperature T of the chamber. The number density of molecules with speeds in the range v to $v + dv$ is thus given by

$$n(v) \propto v^2 \exp(-v^2/\alpha^2) \tag{5B.1}$$

where

$$\alpha = (2kT/m)^{1/2} \tag{5B.2}$$

is the most probable speed in the chamber. Note that the velocity of sound, $(\gamma kT/m)^{1/2}$, where γ is the usual ratio of specific heats $\gamma = C_p/C_v$, is thus comparable to the mean speed. The breadth of the speed distribution (5B.1) is so large that one often requires some "velocity selection" technique to provide a more monoenergetic beam suitable for scattering studies. An alternative beam source is the high-pressure nozzle (or jet). If the gas expands through a nozzle from a high- to a low-pressure region (Fig. 5B.1), one obtains a supersonic beam with a narrow speed distribution. In such a beam the stream velocity is higher than the velocity of sound at the temperature of the beam. The distribution of speed is Maxwellian when measured by an observer moving with the mean stream velocity, v_s. To a laboratory-bound observer (as many of us are), the number density

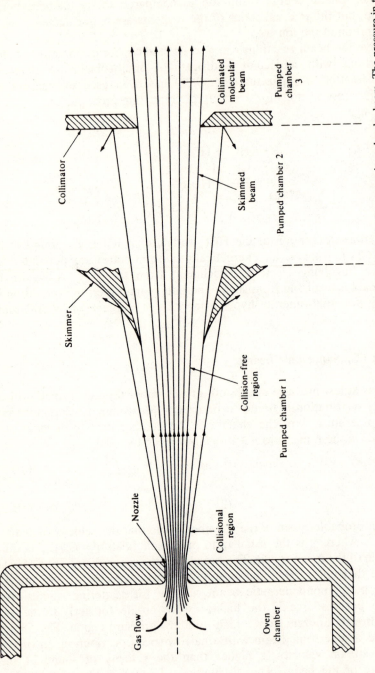

Fig. 5B.1. Schematic representation of the experimental arrangement to produce an intense, collimated supersonic molecular beam. The pressure in the source, or "oven," chamber may be tens of atmospheres, but the pumped chamber (1) should be maintained below 10^{-3} Torr. Pressures in chambers 2 and 3 should be successively lower, typically less than 10^{-6} Torr in chamber 3. The optimum distance from nozzle to skimmer is *ca.* 50–100 nozzle diameters.

distribution will be

$$n(v) \propto v^2 \exp[-(v - v_s)^2/\alpha_s^2] = v^2 \exp[-S^2(v - v_s)^2/v_s^2]. \quad (5B.3)$$

Here, α_s is the width of the speed distribution as measured by a beam-bound observer. It can therefore be used to define the beam temperature by analogy to (5B.2):

$$\alpha_s = (2kT_s/m)^{1/2} \quad (5B.4)$$

The "speed ratio" of the beam is defined $S = v_s/\alpha_s$. The Mach number characterizing the beam is

$$M = v_s/(\gamma kT_s/m)^{1/2} = (2/\gamma)^{1/2}S \quad (5B.5)$$

[Note that the speed ratio is unambiguously determined from the measured $n(v)$, Eq. (5B.3), whereas the Mach number requires a knowledge of γ at the temperature of the beam.]

A supersonic beam has the virtue of a small spread in the speed distribution. Figure 5B.2 shows typical distributions for nozzle beams, for different values of the speed ratio S.

The nozzle-beam technique is also advantageous for producing molecular beams of widely different translational energies. When an expanding light gas (the "carrier gas") contains a small fraction of a heavier gas (the "seed

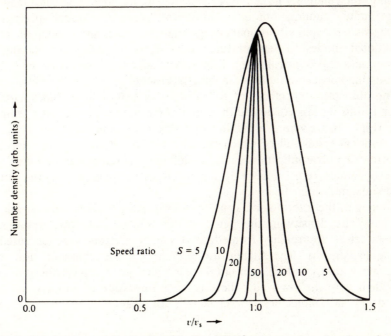

Fig. 5B.2. Calculated speed distributions [via Eq. (5B.3)] in terms of a reduced speed, v/v_s, for supersonic nozzle beams with specified values of speed ratio $S \equiv v_s/\alpha_s$. Distributions with a fractional width as small as 1% have been obtained experimentally.

gas''), the heavy molecules will reach nearly the same mean stream velocity as the light ones, that is, they are accelerated via the collisions in the nozzle. Since they are heavier, however, their kinetic energy will be that much higher. Seeding a heavy reagent in a light inert carrier gas is a practical route to producing reagents with well-defined, high translational energies. By controlling the mole fraction of the seed gas and the nature of the carrier gas, as well as the oven temperature, one can vary the kinetic energy over a wide range from thermal to several eV (for heavy seed molecules).

The sharp velocity distributions of supersonic beams are not their only merit. When a molecular gas is expanded in a jet, substantial cooling of the internal degrees of freedom of the molecules results from the isentropic expansion process, converting random translational motion and internal energy to directed translational motion. Because of the opportunity for many, many intermolecular collisions in the nozzle, and since such collisions will couple the internal degrees of freedom to the translation, such energy-transferring collisions will tend to cool these internal degrees of freedom. The most efficient process (to be discussed in Sec. 6.1) is rotation–translation energy exchange. Measurements of the distribution of rotational energy of molecules in supersonic beams in the collision-free zone show an essentially thermal (Boltzmann) distribution with a temperature not much higher than T_s (and often below 5 K). Regarding vibrational cooling, the relatively large energy spacings in diatomic molecules lead to a more limited cooling; but in polyatomics, where vibrational spacings are much smaller, rapid vibrational energy transfer occurs and cooling of certain vibrational modes to temperatures near T_s is feasible. "Jet-cooling" of polyatomics makes possible much improved spectroscopic measurements by eliminating congestion and minimizing hot bands.

The absorption spectrum of polyatomic molecules in a cold supersonic beam is thus far less congested than that of the bulk vapor at room temperature (Fig. 5B.3), where a significant fraction of molecules are in excited vibrotational states, all of which can also absorb. Preparing (through light absorption) polyatomic molecules in definite excited states and exploring their dynamics (e.g., Fig. 1.5) has been made possible by the use of supersonic molecular beams.†

Cooling of the intermolecular motions eventually leads to condensation of the gas. The density in the beam is typically far too low, but dimers, trimers, etc., are readily formed, and higher clusters are generated by strong expansions. The higher the pressure in the chamber, the more extensive is the "condensation" in the expanded jet. Even for the rare gases with their very shallow wells, there is facile formation of adducts, so-called van der Waals molecules or "clusters", for example, $Ar_2, Ar_3, \ldots, Ar_n$.

---

† At room temperature the internal energy content of a large polyatomic molecule is quite significant. To achieve the best cooling the expanding gas is usually a mixture containing a small mole fraction of the polyatomic in a monatomic carrier gas.

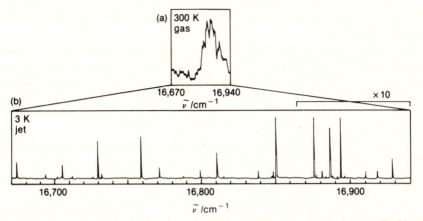

Fig. 5B.3. Jet spectroscopy. (a) Portion of the absorption spectrum of nitrogen dioxide at 300 K. Even at low pressures, the complexity of the spectrum precludes analysis. (b) Jet-cooled spectrum (with an expanded frequency scale). Here, individual spectral lines are resolved and full spectroscopic–structural analysis is possible. [Adapted from D. H. Levy, *Sci. Am.*, Feb. 1984, p. 68).]

When a mixture is expanded, mixed van der Waals molecules are readily produced, for example, ArKr, HCl·Ar, He·I_2, HCl·NO, and even clusters of large organic molecules. The structure, reactivity, and dynamics of such clusters is the subject of Section 7.4.

5.4 The collisional method

As we have already seen, the collisional method is a very powerful one, bringing us closest to our "ideal" experiment discussed in Section 5.1. But some attention is required in order to achieve a useful interpretation of the "raw" experimental measurements. In the following section the essential details involved in the analysis of reactive scattering data are presented.

5.4.1 Angular distribution of reaction products

In our idealized experiment of Section 5.1 we plan to measure the angular distribution of the products for each of the different final internal states. This requires an "internal state analyzer" as a detector, recording the flux of each product in each of its internal states as a function of the scattering angle in the laboratory. We have seen how the use of polarized laser radiation as a probe brings us closer to this ideal.

In the absence of such detailed data, we can still obtain much dynamical information from the "overall" angular distribution of products. However, even to reach this limited objective we must know the products' velocity distribution in order to convert the observed scattered intensity versus angle data measured in the laboratory system to meaningful angular distributions in the c.m. system.

Assuming we know the initial internal energy of the reactants, E_I, and the initial relative translational energy, E_T, then conservation of total energy E gives us a relation between the products' internal energy E'_I and the final relative translational energy E'_T:

$$E = E'_I + E'_T = E_I + E_T - \Delta E_0 \tag{5.5}$$

where ΔE_0 is the reaction endoergicity, that is, the usual zero point to zero-point energy change of the reaction (cf. Fig. 1.1). Thus, given the products' relative velocity, we know the final translational energy and therefore (the sum of) their internal energy E'_I.†

So velocity analysis serves two purposes, first enabling us to transform the laboratory cross-section data to the c.m., and second, determining the distribution of E'_T and thus of products' internal energy as a function of c.m. scattering angle.

But how are we going to represent such a wealth of observations, and in such a way that the major features will be evident? For example, which internal states are most probable? Which directions in space do the products favor? To compact the results, a "flux-velocity" representation is used.

5.4.2 Scattering in velocity space

The most direct presentation of the observed flux distribution of product velocity vectors is by means of flux-velocity contour maps first in the laboratory and then in the c.m. system. Let us develop this concept by means of an example. Consider the elementary reaction

$$K + I_2 \rightarrow KI(v', j') + I$$

Each particular vibrotational state of KI has its own E'_I and hence corresponds to a different final relative translational energy E'_T. However, $E'_T = E - E'_I$, and the final relative velocity is $v' = (2E'_T/\mu')^{1/2}$, where $\mu' = m_{KI}m_I/M$ is the reduced mass of the products and $M = m_K + m_I + m_I$ is the total mass. Now we must relate the measured velocity of the KI in the laboratory system, say $\mathbf{v}'_{KI}$, with the desired final relative velocity in the c.m. system, $\mathbf{v}'$.

First of all, we note that the final c.m. relative velocity can be partitioned between the recoil velocities of the I, say $\mathbf{w}'_I$, and the velocity of the KI, $\mathbf{w}'_{KI}$, with respect to the c.m.:

$$\mathbf{v}' = \mathbf{w}'_{KI} - \mathbf{w}'_I \tag{5.6}$$

---

† In realistic situations there is always a distribution in E_T, albeit often very narrow compared to the total energies E. Also, except for experiments with reagents in selected internal states, there is a distribution in E_I. Thus, in practice, Eq. (5.5) is written in terms of averages, $\langle E \rangle = \langle E_I \rangle + \langle E_T \rangle - \Delta E_0$.

Conservation of momentum requires

$$m_{KI}\mathbf{w}'_{KI} + m_I\mathbf{w}'_I = 0 \tag{5.7}$$

Using Eq. (5.6),

$$\mathbf{w}'_{KI} = \mathbf{v}'(m_I/M) \tag{5.8}$$

But the laboratory velocity of the KI is simply its velocity with respect to the centroid plus the velocity of the c.m. itself:

$$\mathbf{v}'_{KI} = \mathbf{w}'_{KI} + \mathbf{v}_{c.m.} \tag{5.9}$$

where $\mathbf{v}_{c.m.}$ is obtained from the initial velocities

$$m_K\mathbf{v}_K + m_{I_2}\mathbf{v}_{I_2} = M\mathbf{v}_{c.m.} \tag{5.10}$$

Thus, given an observed $\mathbf{v}'_{KI}$, we can calculate $\mathbf{v}'$ and thereby E'_I directly. [Note that only one of the two products need be observed due to momentum conservation, Eq. (5.7).]

We now consider the product distribution in velocity space. A sphere of radius w'_{KI} with the c.m. as origin is the locus of the tips of the velocity vectors of all those KI molecules formed in a particular internal state (and hence given E'_T), irrespective of the angle of scattering. Different KI internal states would each give rise to different spheres in velocity space. The one with the largest radius represents those KI molecules with maximum recoil velocity allowed by energy conservation, that is, $(E'_T)_{max} = E$ [via Eq. (5.5)], so those would be ground-state molecules. All other spheres correspond to internally excited products. The smaller the radius, the larger the internal energy.

The spheres provide a spherical polar coordinate representation of the state-to-state angular distribution. The radius (fixed for given E'_I) is the recoil velocity w'_{KI}, and the polar angle θ is the angle of scattering (with respect to $\mathbf{v}$) in the c.m. system. In the absence of fields, the scattering is cylindrically symmetrical around the initial relative velocity $\mathbf{v}$ (i.e., there is no ϕ dependence) and a given θ gives rise to a cone of equi-intensity scattering in the c.m..

We can take advantage of the above cylindrical symmetry in the absence of fields (i.e., for randomly oriented reactants) and consider the scattering in the plane defined by the velocities of the reactants (Fig. 5.23). The velocity spheres are represented as circles on this plane.

All product KI scattered at a given θ in the c.m. will appear on a given "ray" from the c.m., while all those with a given E'_T (and thus a given E'_I) would be on a given circle (Fig. 5.23). The state-to-state differential solid-angle reaction cross section (Sec. 3.1.6) could be represented on such a plot by "contours" joining all points of equal differential c.m. cross section.[†]

[†] The three-dimensional velocity space distribution is generated from such a planar representation by rotating the figure about the initial relative velocity axis, thereby generating the cones of equi-intensity for a given θ.

(a)

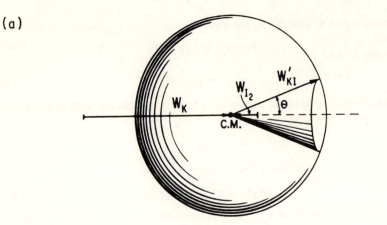

(b)

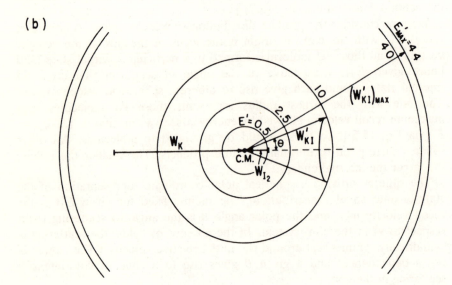

(c)

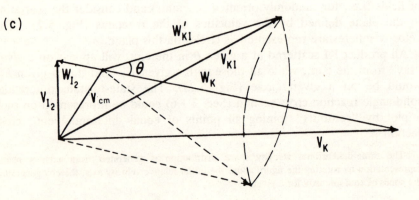

For most diatomics, however, the rotational energy spacings are so small (compared to the energy resolution), that is, the levels so closely spaced, that we treat the final energy (and thus the velocity, v'_{KI}) as a continuous variable. The circles corresponding to all possible E'_T are so very tightly spaced that they meld into a continuum. Thus, instead of trying to determine the differential solid-angle cross section into a definite internal energy state E'_I, we assume a continuous distribution of product energy states. Thus, we consider the number of molecules (of KI) scattered per unit solid angle† and unit time, with the final translational energy in the range E'_T to $E'_T + dE'_T$:

$$d\dot{N}(\theta, E'_T) = \frac{d^3\sigma(\theta, E'_T)}{d^2\omega \, dE'_T} d^2\omega \cdot I_K N_{I_2} \, dE'_T \qquad (5.11)$$

Here, as previously [Eq. (3.17)], I_K is the flux of K atoms (number per unit area per unit time) and N_{I_2} is the number of I_2 molecules in the target volume.

Such a differential cross section is really a sum over many discrete contributions from all final states of the KI with internal energy in the appropriate interval. Thus, formally, we have

$$\frac{d^3\sigma(\theta, E'_T)}{d^2\omega \, dE'_T} = \sum_{v',j'} \frac{d^2\sigma(vj \to v'j'; \theta)}{d^2\omega} \delta(E - E'_T - E'_I) \qquad (5.12)$$

The discrete quantal nature of the internal states is smeared out by any energy-averaging (as in the experiment).

The velocity-analysis experiment itself is usually reported in terms of the scattered flux with recoil velocity in the corresponding range $\mathbf{w}', \mathbf{w}' + d\mathbf{w}'$

† In this section we use $d^2\omega$ for an element of solid angle, $\sin\theta \, d\theta \, d\phi$, to emphasize its two-dimensional nature.

Fig. 5.23. (a) Sphere corresponding to locus of recoil velocities $\mathbf{w}'_{KI}$ of scattered KI from the K + I_2 reaction for a given translational-energy release (constant E'_T and thus w'_{KI}). The vectors $\mathbf{w}_K$ and $\mathbf{w}_{I_2}$ are the initial relative velocities of the reactants with respect to the center of mass (c.m.). The cone shown corresponds to the KI, which recoils at an angle θ with respect to the incident relative velocity of the K. (The recoiling partner, I, always directed opposite to the KI, is not shown). (b) Spheres of different radii w'_{KI} (projected as circles on a plane) corresponding to the indicated E'_T values (in kcal mol^{-1}). The maximum value of the recoil velocity of the KI with respect to the c.m., $(w_{KI})_{max}$, is associated with $E'_{max} = E \simeq 44$ kcal mol^{-1}. The particular value of w'_{KI} yielding the cone shown, and that in (a), corresponds to $E'_T = 10$ kcal mol^{-1}. The initial relative translational energy is $E_T = 2.7$ kcal mol^{-1}. (c) Velocity vector triangle (Newton diagram) showing the relation between the laboratory velocity of the product KI, $\mathbf{v}'_{KI}$ and its recoil velocity with respect to the c.m., $\mathbf{w}'_{KI}$, [Eq. (5.9)]. The initial relative velocity $\mathbf{v} = \mathbf{v}_K - \mathbf{v}_{I_2} = \mathbf{w}_K - \mathbf{w}_{I_2}$ forms the hypotenuse of the Newton triangle. As in (b), $\mathbf{w}'_{KI}$ is the recoil velocity of the KI scattered (here in the crossed-beam plane) at a c.m. angle θ with a given E'_T. The dashed lines denote a second (in-plane) w'_{KI} [(a)] for KI recoiling at the same θ and w'_{KI} but with a quite different laboratory velocity! The cone of equiproduct intensity, given θ and variable ϕ (a), is indicated by the light dashed arcs.

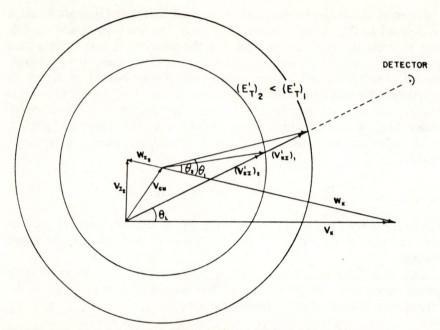

Fig. 5.24. Velocity vector diagram [similar to Fig. 5.23(c)] showing how the KI flux at the detector (set at laboratory angle Θ_L) consists of contributions from more than one c.m. scattering angle θ and recoil velocity w'_{KI} (corresponding to different values of E'_T).

(where $E'_T \propto w'^2$). The corresponding cross section is

$$\frac{d^3\sigma(\theta, w')}{d^2\omega\, dw'} = \frac{d^3\sigma(\theta, E'_T)}{d^2\omega\, dE'_T}\left(\frac{dE'_T}{dw'}\right) \tag{5.13}$$

The usual† flux velocity contour map is of this quantity in a θ, w' polar coordinate system.

It should be emphasized that the c.m. cross section is *not* simply proportional to the observed scattered intensity. A measurement of the product flux at a given laboratory angle, Θ_L, samples different E'_T circles at different θ values (Fig. 5.24). However, for the ideal experiment with velocity analysis and nearly monoenergetic reagent beams, the scattered intensity into a volume element $d\mathbf{v}'_{KI}$ in the laboratory velocity space can be related to the scattered intensity into a volume element $d\mathbf{w}'_{KI}$ in the c.m. velocity space. Thus,

$$d^3\sigma(\theta, w')/d\mathbf{w}' = d^3\sigma(\Theta_L, v')/d\mathbf{v}' \tag{5.14}$$

or

$$\frac{d^3\sigma(\theta, w')}{d^2\omega\, dw'} = \frac{d^3\sigma(\Theta_L, v')}{d^2\Omega\, dv'}\left(\frac{w'}{v'}\right)^2 \tag{5.15}$$

† Some contour maps represent the velocity space distribution $d^3\sigma(\theta, w')/d\mathbf{w}'$, where $d\mathbf{w}' = (w')^2\, dw'\, d^2\omega$. These are called "Cartesian" maps.

where $(w'/v')^2$ is the Jacobian appropriate for the present transformation [cf. Eq. (3A.4) for elastic scattering].

5.4.3 Intensity contour maps: Qualitative aspects of flux–velocity distributions

Most diatomic molecules (and essentially all polyatomics) have rotational energy spacings that are small compared to the collision energy. Hence velocity analysis of the collision products usually cannot resolve the individual final internal states. Rather, what can be measured is the relative intensity of the products† $P(\theta, w')$ scattered into an element of solid angle, with the final product velocity in the range w' to $w' + dw'$. In other words, $P(\theta, w')$ is the product flux-velocity distribution, with the discrete quantal nature of the final internal energy smeared out.

Considered as a function of two variables, $P(\theta, w')$ can be represented as contours of equal product intensity in the polar coordinates w' and θ. The origin is at the c.m., and the distance of every point from the origin is proportional to the final product velocity w'.

Figure 5.25 shows such a contour map for KI from the $K + I_2$ reaction. The zero of θ is taken in the direction of the initial relative velocity $\mathbf{w_K}$ of the atomic reactant. It is immediately seen from the contours that the KI molecules are scattered predominantly in the forward direction and that the typical recoil velocity $\mathbf{w'_{KI}}$ is quite small.‡ Thus, most of the product KI molecules must be highly excited. These findings are in accord with our expectations based on the harpoon mechanism (Sec. 4.2).

Now let us look at another KI-forming reaction:

$$K + ICH_3 \rightarrow KI + CH_3.$$

Figure 5.26 displays the contour map, which shows predominantly backward scattering of the KI [a "rebound" reaction (Sec. 3.3) with a repulsive (or late) energy release].

Figure 5.27 is a map for DI from the reactions

$$D + I_2 \rightarrow DI + I,$$

showing the "side scattering" of product. Similar results are obtained when I_2 is replaced by IBr or ICl. These distributions can be interpreted in terms of a Franck–Condon-type model (Secs. 1.4 and Appendix 4C). During the rapid attack by the light atom, the two heavy atoms will not move much. We can thus consider the $D + IX$ reaction as analogous to a direct photodissociation of IX, in which case the heavy atoms will recede along the

---

† As in Section 3.1.6,

$$P(\theta, w') = [d^3\sigma(\theta, w')/d^2\omega \, dw']/\sigma_R$$

where σ_R is the reaction cross section.

‡ The velocity space distribution is shown in Fig. 3.24.

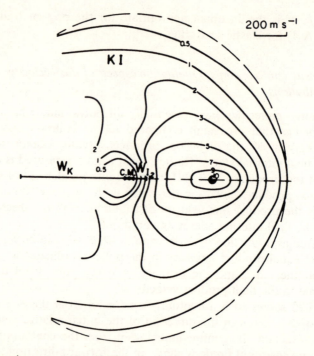

Fig. 5.25. Intensity contour map of $P(\theta, w')$, the flux (velocity–angle) distribution for the KI product of the $K + I_2$ reaction. Plotted are the contours of $P(\theta, w')$. All KI molecules with the same recoil velocity w' will lie on a circle centered at the c.m. (dot). The dashed circle corresponds to the highest recoil velocity w'_{max} allowed by conservation of energy, and hence to internally cold KI molecules. The closer the point is to the c.m., the higher is the internal energy of the KI product. KI scattered at a given direction θ with respect to the direction of the incident K atom will appear on a given ray (from the c.m.). Shown also are the initial relative velocities of K and I_2 ($E_T = 2.7$ kcal mol^{-1}). [Adapted from K. T. Gillen, A. M. Rulis, and R. B. Bernstein, *J. Chem. Phys.*, *54*, 2831 (1971).]

direction of their bond axis. Because of the low mass of the D atom, DI should also take the direction of the IX axis at the moment of the bond rupture. The direction of recoil of the DI product with respect to the direction of the incident atom thus tells us the angle of approach of the D atom for those collisions that favor reaction. From the scattering data we deduce a preferential "sideways" attack of the D atom on the IX molecule.

Figure 5.28 shows a flux-velocity contour map for the reaction

$$O + Br_2 \rightarrow OBr + Br$$

There is marked forward–backward symmetry to the scattered product, suggesting the existence (quite reasonably!) of a long-lived complex (see Sec. 7.2). We will discuss the chemical implications of such results later.

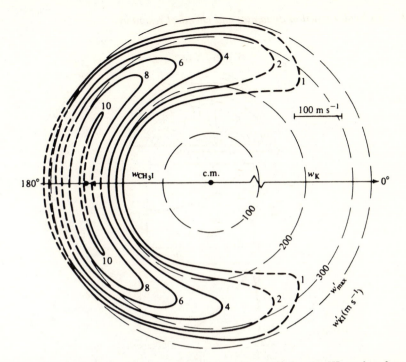

Fig. 5.26. Flux (velocity–angle) contour map (as in Fig. 5.25) for the KI product from the $K + ICH_3$ reaction ($E_T = 2.8\,kcal\,mol^{-1}$). [Adapted from A. M. Rulis and R. B. Bernstein, *J. Chem. Phys.*, **57**, 5497 (1972).]

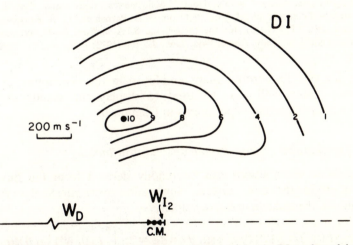

Fig. 5.27. Flux (velocity–angle) contour map (upper half only) for the DI product of the $D + I_2$ reaction, for $E_T \simeq 9\,kcal\,mol^{-1}$. The extensive "sideward" scattering has been interpreted in terms of a preferred bent geometry for the supermolecule in the transition state with the D atom approach sideways to I_2. This is consistent with molecular-orbital considerations ("Walsh's rules" for triatomic structures) and with the DIPR model (Appendix 4C). [Adapted from J. D. McDonald, P. R. LeBreton, Y. T. Lee, and D. R. Herschbach, *J. Chem. Phys.*, **56**, 769 (1972); angular distributions and branching ratios for the reactions of D + IBr and ICl are also presented.]

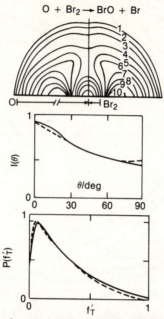

Fig. 5.28. Detailed results on the reactive scattering of $O + Br_2$ at $E_T = 3.0\,kcal\,mol^{-1}$. (Top) C.m. contour map of BrO flux (velocity–angle) distribution. (Center) Angular distribution at fixed BrO recoil velocity. Zero degrees refers to the direction of the incident O atom in the c.m. system. (Bottom) Distribution of product relative translation energy E'_T expressed as the fraction $f'_T \equiv E'_T/E$, where $E \simeq 13\,kcal\,mol^{-1}$. Solid curves experimental; dashed: calculated from a statistical complex model. [Adapted from D. D. Parrish and D. R. Herschbach, *J. Am. Chem. Soc.*, **95**, 6133 (1973); for the theory, see S. A. Safron, N. D. Weinstein, D. R. Herschbach, and J. C. Tully, *Chem. Phys. Lett.*, **12**, 564 (1972).]

The flux-velocity contour maps employed in the description of reactive scattering results will also be used to display the results of inelastic (nonreactive) scattering experiments in Chapter 6.

5.4.4 Translational exoergicity and angular distributions

Less detailed distributions can be readily derived from the flux-velocity distribution in the c.m. system. Integration over all angles yields the distribution of products' relative velocity,

$$P(w') = \iint P(\theta, w') \sin\theta \, d\theta \, d\phi = 2\pi \int_0^\pi P(\theta, w') \sin\theta \, d\theta \quad (5.16)$$

The distribution in translational energy, $P(E'_T)$, is obtained by a change of variable $P(E'_T)\, dE'_T = P(w')\, dw'$ or

$$P(E'_T) = P(w') \, |dw'/dE'_T| \quad (5.17)$$

Figure 5.29 shows the results for two reactions yielding KI, obtained from

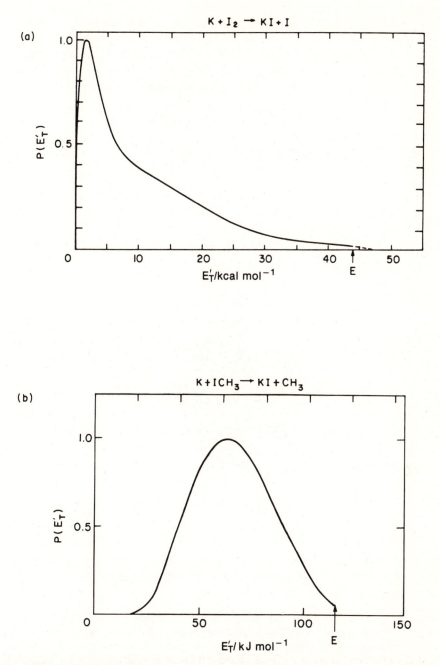

Fig. 5.29. (a) Distribution of final translational energy $P(E'_T)$ for the reaction $K + I_2 \rightarrow KI + I$ (total available energy indicated by the arrow). Note the low average translational release, compared with that for the reaction of $K + CH_3I$ [*cf.* (b)]. (From the data of Fig. 5.25.) (b) Products' relative translational energy distribution for the $CH_3I + K$ reaction; arrow denotes conservation limit. (Based on results of Fig. 5.26.)

Figs. 5.25 and 5.26. The different character of the two reactions is obvious. The $K + I_2$ reaction has a very low translational release typical of a spectator mechanism, while the rebound-type $K + CH_3I$ reaction has a much higher translational exoergicity. Note that such distributions can be further "coarse-grained" simply by evaluating the average value of E_T', that is,

$$\langle E_T' \rangle = \int_0^E E_T' P(E_T') \, dE_T' \tag{5.18}$$

Alternatively, one can integrate over all values of w' to obtain the angular distribution as measured in "primitive" experiments (Sec. 3.3) without velocity resolution:

$$P(\theta) = \int_0^{w_{max}} P(\theta, w') \, dw' \tag{5.19}$$

Here, w_{max} is the maximal value of the product velocity (i.e., for the limit in

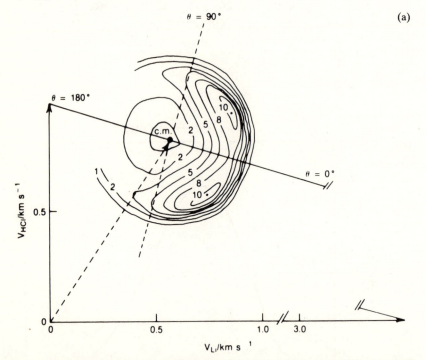

Fig. 5.30. Experimental product flux–velocity distribution for the reaction $HCl + Li \rightarrow LiCl + H$ at $E_T = 9.2 \, kcal \, mol^{-1}$. (a) Flux contour map in the c.m. system superimposed on the laboratory velocity vector triangle. The origin of the product flux–velocity map is at the c.m. (the tip of the centroid vector). The LiCl product is concentrated in the "forward" hemisphere ($\theta < 90°$) but with a peak intensity at *ca.* 45° with a recoil velocity w' about 3/4 of w_{max}'. (b) Integrated results in the form of the products' final relative translational distribution $P(E_T')$ and c.m. angular distribution $P(\theta)$. Solid curve: best fit to data; dashed curves enclose shaded zones of uncertainty. [Adapted from C. H. Becker, P. Casavecchia, P. W. Tiedemann, J. J. Valentini, and Y. T. Lee, *J. Chem. Phys.*, *73*, 2833 (1980).]

which the products are in their ground internal states and the total energy is released as translation, $E'_T = E$).

Figure 5.30 shows $P(E'_T)$ and $P(\theta)$ derived from a c.m. flux-velocity contour map for the crossed-beam reaction

$$\text{HCl} + \text{Li} \rightarrow \text{LiCl} + \text{H}$$

To evaluate the total reaction cross section it is not sufficient to know the relative velocity-angle distribution. Rather, one needs the actual (i.e., absolute) differential cross section, Eq. (5.12). The first integration is over all angles, yielding [cf. Eq. (5.16)] the cross section for scattering into a given final translational energy,

$$\frac{d\sigma}{dE'_T} = \iint d^2\omega \, \frac{d^3\sigma(\theta, E'_T)}{d^2\omega \, dE'_T} = \iint \frac{d^3\sigma}{d^2\omega \, dE'_T} \sin\theta \, d\theta \, d\phi$$

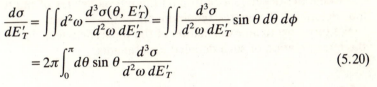

$$= 2\pi \int_0^\pi d\theta \sin\theta \, \frac{d^3\sigma}{d^2\omega \, dE'_T} \tag{5.20}$$

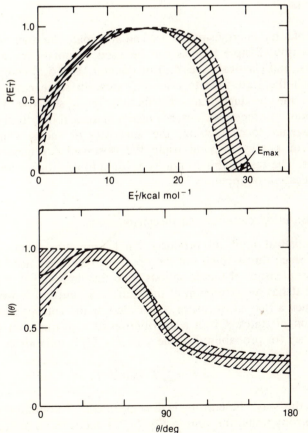

(b)

Fig. 5.30. (continued)

This is the "continuum" analog of the state-to-state total cross section [cf. Eq. (5.12)],

$$\frac{d\sigma}{dE'_T} = \sum_{v'j'} \sigma(vj \rightarrow v'j') \delta(E - E'_T - E'_I)$$

The final integration, over E'_T, yields the reaction cross section, for given reactants,

$$\sigma_R = \int_0^E \frac{d\sigma}{dE'_T} dE'_T = \sum_{v'j'} \sigma(vj \rightarrow v'j') \qquad (5.21)$$

When measurements are made for different initial conditions, one can determine the dependence of σ_R on the internal states of the reactants, and thus determine the energy requirements of the reaction.

We turn now to an examination of the theoretical techniques available for the interpretation of such detailed experiments.

5.5 Surprisal

Given a potential energy surface, we can anticipate the main role of energy in the reaction. Simple models can be used to obtain more quantitative predictions, and classical (Sec. 4.3) or quantal (Sec. 5.6) computations can generate all the state-to-state rate constants. In some cases we obtain a highly specific product state distribution or a selective reagent state consumption; in others, we observe nearly statistical distributions. It would be advanageous to characterize the selectivity of energy consumption or specificity of energy disposal using the observed (or computed) results themselves. The rest of this section is devoted to this topic with some of the details relegated to Appendix 5.C.

5.5.1 Measures of selectivity and specificity

In the statistical limit all products' final (quantum) states are equally probable. Since this is often not the observed result, we need to provide a quantitative measure of the deviations from this limit. What is required is a measure that has an extremum at the statistical limit but is well defined in general. Such a (global) measure is provided by the entropy associated with a population of states. Given a distribution of final quantum states where the frequency (or probability) of the state f is $P(f)$, the entropy is given by

$$S = -\sum_f P(f) \ln P(f) \qquad (5.22)$$

Defined in this way, the entropy is dimensionless. [To obtain entropy in the familiar entropy units, the sum in Eq. (5.22) needs to be multiplied by the gas constant R.] The entropy (5.22) generalizes the standard thermodynamic concept in one sense: So far we have not specified the distribution $P(f)$ over quantum states. It can be *any* distribution, including that of the nascent products of a chemical reaction. If we specify that $P(f)$ is a distribution of quantum states for a system in *equilibrium*, then the entropy (5.22) is just the very one introduced by Clausius. To verify this, recall that at thermal equilibrium $P(f) = \exp(-E_f/\mathbf{k}T)/Q$, where Q is the partition function. Using this result in Eq. (5.22), multiplying the sum by $R = \mathbf{k}N_A$ (where N_A is Avogadro's number), we recover the familiar result,

$$S = R \ln Q + \langle E \rangle / T$$

where

$$\langle E \rangle = N_A \sum_f E_f P(f)$$

is the mean energy per mole.

In principle, Eq. (5.22) is all that we require. In applications it is convenient to recognize that one seldom measures the fully resolved products' quantum state distribution: One often groups together quantum states into larger sets. In an atom–diatom reaction, we can consider for example the set of quantum states corresponding to AB in a given vibrotational state (or in a given vibrational state, etc.). The counting of the number of states in a given group is discussed for a number of simple cases in Appendix 5C. For the moment, let there be $g(v)$ states in the group whose probability is $P(v)$. Then the probability per quantum state f of the group v is $P(f) = P(v)/g(v)$. Hence,

$$S = -\sum_f [P(v)/g(v)] \ln[P(v)/g(v)]$$

Let us now sum over f by summing first within each group on all the states that belong to the group and then over all groups. For each value of v there are $g(v)$ terms in the sum and each term has the same value. Therefore, we are left with an expression for the entropy as a sum over groups

$$S = -\sum_v P(v) \ln[P(v)/g(v)] \qquad (5.23)$$

Note that the sum is no longer of the form of Eq. (5.22).

It is advantageous to rewrite (5.23) as follows. Let $P^0(v)$ be the distribution when all final quantum states are equally probable,

$$P^0(v) = g(v) \Big/ \sum_v g(v) \qquad (5.24)$$

We shall refer to $P^0(v)$ as the *prior distribution* (our prior expectation) for reasons given below. We now write the entropy in terms of the deviance of the actual distribution from the prior one. The result is

$$S = S_{\max} - DS \qquad (5.25)$$

where we have introduced S_{max},

$$S_{max} = \ln\left(\sum_v g(v)\right) = \ln\left(\sum_f 1\right) \qquad (5.26)$$

the logarithm of the number of quantum states. The subscript "max" is because (5.26) is the highest possible value of the entropy, realized when all final quantum states are equally probable. Since $DS = S_{max} - S$, it is a measure of how far the actual entropy is below its maximal value, and is termed the "entropy deficiency",

$$DS = \sum_v P(v)\ln[P(v)/P^0(v)] \geq 0 \qquad (5.27)$$

The entropy deficiency is zero (i.e., the entropy is at its global maximal) if (and only if) the actual distribution is the prior distribution.

Given an observed final distribution $P(v)$, we can consider either the *integral* measure of specificity given by DS or the *local* measure given by $I(v)$,

$$I(v) = -\ln[P(v)/P^0(v)] \qquad (5.28)$$

known as the *surprisal*.

In Section 5.5.3 we shall point out that everything we have said about final states is equally valid for initial states, so much so that the surprisal of energy disposal in the forward reaction is numerically equal to the surprisal of the energy requirements of the reverse reaction at the same total energy E. The same is true for the entropy deficiency.

5.5.2 Surprisal analysis of energy disposal

Consider first a simple situation, the vibrational energy distribution in the elementary reactions:

$$Cl + \begin{cases} HI \\ DI \end{cases} \rightarrow \begin{cases} HCl(v) \\ DCl(v) \end{cases} + I$$

Using the prior distribution (from Appendix 5C),

$$P^0(v) \propto (E - E_v)^{3/2} \propto (1 - f_v)^{3/2} \qquad (5.29)$$

Figure 5.31 shows the surprisal $-\ln[P(v)/P^0(v)]$ plotted versus the reduced variable f_v, where

$$f_v = E_v/E \qquad (5.30)$$

is the fraction of the total energy in the product vibration. While the observed distribution, $P(v)$, is qualitatively different in shape than the prior distribution (both are shown in Fig. 5.31), the surprisal is well represented as a straight line,

$$I(f_v) \equiv -\ln[P(v)/P^0(v)] = \lambda_0 + \lambda_v f_v \qquad (5.31)$$

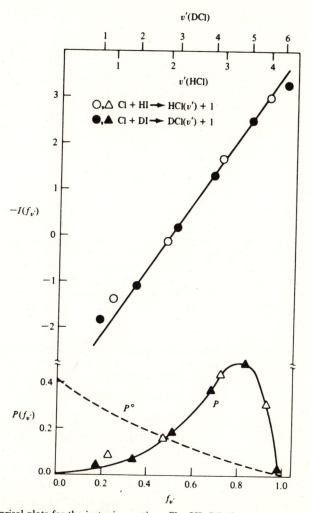

Fig. 5.31. Surprisal plots for the isotopic reactions Cl + HI, DI. Open symbols denote the HCl product; solid symbols, DCl. Note that the use of the reduced energy variable $f_{v'}$ unifies the data for the two isotopic product molecules. The upper graph is a vibrational surprisal plot with $\lambda_v = -8.0$. [Adapted from A. Ben-Shaul, R. D. Levine, and R. B. Bernstein, *J. Chem. Phys.*, *57*, 5427 (1972); data sources therein.]

In Section 5.5.4 we shall interpret f_v as a "constraint" on the vibrational state distribution and show that λ_0 is a function of λ_v, so that Eq. (5.31) is a one-parameter representation of the data. That the representation is an accurate one can be seen from the Fig. 5.31 (which also shows the distribution whose surprisal is exactly linear).

Note also the advantage of representing the distribution and the surprisal in terms of the reduced variable f_v. It has largely eliminated the isotope effects. This is also evident in Fig. 5.32, which shows a plot of the surprisal

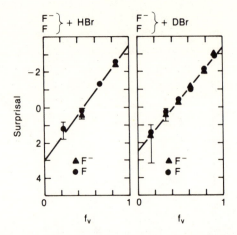

Fig. 5.32. Surprisal plot for the nascent HF vibrational distribution from the $F^- + HBr$ and $F + HBr$ reactions and corresponding results for DF. For the relatively widely spaced HF (or DF) rotational levels, the prior distribution is computed by a discrete summation over rotational states [see Eq. (5C.11)]. [Adapted from A. O. Langford, V. M. Bierbaum, and S. R. Leone, *J. Chem. Phys.*, *83*, 3913 (1985). Nascent distributions for $F + HBr/DBr$ from K. Tamagake, D. W. Setser, and J. P. Sung, *J. Chem. Phys.*, *73*, 2203 (1980).]

of the HF vibrational distribution produced in both the neutral and negative-ion reactions,

$$F + HBr \rightarrow FH + Br$$

$$F^- + HBr \rightarrow FH + Br^-$$

and the corresponding results for DBr. As we shall argue in Section 5.5.4, the functional form (5.31) for the surprisal implies that a measure of the extent of products' vibrational excitation is $\langle f_v \rangle$, the average fraction of the available energy in products' vibration. The general one-to-one relation between $\langle f_v \rangle$ and the magnitude of λ_v will be shown in Fig. 5.39 below. Since $\langle f_v \rangle$ decreases as λ_v increases (from negative to positive values), one speaks of distributions with negative values of λ_v as being "hotter" than the prior and vice versa.

It is the negative value of λ_v for the $Cl + HI$ reaction that ensures that $P(v)$ peaks at a high value of v. Not all reactions channel energy preferentially into vibration. Figure 5.33 shows the results for the CuF vibrational distribution in several distinct electronic states produced in the chemiluminescent reaction

$$Cu + F_2 \rightarrow CuF^*(v) + F$$

As is often the case for reactions involving a multitude of curve crossings (see Sec. 6.6), what is observed is essentially the prior distribution. Another illustration of a prior distribution of the excess energy is the reaction

$$H + DCl \rightarrow HCl(v) + D$$

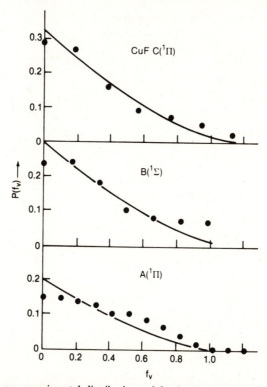

Fig. 5.33. Circles are experimental distributions of f_v, the fraction of energy available to the products appearing in CuF* vibrations, based on detection of product flux, for the reactions $Cu + F_2 \rightarrow CuF(A, B, C) + F$. Solid lines represent prior distributions. [Adapted from R. W. Schwenz and J. M. Parson, *J. Chem. Phys.*, *73*, 259 (1980).]

shown in Fig. 5.34. It can even be that the energy release into vibration is less than that of the prior distribution. An example is shown in Fig. 5.35 for the $H + D_2$ reaction using hot H atoms. However, for exoergic reactions, where the energy release is typically early, a negative λ_v is the norm, as is evident from the sample of results in Table 5.3.

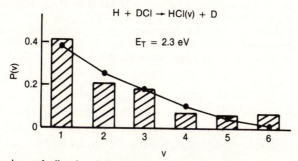

Fig. 5.34. Experimental vibrational distributions of HCl from the reaction of translationally hot H atoms with DCl at $E_T = 2.3$ eV (histogram). Points, prior distribution. [Adapted from C. A. Wight, F. Magnotta, and S. R. Leone, *J. Chem. Phys.*, *81*, 3951 (1984).]

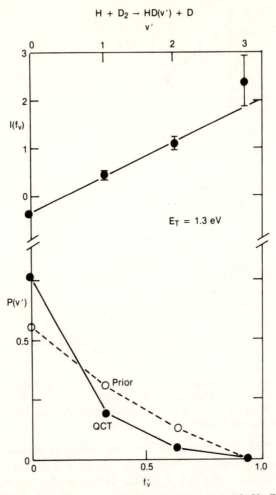

Fig. 5.35. Vibrational surprisal plot for the $H + D_2$ reaction at 1.3 eV. (Upper panel) The surprisal of the classical trajectories (QCT)-computed vibrational state distribution of HD (dots) versus f_v (lower scale) or versus v (upper scale). The slope of the straight line is $\lambda_v = 2.3$. Shown also are the error bars on the surprisal due to the QCT statistical uncertainties. (Lower panel) The QCT-computed vibrational distribution (dots). The solid line connects the points with exact linear surprisal. Prior distribution: open circles. [Adapted from E. Zamir, R. D. Levine, and R. B. Bernstein, *Chem. Phys. Lett.*, 107, 217 (1983).]

Surprisal analysis is not limited to the final vibrational state distribution. Here, we consider the rotational-state distribution. For a given vibrational manifold it is convenient to introduce the reduced variable g_R,

$$g_R = E_R/(E - E_v) \tag{5.32}$$

which varies between 0 and 1 within any one vibrational manifold. The joint

Table 5.3 Surprisal analysis of vibrational energy disposal[a]

Reactants	E (kcal mol^{-1})	λ_v	$\langle f_v \rangle$
HBr + Cl	18.2	-3.0 ± 0.6	0.38
HBr + F[b]	51.0	-5.7 ± 0.1	0.60
DBr + F[b]	51.5	-5.4 ± 0.2	0.58
HI + O	34.4	-6.9 ± 0.1	0.67
H$_2$Se + F	61.3	-3.8 ± 0.5	0.48
H$_2$Se + Cl	29.1	-5.8 ± 0.5	0.40
H$_2$Se + O	29.0	-6.9 ± 0.1	0.47

[a] Adapted from B. S. Agrawalla and D. W. Setser, in *Gas-Phase Chemiluminescence and Chemi-ionization*, A. Fontijn, ed., North-Holland, Amsterdam 1985 (data sources therein).

[b] See also Fig. 5.32.

vibrotational distribution is then represented by

$$-\ln[P(v, j)/P^0(v, j)] = \lambda_0 + \lambda_v f_v + \theta_R g_R \qquad (5.33)$$

where the prior distribution is (see Appendix 5C)

$$P^0(v, j) \propto (2j + 1)(E - E_v - E_R)^{1/2} \qquad (5.34)$$

For the H + D$_2$ reaction such a representation has already been used in Fig. 1.4 and for the O(^1D) + H$_2$ reaction in Fig. 4.30. A lower rotational excitation is shown in Fig. 5.36 for the H + H$_2$ reaction using quantum-mechanical scattering computations at a low energy to generate the final H$_2$ rotational distribution. A colder than prior population of OH rotational states is found for the H + CO$_2 \rightarrow$ HO + CO reaction (Fig. 5.37), in contrast to the hotter than prior ($\theta_R = -3.5$) OH rotational distribution from O(^1D) + H$_2$ shown in Fig. 4.30.

5.5.3 Surprisal analysis of energy requirements

The value of the surprisal of, say, the vibrational energy disposal in the exoergic O + CS reaction equals the value of the surprisal of the relative reactivity of the vibrational states in the reversed (endoergic) reaction

$$S + CO(v) \rightarrow CS + O$$

What requires comment is what we mean by the distribution $P(v)$ for the reverse reaction.

The answer has already been provided in Sections 4.4.3 and 4.4.9. The probabilities we have been talking about are the very same ones defined there. The expression (4.62), symmetric in the two probabilities, was derived on the basis of detailed balance. The same procedure applies to more averaged rates. Consider the exoergic atom–molecule exchange

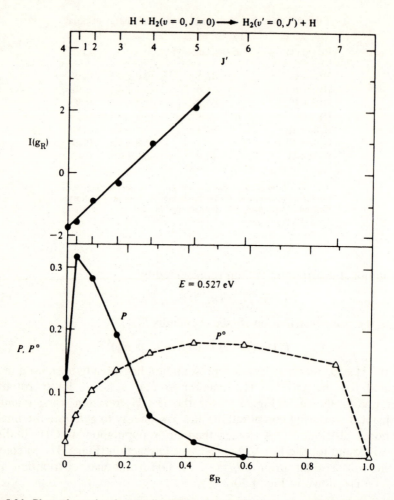

Fig. 5.36. Plots of rotational state distribution for the H_2 ($v' = 0$) product of the hydrogen exchange reaction $H + H_2$ ($v = 0, j = 0$) → H_2 ($v' = 0, j'$) + H at the indicated total energy. (Bottom) Solid circles: $P(j')$ calculated via accurate quantal method; open triangles: prior $P^0(j')$. (Top) Rotational surprisal plot, all as a function of the products' reduced rotational energy g_R. [Adapted from R. E. Wyatt, *Chem. Phys. Lett.*, *34*, 167 (1975).]

reaction

$$A + BC(v) \underset{r}{\overset{f}{\rightleftharpoons}} AB(v') + C$$

The thermal rate constants for the forward (f) and reverse (r) reactions at a temperature T, i.e., $k_f(T)$ and $k_r(T)$, are related by the equilibrium constant $K(T) = k_f(T)/k_r(T)$.

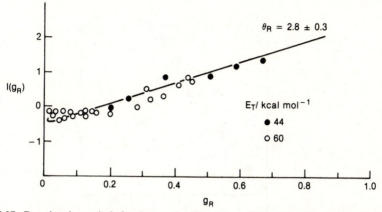

Fig. 5.37. Rotational surprisals for the reaction $H + CO_2 \rightarrow OH + CO$ at the indicated collision energies. [Adapted from J. Wolfrum, *J. Chem. Phys.*, *89*, 2525 (1985).]

From Eq. (4.38) we obtain

$$\frac{k_f(\text{all} \rightarrow v'; T)}{k_r(v' \rightarrow \text{all}; T)} = K(T)p(v') \tag{5.35a}$$

We make use of the operational definition of $P(v')$, the products' vibrational state distribution for the forward reaction,

$$P(v') = k_f(\text{all} \rightarrow v'; T) \Big/ \sum_{v'} k_f(\text{all} \rightarrow v'; T)$$

$$= k_f(\text{all} \rightarrow v'; T)/k_f(T) \tag{5.35b}$$

Substituting (5.35a), we obtain

$$P(v') = K(T)p(v')k_r(v' \rightarrow \text{all}; T)/k_f(T)$$

$$= p(v')k_r(v' \rightarrow \text{all}; T)/k_r(T) \tag{5.35c}$$

With $p(v')$ a thermal distribution, i.e., $p(v') \propto \exp(-E_{v'}/kT)$, Eq. (5.35c) yields

$$k_r(v' \rightarrow \text{all}; T) \equiv k_r(v'; T) = k_r(T)P(v')/p(v') \tag{5.36a}$$

That is,

$$k_r(v'; T) \propto k_r(T)P(v') \exp(E_{v'}/kT) \tag{5.36b}$$

Over and above the exponential increase with $E_{v'}$ from $p(v')$, we have the usual increase of $P(v')$ with $E_{v'}$ to give a strong positive $E_{v'}$ dependence to $k_r(v'; T)$—e.g., Fig. 1.3.

Equation (5.35c) defines $P(v')$, the distribution of vibrational states in the reactive reactants for the (reversed) reaction. The very same surprisal analysis characterizes both the energy disposal in one direction and the

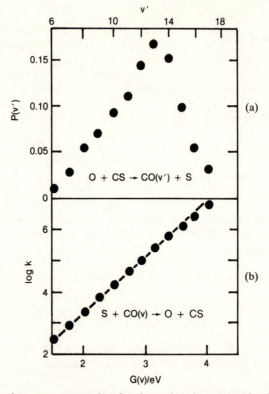

Fig. 5.38. Vibrational energy consumption for the endoergic reaction $S + CO(v) \rightarrow O + CS$ deduced from the product vibrational distribution for the reverse, exoergic reaction $O + CS \rightarrow CO(v') + S$. (a) Experimental data on relative rates of formation [expressed as $P(v')$] of $CO(v')$ from the $O + CS$ reaction at 300 K. By detailed balance, $P(v)$ is also the reaction probability of the different vibrational states of the reverse reaction. (b) Plot of the log of the relative rate constant for the $S + CO(v)$ reaction as a function of the vibrational energy of the CO, i.e., $G(v)$. [Adapted from H. D. Kaplan, R. D. Levine, and J. Manz, *Chem. Phys.*, *12*, 447 (1976).]

energy requirements in the other direction. Typical results are shown in Fig. 5.38.

5.5.4 The maximum entropy formalism

The maximum entropy formalism provides an interpretation of the significance of the results of surprisal analysis. The empirical observation is that distributions can deviate from the (prior) expectations where all final quantum states are equally probable. To account for such a deviance it is necessary to reformulate the problem so that the statistical distribution, where all states are equally probable, is but a limiting case. In Section 5.5.1 we noted that the prior distribution is the one of maximal entropy (and we shall offer a formal proof below). If there are deviations from the prior distribution, the entropy of the distribution will not reach its maximal possible value (at the given total energy). Let us adopt the premise that the distribution is as statistical as it can be. If it is not equal to the prior distribution, there must be some constraints that preclude this. The technical premise is then that the distribution of quantum states of the system is of maximal entropy *subject to constraints*. That the maximum is subject to constraints is the central point. The "physics" of the problem enters through the choice of constraints. Before we consider this point further, let us spell out the formalism itself.

A constraint on a distribution is an observable A, say the vibrational energy, which has the value of $A(f)$ when the system is in the quantum state f and *whose average value,*

$$\langle A \rangle = \sum_f A(f)P(f) \tag{5.37}$$

is specified. Since there are many final states f and only one equation, Eq. (5.37), the numerical value of $\langle A \rangle$ does not suffice to determine a unique distribution $P(f)$.

When the prior distribution is used in (5.37), the resulting numerical value of $\langle A \rangle$ will be different. Thus, while (5.37) does not suffice to determine the sought-after distribution $P(f)$, it does suffice to imply that it differs from the prior one. There is, however, an additional constraint on the distribution, one that is always present, namely, that the distribution is normalized:[†]

$$1 = \sum_f P(f) \tag{5.38}$$

The condition of normalization, (5.38), cannot be used to distinguish between the prior and the actual distributions since it applies to both. Not so for condition (5.37). Not all normalized distributions will lead to the same value for $\langle A \rangle$. When we specify the numerical value of $\langle A \rangle$ we exclude many of the distributions that are allowed if the only condition imposed is normalization. Say we now impose yet another constraint: The mean value of some observable B, i.e., $\langle B \rangle$, is also specified. Among all the normalized distributions that have a given value for $\langle A \rangle$, some will and some will not have the given value for $\langle B \rangle$. Any time we add a constraint we narrow the range of possible distributions.[‡] One can now go on, adding more and more constraints until the distribution is pinned down. We want to impose the *minimal* number of constraints and, among the set of distributions that satisfy those constraints, choose the one whose entropy is maximal.

---

† Note that the normalization constraint (5.38) is of the form (5.37), with $A(f) = 1$ for all states f.

‡ At worst, the range is unchanged if all previously allowed distributions have the same value for the new constraint. What is critical to the argument is that the range cannot increase by the addition of new constraints.

Let us do so by stages. In the 0th stage let normalization be the only constraint. The distribution whose entropy is maximal is then the uniform one,

$$P^0(f) = \exp(-\lambda_0) \tag{5.39}$$

The quantity λ_0 is a number, known as a Lagrange multiplier, whose value is to be chosen such that the constraint [in this case, normalization, (5.38)] is satisfied, that is,

$$1 = \sum_f P(f) = \exp(-\lambda_0) \sum_f 1 \tag{5.40}$$

Hence, another way of writing (5.39) is

$$P^0(f) = 1 \Big/ \sum_f 1 \tag{5.41}$$

The probability, $P^0(v)$, of a group of $g(v)$ quantum states can be found from (5.41):

$$P^0(v) = g(v) \Big/ \sum_f 1 = g(v) \Big/ \sum_v g(v) \tag{5.42}$$

With (5.42) we have recovered the interpretation of the prior distribution: It is the distribution of maximal entropy subject to the ever-present constraints (e.g., normalization, conservation of energy).

The prior distribution does not always represent the data (Sec. 5.5.2). One or more specific constraints may be required. Hence in the first stage we take for the first constraint A the vibrational energy:

$$\langle E_v \rangle = \sum_f E_v P(f) \tag{5.43}$$

Among all normalized distributions with a given numerical value of $\langle E_v \rangle$, the (unique) one of maximal entropy can be shown to be

$$P(f) = \exp(-\lambda_0 - \lambda_v E_v) \tag{5.44}$$

Here there are two constraints (normalization and a given $\langle E_v \rangle$) and hence two Lagrange multipliers, λ_0 and λ_v. Their numerical values, as always, are to be determined in terms of the values of the constraints. Let us determine λ_0 first:

$$1 = \sum_f P(f) = \exp(-\lambda_0) \sum_f \exp(-\lambda_v E_v) \tag{5.45}$$

Normalization makes λ_0 an explicit function of λ_v:

$$\exp(\lambda_0) = \sum_f \exp(-\lambda_v E_v) \tag{5.46}$$

The distribution of maximal entropy as given by (5.44) is now normalized and contains only one undetermined Lagrange multiplier, namely, λ_v. Its value can be determined in terms of $\langle E_v \rangle$. Using (5.43), (5.44), and (5.46), the equation for λ_v is

$$\langle E_v \rangle = \sum_f E_v \exp(-\lambda_v E_v) \Big/ \sum_f \exp(-\lambda_v E_v) \qquad (5.47)$$

This is an implicit equation (since it is $\langle E_v \rangle$ which is given). Figure 5.39 provides a graphical representation of the relation between $\langle f_v \rangle = \langle E_v \rangle / E$ and λ_v for an A + BC reaction.

What can we say about the solution (5.44)? One thing is clear. The introduction of the constraint means that not all quantum states are equally probable. Note, however, that there are still groups of quantum states which are of the same probability. These are all the quantum states where BC is in a given vibrational state v. Thus, we can easily compute $P(v)$ by summing (5.44) over all final quantum states that correspond to a given vibrational level of BC;

$$P(v) = g(v) \exp(-\lambda_0 - \lambda_v E_v) \qquad (5.48)$$

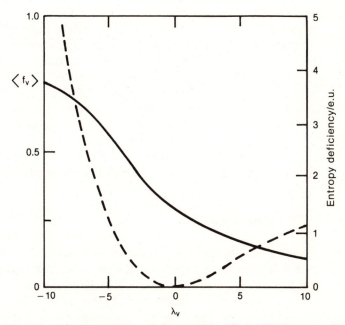

Fig. 5.39. The role of λ_v as a measure of the specificity of vibrational energy disposal. Left scale: The mean fraction of the available energy released as product vibrational excitation $\langle f_v \rangle$ versus λ_v (solid curve: monotonic dependence). Right scale: The entropy deficiency of the product vibrational state distributions versus λ_v (dashed curve). In the prior limit ($\lambda_v = 0$), the entropy deficiency is zero and $\langle f_v \rangle = 2/7$. [Adapted from R. D. Levine and A. Ben-Shaul, in Moore (1976).]

or by dividing by $\Sigma g(v)$ and redefining λ_0

$$P(v) = P^0(v) \exp(-\lambda_0 - \lambda_v E_v) \qquad (5.49)$$

The distribution (5.49) is the one often observed (e.g., Fig. 5.31). Surprisal analysis identifies the constraints on the distribution in question.

Equation (5.49) is not always a good representation of the vibrational energy disposal. "Repulsive release" of the exoergicity often requires a modified form of the constraint. That is just as it should be. The "dynamics" come in through the choice of constraints.

Rather than starting from a potential and computing all the way to the observed distribution, the maximum entropy formalism starts in the middle and works out in both directions. The reasoning is as follows.The complexity of the problem *and* the extensive averaging (over initial states) and summing (over final states) required means that some, or many, of the details of the potential and the initial state become largely irrelevant to the final, often coarse, distribution of interest. What remains relevant are the constraints. It makes sense, therefore, to start from the constraints. Then, on the one hand, the constraints required to account for the observations are identified, and, on the other hand, dynamics are used to specify the appropriate constraints.

Why use maximum entropy to go from the constraints to the distribution? There are several reasons. The most familiar, based on information theory, is that this leads to the least biased and therefore most reasonable inference. In the natural sciences and particularly for our problem, where we are dealing with very many repetitions of the "isolated collision," we can give a stronger reason. For reproducible experiments, maximum entropy is the only consistent inference that is independent of the number of repetitions. In other words, the maximum entropy procedure corresponds to the law of large numbers. It is a procedure of probability theory, not unique to any particular problem. The "science" enters through the choice of constraints.

Appendix 5C: The prior distribution

For reactions that populate the various internal states of the products according to statistical (phase space) considerations alone (i.e., no dynamical effects), the product state distribution and its dependence on reagent states and upon total energy E are simply those of the "prior". Thus, before evaluating the surprisal and the entropy deficiency, it is necessary to calculate this prior distribution. This requires only quantum state-counting since the prior expectation is to populate every energetically allowed group of product states with a probability proportional to the number of such states in that group.

First we note that the density of translational states is given by the expression

$$\rho_T(E_T) = \frac{\mu^{3/2}}{2^{1/2}\pi^2\hbar^3} E_T^{1/2} \tag{5C.1}$$

with ρ the number of states per unit volume per unit interval in E_T and μ is the reduced mass. The density of quantum states of the $AB(v'j') + C$ products at a given total energy E is thus given by

$$\rho(v'j', E) = (2j' + 1)\rho_T(E - E'_I) \tag{5C.2}$$

Here, E'_I is the internal (i.e., vibrotational plus electronic, if any) excitation energy of AB, so that $E'_T = E - E'_I$. The *total* density of states, $\rho(E)$, is obtained by summing over the various possible final states,

$$\rho(E) = \sum_{v'=0}^{v'^*} \sum_{j'=0}^{j'^*(v')} (2j' + 1)\rho_T(E - E'_I) \tag{5C.3}$$

where $j'^*(v')$ is the largest energetically allowed j' (for the given v'), and v'^* the largest v' allowed for given E.

The summations in (5C.3) are over a finite range and can always be carried out explicitly given the vibrotational energy levels of $AB(v'j')$. The fraction of the total number of quantum states in a particular group of products' states is obtained with $\rho(E)$ serving as the normalization factor. Thus, the prior-expectation probability of an individual vibrotational $P^0(v', j')$ is

$$P^0(v', j') = (2j' + 1)\rho_T(E - E'_I)/\rho(E) \tag{5C.4}$$

The prior distribution of vibrational states (*irrespective* of j'), $P^0(v')$, is therefore the sum over j':

$$P^0(v') = \sum_{j=0}^{j^*(v')} P^0(v', j') \tag{5C.5}$$

where for the $AB(v', j') + C$ products

$$P^0(v', j') \propto (2j' + 1)(E - E_{v'} - E_{j'})^{1/2} \tag{5C.6}$$

In the "ideal" limiting case in which the diatomic energy levels are approximated by the rigid-rotor harmonic oscillator (RRHO) level scheme, simple formulas for the priors have been worked out in terms of the reduced energy variables. One obtains compact approximate (RRHO) expressions such as

$$P^0(f_v) = \tfrac{5}{2}(1 - f_v)^{3/2} \tag{5C.7}$$

$$P^0(f_T) = \tfrac{15}{4}f_T^{1/2}(1 - f_T) \tag{5C.8}$$

$$P^0(f_R \mid f_v) = \frac{3}{2}\frac{(1 - f_v - f_R)^{1/2}}{(1 - f_v)^{3/2}} = \tfrac{3}{2}(1 - g)^{1/2}/(1 - f_v) \tag{5C.9}$$

where we have dropped the primes. The reduced rotational energy variable g is defined implicitly in (5C.9), i.e., $g = f_R/(1 - f_v)$.

Thus, the vibrational surprisal (for the RRHO case), is given by

$$I(f_v) = -\ln[P(f_v)/P^0(f_v)] = -\ln\left(\frac{2P(f_v)}{5(1-f_v)^{3/2}}\right) \qquad (5C.10)$$

The RRHO expressions (5C.7)–(5C.9) are derived by regarding the internal energy as a continuous variable. Sometimes (especially for hydrides) the vibrational-energy spacings are too wide to make this a reasonable approximation. Hence an intermediate procedure is to integrate over $E_{j'}$ but to sum over v'. If we take the v' dependence of $E_{j'}$ to be that of a vibrating rotor (VR), the result is

$$P^0(f_v) = B_v^{-1}(1-f_v)^{3/2} \Big/ \sum_v B_v^{-1}(1-f_v)^{3/2} \qquad (5C.11)$$

Here, B_v is the v-dependent rotational constant.

For more general processes, such as an atom–polyatom or diatom–polyatom product pairs, the state-counting proceeds in the same fashion except, of course, that there are more quantum numbers. In the RRHO limit, simple explicit results can be worked out, but when the energy can be partitioned over more degrees of freedom, the classical-like RRHO approximation is no longer quantitatively accurate.

5.6 Quantum dynamics

So far we have largely implied that classical trajectories are the way to perform exact dynamical computations. Where and why do we expect quantal effects. One very obvious deficiency of the classical mechanics approach is the description of internal states. Why not go all the way and describe the internal states quantum mechanically? The problem is that, as usually formulated, quantum mechanics does too good a job. To describe the internal states of HCl by quantum mechanics means describing not only the vibrational states but also the rotational states. Now a rotational *state* means not only the magnitude j of the rotational angular momentum but also its projection, m_j. There are $2j + 1$ distinct quantum states for each value of j. If we probe the orientation of the product then, of course, the distribution of m_j's is what we are after. Otherwise, however, they need all be summed over. There are, for example, over 5000 energetically allowed vibrotational quantum states of HCl produced in the Cl + HBr thermal reaction. The development of computational techniques that avoid the need to compute such details (which will be summed or averaged over) is a current research topic.

The second quantum-mechanical aspect is that of the relative motion. Now, except possibly near the barrier to reaction, the de Broglie wavelength for the relative motion is usually short compared to the range of variation of the potential.† At the barrier (or at the turning point for inelastic collisions), the system can penetrate the classically inaccessible region. Although the probability of tunneling through the barrier is exponentially small, for thermal reagents at lower temperatures, the probability of having energy in excess of the barrier height is also small. The tunneling correction is thus primarily of interest at lower temperatures. Should one not therefore restrict quantum mechanics to the internal states and treat the relative motion classically? This is the solution often adopted.

The third important quantal aspect is due to the principle of superposition (recall Sec. 3.1.7). Quantum mechanical probabilities are evaluated by adding together all the contributing probability amplitudes and squaring their sum. The resulting interference pattern is resolvable, at least for elastic scattering. Such interference should, however, be present whenever the transition of interest can be realized by more than one classical path. Of course, any averaging tends to smear out interferences. Since the average behavior is the classical result, this effect, while always present, is evident only in very high resolution experiments. The very multitude of final internal states also tends to smear out the interference pattern. That is not to say that it cannot be observed. After all, Fig. 3.13 shows a rainbow in a final state distribution.

Beyond these general remarks we also need to consider a number of more technical aspects of quantal scattering theory. The discussion below is aimed at delineating the key points rather than provide "working equations".

5.6.1 Coupled channels

For elastic scattering (see Sec. 3.1.7) the wave function for relative motion behaves, at large separation, as an incident wave plus a scattered wave. When the colliding particles have internal structure we may wish to try a similar approach. There is no problem about the incident part. We simply multiply the incident wave by the wave function $\phi_n(r)$ for the internal initial state. But in the scattered part many final internal states are possible. The scattered part is then a sum over many final states, each term being a product of a scattered wave for the relative motion and an internal state wave function. Equation (3.28) is thus replaced by

$$G_{lm}(R) \to \sum_n k_n^{-1/2} \{\delta_{nm} \exp[-i(kR - \tfrac{1}{2}l\pi)] - S_{nm}^l \exp[i(kR + \tfrac{1}{2}l\pi)]\} \phi_n(r)$$

$$(5.50)$$

---

† This condition not only suggests that classical mechanics will be a reasonable approximation but also signals the need for care in computing the wave function for relative motion. The reason is that when the wavelength is short, the wave function is highly oscillatory. Small integration steps are thus required. One solution is to solve separately for the amplitude and the phase of the wavefunction. When that is done, the leading term is the semiclassical limit.

Here, n is an index of the internal state and the first term in the curly brackets corresponds to the initial state m. For simplicity we have assumed that the orbital angular momentum is conserved, which, in general, will not be the case. Each term in (5.50) is usually referred to as a channel. Note that we are still not facing the full complexity of the problem of reactive scattering. In (5.50) the incident and scattered waves are along the same coordinate R. Thus, expression (5.50) is really applicable only to inelastic scattering. Rearrangement collisions (and thus all of chemistry) require a scattered wave also along the coordinate of relative separation of the products. Either we add such terms to (5.50) or we need go to the "natural" coordinates discussed in Section 5.6.6.

Even for inelastic collisions there can be an incident wave accompanied by any energetically allowed internal state.† Thus, if N internal states are allowed at energy E, there will be N distinct wave functions of the type (5.50), one for each initial internal state. Thus, $S_{n'n}^l$, defined by (5.50) is an element of an $N \times N$ matrix, known as the scattering matrix. Note that for each value of the initial angular momentum l we shall have a different **S** matrix.

To determine the scattering matrix one needs to solve the Schrödinger equation. While the details are beyond the scope of this primer, one aspect common to essentially all available numerical methods should be noted: The computations determine the entire $(N \times N)$ matrix, and the procedures require matrix manipulations. This severely limits the number, N, of internal quantum states that can be handled. The high degeneracy of rotational states implies that most realistic problems are outside the capabilities of current computers. It also means that *much more detail than necessary* is generated. For most problems we wish to examine one or a few initial states. Yet we have to solve for all possible initial states.‡

Feasible approximation schemes for the **S** matrix have been, and continue to be, the subject of active study. Advantage is taken of the semiclassical nature of the relative motion of the heavy atoms and of the relatively slow rotation of molecules to simplify the problem. A few such methods will be outlined below.

The S_{nm}^l element of the scattering matrix determines the amplitude of the outgoing (i.e., scattered) wave in the nth channel for an incoming (incident) wave of unit amplitude in the mth channel. Microscopic reversibility means that **S** is a symmetric matrix:

$$S_{mn}^l = S_{nm}^l \tag{5.51}$$

Conservation of flux requires that

$$\sum_n |S_{nm}^l|^2 = 1, \text{ for all } m \tag{5.52}$$

---

† At a given total energy E, the internal energy E_n must satisfy $E = E_n + \hbar^2 k_n^2 / 2\mu$, where k_n must be real.

‡ In a reactive collision, the states of the products are also possible initial states.

More generally, one can show that conservation of flux requires that the **S** matrix is unitary, $S^\dagger S = I$, of which (5.52) is a special case.

5.6.2 From the S matrix to the cross section

Computing the scattering amplitude for the case where the orbital angular momentum is conserved proceeds just as in the elastic case, and the final result looks deceptively similar:

$$f_{mn}(\theta) = (2ik_n)^{-1} \sum_{l=0}^{\infty} (2l+1)(S_{mn}^l - \delta_{mn})P_l(\cos\theta) \tag{5.53}$$

The two points of essential difference are that the wave vector in (5.53) is the one for the incident channel, and the subtraction of the contribution of the incident wave is relevant only for the elastic case. Integrating over angles, we have for the cross section

$$\sigma_{mn} = (\pi/k_n^2) \sum_{l=0}^{\infty} (2l+1) |S_{mn}^l - \delta_{mn}|^2 \tag{5.54}$$

The sum in (5.54) converges since the **S** matrix (which needs to be computed anew for each l) tends to the identity matrix as l increases. The reason, just as in the elastic case, is the centrifugal barrier contribution to the kinetic energy for the relative motion. As l increases, the classical turning point is further out. In the nonclassical region, the amplitude of the wave functions for the relative motion becomes exponentially small. To couple different channels the intermolecular potential $V(R, r)$ need be large for such R values that the wave functions can sample. Hence the effectiveness of the coupling diminishes once l is large.†

The orbital angular momentum is strictly conserved only for a central potential. Since, typically, the intermolecular potential does depend on the mutual orientation, l is not conserved, either in magnitude or direction. The scattering amplitude will thus contain a threefold sum (a sum over l' and its projection $m_{l'}$ and an average over l). The number of indices carried by any **S** matrix element becomes considerable. Even in the simplest case of a nonreactive atom–rigid-rotor collision, an incident channel needs to be specified by four quantum numbers (j, m_j, l, m_l); the same is true for any

---

† Another factor, and one that will occupy much of our attention in Chap. 6, is the change in the relative kinetic energy due to internal excitation. Say the incident molecule is in its ground internal state while the outgoing molecule is in a highly excited internal state. The kinetic energy for relative motion is then high in the incident channel but low in the exit channel. The turning points in the two channels will thus occur at significantly different values of R. Near the turning point in the exit channel, where the radial wave function is large, the radial wave function in the entrance channel is still highly oscillatory (since the wave vector is large). Near the turning point of the entrance channel, the radial wave function in the exit channel is already exponentially small: clearly a no-win situation. In Chap. 6 we shall argue that under such circumstances the **S** matrix elements are exponentially small.

exit channel. The number of coupled equations and the size of the scattering matrix are then unmanageable. We can reduce the problem by replacing the computation of a single very large matrix by computing many matrices of more reasonable size. For the simple $A + BC$ collision this is achieved by taking into account the conservation of the *total* angular momentum $\mathbf{J}$

$$\mathbf{J} = \mathbf{j} + \mathbf{L} = \mathbf{j}' + \mathbf{L}' \tag{5.55}$$

Therefore, at a given eigenvalue J of $\mathbf{J}$ and for a given $j \rightarrow j'$ transition, the ranges of both l and l' are restricted,

$$|J - j| \leq l \leq J + j \tag{5.56}$$

and similarly for l'. We now need to compute a separate $\mathbf{S}$ matrix for each value of J, i.e., $S^J_{jl,j'l'}$. Summing over all values of J we have for (the degeneracy-averaged) $j \rightarrow j'$ cross section

$$\sigma(j \rightarrow j') = (\pi/k_j^2)(2j + 1)^{-1} \sum_{J=0}^{\infty} (2J + 1) \sum_{l=|J-j|}^{J+1} \sum_{l'=|J-j'|}^{J+j'} |S^J_{jl,j'l'}|^2 \tag{5.57}$$

If we sum first over l' and denote the sum by $P^J_{j,j'}(l)$ and interchange the order of summation over l and J, we end up with

$$\sigma(j \rightarrow j') = (\pi/k_j^2) \sum_{l=0}^{\infty} (2l + 1) \sum_{J=|l-j|}^{l+j} \frac{(2J + 1)}{(2j + 1)(2l + 1)} P^J_{j,j'}(l) \tag{5.58}$$

There are $(2j + 1)(2l + 1)$ initial states of a given j and l. Of these, $(2J + 1)$ states have the same J value,

$$\sum_{J=|j-l|}^{j+l} (2J + 1) = (2j + 1)(2l + 1) \tag{5.59}$$

The fraction $(2J + 1)/(2j + 1)(2l + 1)$ is the fraction of initial states with a given J, while $P^J_{j,j'}(l)$ is the fraction (of those with a given J) that proceed to j'. Hence (5.58) can be written in the manner made familiar in our study of the optical model:

$$\sigma(j \rightarrow j') = (\pi/k_j^2) \sum_{l=0}^{\infty} (2l + 1)P_{j,j'}(l) \tag{5.60}$$

Here $P_{j,j'}(l)$ is the opacity function for the $j \rightarrow j'$ transition,

$$P_{j,j'}(l) = \sum_{J=|l-j|}^{l+j} \frac{(2J + 1)}{(2j + 1)(2l + 1)} P^J_{j,j'}(l) \tag{5.61}$$

In Section 2.4.2 we used the semiclassical correspondence $bk \approx l + \frac{1}{2}$ to transform (2.47) to (2.45). The same correspondence can be used here to rewrite (5.60) as

$$\sigma(j \rightarrow j') = \int_0^{\infty} db \, 2\pi b P_{j,j'}(b) \tag{5.62}$$

where $P_{j,j'}(b)$ is evaluated from $P_{j,j'}(l)$.

5.6.3 The classical path limit

The classical path approximation is very commonly applied when quantal effects in the relative motion are not very important. This requires that the de Broglie wavelength be short. The usual formulation of the method also requires that the kinetic energy be high compared to the spacings of internal energy levels. The reason for this second condition is that the relative motion is treated completely classically. This implies that at every relative separation one can assign a unique velocity to the relative motion. If exchange of energy between translational and internal energy takes place, that is not going to be the case. In particular, after the collision, each internal state of the products can correspond to a different final translational energy as determined by conservation of total energy.† The classical path approximation therefore requires that the fractional change in the initial translational energy be small.

Turning now to a brief derivation, consider an atom (A)–molecule (BC) collision. To simplify the notation we shall regard the molecule as a harmonic oscillator. Assume that we know the relative separation, as a function of time, $R(t)$, throughout the collision. The interaction potential can then be regarded as a function of time, rather than as a function of the relative separation: Thus $V(R)$ becomes $V[R(t)]$. The appropriate equation of motion is thus Schrödinger's *time-dependent* equation

$$i\hbar \frac{\partial \psi(t)}{\partial t} = H(t)\psi(t) \tag{5.63}$$

where $H(t)$ is the Hamiltonian for the system. Well before $(t \to -\infty)$ or after $(t \to +\infty)$, the collision $H(t)$ is just the Hamiltonian of the *isolated* harmonic oscillator, H_0. This Hamiltonian and its eigenfunctions ϕ_n and eigenvalues E_n are well-known and satisfy the Schrödinger *time-independent* equation

$$H_0\phi_n = E_n\phi_n \tag{5.64}$$

and the orthonormality relation

$$\int \phi_n \phi_{n'} \, dR_{BC} = \delta_{nn'} \tag{5.65}$$

where δ is the Kronecker delta.

To solve equation of motion (5.63) we use a "trial function" of the form

$$\psi(t) = \sum_n a_n(t)\phi_n \exp(-iE_n t/\hbar) \tag{5.66}$$

where the (complex) coefficients $a_n(t)$ are to be determined from the

---

† Indeed, measuring the distribution of kinetic energy of the products serves to characterize their internal state distribution, e.g., Section 5.3.

condition that $\psi(t)$ satisfy the equation of motion. The physical significance of the coefficients follows when we consider the probability that at the time t the oscillator will be found in the state n'. According to the usual quantal procedure, this probability is given by†

$$\left| \int \phi_{n'} \psi(t) \, dR_{BC} \right|^2 = |a_{n'}(t)|^2 \tag{5.67}$$

In particular, the probability that after the collision the oscillator is in state n' is $|a_{n'}(\infty)|^2$.

The time-dependent wave function $\psi(t)$ provides a *complete description of the dynamics throughout the collision*. In particular, for $t \to -\infty$, it describes the initial, precollision state of the oscillator. That initial state is not arbitrary but is specified beforehand when the experimentalist "prepares" the initial state. The wave function at $t \to -\infty$ is thus known. The equation of motion (5.63) is a first-order differential equation; hence, when $\psi(-\infty)$ is specified, we obtain a *unique* solution for $\psi(\infty)$ at all subsequent times and, in particular, for $t \to +\infty$. In other words, $\psi(\infty)$, the final postcollision wave function, for a given intermolecular potential, is determined‡ by the initial state.

To put these abstract considerations into a more concrete form, we return now to the concept of the scattering matrix. The wave function (5.66) is given in terms of the coefficients $a_{n'}(t)$. It is convenient to regard the collection of coefficients at a time t as elements of a column vector $\mathbf{a}(t)$. Then, the nth component of this vector is just the coefficient $a_n(t)$. As an example, consider the precollision condition that the oscillator was in a particular state n. Then, we have that

$$\mathbf{a}(t) = \begin{pmatrix} a_1(t) \\ a_2(t) \\ \vdots \\ a_n(t) \\ \vdots \end{pmatrix} \xrightarrow[t \to -\infty]{} \mathbf{a}(-\infty) = \begin{pmatrix} 0 \\ 0 \\ \vdots \\ 1 \\ \vdots \\ 0 \end{pmatrix} \tag{5.68}$$

i.e., there is only one nonvanishing element of $\mathbf{a}(-\infty)$, the nth element in the column, $a_n(-\infty)$, and its value is unity. In general, the precollision wave function is specified by the vector $\mathbf{a}(-\infty)$. The solution of the equation of motion determines $\mathbf{a}(t)$ for all subsequent times t and, in particular, for $t \to \infty$. The *probability* of any particular final vibrational state n' after the collision is then, via Eq. (5.67), $|a_{n'}(\infty)|^2$.

---

† To derive (5.67), multiply (5.66) by $\phi_{n'}$ and integrate over R_{BC}. From the orthogonality relation (5.65), only the n' term survives.

‡ Moreover, since the equation of motion is linear, there is a linear transformation from the initial to the final state. In other words, if ψ_1 and ψ_2 are solutions of the equation, so is $\psi = a\psi_1 + b\psi_2$, where a and b are constants.

The linear transformation from the initial to the final state can now be written as a matrix equation

$$\mathbf{a}(+\infty) = \mathbf{S}\mathbf{a}(-\infty) \tag{5.69}$$

or, explicitly, in terms of the rules for matrix multiplication:

$$a_{n'}(+\infty) = \sum_n S_{n'n} a_n(-\infty) \tag{5.70}$$

Thus, if we know the initial state, we can find the final state, provided we know the elements of the matrix **S**. Therefore, **S** must contain all possible information about the dynamic processes going on between $t = -\infty$ and $t = +\infty$.

Equation (5.69) has the structure

$$\begin{pmatrix} a_1(+\infty) \\ a_2(+\infty) \\ \vdots \end{pmatrix} = \begin{pmatrix} S_{11} & S_{12} & \cdots \\ S_{21} & \\ \vdots \end{pmatrix} \begin{pmatrix} a_1(-\infty) \\ a_2(-\infty) \\ \vdots \end{pmatrix} \tag{5.71}$$

The matrix **S** must be such that it *ensures the conservation of probability*:

$$\sum_{n'} |a_{n'}(t)|^2 = 1 \tag{5.72}$$

It follows from (5.70) that S must satisfy

$$\mathbf{S}^\dagger \mathbf{S} = \mathbf{I} \tag{5.73}$$

where **I** is the unit matrix and the superscript dagger (†) denotes the Hermitian adjoint. Explicitly [from Eq. (5.70)] this condition is

$$\sum_{n'} S_{n'n} S^*_{n'n''} = \delta_{nn''} \tag{5.74}$$

The matrix **S** is the *scattering matrix*.† For any initial state $\mathbf{a}(-\infty)$, and for a given intermolecular potential, the scattering matrix specifies a unique final state $\mathbf{a}(+\infty)$.

Once more we note the excess computational effort in obtaining the global **S** matrix. Experiments require cross sections for a very limited number of initial states. We therefore turn to procedures that share (with the method of classical trajectories) the advantage that attention can be limited only to the desired initial state.

---

† In the classical path approximation, since the path, $R(t)$, depends on the impact parameter [see Eq. (2.31)], so does the **S** matrix. This b dependence is the semiclassical limit of the **S** matrix having a separate set of entries for each l (Sec. 5.6.1).

5.6.4 The sudden approximation for "fast" collisions

The motivation for introducing the sudden approximation is the large number of final rotational states. What makes it possible is the comparatively long rotational period, t_R. To compute the classical rotation time t_R, note that in terms of the angular velocity ω, the rotational energy of a diatomic rigid rotor is given by $E_R = (\frac{1}{2})I\omega^2$, where I is the moment of inertia. The quantal result is $E_R = \hbar^2 j(j+1)/2I$. Hence, $\hbar j \approx I\omega$. (Compare $L \approx \mu b v$). In terms of the rotational constant B (with dimensions of energy), $B = \hbar^2/2I$, $\omega = jB/\hbar$, or $t_R = \hbar/jB$. With the exception of hydrides, B is often below $1 \, \text{cm}^{-1}$ or $2 \times 10^{-23} \, \text{J}$ (see general Appendix). Thus, $t_R \geq (5/j)$ ps. The vibrational period $t_v = 1/v$ is often below 0.1 ps and comparable to the duration of a collision.† The conclusion that (for low j's) the rotational period is long compared to the duration of the collision is essentially equivalent to the conclusion that the rotational energy spacings are small compared to the collision energy.‡

Under such circumstances it is reasonable to assume that the rotational motion is "frozen" while the collision takes place (Fig. 5.40). The scattering can now be computed for different, fixed, initial orientations γ of the diatomic molecule. The computation is far simpler since the rotational states are not included. This is especially the case for an inelastic atom–rigid-rotor collision for which the computation then reduces to that of elastic scattering. Of course, for each value of γ ($\gamma \equiv \alpha, \beta$) the potential will be different, so that the phase shift needs to be computed as a function of γ. Given $S(\gamma) = \exp[2i\delta(\gamma)]$, we compute the **S** matrix by

$$S_{jm_j j'm_j'} = \langle j'm_{j'}| S(\gamma) |jm_j\rangle \qquad (5.75)$$

where the angular brackets denote the usual quantum-mechanical scalar product.

As long as we are approximating, we note that the validity conditions for the sudden approximation also imply that conservation of total angular momentum is not critical.§ Thus, we might as well compute the scattering amplitude using (5.53):

$$f_{j'm_{j'},jm_j}(\theta) = (2ik)^{-1} \sum_{l=0}^{\infty} (2l+1)\langle j'm_{j'}| \exp[2i\delta_l(\gamma)] - 1 |jm_j\rangle P_l(\cos\theta) \qquad (5.76)$$

which can be written as

$$\langle j'm_{j'}| f(\theta \mid \gamma) |jm_j\rangle \qquad (5.77)$$

---

† As a rough estimate for the duration of the collision, $t_c = a/v$, where a is the range of the potential.

‡ Take the range a of the potential to be equal to bond length of the diatom. Then, $B = \hbar^2/2\mu a^2$. $t_c < \tau_R$ implies $(v/a)^2 > j(j+1)B^2/\hbar^2$ or $E_T = (\frac{1}{2})\mu v^2 > Bj(j+1) = E_R$.

§ Take the range a of the potential as a typical value of the impact parameter. Then, $\hbar l = \mu v a = \mu a^2/t_c = I/t_c$. In the sudden regime, $\hbar l = I/t_c > I/t_R = \hbar j$, i.e., $l > j$ or $l \pm j \approx l$.

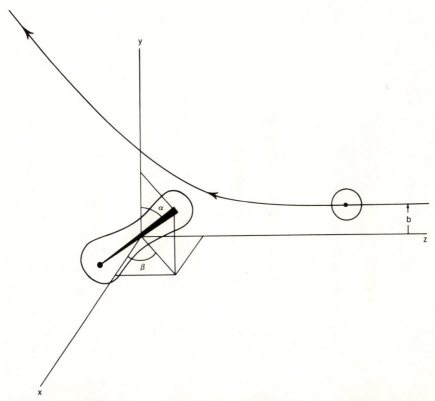

Fig. 5.40. Schematic illustration of a "sudden" collision of an atom with a (rigid-body-modeled) molecule at fixed orientation (angles α, β) with specified impact parameter b. In general, the trajectory will *not* remain in the yz plane.

where $f(\theta \mid \gamma)$ is the elastic scattering amplitude at a given value of γ,

$$f(\theta \mid \gamma) = (2ik)^{-1} \sum_{l=0}^{\infty} (2l + 1)\{\exp[2i\delta_l(\gamma)] - 1\}P_l(\cos \theta) \qquad (5.78)$$

The sudden approximation for reactive scattering is, of course, more complicated; there one needs to consider the orientation angles of both reactants and products.

5.6.5 Wave packets

The classical path approximation has the conceptual advantage that one can think of the collision proceeding in time. (At each instant t, the probability of being in state n is $|a_n(t)|^2$). In classical mechanics we can exactly follow the collision in time. Is there an exact quantum-mechanical analog? It might appear that the Heisenberg time–energy uncertainty relation precludes such a description. However, if we are willing to forego results at sharp values of

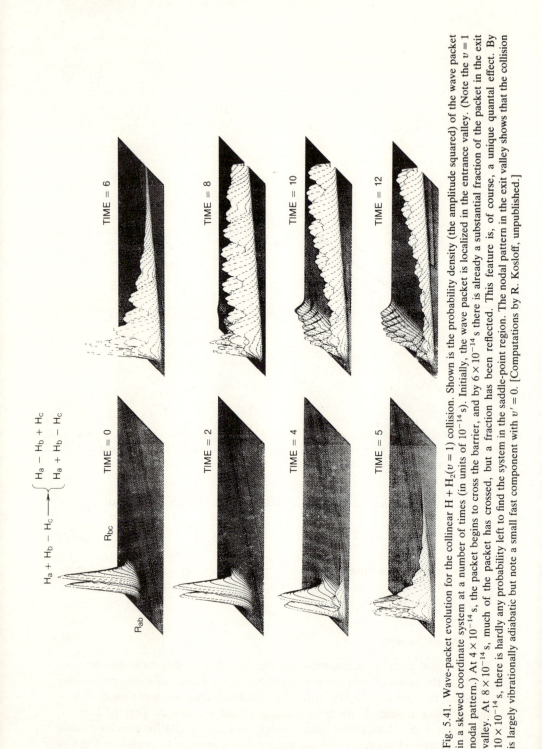

Fig. 5.41. Wave-packet evolution for the collinear $H + H_2(v = 1)$ collision. Shown is the probability density (the amplitude squared) of the wave packet in a skewed coordinate system at a number of times (in units of 10^{-14} s). Initially, the wave packet is localized in the entrance valley. (Note the $v = 1$ nodal pattern.) At 4×10^{-14} s, the packet begins to cross the barrier, and by 6×10^{-14} s there is already a substantial fraction of the packet in the exit valley. At 8×10^{-14} s, much of the packet has crossed, but a fraction has been reflected. This feature is, of course, a unique quantal effect. By 10×10^{-14} s, there is hardly any probability left to find the system in the saddle-point region. The nodal pattern in the exit valley shows that the collision is largely vibrationally adiabatic but note a small fast component with $v' = 0$. [Computations by R. Kosloff, unpublished.]

the energy, e.g., allowing a finite spread in the initial momentum, one can localize the projectiles in the initial state about some mean separation and then follow the exact quantal time evolution. This will also have the advantage that we can solve the scattering problem directly for the initial state of interest.

Figure 5.41 shows such a wave packet for a reactive $H + H_2$ (collinear) collision at a number of times during the course of a collision. An important feature, clearly evident from the figure, is that starting with a given (pure) initial state, one obtains a finite probability both for reaction and for a nonreactive encounter. This is, of course, not possible for a single classical trajectory, which must exit either in the entrance or in the products' valley. A wave packet corresponds to a swarm of classical trajectories (Sec. 4.3.2), some of which do and some of which do not react.

5.6.6 Natural collision coordinates

With Fig. 5.41 we finally face the full complexity of quantal reactive scattering computations. A major problem is that of the boundary conditions on $\psi(R, r)$ requiring outgoing waves not only in the R but also in the R' direction. A possible solution is to introduce a coordinate, say s, that transforms continuously from being the relative separation (R) of the reactants, when these are far apart to the relative separation (R') of the products, as they move out. It is easy to visualize such a coordinate, as shown in Fig. 5.42. The problem is also evident, i.e., s is not a straight-line coordinate. The kinetic energy for motion along s is *not* simply given by $-(\hbar^2/2\mu) \, \partial^2/\partial s^2$.

One aspect of this that we have already noted (Fig. 4.23a) is the "bobsled" effect, that is, the tendency of the motion along s to "climb the sidewall" rather than to follow the minimum-energy path along the reaction coordinate.

The change in the physical interpretation of the coordinate s is accompanied by changes in the other coordinates. The coordinate ρ, orthogonal to s (Fig. 5.42) starts (where $s \to -\infty$) as the vibration of BC, becomes (near $s = 0$) the symmetric vibration of ABC and ends up (as $s \to \infty$) as the vibration of AB, and similarly for the other coordinates. Computational schemes based on expressing the Hamiltonian as a function of the natural collision coordinates have been implemented. Such schemes are attractive because the coordinates center attention on those regions of the potential most relevant to the dynamics. For the same reason such coordinates are useful for qualitative discussions and for developing models.

5.6.7 Statistical dynamics

The labor of performing exact dynamic computations is such that any simplifications are always welcome. One type of approximation makes use

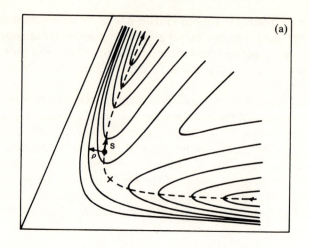

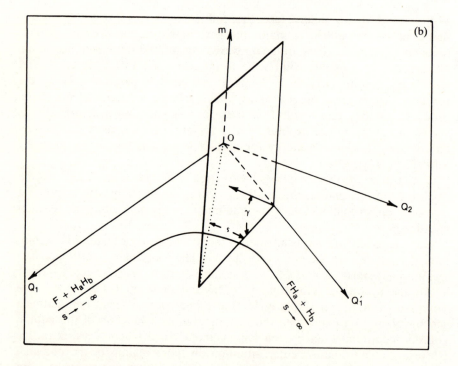

Fig. 5.42. Potential energy surface contours, showing the minimum-energy path. The natural collision coordinates are referred to as s (reaction coordinate) and ρ (orthogonal coordinate). (a) Collinear collision. [Based on R. A. Marcus, *J. Chem. Phys.*, *45*, 4493 (1966).] (b) Three-dimensional collision shown for the $F + H_aH_b \rightarrow FH_a + H_b$ reaction. Q_1 and Q_2 are the (mass-scaled) relative separations of F—H_2 and H—H, respectively. Q_1' is the FH—H separation. Nonlinear geometries are measured by m, and the $m = 0$ plane is the one for collinear collisions. At large relative separations γ is the orientation angle of the reactants. [Adapted from J. F. McNutt, R. E. Wyatt, and M. J. Redmon, *J. Chem. Phys.*, *81*, 1693 (1984).]

of the statistical nature of the dynamics, when applicable. For example, let us consider the configuration of no return. Even though the approximation of no return is not perfect, it may be quite reasonable. Hence to improve upon it we can start the (classical or quantal) solution at the configuration of no return and propagate it for a while. Of course, if we propagate all the way to the reactant or product region, we have not gained over the usual procedures. We can, however, propagate it just enough to be able to better decide whether it will go out or will return. Such a procedure simplifies the computation of the reaction rate $k(E)$.

In both quantum and classical dynamics much of the effort is spent on the computation of the distribution of products in their different accessible states.† Yet in Section 5.5 we have argued that the distributions of final states are often characterized by a small number of constraints. A more efficient route would be to compute the mean values of the constraints directly and to determine the distribution from them via the maximum entropy formalism.

5.6.8 Concluding remarks

The quantum theory of reactive molecular scattering is (formally) fully developed, but difficult to implement computationally in three dimensions (3D). Exact numerical solutions (for the wave functions and S-matrix elements) are possible for the collinearly constrained problem. Approximate methods for the 3D scattering computations have been developed and applied to a few simple systems, for example, the $H + H_2$ exchange reaction at low E_T and the $F + H_2$ reaction near threshold. Reactive infinite-order, sudden-approximation computations are often useful under appropriate conditions. Quasiclassical trajectory calculations still represent the "workhorse method" of obtaining coarse dynamical features. Naturally, such results cannot deal with quantal phenomena such as interferences and "resonances."

In conclusion, reactive scattering theory (and computation) is still a growing research topic with profound implications in the field of molecular reaction dynamics. The more technical details of the subject are, however, largely beyond the scope of this primer.

Suggested reading

Section 5.1

Book

Bernstein, R. B., *Chemical Dynamics via Molecular Beam and Laser Techniques*, Oxford, New York, 1982.

---

† Recall that in the Monte Carlo version of the classical trajectory approach we need essentially as many "successful" trajectories to compute the detailed rate constant into a specific final state as for the computation of the overall reaction rate, to the same accuracy.

Review Article

Leone, S. R., "State-Resolved Molecular Reaction Dynamics." *Annu. Rev. Phys. Chem.*, *35*, 109 (1984).

Section 5.2

Books

Corney, A., *Atomic and Laser Spectroscopy*, Clarendon, Oxford, 1977.

Crosley, D. R. (ed.), *Laser Probes for Combustion Chemistry*, American Chemical Society, Washington, D.C., 1980.

Demtröder, W., *Laser Spectroscopy—Basic Concepts and Instrumentation*, Springer-Verlag, Berlin, 1981.

Glorieux, P., Lecler, D., and Vetter, R. (eds.), *Chemical Photophysics*, Editions du CNRS, Paris, 1979.

Harper, P. G., and Wherret, B. S. (eds.), *Nonlinear Optics*, Academic, New York, 1977.

Hinkley, E. D. (ed.), *Laser Monitoring of the Atmosphere*, Springer-Verlag, Berlin, 1976.

Jacobs, S. F., Sargent, M., Scully, M. O., and Walker, C. T., *Laser Photochemistry, Tunable Lasers and Other Topics*, Addison-Wesley, Reading, MA, 1976.

Jackson, W. M., and Harvey, A. B. (eds.), *Lasers as Reactants and Probes in Chemistry*, Howard University Press, Washington, D.C., 1985.

Kompa, K. L., and Smith, S. D. (eds.), *Laser-induced Processes in Molecules*, Springer-Verlag, Berlin, 1979.

Lengyel, B. A., *Lasers*, Wiley, New York, 1971.

Letokhov, V. S., and Chebotayev, V. P., *Nonlinear Laser Spectroscopy*, Springer-Verlag, Berlin, 1977.

Mooradian, A., Jaeger, T., and Stockseth, P. (eds.), *Tunable Lasers and Applications*, Springer-Verlag, Berlin, 1976.

Moore, C. B. (ed.), *Chemical and Biochemical Applications of Lasers*, Vols. 1–5, Academic, New York, 1974-1980.

Okabe, H., *Photochemistry of Small Molecules*, Wiley, New York, 1978.

Pao, Y. H. (ed.), *Optoacoustic Spectrometry and Detection*, Academic, New York, 1977.

Picosecond Phenomena, Vols. I–IV, 1979-84

Rhodes, C. K. (ed.), *Excimer Lasers*, 2nd ed., Springer-Verlag, Berlin, 1984.

Schafer, F. P., *Dye Lasers*, Springer-Verlag, Berlin, 1977.

Shank, C. V., Ippen, E. P., and Shapiro, S. L. (eds.) *Picosecond Phenomena*, Vol. 1, Springer-Verlag, Berlin, 1978.

Shen, Y. R. (ed.), *Nonlinear Infrared Generation*, Springer-Verlag, Berlin, 1977.

Shen, Y. R., *The Principles of Nonlinear Optics*, Wiley, New York, 1984.

Steinfield, J. I., *Molecules and Radiation*, M.I.T. Press, Cambridge, MA, 1985.

Steinfeld, J. I. (ed.), *Laser and Coherence Spectroscopy*, Plenum, New York, 1981.

Yariv, A., *Quantum Electronics*, 2nd ed., Wiley, New York, 1975.

Review Articles

Altkorn, R., and Zare, R. N., "Effects of Saturation on Laser-Induced Fluorescence Measurements of Population and Polarization." *Annu. Rev. Phys. Chem. 35,*

265 (1984).

Baggott, J. E., "Gas Phase Photoprocesses," in *Photochemistry,* Vol. 16 (Specialist Periodical Report, 1985).

Balint-Kurti, G. G., and Shapiro, M., "Quantum Theory of Molecular Photodissociation," in Lawley (1985).

Baronavski, A. P., "Laser Ultraviolet Photochemistry," in Jackson and Harvey (1985).

Baronavski, A. P., Umstead, M. E., and Lin, M. C., "Laser Diagnostics of Reaction Product Energy Distributions." *Adv. Chem. Phys. 47* (Pt. 2), 85 (1981).

Berry, M. J., "Laser Studies of Gas Phase Chemical Reaction Dynamics." *Annu. Rev. Phys. Chem. 26,* 259 (1975).

Beswick, J. A., and Durup, J., "Half Collisions Induced by Lasers," in Glorieux et al. (1979).

Brumer, P. and Shapiro, M., "Theoretical Aspects of Photodissociation and Intramolecular Dynamics," in Lawley (1985).

Bondybey, V. E., "Relaxation and Vibrational Energy Redistribution Processes in Polyatomic Molecules." *Annu. Rev. Phys. Chem. 35,* 591 (1984).

Clyne, M. A. A., and McDermid, I. S., "Laser-Induced Fluorescence: Electronically Excited States of Small Molecules." *Adv. Chem. Phys. 50,* 1 (1982).

Cool, T. A., "Chemical Lasers," in Smith (1980).

Craig, D. P., and Thirunamachandran, T., "Radiation-Molecule Interactions in Chemical Physics." *Adv. Quant. Chem. 16,* 98 (1982).

Demtröder, W., "High Resolution Laser Spectroscopy of Molecules," in Woolley (1980).

Dupont-Roc, J., "Interaction Between Radiation and Matter," in Glorieux et al. (1979).

Durup, J., "On the Extraction of Time Information from Energy-Resolved Experiments." *Laser Chem. 3,* 85 (1983).

Eckbreth, A. C., Use of Lasers in Nonlinear Raman Spectroscopy," in Jackson and Harvey (1985).

Eisenthal, K. B., "Studies of Chemical Physical Processes with Picosecond Lasers." *Acc. Chem. Res. 8,* 118 (1975).

Eisenthal, K. B., "Picosecond Spectroscopy." *Annu. Rev. Phys. Chem. 28,* 207 (1977).

Ewing, J. J., "New Laser Sources," in Moore (1974).

Ewing, J. J., "Rare-Gas Halide Lasers." *Phys. Today 31,* 32 (1978).

Fleming, G. R., "Applications of Continuously Operating, Synchronously Mode-Locked Lasers." *Adv. Chem. Phys. 49,* 1 (1982).

Flynn, G. W., and Weston, R. E., Jr., "Hot Atoms Revisited: Laser Photolysis and Product Detection," *Annu. Rev. Phys Chem. 37,* 551 (1986).

Gelbart, W. M., "Photodissociation Dynamics of Polyatomic Molecules." *Annu. Rev. Phys. Chem. 28,* 323 (1977).

Gelbart, W. M., Elert, M. L., and Heller, D. F., "Photodissociation of the Formaldehyde Molecules: Does It or Doesn't It?" *Chem. Rev. 80,* 403 (1980).

Haas, Y., and Asscher, M., "Two-Photon Excitation as a Kinetic Tool: Application to Nitric Oxide Fluorescence Quenching." *Adv. Chem. Phys. 47* (Pt. 2), 17 (1981).

Harris, F. M., and Beynon, J. H., "Photodissociation in Beams: Organic Ions," in Bowers (1984).

Hilinski, E. F., and Rentzepis, P. M., "Chemical Applications of Picosecond Spectroscopy." *Acc. Chem. Res. 16,* 224 (1983).

Hirota, I., and Kawaguchi, K., "High Resolution Infrared Studies of Molecular Dynamics." *Annu. Rev. Phys. Chem. 36,* 53 (1985).

Hochstrasser, R. M., and Trommsdorff, H. P., "Nonlinear Optical Spectroscopy of Molecular Systems." *Acc. Chem. Res. 16,* 376 (1983).

Huppert, D., and Rentzepis, P. M., "Picosecond Kinetics," in Levine and Jortner (1976).

Ippen, E. P., and Shank, C. V., "Sub-Picosecond Spectroscopy," *Phys. Today 31,* 41 (1978).

Kaufmann, K. J., and Rentzepis, P. M., "Picosecond Spectroscopy in Chemistry and Biology." *Acc. Chem. Res., 8,* 407 (1975).

Kenney-Wallace, G. A., "Picosecond Spectroscopy and Dynamics of Electron Relaxation Processes in Liquids." *Adv. Chem. Phys. 47* (Pt. 2), 535 (1981).

Kimel, S., and Speiser, S., "Lasers and Chemistry." *Chem. Rev. 77,* 437 (1977).

Kinsey, J. L., "Laser Induced Fluorescence." *Annu. Rev. Phys. Chem. 28,* 349 (1977).

Kinsey, J. L., "Fourier Transform Doppler Spectroscopy: A New Tool for State-to-State Chemistry," in Brooks and Hayes (1977).

Kresin, V. Z., and Lester, W. A., Jr., "Theory of Polyatomic Photodissociation. Adiabatic Description of the Dissociative State and the Translation-Vibration Interaction." *J. Phys. Chem. 86,* 2182 (1982).

Laubereau, A., and Kaiser, W., "Picosecond Spectroscopy of Molecular Dynamics in Liquids." *Annu. Rev. Phys. Chem. 26,* 83 (1975).

Laudenslager, J. B., "Ion-Molecule Processes in Lasers," in Ausloos (1979).

Lawley, K. P. (ed.), "Photodissociation and Photoionizaton," *Adv. Chem. Phys. 60* (1985).

Leone, S. R., "Photogragmentation Dynamics." *Adv. Chem. Phys. 50,* 255 (1982).

Leone, S. R., and Moore, C. B., "Laser Sources," in Moore (1974).

Letokhov, V. S., "Laser Selective Detection of Single Atoms," in Moore (1980).

Letokhov, V. S., and Ryabov, E. A., "Time-Resolved Raman Spectroscopy of Highly Excited Vibrational States of Polyatomic Molecules," in Lauberau and Stockburger (1985).

Letokhov, V. S., and Moore, C. B., "Laser Isotope Separation." in Moore (1977).

Lin, M. C., Umstead, M. E., and Djeu, N., "Chemical Lasers." *Annu. Rev. Phys. Chem. 34,* 557 (1983).

Lineberger, W. C., "Laser Spectroscopy of Gas Phase Ions, in Moore (1974).

Mead, R. D., Stevens, A. E., and Lineberger, W. C., "Photodetachment in Negative Ion Beams," in Bowers (1984).

Miller, T. A., "Light and Radical Ions." *Annu. Rev. Phys. Chem. 33,* 257 (1982).

Moore, C. B., and Weisshaar, J. C., "Formaldehyde Photochemistry." *Annu. Rev. Phys. Chem. 34,* 525 (1983).

Moseley, J. T., "Half-Collision Aspects of Ion Photofragment Spectroscopy." *J. Phys. Chem. 86,* 3282 (1982).

Moseley, J., and Durup, J., "Fast Ion Beam Photofragment Spectroscopy." *Annu. Rev. Phys. Chem. 32,* 53 (1981).

Pimentel, G. C., and Kompa, K. L., "What is a Chemical Laser?," in Gross and Bott (1976).

Reintjes, J. F., "Coherent Ultraviolet Vacuum Ultraviolet Sources," in Laser Handbook, Vol. 5, Bass, M., and Stitch, M. L., eds., North-Holland, Amsterdam, 1985.

Reisler, H., Mangir, M., and Wittig, C., "Laser Kinetics Spectroscopy of Elementary Processes," in Moore (1980).

Rentzepis, P. M., "Picosecond Spectroscopy and Molecular Relaxation." *Adv. Chem. Phys. 23*, 189 (1973).

Robinson, G. W., and Jalenak, W. A., "Chemical Reactions in Solution: The New Photochemistry." *Laser Chem. 3*, 163 (1983).

Schafer, F. P., "New Developments in Laser Dyes." *Laser Chem. 3*, 265 (1983).

Shank, C. V., "Advances in Femtosecond Optical Spectroscopy Techniques." *Laser Chem. 3*, 133 (1983).

Shank, C. V., and Greene, B. I., "Femtosecond Spectroscopy and Chemistry." *J. Phys. Chem. 87*, 732 (1983).

Simon, J. D., and Peters, K. S., "Picosecond Studies of Organic Photoreactions." *Acc. Chem. Res. 17*, 277 (1984).

Simons, J. P., "The Dynamics of Photodissociation." *Spec. Per. Rep. 2*, 58 (1976).

Simons, J. P., "Photodissociation: A Critical Survey." *J. Phys. Chem. 88*, 1287 (1984).

Steinfield, J. I., "Laser-Induced Chemical Reactions: Survey of the Literature, 1965–1979, in Steinfield (1981).

Swofford, R. L., and Albrecht, A. C., "Nonlinear Spectroscopy," *Annu. Rev. Phys. Chem. 29*, 421 (1978).

Tramer, A., "Relaxation of Photo-Excited Molecules," in Glorieux et al. (1979).

Valentini, J. J., "Laser Raman Techniques" in Radziemski et al. (1986).

Wallace, S. C., "Nonlinear Optics and Laser Spectroscopy in the Vacuum Ultraviolet." *Adv. Chem. Phys. 47* (Pt. 2), 153 (1981).

Walther, H. (ed.), *Laser Spectroscopy of Atoms and Molecules*, Springer-Verlag, Berlin (1976).

Welge, K. H., and Schmiedl, R., "Doppler Spectroscopy of Photofragments," *Adv. Chem. Phys. 47* (Pt. 2), 133 (1981).

Wolfrum, J., "Laser Stimulation and Observation of Bimolecular Reactions," in El-Sayed (1986).

Wolfrum, J., "Laser Induced Chemical Reactions in Combustion and Industrial Processes," *Laser Chem. 6*, 125 (1986).

Zare, R. N., "Interference Effects in Molecular Fluorescence." *Acc. Chem. Res. 4*, 361 (1971).

Zare, R. N., "Photoejection Dynamics." *Mol. Photochem. 4*, 1 (1972).

Zare, R. N., "Laser Techniques for Determining State-to-State Reaction Rates," in Brooks and Hayes (1977).

Zare, R. N., "Laser Chemical Analysis." *Science 226*, 298 (1984).

Zare, R. N., and Dagdigian, P. J., "Tunable Laser Fluorescence Method for Product State Analysis." *Science 185*, 739 (1974).

Zewail, A. H., "Picosecond Laser Chemistry in Supersonic Jet Beams." *Laser Chem. 2*, 55 (1983).

Section 5.3

Books

Ausloos, P. (ed.), *Interaction Between Ions and Molecules*, Plenum, New York, 1975.

Ausloos, P. (ed.), *Kinetics of Ion–Molecule Reactions,* Plenum, New York, 1979.

Drukarev, G. F., *Collisions of Electrons with Atoms and Molecules,* Plenum, New York, 1985.

Eichler, J., Hertel, I. V., and Stolterfot, N. (eds.), *Electronic and Atomic Collisions,* North-Holland, Amsterdam, 1984.

Faraday Discussions of the Chemical Society, 55, Molecular Beam Scattering, (1973).

Ferreira, M. A. A. (ed.), *Ionic Processes in the Gas Phase,* Reidel, Dordrecht, 1982.

Kleinpoppen, H., Briggs, J. S., and Lutz, H. O. (eds.), *Fundamental Processes in Atomic Collision Physics.* Plenum, New York, 1985.

Leach, S. (ed.), *Molecular Ion Studies, J. Chim. Phys.,* 77, 7/8 (1980).

Ramsey, N. F., *Molecular Beams,* Oxford University Press, New York, 1986.

Setser, D. W. (ed.), *Gas Phase Intermediates, Generation and Detection,* Academic, New York, 1980.

Review Articles

Baer, T., "State Selection by Photoion-Photoelectron Coincidence," in Bowers (1979).

Brooks, P. R., "Molecular Beams," in Moore, Vol. 1 (1974).

Dubrin, J., and Henchman, M. J., "Ion–Molecule Reactions." *Int. Rev. Sci. Phys. Chem. 9,* 213 (1972).

Dunbar, R. C., "Ion Photodissociation," in Bowers (1979).

Ferguson, E. E., "Ion–Molecule Reactions." *Annu. Rev. Phys. Chem. 26,* 17 (1975).

Ferguson, E. E., and Arnold, F., "Ion Chemistry of the Stratosphere." *Acc. Chem. Res. 14,* 327 (1981).

Gentry, R. W., "Molecular Beam Studies of Ion-Molecule Reactions," in Ausloos (1979).

Gentry, R. W., "Molecular Beam Techniques: Applications to the Study of Ion-Molecule Collisions," in Bowers (1979).

Grice, R., "Reactions Studied by Molecular Beam Technques." *Sp. Per. Rep. 4,* 1 (1981).

Hack, W., "Detection Methods for Atoms and Radicals in the Gas Phase." *Int. Rev. Phys. Chem. 4,* 165 (1985).

Kosko, W. S., "Scattering of Positive Ions by Molecules." *Adv. Chem. Phys. 30,* 185 (1975).

Levy, D. H., Wharton, L., and Smalley, R. E., "Laser Spectroscopy in Supersonic Jets," in Moore (1974).

Mahan, B. H., "Ion-Molecule Collision Processes." *Acc. Chem. Res. 3,* 393 (1970).

Mahan, B. H., "Ion-Molecule Collision Phenomena." *Int. Rev. Sci. Phys. Chem. 9,* 25 (1975).

Moran, T. F., "State-to-State Ion-Molecule Reactions," in Brooks and Hayes (1977).

Moseley, J. T., "Ion Photofragment Spectroscopy." *Adv. Chem. Phys. 60,* 245 (1985).

Powis, I., "Ionization Processes and Ion Dynamics," in *Mass Spectrometry,* Vol. 8 (Specialist Periodical Report, 1985).

Smalley, R. E., Wharton, L., and Levy, D. H., "Molecular Optical Spectroscopy with Supersonic Beams and Jets." *Acc. Chem. Res. 10,* 139 (1977).

Tiernan, T. O., and Lifshitz, C., "Role of Excited States in Ion-Neutral Collisions." *Adv. Chem. Phys. 52,* 82 (1980).

Section 5.4

Books

Fluendy, M. A. D., and Lawley, K. P., *Chemical Applications of Molecular Beam Scattering,* Chapman and Hall, London, 1973.

Review Articles

Anderson, J. B., Andres, R. P., and Fenn, J. B., "Supersonic Nozzle Beams." *Adv. Chem. Phys. 10,* 275 (1966).

Bernstein, R. B., "Reaction Dynamics by Molecular Beams," in Zewail (1978).

Campargue, R., "Progress in Overexpanded Supersonic Jets and Skimmed Molecular Beams in Free-Jet Zones of Silence," in El-Sayed (1984).

Gentry, W. R., "Low Energy Pulsed Beam Sources," in Scoles (1986).

Lee, Y. T., "Reactive Scattering: Non-Optical Methods," in Scoles (1986).

Neynaber, R. H., "Merging Beams," in Bederson and Fite (1968).

Pauly, H., and Toennies, J. P., "Beam Experiments at Thermal Energies," in Bederson and Fite (1968).

Rettner, C. T., Marinero, E. E., Zare, R. N., and Kung, A. H., "Pulsed Free Jets: Novel Nonlinear Media for Generation of VUV and XUV Radiation," in El-Sayed (1984).

Smalley, R. E., "Lasers and Pulsed Beams," in Jackson and Harvey (1985).

Section 5.5

Books

Katz, A., *Principles of Statistical Mechanics: The Information Theory Approach,* Freeman, San Francisco, 1967.

Levine, R. D., and Tribus, M. (eds.), *Maximum Entropy Formalism,* MIT Press, Cambridge, Mass., 1979.

Review Articles

Bernstein, R. B. and Levine, R. D., "Role of Energy in Reactive Molecular Scattering: An Information Theoretic Approach." *Adv. At. Mol. Phys. 10,* 216 (1975).

Jaynes, E. T., *Statistical Physics,* in 1962 Brandeis Lectures, K. Ford (ed.), Benjamin, New York, 1963.

Levine, R. D., "Information Theory Approach to Molecular Reaction Dynamics." *Annu. Rev. Phys. Chem. 29,* 59 (1978).

Levine, R. D., "The Information Theoretic Approach to Intramolecular Dynamics." *Adv. Chem. Phys. 47* (Pt. 1), 239 (1981).

Levine, R. D., "Statistical Dynamics," in Baer (1985).

Levine, R. D., and Ben-Shaul, A., "Thermodynamics of Molecular Disequilibrium," in Moore (1974).

Levine, R. D., and Bernstein, R. B., "Energy Disposal and Energy Consumption in Elementary Chemical Reactions: The Information Theoretic Approach." *Acc. Chem. Res. 7*, 393 (1974).

Levine, R. D., and Bernstein, R. B., "Thermodynamic Approach to Collision Processes," in Miller (1976).

Levine, R. D., and Kinsey, J. L., "Information-Theoretic Approach: Application to Molecular Collisions," in Bernstein (1979).

Nesbet, R. K., "Surprisal Theory," in Henderson (1981).

Section 5.6

Books

Baer, M. (ed.), *The Theory of Chemical Reaction Dynamics,* CRC Press, Boca Raton, Fla. 1985.

Bernstein, R. B. (ed.), *Atom–Molecule Collision Theory. A Guide for the Experimentalist,* Plenum, New York, 1979.

Eu, B. C., *Semiclassical Theories of Molecular Scattering,* Springer-Verlag, Berlin, 1984.

Fain, B., *Theory of Rate Processes in Condensed Media,* Springer-Verlag, Berlin, 1980.

Gianturco, F. A., *The Transfer of Molecular Energies by Collision: Recent Quantum Treatments,* Springer-Verlag, Berlin, 1979.

Gianturco, F. A. (ed.), *Atomic and Molecular Collision Theory,* Plenum, New York 1982.

Levine, R. D., *Quantum Mechanics of Molecular Rate Processes,* Clarendon, Oxford, U.K. 1969.

Newton, R. G., *Scattering Theory of Waves and Particles,* 2nd ed., Springer-Verlag, Heidelberg, 1982.

Woolley, R. G. (ed.), *Quantum Dynamics of Molecules,* Plenum, New York (1980).

Review Articles

Adelman, S. A., and Doll, J. D., "Brownian Motion and Chemical Dynamics on Solid Surfaces." *Acc. Chem. Res. 10*, 378 (1977).

Baer, M., "A Review of Quantum-Mechanical Approximate Treatments of Three-Body Reactive Systems." *Adv. Chem. Phys. 49*, 191 (1982).

Baer, M., "The General Theory of Reactive Scattering: The Differential Equation Approach," in Baer (1985).

Balint-Kurti, G. G., "The Theory of Rotationally Inelastic Molecular Collisions." *Int. Rev. Sci. Phys. Chem. 1*, 285 (1975).

Basilevsky, M. V., and Ryaboy, V. M., "Quantum Dynamics of Linear Triatomic Reactions." *Adv. Quant. Chem. 15*, 1 (1982).

Bottcher, C., "Numerical Solution of the Few-Body Schrödinger Equation," in Eichler et al. (1984).

Bowman, J. M., "Reduced Dimensionality Theories of Quantum Reactive Scattering." *Adv. Chem. Phys. 61*, 115 (1985).

Bowman, J. M., Ju, G.-Z., and Lee, K. T., "Incorporation of Collinear Exact

Quantum Reaction Probabilities into Three-Dimensional Transition-State Theory." *J. Phys. Chem. 86,* 2232 (1982).

Brumer, P., and Shapiro, M., "Theoretical Aspects of Photodissociation and Intramolecular Dynamics." *Adv. Chem. Phys. 60,* 371 (1984).

Casassa, M. P., Western, C. M., and Janda, K. C., "Photodissociation of van der Waals Molecules: Do Angular Momentum Constraints Determine Decay Rates?," in Truhlar (1984).

Cerjan, C. J., Shi, S., and Miller, W. H., "Applications of a Simple Dynamical Model to the Reaction Path Hamiltonian: Tunneling Corrections to Rate Constants, Product State Distributions, Line Widths of Local Mode Overtones, and Mode Specificity in Unimolecular Decomposition." *J. Phys. Chem. 86,* 2244 (1982).

Child, M. S., "Semiclassical Reactive Scattering," in Baer (1985).

Chu, S.-I., "Complex-Coordinate Coupled-Channel Methods for Predissociating Resonance in van der Waals Molecules," in Truhlar (1984).

DePristo, A. E., and Rabitz, H., "Vibrational and Rotational Collision Processes." *Adv. Chem. Phys. 42,* 271 (1980).

Diestler, D. J., "Collision-Induced Dissociation: Quantal Treatment," in Bernstein (1979).

Eno, L., and Rabitz, H., "Sensitivity Analysis and Its Role in Quantum Scattering." *Adv. Chem. Phys. 51,* 177 (1982).

Freed, K. F., Metiu, H., Hood, E., and Jedrzejek, C., "Quantum Mechanical Model of the Dynamics of Desorption Processes," in Jortner and Pullman (1982).

Garrett, B. C., Schwenke, D. W., Skodje, R. T., Thirumalai, D., Thompson, T. C., and Truhlar, D. G., "Bimolecular Reactive Collisions: Adiabatic and Nonadiabatic Methods for Energies, Lifetime, and Branching Probabilities," in Truhlar (1984).

George, T. F., and Ross, J., "Quantum Dynamical Theory of Molecular Collisions." *Annu. Rev. Phys. Chem. 24,* 263 (1973).

Gordon, R. G., "Rational Selection of Methods for Molecular Scattering Calculations," in *Faraday Discuss. Chem. Soc.* (1973).

Heller, E. J., "Potential Surface Properties and Dynamics from Molecular Spectra: A Time-Dependent Picture," in Truhlar (1981).

Heller, E. J., "The Semiclassical Way to Molecular Spectroscopy." *Acc. Chem. Res. 14,* 368 (1981).

Jellinek, J., and Kouri, D. J., "Approximate Treatments of Reactive Scattering: Infinite Order Sudden Approximation," in Baer (1985).

Jordan, K. D., "Theoretical Studies of the Reactions of Atoms with Small Molecules," in Fontijn (1985).

Kouri, D. J., "General Theory of Reactive Scattering: The Integral Equation Approach," in Baer (1985).

Kouri, D. J., and Fitz, D. E., "Angular Momentum Decoupling Approximations. Current Status, Successes and Difficulties." *J. Phys. Chem. 86,* 2224 (1982).

Kuppermann, A., "Theoretical Aspects of the Mechanism of Simple Chemical Reactions," in Glorieux et al. (1979).

Kuppermann, A., "Accurate Quantum Calculations of Reactive Systems," in Henderson (1981).

Lester, W. A., Jr., "Calculation of Cross Sections for Rotational Excitation of

Diatomic Molecules by Heavy Particle Impact: Solution of the Close-Coupled Equations." *Methods Comput. Phys. 10,* 211 (1971).

Lester, W. A., Jr., "Coupled-Channel Studies of Rotational and Vibrational Energy Transfer by Collision." *Adv. Quant. Chem. 9,* 199 (1975).

Lester, W. A., Jr., "The N Coupled-Channel Problem," in Miller (1976).

Levine, R. D., "Quasi-Bound States in Molecular Collisions." *Acc. Chem. Res. 3,* 273 (1970).

Light, J. C., "Quantum Theories of Chemical Kinetics." *Adv. Chem. Phys. 19,* 1 (1971).

Light, J. C., "Inelastic Scattering Cross Sections: Theory," in Bernstein (1979).

Light, J. C., "Reactive Scattering Cross Section: General Quantal Theory," in Bernstein (1979).

Manz, J., "Molecular Dynamics Along Hyperspherical Coordinates," *Comments At. Mol. Phys. 17,* 91 (1985).

Marcus, R. A., "The Theoretical Approach," in *Faraday Discuss. Chem. Soc.* (1973).

McCurdy, C. W., and Miller, W. H., "A New Helicity Representation for Reactive Atom–Diatom Collisions," in Brooks and Hayes (1977).

Micha, D. A., "Long-Lived States in Atom–Molecule Collision." *Acc. Chem. Res. 6,* 138 (1973).

Micha, D. A., "Quantum Theory of Reactive Molecular Collisions." *Adv. Chem. Phys. 30,* 7 (1975).

Micha, D. A., "Optical Models in Molecular Collision Theory," in Miller (1976).

Micha, D. A., "Overview of Non-Reactive Scattering," in Truhlar (1981).

Micha, D. A., "General Theory of Reactive Scattering: A Many-Body Approach," in Baer (1985).

Miller, W. H., "Classical-Limit Quantum Mechanics and The Theory of Molecular Collisions." *Adv. Chem. Phys. 25,* 69 (1974).

Miller, W. H., "Classical S-Matrix in Molecular Collisions." *Adv. Chem. Phys. 30,* 77 (1975).

Miller, W. H., "Reaction Path Hamiltonian for Polyatomic Systems: Further Developments and Applications," in Truhlar (1981).

Miller, W. H., "Reaction-Path Dynamics for Polyatomic Systems." *J. Phys. Chem. 87,* 3811 (1983).

Rabitz, H., "Effective Hamiltonians in Molecular Collisions," in Miller (1976).

Reinhardt, W. P., "Complex Coodinates in the Theory of Atomic and Molecular Structure and Dynamics." *Annu. Rev. Phys. Chem. 33,* 223 (1982).

Secrest, D., "Amplitude Densities in Molecular Scattering." *Methods Comput. Phys. 10,* 243 (1971).

Secrest, D., "Vibrational Excitation: The Quantal Treatment," in Bernstein (1979).

Secrest, D., "Inelastic Vibrational and Rotational Quantum Collisions," in Bowman (1983).

Shapiro, M., and Bersohn, R., "Theories of the Dynamics of Photodissociation." *Annu. Rev. Phys. Chem. 33,* 409 (1982).

Singer, S. J., Freed, K. F., and Band, Y. B., "Photodissociation of Diatomic Molecules to Open Shell Atoms." *Adv. Chem. Phys. 61,* 1 (1985).

Tang, K. T., "Approximate Treatments of Reactive Scattering: The T Matrix Approach," in Baer (1985).

Truhlar, D. G., Abdallah, J., Jr., and Smith, R. L., "Algebraic Variational Methods in Scattering Theory." *Adv. Chem. Phys. 25,* 211 (1974).

Truhlar, D. G., Mead, C. A., and Brandt, M. A., "Time-Reversal Invariance, Representations for Scattering Wave Functions, Symmetry of the Scattering Matrix, and Differential Cross Sections." *Adv. Chem. Phys. 33,* 295 (1975).

Walker, R. B., and Light, J. C., "Reactive Molecular Collisions." *Annu. Rev. Phys. Chem. 31,* 401 (1980).

Wyatt, R. E., "Reactive Scattering Cross Sections: Approximate Quantal Treatment," in Bernstein (1979).

Wyatt, R. E., "Direct-Mode Chemical Reactions: Methodology for Accurate Quantal Calculations," in Bernstein (1979).

6
Molecular energy transfer

6.1 A macroscopic description of energy transfer

Inelastic, energy-transferring collisions provide the mechanism for bringing a system to thermal equilibrium. This is but one reason for examining the molecular level description of energy exchange. Determining the inter*molecular* potential is another reason for studying inelastic collisions. An aspect not stressed before but which can no longer be ignored is that a collision need not occur on a single potential energy surface. The concept of a unique potential as a function of the interatomic distances requires that the system be in a single electronic state. But the very phenomenon of electronic energy transfer, for example, the quenching of excited Hg atoms

$$Hg^* + CO(v = 0) \rightarrow Hg + CO(v')$$

or the pumping process of the iodine chemical laser

$$O_2^*(^1\Delta) + I(^2P_{3/2}) \rightarrow O_2(^3\Sigma) + I(^2P_{1/2}),$$

implies the participation of more than one electronic state of the system. Even when the collision occurs on a single surface, the electronically excited states are still there and can be accessed by visible or ultraviolet radiation interacting with the bimolecular colliding pair. The complementary process, that of emision down to the ground electronic state from a collision taking place on an excited surface, is equally an important probe of the dynamics. From the theoretical viewpoint, the boundary conditions on the scattering wave function are simpler for inelastic than for reactive collisions (Sec. 5.5), so that the theory of energy-transfer collisions is simpler and further developed. There is also more scope here for simple but realistic models.

6.1.1 Equilibrium and disequilibrium

Equilibrium in the gas phase can be characterized by a constant (time-independent) fraction of molecules in any given energy level. At a temperature T the relative population in the ith energy level is given by the *Boltzmann distribution*:

$$p_i = n_i/N = g_i \exp(-E_i/\mathbf{k}T)/Q_I \tag{6.1}$$

Here, n_i is the number density of molecules in the level i, and N is the total number of molecules. The *degeneracy* g_i is the number of possible quantum states of the molecule corresponding to the energy level E_i. Q_I is the internal partition function.

Various processes can lead to a transient population distribution that deviates from Eq. (6.1). Such a state of disequilibrium can be achieved, for example, by very rapid heating in a shock tube. Alternatively, and more selectively, a preferential population of excited states can be created either by physical activation (e.g., absorption of light) or by chemical reactions that are selective in their energy disposal. Following a transient disturbance the gas "relaxes" into an equilibrium distribution. One can follow this relaxation process by observing the changes in the population of the different energy levels. Such changes are the manifestations of *energy-transferring* molecular collision.† For example, a vibrationally excited HCl molecule can lose its excess energy in a collision where the initial *vibrational energy* is converted upon collision with some other molecule M to *translational energy* of the "products":

$$HCl(v = 1) + M \rightarrow HCl(v = 0) + M$$

We shall refer to such an inelastic collision as a *V–T transfer*. The reverse collision, where HCl gains vibrational excitation at the expense of the initial translational energy is a *T–V transfer* collision.

The *V–T* transfer from the lowest excited state to the ground vibrational state is usually the slowest, or rate determining, step in the complete return to thermal equilibrium. In a real system, and particularly for polyatomic molecules, many additional processes are possible, and will be discussed below.

6.1.2 The relaxation rate equations

To determine the rate at which the disturbed population relaxes towards equilibrium we consider a simple example where the only two important inelastic processes are the vibrational deexcitation and (reverse) excitation collisions:

$$HCl(v = 1) + M \rightleftarrows HCl(v = 0) + M$$

This is the case when we perturb an equilibrium mixture of HCl and an inert

---

† The relaxation can also take place via a unimolecular mechanism by which excited molecules either emit light or (if they are sufficiently energized) dissociate. In general, the lifetime for losing vibrotational energy as radiation is far longer (e.g., *ca.* 1 ms) than typical observed relaxation times of the gas. On the other hand, in interstellar space, where the density of molecules, and hence the collision frequency, is very small, the unimolecular mechanism dominates the approach to equilibrium. Another condition that can enhance the rate of photon emission is the presence of photons of the same frequency. This stimulated emission process is central to the operation of a laser, as discussed in Appendix 6A.

buffer gas at low temperatures, where, at equilibrium, most of the HCl molecules are in the ground, $v = 0$, state. We now write conventional chemical kinetic second-order rate expressions for the forward and reverse bimolecular processes of $V–T$ and $T–V$ transfer,

$$-\frac{dn_0}{dt} = \frac{dn_1}{dt} = -k_{10}n_Bn_1 + k_{01}n_Bn_0 \tag{6.2}$$

Here, n_1 and n_0 are the number densities of HCl molecules in the $v = 1$ and $v = 0$ states and n_B is the number density of the buffer gas; k_{10} is the rate constant for the $V–T$ process, which depletes the population in the $v = 1$ state, and k_{01} is the rate constant for the reverse $T–V$ process, which populates the $v = 1$ state. At equilibrium there is no further change in the populations, but the energy-transfer processes do not stop. Rather, the populations have adjusted themselves so as to ensure that the rate of the forward and reverse process is the same. At equilibrium, (6.2) yields

$$\frac{k_{01}}{k_{10}} = \left[\frac{n_1}{n_0}\right]_{eq} = \exp[-(E_1 - E_0)/kT] = \exp(-\theta/T). \tag{6.3}$$

Here, the subscript "eq" denotes (the population ratio at) equilibrium. The magnitude of this ratio is known from the Boltzmann distribution [Eq. (6.1)], $\theta = \hbar\omega/k$, while $E_1 - E_0 = \hbar\omega$, so that θ is the characteristic vibrational temperature. The result, Eq. (6.3), is a particular example of detailed balance as discussed in Section 4.4.3.

6.1.3 The relaxation time

When the mixture is suddenly displaced from equilibrium (say by a pulse of light of the resonant frequency ω), the population of HCl in the $v = 1$ state will temporarily increase from its equilibrium value, $(n_1)_{eq}$, to $(n_1)_{eq} + \Delta n_1$. The population of the $v = 0$ state will decrease by the corresponding amount. From (6.2),

$$\frac{dn_1}{dt} = \frac{d}{dt}[(n_1)_{eq} + \Delta n_1] = -k_{10}[(n_1)_{eq} + \Delta n_1]n_B + k_{01}[(n_0)_{eq} - \Delta n_1]n_B$$

or, using (6.3),

$$\frac{d\,\Delta n_1}{dt} = -(k_{10} + k_{01})n_B\,\Delta n_1 = -\Delta n_1/\tau \tag{6.4}$$

Thus,

$$\Delta n_1(t) = n_1(t) - (n_1)_{eq} = \Delta n_1(t = 0)\exp(-t/\tau) \tag{6.5}$$

where

$$\tau = 1/n_B(k_{10} + k_{01}) = 1/n_Bk_{10}[1 + \exp(-\theta/T)] \approx 1/n_Bk_{10} \tag{6.6}$$

is the *relaxation time* for restoring the equilibrium. Since the number density

n_B of the inert buffer gas is proportional to its pressure, the results of bulk relaxation studies are often given in terms of the *product* $P\tau$, where $(P\tau)^{-1}$ is a second-order rate constant in units of $(\text{atm s})^{-1}$. To convert $(P\tau)^{-1}$ to the more familiar units (liters $\text{mol}^{-1}\,\text{s}^{-1}$), we note that the mole density (for an ideal gas) is P/RT, so that $RT/P\tau = k$ is the rate constant in the familiar units.†

6.1.4 Overview of relaxation rates

A summary of the possible energy-transfer processes (on a single potential energy surface) is shown in Fig. 6.1 together with typical $P\tau$ values. An example of a $V-V$ process is

$$\text{HCl}(v = 1) + \text{HCl}(v = 1) \rightarrow \text{HCl}(v = 2) + \text{HCl}(v = 0)$$

Here the translational exoergicity will be small (entirely due to the anharmonicity of the HCl vibration), even though a large amount of energy is exchanged. A fairly efficient $R-V$ transfer is

$$\text{I}_2(v = 0) + \text{H}_2(j = 2) \rightarrow \text{I}_2(v = 1) + \text{H}_2(j = 0)$$

Both collisions mentioned can lead to other final states besides the ones shown. For example, in the first case we can also have a $V \rightarrow T$ transfer,

$$\text{HCl}(v = 1) + \text{HCl}(v = 1) \rightarrow \text{HCl}(v = 1) + \text{HCl}(v = 0) + Q$$

with a large translational exoergicity, Q. Alternatively, one of the HCl molecules could be excited ($T-V$ transfer). Both of these processes are far less efficient than "near resonant" $V-V$ transfer in which there is little change in the translational energy.

One of our aims is to learn to distinguish between the more probable and the less probable processes. As a rule, we shall learn to recognize that a *large translational energy "defect"* is an indication of an *inefficient transfer* process. This generalization is supported by the summary of relaxation times shown in Fig. 6.1. Rotational-energy spacings are small, and so both $R-R'$ and $R-T$ transfers are very efficient. Least efficient of all are $V-T$ transfers, particularly for the diatomic molecules with the larger vibrational spacings.

A measure of the efficiency of energy-transfer collisions is the *collision*

---

† In practical units,

$$k(\text{cm}^3\,\text{mol}^{-1}\,\text{s}^{-1}) = 2.445 \times 10^4 (T/298)/P\tau$$

where P is in atm, τ in s. Note also that the "effective" cross section σ related by $k = \langle v \rangle \sigma$ is, in practical units,

$$k(\text{cm}^3\,\text{mol}^{-1}\,\text{s}^{-1}) = 8.76 \times 10^{11} (T/\mu)^{1/2}\sigma$$

where μ is the reduced mass in atomic mass units ($C = 12$) and σ is in Å^2.

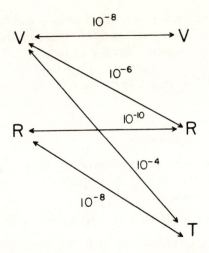

Fig. 6.1. Schematic diagram showing the energy transfer processes occurring in thermal molecular collisions. *V, R,* and *T* refer to vibrational, rotational, and translational energy, respectively. The numbers are typical values of $P\tau$ (atm s), the bulk "relaxation time" characterizing the particular mode of energy transfer. [Adapted from W. H. Flygare, *Acc. Chem. Res., 1,* 121 (1968).]

number, Z_{A-B}, defined as the ratio of the relaxation time to the time interval between collisions† [ω^{-1}, Eq. (2.8)],

$$Z_{A-B} = \omega\tau_{A-B} \tag{6.7}$$

for a particular $A-B$ relaxation process. For $R-R'$ transfer, $Z_{R-R'}$ is typically between 1 and 10, showing that $R-R'$ transfer occurs on nearly every "collision," while for $V-T$ transfer, of, say O_2 in Ar at room temperature, $Z_{V-T} \sim 10^6$. In qualitative terms, we can regard Z_{A-B}^{-1} as the probability of inducing the particular inelastic transition in a single collision. Thus, the expected transition will occur, on the average, once out of every Z_{A-B} collisions.

The studies of relaxation in the bulk have yielded valuable information, summarized in Fig. 6.1, on the rates of energy–transfer processes. Several questions are still unanswered. On the practical side we still need to know not just the overall rate of relaxation but also the rates of specific excitation transfer processes (e.g., as in Fig. 1.5). We address this question in the next section. There, we also demonstrate the practical application of such knowledge. The nature of the intermolecular forces that lead to effective energy transfer and the physical reasons that favor low translational exoergicity also need clarification. Finally, we need to examine electronic

---

† One should distinguish the (dimensionless) collision number Z_{A-B} from the rate constant for bimolecular collisions, the Z of (2.10). In terms of Z, $Z_{A-B} = Zn_B\tau_{A-B} = Z/k_{A-B}$, where k is the relaxation rate constant.

energy transfer. In Appendix 6A we consider the question of equilibrium in the presence of photons and examine the condition for lasing.

6.1.5 A hierarchy of relaxation times

Many different collision-induced energy-transfer processes occur with widely differing rates as a gas relaxes to equilibrium. Yet bulk relaxation studies can often be characterized by a single relaxation time. Some indication of why that should be the case is provided by Fig. 6.1. Precisely because some processes are very much faster than others,

$$\tau_{R-R'} < \tau_{V-V'} < \tau_{V-T} \tag{6.8}$$

they are essentially complete on the time scale necessary to monitor the slow process. In other words, when we propose to measure $V-T$ relaxation we need time resolution of the order of τ_{V-T}, say 10^{-3} s at a pressure of 1 Torr. This is a very long time as far as the $R-R'$ transfer process are concerned. As a result, there is ample time for the entire manifold of rotational states, belonging to a given vibrational state, to come to equilibrium among themselves.† Then $V-V'$ processes equilibrate among the excited vibrational states. (Ultimately we come to the slowest process, the $v = 1 \rightarrow 0$ deactivation of the lowest excited vibrational state by $V-T$ transfer.) The observed relaxation time is thus the relaxation time of the slowest or *rate-determining step*. The population of all the excited states relaxes with this longest relaxation time, since the other transfer processes are much faster and, at any point, rapidly equilibrate among all excited states. The lowest excited state must, however, relax by a *slow $V-T$* (or $V-R$) transfer (where else can the energy go?), and its relaxation determines the net rate of depopulation of all other excited levels.‡ Before we accept this picture, we should try to validate it by a direct experimental observation.

6.1.6 Laser-induced fluorescence spectroscopy

When the laser emission frequency matches a vibrotational absorption line of a molecule, it is possible to excite selectively a significant fraction of the molecules in the bulk into the upper level and then to measure the collisional relaxation of this particular level. The changes in the population

---

† In Appendix 6A we explore some of the consequences of this fast relaxation.

‡ For reactions with an activation energy, the relaxation time towards *chemical* equilibrium is usually longer than even the slowest relaxation time towards thermal (i.e., physical) equilibrium. It is for this reason that, despite the selective energy requirements of chemical reactions, we can often make the assumption that the reactants are in thermal equilibrium. This assumption does fail in those cases (e.g., collision-induced dissociation) in which the rate of depletion, by collisional dissociation, of high vibrational states, is large enough to compete with $T-V$ and $V-V'$ transfer rates.

are monitored either via the infrared fluorescence from the excited levels†
or by absorption spectroscopy.

An application to the simplest case, V–V' transfer in diatomics, is shown
in Fig. 6.2(a). A mixture of HCl and DCl in an excess of Ar buffer gas is
irradiated by a brief pulse of infrared light (from an HCl chemical laser).
HCl molecules are thereby selectively excited from the $v = 0$ to the $v = 1$
level. Thereafter, the population of the HCl($v = 1$) level is depleted both by
V–T transfer (mainly due to collisions with the excess Ar buffer gas) and by
the nonresonant V–V' process:

$$\text{HCl}(v = 1) + \text{DCl}(v = 0) \rightarrow \text{HCl}(v = 0) + \text{DCl}(v = 1), \qquad Q = 775 \text{ cm}^{-1}$$

The $v = 1$ population of DCl is depleted again, mainly by V–T transfer.

The time-resolved fluorescence of DCl shows [Fig. 6.2(b)] both a
fast-rising exponential due to the rapid V–V' transfer and a slower
decaying exponential due to V–T (and V–R) transfer. The measured V–V'
rates ($P\tau \sim 10^{-8}$ atm s) are about two orders of magnitude faster than the
V–T rates.

6.1.7 V-V' processes in polyatomics

In polyatomic molecules, additional types of V–V' transfer processes are
possible. Figure 6.3 shows the low-lying vibrational states of CO_2 together
with their spectroscopic designation. The triplet of numbers (m, n, p)
indicates the numbers of vibrational quanta in the symmetric stretching (m),
the bending (n), and the asymmetric stretching (p) modes, respectively. It
was found that the *intra*molecular collision-induced transfer

$$\text{CO}_2(020) + M \rightarrow \text{CO}_2(100) + M, \qquad Q = 103 \text{ cm}^{-1} \qquad (6.9)$$

is very efficient ($Z_{V-V'} \sim 10$). Next are the two *inter*molecular sharing
processes

$$\text{CO}_2(100) + \text{CO}_2(000) \rightarrow 2\text{CO}_2(010), \qquad Q = 54 \text{ cm}^{-1} \qquad (6.10)$$

and

$$\text{CO}_2(020) + \text{CO}_2(000) \rightarrow 2\text{CO}_2(010), \qquad Q = -49 \text{ cm}^{-1} \qquad (6.11)$$

for which $Z_{V-V'} \sim 50$. [The V–V' process (6.11) is also possible in
diatomics; it is the major mechanism for bringing excited vibrational levels
of diatomics into equilibrium with each other.] In CO_2, intermolecular

---

† Radiative lifetimes in the infrared are sufficiently long so that one can neglect the depletion
of excited states by fluorescence as compared to collisional loss (or gain). However, though
only a small fraction of excited molecules lose their excitation by the (unimolecular) radiative
decay mode, *enough* of them do so that we can use the fluorescence intensity as a monitor of
the concentration of the excited molecules. Radiative lifetimes in the visible or ultraviolet
region are much shorter, and the unimolecular radiative decay *can* effectively complete with
the bimolecular energy-transfer processes (Sec. 6.6.2).

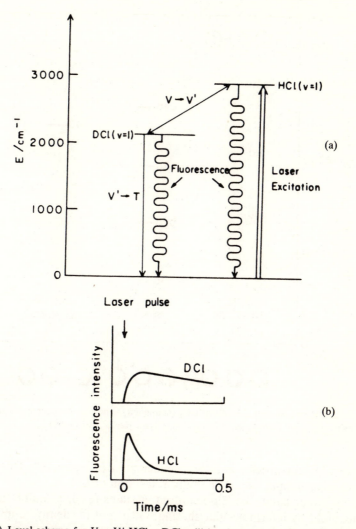

Fig. 6.2. (a) Level scheme for $V \to V'$ HCl $\to$ DCl collisional energy transfer. Following laser excitation of HCl ($v = 1$), the population can be monitored by the infrared fluorescence. (b) Fluorescence intensity from HCl ($v = 1$) following a laser excitation pulse (bottom) compared to that (top) from DCl ($v = 1$) formed in $V \to V'$ transfer. [Adapted from H. L. Chen and C. B. Moore, *J. Chem. Phys.*, *54*, 4072 (1971).]

overtone sharing of the type (6.11) "equilibrates" among the levels of the $0n0$ manifold, while the intra- and intermolecular transfer processes (6.9) and (6.10) couple the $m00$ and $0n0$ levels. Thus, the whole $mn0$ manifold of states is rapidly equilibrated and is then depleted by the slow V–T transfer out of the lowest possible vibrationally excited state

$$CO_2(010) + M \to CO_2(000) + M, \qquad Q = 667 \text{ cm}^{-1} \qquad (6.12)$$

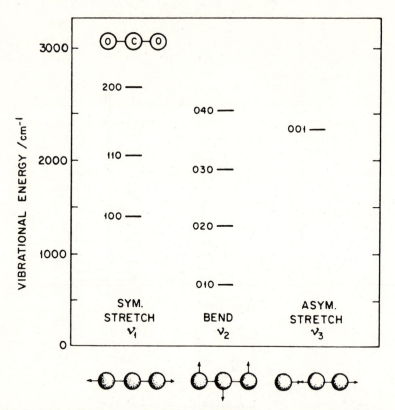

Fig. 6.3. Low-lying vibrational levels of CO_2 (simplified). Their spectroscopic designations are also shown. [Adapted from C. B. Moore, *Acc. Chem. Res.* **2**, 103 (1969).]

In pure CO_2 this process is rather inefficient: $Z_{V-T} \sim 5 \times 10^4$. This, then, is the rate-determining step in the bulk relaxation of CO_2.

On the other hand, the $00p$ manifold of states reaches equilibrium within the manifold by intermolecular sharing, but is not efficiently coupled to the $mn0$ manifold, as there are no states near enough in energy to the 001 state (Fig. 6.3). Thus, $V-V'$ transfer processes out of the 001 state are comparatively inefficient, with $Z_{V-V'} \sim 2 \times 10^4$. We thus see a somewhat uncommon case of a slow $V-V'$ step that is nearly rate determining. In fact, $V-V'$ transfer out of the 001 state does become the rate-determining step when the collision partner M in the deactivation step (6.12) is not CO_2 but, say, H_2 (for which $Z_v \sim 150$). Hence, in the presence of H_2 buffer, vibrationally excited CO_2 can be viewed as a mixture of two species, each species being in thermal equilibrium! One species is CO_2 molecules in vibrational states of the $00p$ manifold and the other is CO_2 molecules in $mn0$ manifold. Collisional transfer between the two manifolds is inefficient.

Another well-characterized example of intermode $V-V'$ processes is that of CH_3F. This five-atom molecule has 9 vibrational modes, of which three are degenerate. The CF stretch vibration (designated v_3, frequency 1050 cm^{-1}) is collisionally strongly coupled to the CH$_3$ rocking mode (v_6, 1190 cm^{-1}). The CH stretches (v_1 and v_4, ~2980 cm^{-1}) are collisionally coupled to bending modes (v_2, v_5, ~1470 cm^{-1}) by a 1:2 quantum exchange. The weakest $V-V'$ coupling is between v_6 and v_2 and v_5. Here, too, the conclusion is that on short time scales (i.e., less than the $V-T$ relaxation time, which is about 1 ms for CH$_3$F at 1 Torr), energy can be localized in a subset of vibrational modes of a polyatomic molecule. Figure 6.4 contrasts the energy content in the different vibrational modes of CH$_3$F when only $V-V'$ (but not $V-T$) equilibrium is maintained with the distribution at complete thermal equilibrium.

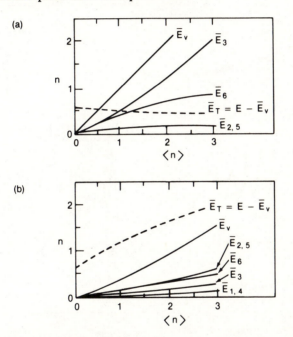

Fig. 6.4. The energy per mode of CH$_3$F versus the mean total energy per molecule. Energy is plotted as n, i.e., energy in units of CO$_2$ laser photons (0.13 eV), so that the abscissa is the mean number, $\langle n \rangle$, of CO$_2$ photons absorbed per CH$_3$F molecule. (a) Laser pumping of CH$_3$F at the v_3 frequency with fast $V-V'$ equilibration only. (b) At complete thermal equilibrium (which is equivalent to slow "Bunsen burner" heating). [Adapted from I. Shamah and G. W. Flynn, *J. Chem. Phys.*, *69*, 2474 (1978). See also E. Weitz and G. W. Flynn, *Adv. Chem. Phys.*, *47*, 185 (1979).]

In larger polyatomic molecules, where the density of vibrational states exceeds a critical value (see Fig. 4.39), the intramolecular vibrational energy redistribution need no longer be collision induced.† The (optically or

---

† In other words, the high density of states implies that there are several (or many) optically inactive levels that are quasi resonant in energy with the optically prepared initial level.

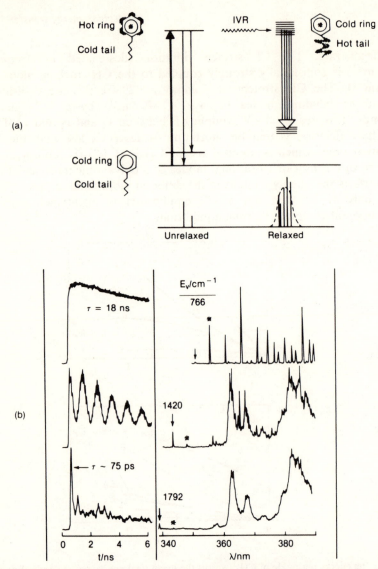

Fig. 6.5. (a) Schematic principles of a time-resolved fluorescence experiment for large polyatomic molecules (e.g., an alkyl benzene). An optically active mode (usually, in an electronically excited state) is pumped from the (jet-cooled) ground state. It can either resonance fluoresce down or undergo intramolecular vibrational-energy redistribution. After this relaxation the fluorescence will be to the red of the original excitation since the down transitions must lead to vibrationally excited levels of the ground electronic state. (Shown pictorially are the relevant modes for an alkyl benzene; see Sec. 4.4.11.) (b) Experimental (picosecond molecular beam) results on IVR of anthracene. Left, time-resolved data for different excess vibrational energies. Right, corresponding fluorescence spectra. [From P. M. Felker and A. H. Zewail, *J. Chem. Phys.* *82*, 2975 (1985).] Similar real-time results on alkyl anilines have been reported [J. S. Baskin, M. Dantus, and A. H. Zewail, *Chem. Phys. Lett.* *130*, 473 (1986).] For example, in *p*-propyl aniline, although the dispersed fluorescence is highly congested [see, e.g., S. M. Beck, J. B. Hopkins, D. E. Powers, and R. E. Smalley, *J. Chem. Phys.*, *74*, 43 (1981)], coherent vibrational motion (quantum beats) were observed, implying restricted, non-dissipative IVR.

otherwise) excited state can couple to other states [either due to anharmonicity or via rotation-mediated (Coriolis) coupling], leading to a unimolecular process that is detectable via time resolved fluorescence (Fig. 6.5). Such intramolecular energy-randomization processes destroy the selectivity of the initial excitation.

6.1.8 The CO_2 laser

The CO_2 laser is frequently used in the infrared. It can be line-tuned in a region (920–1080 cm^{-1}) where many molecules absorb and can provide high pulse energies or power densities (peak powers in a pulsed mode of 10^3–10^4 MW cm^{-2}). This device is possible because of the selective intermode V–V coupling discussed in the previous section.

Nitrogen molecules, vibrationally excited by electron impact

$$N_2(v = 0) + e^- \rightarrow N_2(v = 1) + e^-$$

undergo an efficient V–V' transfer with CO_2,

$$N_2(v = 1) + CO_2(000) \rightleftarrows CO_2(001) + N_2(v = 0), \qquad Q = -18 \text{ cm}^{-1} \quad (6.13)$$

and promote a significant fraction of CO_2 molecules into the 001 level. As a basis for lasing, induced emission from the 001 level (Fig. 6.6)

$$CO_2(001) \rightarrow CO_2(100) + hv$$

or

$$CO_2(001) \rightarrow CO_2(020) + hv$$

could be used, provided that (1) collisional loss of $CO_2(001)$ is kept to a minimum and that (2) collisional loss of $CO_2(100)$ and $CO_2(020)$ can be enhanced so that the upper level population exceeds that of the lower levels.

The first condition is obtained as a result of the inefficient V–V' transfer out of the 001 state. The second condition is fulfilled by depletion of $CO_2(020)$ and $CO_2(100)$ by the efficient processes (6.10) and (6.11).

The CO_2 can also be pumped by chemical means. As an example, vibrationally excited DF molecules, produced in the fast exoergic reaction

$$F + D_2 \rightarrow D + DF(v')$$

can undergo efficient V–V' transfer with CO_2:

$$DF(v') + CO_2(000) \rightarrow CO_2(001) + DF(v' - 1)$$

From here on, the fate of the CO_2 is as before. The necessary F atoms can be produced by the precursor reaction

$$NO + F_2 \rightarrow NOF + F$$

In this way, by mixing tank gases (NO, F_2, D_2, CO_2) it is possible to effect a

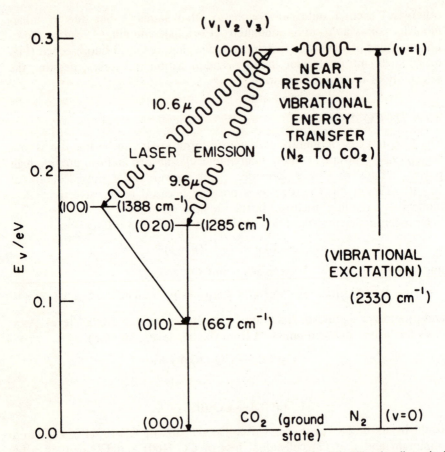

Fig. 6.6. Level diagram appropriate for the N_2–CO_2 laser (*cf.* Fig. 6.3). Vibrationally excited N_2 molecules are the source of the excitation of the 001 vibration of the CO_2 molecules. The final states $CO_2(100)$ and $CO_2(020)$ are depleted by V–V' transfer.

direct conversion of chemical energy to coherent laser radiation energy in the infrared.

At the center of our discussion in this section has been the concept of a state-to-state rate constant for energy transfer, and the wide variations possible in such rates. In the following section we explore the basic dynamical factors that determine these rates.

6.2 Simple models of energy transfer

6.2.1 *Two extremes of vibrational energy transfer*

The simplest inelastic event, conceptually, is the collinear collision between an atom A and a "harmonic oscillator molecule" BC. In such a collision all three atoms are confined to a line, and atom A hits the "near atom" of the

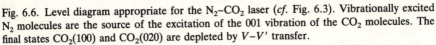

oscillator, say atom B. Two limiting situations can be considered. If the oscillator *spring* is extremely *stiff*, BC is practically a rigid body. When A collides with the oscillator, any kinetic energy lost by A is transferred to the "structureless" molecule BC as kinetic energy of its center of mass. In other words, after A hits B, atoms B and C recoil synchronously without any change in their relative velocity. In the limit of a stiff bond, the collision is purely elastic, as the total kinetic energy has not changed.

The opposite extreme is that of a very *weak* oscillator spring. Here, atom C is so very *weakly coupled* to B that it hardly responds to a change in the velocity of atom B. In the limit of a vanishingly small oscillator force constant, atom C is a *spectator* to the A–B collision. Any change in the velocity of B is thus a change in the relative velocity of B and C and hence a change in the *vibrational energy* of BC.

We consider the collinear A + BC collision, where BC is initially at rest. In the spectator limit, the velocity v'_B acquired by B after it is hit by A is the final relative velocity of B and C. The vibrational energy acquired by BC is then (Sec. 2.2)

$$\Delta E_{vib} = \tfrac{1}{2}\mu_{BC}(v'_B)^2 \qquad (6.14)$$

where $\mu_{BC} = m_B m_C/(m_B + m_C)$ is the reduced mass of BC. Figure 6.7 shows the momentum balance when C is a spectator. The initial momentum of A,

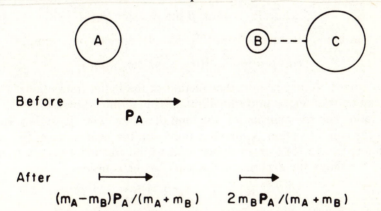

Fig. 6.7. Linear momentum conservation for the spectator model. Here, $\mathbf{P}_A$ is the initial momentum of A (while $\mathbf{P}_B = 0$). After the collision $\mathbf{P}'_A = (m_B - m_A)P_A/(m_A + m_B)$ and $\mathbf{P}'_B = 2m_B P_A/(m_A + m_B)$, summing to $\mathbf{P}_A$. [Adapted from B. H. Mahan, *J. Chem. Phys.*, *52*, 5221 (1970).]

$\mathbf{P}_A$, is partitioned after the elastic A + B collision, between A and B so as to conserve energy and momentum.† Thus, we can solve for the final velocity of B: $v'_B = 2P_A/(m_A + m_B)$. The inital relative translational energy, E_T, is the kinetic energy of the relative motion of A and BC,

$$E_T = \tfrac{1}{2}\mu v_A^2 = \tfrac{1}{2}(\mu/m_A^2)P_A^2 \qquad (6.15)$$

† $\mathbf{P}_A = m_B v'_B + m_A v'_A$, $P_A^2/2m_A = \tfrac{1}{2}(m_B v'^2_B + m_A v'^2_A)$.

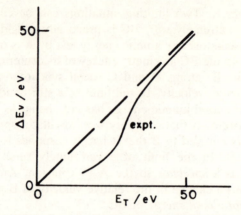

Fig. 6.8. Experimental results for the most probable value of the vibrational energy transfer as a function of the collision energy, E_T, for the K^+—H_2 system. The dashed straight line is the prediction of the spectator model. [Adapted from H. van Dop, A. J. H. Boerboom, and J. Los, *Physica*, *54*, 223 (1971).]

where μ is the reduced mass, $\mu = m_A m_{BC}/(m_A + m_{BC})$, $m_{BC} = m_B + m_C$. From (6.15) and (6.14) we thus obtain, in a compact form,

$$\Delta E_{vib}/E_T = 4\cos^2\beta \sin^2\beta = \sin^2 2\beta \leq 1 \qquad (6.16)$$

where our ubiquitous mass factor [Eq. (4A.4)] appears once more,

$$\cos^2\beta = m_A m_C/[(m_A + m_B)(m_B + m_C)] \qquad (6.17)$$

When atom C is a spectator, the fraction of the initial translational energy that can be transferred into the vibration is determined by a dimensionless mass ratio. The *spectator limit* is the limit of a *very loose BC spring*. Hence, when the colliding atom A interacts only with the nearest atom (B) of the oscillator, Eq. (6.16) sets an upper limit to the fractional energy transfer. Figure 6.8 shows the experimental results for the collision

$$K^+ + H_2 \rightarrow K^+ + H_2(v)$$

At higher collision energies, the most probable fractional energy transfer observed is in good agreement with the upper bound (6.16). But why is it only at fairly high energies that the spectator limit is reached? Is this limit not the very one termed "a sudden collision" when discussing rotational excitation? Did we not apply it to R–T transfer already at thermal energies (Sec. 5.6.4)? We clearly need to know more than just E_T to estimate the inelastic energy transfer.

6.2.2 The adiabaticity parameter

In physical terms, the limit of a weak oscillator spring is the limit where the duration of the collision, t_c, is short compared to the period t_v of the oscillator†

$$t_c < t_v, \quad \text{or} \quad vt_c < 1 \tag{6.18}$$

where v is the oscillator frequency. Since the duration of the collision decreases with increasing relative velocity, (6.18) is a high-velocity, or a sudden, limit. The opposite extreme,

$$t_c > t_v \quad \text{or} \quad vt_c > 1 \tag{6.19}$$

is a low-velocity, or *adiabatic,* limit. Energy transfer is very inefficient in the adiabatic limit. In this limit the duration of the collision is long compared to the period of oscillation, so that the perturbation by the collision is "slow." The oscillator can accommodate itself to the perturbation, or in our earlier language, it presents a rigid, unyielding front.‡

To determine the range of velocities that correspond to the adiabatic limit, we use as an estimate for the duration of the collision

$$t_c = a/v \tag{6.20}$$

Here, a is the "range" of the intermolecular force and v is the relative velocity during the collision. In the adiabatic range, $v \lesssim a/t_v$, or, in terms of the vibrational energy spacing ΔE, $\Delta E = hv = h/t_v$,

$$v \lesssim a \, \Delta E/h = av \tag{6.21}$$

For many diatomic molecules, $v \approx 10^{13} \, \text{s}^{-1}$, and taking as an estimate $a = 2 \, \text{Å}$, we obtain $v < 2 \, \text{km s}^{-1}$. At ordinary temperatures, where velocities in the gas phase are smaller by an order of magnitude, one is in the adiabatic range for vibrational energy transfer.

We can summarize our qualitative considerations by the introduction of an *adiabaticity parameter*, ξ,

$$\xi = t_c/t_v = a \, |\Delta E|/hv \tag{6.22}$$

where $|\Delta E|$ is a measure of a typical amount of energy transferred into (or out of) the translation, so that ΔE is roughly the translational exoergicity. *Large values of ξ correspond to adiabatic collisions* when energy transfer is *inefficient.* Below $\xi \approx 1$, one enters the region of efficient transfer, and for $\xi \ll 1$, we are in the *sudden or spectator regime*.

Similar considerations apply to *rotational energy transfer* where the adiabaticity parameter is defined by

$$\xi = t_c/t_r \tag{6.23}$$

Here, t_r is the rotational period.

---

† The period, or inverse frequency, is an inverse measure of the oscillator force constant, $t_v = v^{-1} = 2\pi(\mu_{BC}/k)^{1/2}$.

‡ Since $t_c > t_v$, the actual oscillations are very rapid, and the colliding atom A, which measures time on a t_c scale, sees only the average position of the oscillator and cannot probe the oscillations about the average.

The spacings between low rotational levels are (with the exception of hydrides) two to three orders of magnitude smaller than vibrational energy spacings. Hence, for $T-R$ transfer, the adiabatic velocity range is usually below $100 \, \mathrm{m \, s^{-1}}$. At *ordinary temperatures, $R-T$ transfer is in the *sudden range,* and hence rotational relaxation should occur quite readily (Fig. 6.1).

The estimate of the adiabaticity parameter for rotational energy transfer is based on the considerations of Section 5.6.4, which lead to

$$\xi = |\Delta E|/h\omega \tag{6.24}$$

where ω is the orbital angular velocity, $v = \omega a$. Since for rotational energy transfer $\Delta E = h\omega_R$, where ω_R is the angular velocity of the molecule, $\xi = \omega_R/\omega$ is the ratio of internal to relative angular velocity [see Eq. (6.22)].

6.2.3 The exponential gap

A more quantitative approach is to introduce a resonance function, $R(\xi)$, as measure of the efficiency of energy transfer at a given value of ξ. Thus, if $\langle \Delta E \rangle$ is the average energy transfer in the collision, $R(\xi)$ is defined by

$$\frac{\langle \Delta E \rangle}{E_T} = \frac{\langle \Delta E \rangle_{\xi=0}}{E_T} R(\xi) \tag{6.25}$$

where $\langle \Delta E \rangle_{\xi=0}$ is the average energy transfer in the sudden limit. The computation of the resonance (or "energy mismatch") function requires a solution of the collision dynamics. We shall see shortly that, at least for $\xi > 1$, $R(\xi)$ can be approximated by

$$R(\xi) \approx \exp(-\xi) \tag{6.26}$$

The exponential decline of the efficiency of energy transfer with increasing ξ is an important rule of thumb in the field of energy transfer. Deviations from the rule (e.g., Sec. 6.6) are taken as a diagnostic indication that some special features are present.

Some guidance to the physical significance of the resonance function is provided by our model of a collinear $A + BC$ collision. $R(\xi)$ was very small when it was difficult to change the relative momentum of B with respect to C (a stiff bond), while it was large in the opposite extreme. $R(\xi)$ is a measure of the readiness of the BC bond to change its momentum. In other words, $R(\xi)$ is a quantitative measure of a Franck–Condon-like principle, i.e., the reluctance of the nuclei to undergo significant changes in their momenta. Thus, if p and p' are the initial and final momenta, we can write the adiabaticity parameter as

$$\xi = (p - p')a/h = |\Delta E| \, a/hv \tag{6.27}$$

To show the equivalence to (6.22), we put $p - p' = [p^2 - (p')^2]/(p + p')$ or $p - p' = \Delta E/v$, where $v = (p + p')/2\mu$. The larger the change in the momentum of the nuclei, the larger is ξ.

We can thus summarize the necessary conditions for an efficient energy transfer. At a given energy gap ΔE, $R(\xi)$ *will be exponentially small at low velocities*, $v \ll a\,\Delta E/h$ *and will increase with increasing velocities*. This version is the form most useful for collision theory. An indirect illustration of this "exponential gap" rule is the strong positive temperature dependence of the V–T relaxation rate. For most diatomic molecules the temperature dependence is best fitted by a so-called *Landau-Teller* form (Fig. 6.9):

$$\ln(P\tau)^{-1} = A - BT^{-1/3}, \qquad V\text{–}T \quad \text{transfer} \qquad (6.28)$$

which is quite different from the Arrhenius temperature dependence of bimolecular reaction rate constants. Theory shows that (6.28) is the expected result for an inelastic cross section that increases near-exponentially with increasing translational energy,

$$\sigma_{VT} \propto \exp(-v_0/v) \qquad (6.29)$$

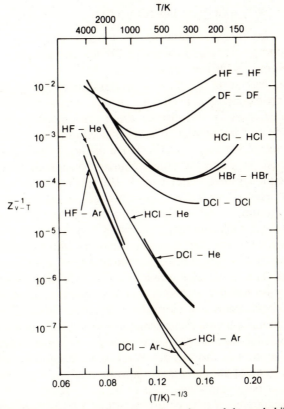

Fig. 6.9. Landau–Teller plot of the temperature dependence of the probability of vibrational relaxation for isotopic hydrogen halides. Linear behavior obtains only at higher temperatures. Shown is Z_{V-T}^{-1} versus $T^{-1/3}$. At low temperature, the long-range attractive dipole–dipole force enhances the relaxation process. [Adapted from Ben-Shaul et al. (1981).]

Indeed, the parameter v_0 is found to be a nearly linear function of the vibrational frequency, as expected from Eq. (6.21). Similarly, the collision number Z_{VT} for $V-T$ transfer in polyatomics is found to be an approximate exponential function of the *lowest*† normal vibrational frequency v_{min},

$$\log Z_{VT} \propto v_{min} \tag{6.30}$$

At low temperatures one observes deviations from the simple predictions based on the direct $V-T$ mechanism. At lower collision velocities the attractive part of the intermolecular force begins to play a more important role. The stronger the attraction, the higher the temperature below which such deviations will occur. Particularly important here are the long-range dipole–dipole forces (Sec. 3.2.4). Even molecules without permanent dipole but that are readily polarizable can undergo efficient $V-T$ transfer via a mechanism involving a dipole-induced dipole forces. An example of the importance of these long-range attractive forces is the efficient $V-V'$ process,

$$CO_2(002) + CO_2(000) \rightarrow 2CO_2(001), \qquad Q = -25 \, cm^{-1}$$

which equilibrates the 00p manifold of states in CO_2 (Sec. 6.1.6) with a $Z_v \approx 2$. The energy transfer process (6.13) for the nonpolar N_2 molecule with $Q = -18 \, cm^{-1}$ is far less efficient.

Another manifestation of these considerations is *intra*molecular $V-T$ transfer. As an example, consider a van der Waals molecule, say $I_2 \cdot He$. Using a tunable laser in the visible, the I_2 chromophore in the van der Waals molecule can be excited to a specific vibrational level of a (bound) electronically excited state of I_2. The $I_2 \cdot He$ attractive well is a very shallow one, hence there is a considerable frequency mismatch between the I—I vibrational frequency in I_2^* and that of He relative to I_2^*. The persistent "collisions" of He against I_2^* do, however, cause the occasional loss of a vibrational quantum of I_2^* which is transferred to the relative I_2^*—He motion. The $I_2^* \cdot He$ molecule then dissociates (since the vibrational quantum of I_2^* far exceeds the van der Waals well depth), and the event is detected by fluorescence of I_2^* from a lower vibrational level than the one to which it was pumped. The (inverse) lifetime of the $I_2^* \cdot He$ can be estimated as the number of "collisions" per second (i.e., the vibrational frequency) times the probability (per collision) of vibrational deactivation of I_2^*. Figure 6.10 shows a logarithmic plot of the calculated lifetimes of such van der Waals molecules versus the adiabaticity parameter $\xi = a |\Delta p|/h$, where $|\Delta p|$ is the change in momentum of the van der Waals bond upon dissociation.

Other fairly general statements can also be made; for example, collisions at higher impact parameters will tend to be more adiabatic (because of the larger range parameter a) than collisions at lower impact parameters. This is exemplified in Fig. 6.11 for the inelastic collision of SF_6 with Ar. The high

---

† The $V-T$ relaxation of the lowest vibrational level is rate determining (Sec. 6.1.5).

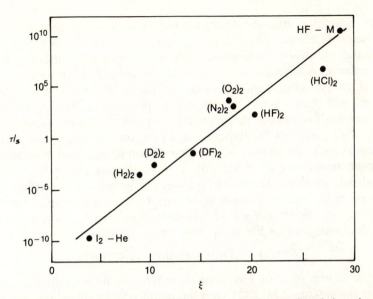

Fig. 6.10. Momentum gap correlation of lifetimes for unimolecular dissociation of van der Waals molecules following vibrational excitation of the chemical (not the van der Waals) bond. [Adapted from G. E. Ewing, *J. Chem. Phys. 71*, 3143 (1979).] Calculated lifetimes [see also J. A. Beswick and J. Jortner, *Adv. Chem. Phys. 47*, 363 (1981)] are plotted as a function of ξ; $\xi = p'/\alpha\hbar$, where $p' = 2\mu'E'$ is the final relative momentum of the fragments and α ($\equiv 1/a$) is the Morse (range) parameter for the van der Waals bond. [Note that the range parameter a is often denoted by ρ in the actual potential function, e.g., Eq. (6.32).]

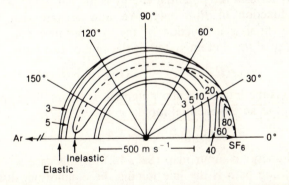

Fig. 6.11. Flux-velocity contour map, in the c.m. system, for the scattering of SF_6 by Ar, showing the increase in inelastic energy transfer (vibrotational excitation) with increasing scattering angle θ. The contours are constructed from observed velocity distributions at different laboratory angles, followed by conversion to the c.m. system. The contours show the range from a maximum intensity of 100 (for forward scattering) down to 3. The outer circle is the locus of velocities for elastic scattering. The dashed curve is the locus of the maxima (i.e., it connects the inelastic peaks). The difference is an indication of the most probable energy transfer, which increases with θ. For this experiment, which used a supersonic SF_6 beam seeded in He, $E_T = 0.73$ eV. [Adapted from J. Eccles, G. Pfeffer, E. Piper, G. Ringer, and J. P. Toennies, *Chem. Phys., 89*, 1 (1984).]

impact parameter, grazing collisions, which lead to nearly forward scattering are essentially elastic. The inelasticity increases at larger scattering angles. In general, "head-on" collisions, leading to backward scattering, are the most inelastic. A corollary is that collisions resulting in a large change in the internal energy (and hence a large change in the vibrational quantum number) will occur mostly at low impact parameters. The corresponding cross sections (and hence the corresponding rate constants) would then be rather small (Sec. 3.3.5). Experimental evidence for large changes in the vibrational quantum number in a single collision comes from molecular- and ion-beam measurements (Chap. 5). A less obvious but valid qualitative generalization is that it is easier to transfer energy into (or out of) an excited oscillator than into a ground state one.† An experimental test is to determine the efficiency of a $V-T$ transfer process as a function of the initial vibrational energy. This has been done for the process

$$CO(v) + CO(0) \rightarrow CO(v-1) + CO(1)$$

A rough linear relation has been found between the efficiency of the transfer (as measured by Z_{VT}^{-1}), and the initial vibrational energy.

6.2.4 Intermolecular potential for vibrational excitation

The simplest illustration is the potential appropriate to describe the collinear collision of an atom A and a harmonic oscillator BC. Since the system is collinear, the potential energy depends upon two coordinates only, for example, the two bond distances R_{BC} and R_{AB}. When the atom is far away, the only term in the potential energy is the harmonic potential of the oscillator, a function of R_{BC} only. We now assume that the A + BC interaction potential is a function of the AB distance only, so that the potential energy of the 3-atom system is the sum of two parts (Fig. 6.12),

$$V_{ABC}(R_{AB}, R_{BC}) = V_{BC}(R_{BC}) + V(R_{AB}) \tag{6.31}$$

Here, $V(R_{AB})$ is the intermolecular potential. It is a realistic approximation to take $V(R_{AB})$ as an exponential repulsion (Sec. 2.1.7),

$$V(R_{AB}) = A \exp(-R_{AB}/\rho) \tag{6.32}$$

A potential energy contour map (Sec. 4.2) for this case is shown in Fig. 6.13. The collision path is the line joining the equilibrium distance of the oscillator for different values of R_{AB}. As the atom A approaches, the oscillator is compressed. If the atom moves very slowly, the compression and subsequent release of the oscillator spring are carried out very gently, with the result that the oscillator is left unchanged by the collision. This is the adiabatic limit. When the approach of the atom is very fast, the rapid compression leaves the oscillator vibrating strongly. This is the sudden limit.

---

† The physical reason is that the amplitude of oscillation of the bond is greater for an increased initial vibrational energy.

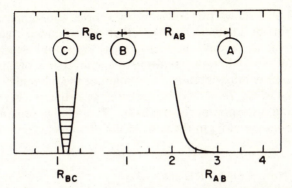

Fig. 6.12. Potentials for a collinear atom (A) harmonic oscillator (BC) collision. At left is the harmonic B–C potential (with vibrational levels shown); at right is the exponential (Born–Mayer) A–B repulsion.

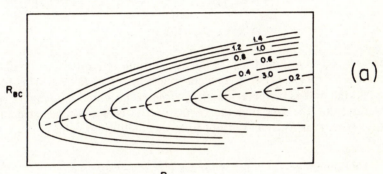

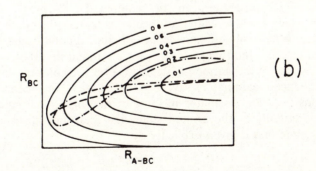

Fig. 6.13. (a) Potential energy contour map corresponding to Fig. 6.12 (energies in arbitrary units). The dashed line is the minimum energy path. Note the "compression" of the oscillator bond during the approach of the atom. (b) Two collision trajectories superimposed on a Landau–Teller potential energy surface (Fig. 6.12). For both trajectories BC has no initial vibrational excitation. The low-energy "adiabatic" trajectory (dashed curve) retraces itself on the way out. The high-energy "sudden" trajectory (dot–dashed curve) shows vibrational excitation (i.e., BC oscillation) induced by the collision. [Adapted from M. Attermeyer and R. A. Marcus, *J. Chem. Phys.*, *52*, 393 (1970).]

Figure 6.13 also shows two collision trajectories superimposed upon the surface, one for an adiabatic and the other for a sudden collision. While both begin in the same way, the high-velocity trajectory clearly acquires a large amplitude for vibration, showing the conversion of initial relative translational energy into vibrational excitation of the diatomic.

The potential of Eq. (6.32) can be rewritten as a sum of a central part and a term involving distortion of the oscillator. The relative separation R, i.e., the distance from the atom to the c.m. of the diatomic, can be written as

$$R = R_{AB} + \gamma R_{BC} \tag{6.33}$$

where $\gamma = m_C/(m_B + m_C) < 1$. Thus,

$$
\begin{aligned}
V(R_{AB}) &= A \exp[-(R - \gamma R_{BC})/\rho] \\
&= A \exp[-(R - \gamma R^e_{BC})/\rho] \exp[+\gamma x/\rho] \\
&= A \exp[\gamma R^e_{BC}/\rho] \exp[-R/\rho] \exp[+\gamma x/\rho] \\
&= A' \exp[-R/\rho]\left\{1 + \frac{\gamma x}{\rho} + \cdots\right\}
\end{aligned} \tag{6.34}
$$

where $x \equiv R_{BC} - R^e_{BC}$ and R^e_{BC} is the equilibrium separation of the diatomic BC. $A' = A \exp(\gamma R^e_{BC}/\rho)$, a constant independent of R or x. This form of the potential makes it clearer that it is a *central* potential *modified* by a term that allows for the fact that the oscillator BC is *not* a rigid, structureless particle (i.e., that R_{BC} can deviate from its equilibrium value).

6.2.5 Vibrational energy transfer for the Landau–Teller model

Both the quantum-mechanical and the classical dynamics for such a "Landau–Teller" potential can be computed. A more approximate procedure, which leads to analytical results, is the classical path approximation, where one assumes knowledge of the relative separation, $R(t)$ throughout the collision (Sec. 5.6.3). It is then found that the distribution of vibrational states of the harmonic oscillator after the collision depends upon only one reduced parameter, say, $\bar{v} = \langle \Delta E_{vib} \rangle / h\nu$, the mean vibrational energy transfer to the oscillator in units of the vibrational spacing. If the oscillator is initially in the ground state, the final-state distribution has a single maximum and is given explicitly by

$$P_v = (\bar{v}^v/v!) \exp(-\bar{v}) \tag{6.35}$$

One readily verifies that

$$\sum_v P_v = 1 \tag{6.36}$$

$$\sum_v E_v P_v = \langle \Delta E_{vib} \rangle$$

Distributions of the type (6.35) have indeed been observed in higher-

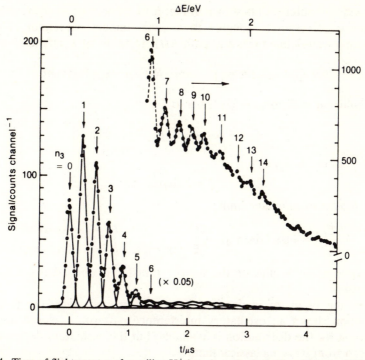

Fig. 6.14. Time-of-flight spectra of recoiling H^+ ions from the inelastic collisions of $H^+ + CF_4$, scattered at $\Theta = 5°$. Here, $E_T = 18.5\,\text{eV}$. Lower abscissa scale, time-of-flight; upper, energy loss, ΔE. Points are experimental, curves calculated assuming the indicated number n_3 of v_3 quanta of excitation of the CF_4, with a relative intensity distribution given by Eq. (6.35). [Adapted from M. Noll and J. P. Toennies, *Chem. Phys. Lett.*, *108*, 297 (1984).]

energy atom–molecule collisions. Figure 6.14 shows a fit based on (6.35) to the observed time-of-flight spectra in the

$$H^+ + CF_4 \rightarrow H^+ + CF_4(nv_3)$$

inelastic collision at 18.3 eV collision energy. The excited mode is v_3, which is the most active mode in the infrared spectrum of CF_4. Excitation of the v_4 mode, which is also infrared active, has also been observed.

A prerequisite for the validity of (6.35) is the classical path approximation, which implies that the fractional loss of kinetic energy $\Delta E/E_T$ should be small. This can be seen to be the case for the example of Fig. 6.14. Other situations in which such an assumption would be warranted include photodissociation, where a triatomic molecule is promoted to a repulsive state. During the "half-collision" where the atom recedes from the diatom, translation–vibration coupling is operative. Another example, discussed in Section 6.6.2, is the quenching of electronically excited atoms by diatomic molecules.

The experimental results shown in Fig. 6.14 are at a given scattering angle and hence correspond† roughly to a given impact parameter. There is therefore a well-defined classical path $R(t)$. In terms of $R(t)$ and (6.34)

$$\bar{v} = \langle \Delta E_{vib} \rangle / h\nu = \left| \int_{-\infty}^{\infty} \exp[-R(t)/\rho] \exp(2\pi i \nu t) \, dt \right|^2 \qquad (6.37)$$

For a collinear collision, (6.37) leads to

$$\frac{\langle \Delta E_{vib} \rangle}{E_T} = 4 \cos^2 \beta \sin^2 \beta \; \left(\frac{\xi}{2} \operatorname{cosech} \frac{\xi}{2} \right)^2 \qquad (6.38)$$

where ξ is here given by‡

$$\xi = 2\pi \omega \rho / v \qquad (6.39)$$

In the *sudden* or *adiabatic* limit.

$$\frac{\langle \Delta E_v \rangle}{E_T} = 4 \cos^2 \beta \sin^2 \beta \; \begin{cases} 1, & \xi \to 0 \quad \text{(sudden)} \\ \xi^2 \exp(-\xi), & \xi \to \infty \quad \text{(adiabatic)} \end{cases} \qquad (6.40)$$

The exponential decline of the energy transfer in the adiabatic range is evident.§

When the oscillator is not initially in the ground state, the final state distribution shows an oscillatory interference pattern. A closed analytical expression for the distribution is available, but the oscillatory dependence is best seen from its semiclassical limit

$$P(v \to v') = |J_{|v-v'|}(\bar{v})|^2 \qquad (6.41)$$

Here, $J_r(\bar{v})$ is the Bessel function of order r and argument $\bar{v}$ (Fig. 6.15). Oscillations occur for $\bar{v} > r$ (i.e., for transitions where the final quantum number is within $\bar{v}$ of the initial vibrational state of the oscillator). The most probable transition (first and largest maximum) is at $|v - v'| = (2/\pi)\bar{v} - \frac{1}{2}$. For larger values of Δv, the decline is exponential, as can also be seen from Eq. (6.35).

The origin of the (interference) oscillations can best be seen from considerations of classical trajectories, as discussed in the following subsection.

6.2.6 Classical trajectories for the Landau–Teller model

For the collinear atom–oscillator collision we have two degrees of freedom. To compute classical trajectories (Sec. 4.3.2) we need to assign four initial

---

† Roughly, rather than exactly, since the intermolecular potential is not quite central [see Eq. (6.34)].

‡ Note that (6.39) differs from (6.22) by a factor of 2π. The origin of the factor is as follows. For harmonic motion, $R_{BC} = A \sin(2\pi t/t_v)$, where A is the amplitude. If the oscillator is hardly to move over the time interval t_c, $2\pi t_c/t_v \ll 1$. Hence, ξ is really $2\pi t_c/t_v$ and not simply t_c/t_v.

§ Note that $\xi \operatorname{cosech} \xi = 2\xi/[\exp(\xi) - \exp(-\xi)]$. For small ξ, we expand $\exp(\xi) \to 1 + \xi$. For large ξ, we neglect $\exp(-\xi)$.

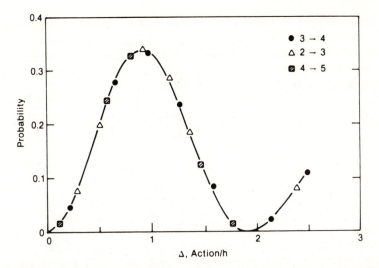

Fig. 6.15. Transition probability for a $\Delta v = 1$ vibrational excitation of a harmonic oscillator via a collinear collision with an atom, plotted as a function of the action transferred, Δ. Note that $\bar{v}$ [Eq. (6.41)] is $2\Delta_{v,v+1}$. The points are from exact quantal calculations; the curve is the semiclassical Eq. (6.41). [Adapted from R. D. Levine and B. R. Johnson, *Chem. Phys. Lett.*, **8**, 501 (1971).]

values (coordinates and momentum for each degree of freedom). Taking the relative separation, we can choose as initial values the initial magnitude of the relative separation R and the initial value of the relative velocity v. It is clear on physical grounds that if the initial value of R is large enough, the collision dynamics will not depend upon the precise value chosen. On the other hand, the magnitude of v is of considerable influence!

For the oscillator, one initial value is chosen to be the initial vibrational energy of the oscillator, corresponding to the experimental practice of specifying the initial vibrational state. What do we choose as the second initial value? We need a quantity that together with the vibrational energy completely specifies the state of the oscillator in classical mechanics. This is the phase angle of the oscillator. In physical terms, the energy determines the amplitude of the oscillation, while the phase determines the partitioning of the energy between kinetic and potential energies. A phase angle $\phi = 0$ corresponds to the oscillator at the equilibrium position where all the energy is kinetic, while $\phi = \pm\pi/2$ corresponds to the turning points where all the energy is potential.

A given trio of initial values v, E_v, ϕ specifies completely a unique collision trajectory and hence determines uniquely a definite amount of energy, ΔE_{vib}, gained (or lost) by the oscillator upon collision. On the other hand, an experimental specification of the initial state in a collinear collision would require only the specification of v and E_v. This is due to the fact that the quantum mechanical state of the oscillator is uniquely specified by the vibrational energy. At a given vibrational state the phase of the oscillator is completely random. To use the classical trajectory method we simulate this state of affairs by calculating many trajectories, at given v and E_v but with different initial phases ϕ, and then performing an average of the computed energy transfer over all values of the initial phase

$$\langle \Delta E_{vib} \rangle = (2\pi)^{-1} \int_0^{2\pi} \Delta E_{vib}(\phi)\, d\phi \tag{6.42}$$

Here, $\Delta E_{vib}(\phi)$ is the energy transfer for a particular classical trajectory, with a given value of ϕ (and v and E_v). As already mentioned, the sampling of initial phases is often done by a Monte-Carlo method.

The final state of the oscillator is, of course, not quantized in a classical trajectory calculation. (Nor was it in the initial state.) We can, however, establish a rough correspondence between the vibrational energy and the quantum state, by, say, regarding all trajectories with a final vibrational energy in the range $v'hv$ to $(v'+1)hv$ as leading to the final vibrational state v' $[E_{v'} = (v' + \tfrac{1}{2})hv]$. As discussed in the previous paragraph, at a given initial relative energy and vibrational energy, the energy transfer is a function of the initial phase only. Hence the usual method of classical trajectories computes the probability of a given final vibrational state as the fractional range of possible initial phases that lead to a final vibrational energy in the desired range (Fig. 6.16),

$$P(v') = (\Delta\phi)_{v'}/2\pi \tag{6.43}$$

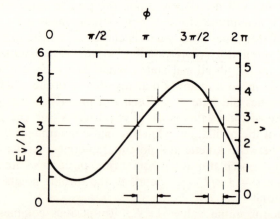

Fig. 6.16. Relation between the vibrational energy of BC acquired by collision with A (at a given collision energy and vibrational energy of BC) and the initial phase of vibration of the oscillator. Note that two ranges of phase angles ϕ (dashed) correspond to excitation to the same energy range, E'_v between $3hv$ and $4hv$. [For a discussion of such plots as means of constructing the proper classical and semiclassical limits, see W. H. Miller, *Acc. Chem. Res.*, **4**, 161 (1971).]

Let us, however, examine Fig. 6.16 in more detail. For the present simple case we can do better than (6.43). Since there is but one initial condition to be varied (i.e., the phase), we can determine the precise value(s) of the phase that lead to any final vibrational quantum number.† The periodicity of the plot versus ϕ means that there will generally be two (or more) distinct initial phases, which will correspond to the same final vibrational quantum number. In quantum mechanics these distinct classical trajectories leading to the same final result can interfere. The origin of the oscillatory interference pattern in the quantal transition probabilities (Fig. 6.15) is thus the same as in the differential cross section versus angle (Sec. 3.1.9).

6.2.7 Rotational energy transfer in the sudden limit

We consider the simple (and not always the most realistic) case where the anisotropic part of the atom–rotor potential, given in general by the Legendre expansion Eq. (3.43), is dominated by a single term, say, $n = 1$ [atom–heteronuclear diatomic case, see Eq. (3.46)] or $n = 2$ (atom–homonuclear diatomic). In the classical path approximation and the sudden limit it can be shown that the opacity function for a $j \to j'$ transition is given by

$$P(j \to j') = (2j' + 1)\, |J_{|j-j'|/n}(q_n)|^2 \qquad (6.44)$$

where q_n is the so-called action integral,

$$q_n = (a_n/\hbar) \int_{-\infty}^{\infty} V_n[R(t)]\, dt \qquad (6.45)$$

with $a_n = 1$ and $3/4$ for $n = 1$ and 2, respectively. The results (6.44) and (6.45) can be compared with expression (6.41) for vibrational excitation. In both cases the argument of the Bessel function is an integral of the intermolecular force along the collision trajectory. For vibrational excitation, the integrand is modulated by the $\exp(i\omega t)$ factor because of the natural oscillation of the vibrating molecule. Here, however, we are in the sudden limit. On the time scale over which $V_n(R)$ varies significantly, the molecule has had hardly any time to rotate. Hence this modulation factor is essentially unity.

The (action) integral q_n can be explicitly evaluated when the R dependence of the intermolecular potential is specified. The trajectory will depend on the initial impact parameter and hence so will q_n. As b increases, the trajectory samples less and less of the intermolecular potential, and hence q_n must decline as $b \to \infty$. As an explicit example consider the long-range intermolecular potential (Sec. 3.2.4). Then, $V_n(R) = -C_s R^{-s}$.

---

† To compute the transition probability, we put $P(v')\, dv' = P(\phi)\, d\phi$, where $P(\phi)$ is the probability density of the initial phase, $P(\phi) = 1/2\pi$. Thus, $P(v') = 1/2\pi\, |dv'/d\phi|$. Note that for high quantum numbers we can replace $dv'/d\phi$ by $\Delta v'/\Delta \phi$, where $\Delta v' = 1$, thereby recovering (6.43).

Since we are dealing with the long-range potential, the trajectory can be approximated by a straight line (unperturbed motion), where $R^2 = b^2 + v^2t^2$, with v the initial relative velocity. Then, changing variables from t to R, (6.45) gives

$$q_n = 2(a_n/\hbar) \int_b^\infty V(R) \frac{dt}{dR} dR$$

$$= (2C_s a_n/\hbar v) \int_b^\infty R^{-s}[R/(R^2 - b^2)^{1/2}] dR$$

$$= 2a_n(C_s/\hbar v)b^{-s+1} \int_1^\infty y^{-s+1}(y^2 - 1)^{-1/2} dy \qquad (6.46)$$

where in the last line we have introduced the reduced variable $y = R/b$. The last integral is a pure number; the entire dependence of q_n on all physical variables is now explicit and is expressed in terms of dimensionless variables. In particular, the rapid decline with b is self-evident.

There are three aspects of the opacity function (6.44) that call for a special comment. The first is the angular distribution. In the sudden limit,

$$\frac{d^2\sigma(j \to j'; \theta)}{d^2\omega} = \left(\frac{d^2\sigma^0(\theta)}{d^2\omega}\right) P(j \to j' \mid b) \qquad (6.47)$$

Here, $d^2\sigma^0(\theta)/d^2\omega$ is the differential cross section for the elastic scattering as determined by the classical path $R(t)$. $P(j \to j' \mid b)$ is the opacity function (6.44) evaluated for that particular b value that corresponds to the scattering angle θ. The first and dominant maximum of the Bessel function $J_r(x)$ versus x means that either in a plot of the differential cross section for a fixed j' versus θ or in a plot of the differential cross section for a fixed θ versus j' there will be a dominant maximum. Both plots reflect the maximum given by the condition $|j - j'|/n = (2/\pi)q_n - \frac{1}{2}$, where q_n is a function of b and hence of θ. Such maxima will be present even if the elastic scattering is due to a purely repulsive potential, so that $(d^2\sigma^0(\theta)/d^2\omega)$ is a monotonically decreasing function of θ.

Another evident feature is the so-called dominant coupling regime. As long as $x > 2r$, the *average* value of $J_r^2(x)$ is independent of r. (Recall the related result in elastic scattering, i.e., the random-phase approximation, Sec. 3.1.8.) Hence the opacity function (6.44) for a $j \to j'$ transition is, on the average, simply proportional to $2j' + 1$. Since there are $2j' + 1$ final quantum states for a given j', in the dominant coupling regime (i.e., large q_n), all final quantum states are equally probable.

Finally, we come to the b dependence of the opacity function. Since q_n decreases as b increases, such a plot, as in Fig. 6.17, reflects the oscillation of the Bessel function $J_r(x)$. The first maximum mentioned earlier is quite evident for large b's. The larger $|\Delta j|$, the smaller the b value at which the

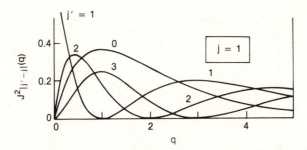

Fig. 6.17. Plot of the quantity $J^2_{|j'-j|}(q)$ versus q for $j = 1$ and $j' = 0, 1, 2, 3$. Note that the ordinate is identified with $P_{j \to j'}/(2j' + 1)$ and the abscissa with an inverse function of b. The behavior characteristic of the dominant coupling regime (obtaining for large q, i.e., small impact parameter collisions) is evident, as the individual P's oscillate rapidly. [Adapted from Y. Alhassid and R. D. Levine, *Phys. Rev.*, *A18*, 89 (1978); for early, sudden approximation numerical calculations, see R. B. Bernstein and K. H. Kramer, *J. Chem. Phys.*, 44, 4473 (1966)].

maximum occurs. The transitions that occur at the largest b values are $j \to j \pm 1$ for $n = 1$ and $j \to j \pm 2$ for $n = 2$.

To compute the state-to-state cross section it is necessary to integrate the opacity function for the specific transition over b. Here we consider the special case of the total inelastic cross section from the initial state j,

$$\sigma(j) = \sum_{j'}{}' \sigma(j \to j') \qquad (6.48)$$

The prime on the sum denotes summation over all j' except $j' = j$. Using a standard identity for Bessel functions, the opacity function for the total inelastic cross section is

$$P(b) = 1 - J_0^2(q) \qquad (6.49)$$

Unless $q \to 0$, $J_0^2(q)$ is very much below unity. Hence, $P(b) \approx 1$ up to such high b's ($\equiv$ low q's) that $J_0(q)$ begins its increase towards unity [*cf.* the $j' = 1$ ($= j$) curve of Fig. 6.17], which happens roughly at $q = \pi/2$. Hence, we can approximate $P(b)$ as a unit step function, with a drop to zero at the critical b value, say, $b = b_c$, which is the solution of $q(b) = \pi/2$. Then,

$$\sigma(j) \approx \pi b_c^2 \qquad (6.50)$$

It is generally the case that for $b > b_c$ only those Δj transitions such that $|\Delta j| = n$ contribute, Fig. 6.18; i.e., when $J_0(x) \to 1$, at most only $J_1(x)$ has a significant value. As b decreases (q_n increases), more and more final j' states can be populated. However, since there are many such states, any individual state-to-state cross section will be small compared to $\sigma(j)$. The only individual cross sections that will be comparable to $\sigma(j)$ are those occurring at the highest b values, i.e., those for which $|\Delta j| = n$. Such transitions are those allowed by first-order perturbation theory.

In Section 7.6 we shall probe in even more detail, considering the role of

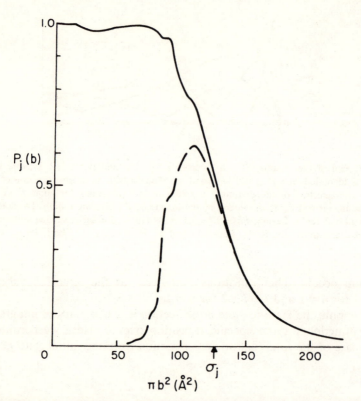

Fig. 6.18. Relative contribution of different transitions to the cross section for rotational excitation, calculated by the sudden approximation. Shown is the opacity function $P_j(b)$ versus πb^2, so that the area under the curve is directly the cross section. The arrow indicates the (computed) total inelastic cross section. The solid curve is the probability for *all* inelastic transitions out of the initial state j. The dashed curve is the contribution of all the "long-range allowed" transitions $j \rightarrow j \pm \Delta j$ ($\Delta j = 1, 2, 3$ in this case), which require only "small" torques and hence are due to the long-range part of the potential. [Adapted from R. W. Fenstermaker and R. B. Bernstein, *J. Chem. Phys.*, *47*, 4417 (1967).]

the orientation of $\mathbf{j}$ (as characterized by the "magnetic" quantum number m_j) or the final orientation of $\mathbf{j}'$. Here, we point out only that in the sudden limit, and, in particular, when the potential leads to fairly impulsive collisions (Fig. 3.11a), there is a simple limiting behavior. The conservation of total angular momentum $\mathbf{j} + \mathbf{l} = \mathbf{j}' + \mathbf{l}'$ implies that $\Delta \mathbf{j} = -\Delta \mathbf{l}$ and hence that [see Eq. (2.27)]

$$\Delta \mathbf{j} = -\Delta(\mathbf{R} \times \mu \dot{\mathbf{R}}) \tag{6.51}$$

For a sudden, impulsive collision, the relative separation $\mathbf{R}$ does not change appreciably during the angular momentum transfer. Hence (see Fig. 3.11a),

$$\Delta \mathbf{j} = -\mathbf{R} \times (\mathbf{p}' - \mathbf{p}) \tag{6.52}$$

i.e., $\Delta \mathbf{j}$ is perpendicular to the momentum transfer, and hence, if the

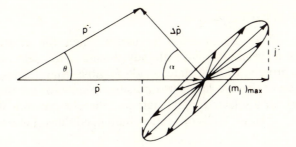

Fig. 6.19. Classical orientation of **j**' in an impulsive collision (*cf.* Fig. 3.11). Taking **j** = 0, we have from Eq. (6.52) that **j**' is orthogonal to the momentum transfer. Hence the vectors **j**' are constrained to be confined to the plane orthogonal to **Δp**, and, for a given *j*', they span the circle as shown. The maximal possible projection of **j**' on, say, the **p** axis is *j*' sin *α*, where angle *α* depends on the scattering angle *θ*. In the sudden ($p = p'$) limit, $\alpha = (\pi - \theta)/2$. [Adapted from R. Schinke and H. J. Korsch, *Chem. Phys. Lett.*, **74**, 449 (1980).]

orientation of **j** is measured with respect to the direction of the momentum transfer, it will be conserved. This has both experimental and computational implications. For example, even if m_j is not selected, then for excitation ($j' > j$) collisions, the conservation of m_j implies that the range of final ($m_{j'}$) orientations will be restricted and cannot reach its ±*j*' limit (Fig. 6.19). Computationally, such conservation can be invoked to make a drastic reduction in the number of coupled channels (Sec. 5.6.2) that need be included.

Such approximate considerations can always be complemented by more exact computations (Sec. 5.6). We have enough theory to interpret *and* to make predictions of state-to-state inelastic cross sections. Now what are the experimental results?

6.3 State-to-state inelastic collisions

State-to-state experiments are the analogs in collision dynamics of the spectroscopic measurements in structural chemistry. As in spectroscopy, such experiments are intended to characterize the potential. However, they do provide much more, for unlike traditional spectroscopy,† the inter-molecular potential during a collision is typically much stronger than the external electrical field used in spectroscopy. Indeed, note the multiquantum transitions revealed in a collision experiment (Fig. 6.14). Moreover, we can vary the nature of the collision partners, the initial internal and translational energy, etc. This section is a survey of some of the results. Inelastic collisions with surfaces will be discussed in Sec. 6.4.

---

† A high-power laser provides an electromagnetic field that can no longer be regarded as a weak perturbation. In Sections 6.6 and again in 7.3 we shall examine some of the implications.

6.3.1 State-to-state rate constants

The concept of a rate constant is useful when the relative velocity distribution is thermal. We have already discussed the selection of initial state and the probing of final states made feasible using lasers. By suitable synchronization of the two lasers it is possible to monitor the distribution of the final states for a selected initial state and thereby measure state-to-state rate constants for single collisions under bulk conditions. Figure 6.20 shows

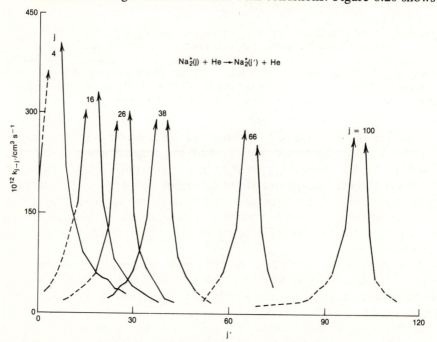

Fig. 6.20. Rate coefficients $k_{j \to j'}$ for rotational excitation and deexcitation of $Na_2^*(j)$ by collision with He. Solid lines connect experimental points. Dashed extrapolations are via exponential gap method. [From data of T. A. Brunner, N. Smith, A. W. Karp, and D. E. Pritchard, *J. Chem. Phys.*, *74*, 3324 (1981).]

the specific rates for $j \to j'$ transitions in the

$$He + Na_2^*(v, j) \to He + Na_2^*(v, j')$$

collision for several initial j values. Note the rapid decline of the rate as $|j - j'|$ increases, a qualitative manifestation of the exponential gap behavior (Sec. 6.2.3). The quantitative dependence of the inelastic rate constant on the energy gap is shown in Fig. 6.21. Plotted is the surprisal, introduced[†] in

---

[†] $k^0(j \to j')$ is the prior rate constant, determined by the number of accessible final quantum states. The experiment samples molecules from a bulk, thermal, distribution which is, hence, not at a sharp value of the total energy; therefore, proper thermal averaging need be performed. Also, rather than $2j' + 1$, a restricted degeneracy of the final j' level is used in Fig. 6.21 [*cf.* Fig. 6.19 and Eq. (6.52)].

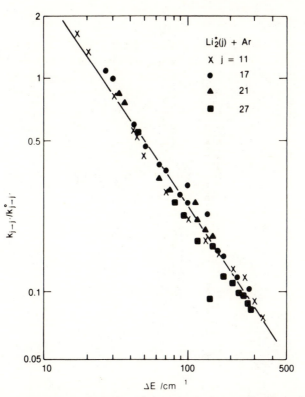

Fig. 6.21. "Power gap law" representation of the state-to-state rotational transition rates for the collision of $Li_2^*(j)$ with Ar. Plotted is the surprisal versus the log of the energy gap $|\Delta E| \equiv |E_{j'} - E_j|$. [Adapted from T. A. Brunner and D. E. Pritchard, *Adv. Chem. Phys.*, **50**, 589 (1982).]

Section 5.5, $\ln[k(j \to j')/k^0(j \to j')]$ versus $\ln|\Delta E|$ for

$$Ar + Li_2^*(v, j) \to Ar + Li_2^*(v, j')$$

at a number of j values.

It is also possible to monitor molecules through their stimulated emission. Figure 6.22 shows a number of observed laser lines in HF operating on pure rotational transitions. Normally, rotational relaxation is fast enough to preclude population inversion between two adjacent rotational states of the same vibrational manifold. However, the transitions shown in Fig. 6.22 originate at uncommonly high initial j values. The spacings among adjacent j states increase with j (or, equivalently, the rotational period decreases with j), so that at high j values a diatomic hydride is no longer in the sudden regime. The inelastic $R–T$ rates are therefore exponentially slower. But how did the HF molecules get to such high j states? Inspection of Fig. 6.22 suggests a $V \to R, T$ process. That is, a molecule at a high v and low

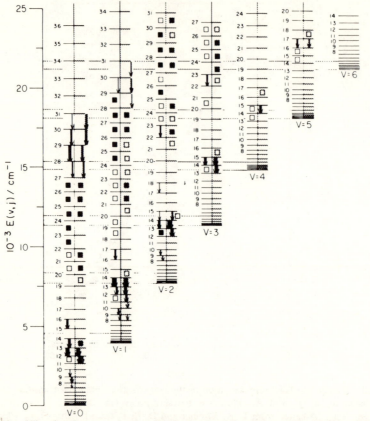

Fig. 6.22. Vibrational energy levels of HF showing observed rotational lasing transitions $(v, j \rightarrow v, j-1)$ for $v = 0$–5 (vertical arrows, whose width and placement denote intensity and timing). From the HF photoelimination reactions of $H_2C{=}CHF$ (left columns) and $H_2C{=}CF_2$ (right) in an Ar buffer. The horizontal cross-hatching shows levels in near-resonance with the $j = 0$–5 levels of each v-state. Considering the pattern for the highest-gain transitions one notes a propensity for $V \rightarrow R$ energy transfer of the type $HF(v, j) + Ar \rightarrow HF(v', j') + Ar$ with $\Delta E_{\text{rot}} \simeq -\Delta E_{\text{vib}}$. [Reproduced with permission from E. R. Sirkin and G. C. Pimentel, *J. Chem. Phys.*, **75**, 604 (1981).]

(thermally populated) j can, upon collision, transfer to a $v-1$, high j state with little energy transfer to translation $(Q \approx 0)$.

Other examples of efficient collision-induced intramolecular energy transfer have already been noted. Intermolecular energy transfer with a low net energy gap is exemplified by V–V transfer in diatomic and small polyatomic molecules.

Thus far we have discussed "degeneracy averaged" rates and cross sections (summed over $m_{j'}$, averaged over m_j). However, using polarized lasers it is possible both to select m_j and to probe $m_{j'}$. This allows measurement of reorientation cross sections, i.e., Δm transitions induced by

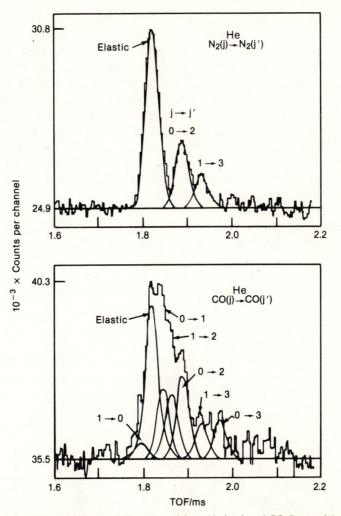

Fig. 6.23. Time-of-flight spectrum of He scattered from N_2 (top) and CO (bottom) in a crossed molecular beam experiment; the laboratory angle $\Theta = 39.5°$ and $E_T = 0.0277$ eV in both cases. Arrows denote expected peak positions corresponding to the designated rotational state transitions $j \rightarrow j'$; the peaks have been deconvoluted (Gaussian shapes) from the data. The strong elastic peaks are dominant in both cases. From symmetry considerations the selection rule $\Delta j =$ even applies to the homonuclear N_2 molecule; for CO, $\Delta j = 1, 2$, and 3 transitions are allowed (and observed). [Adapted from M. Faubel, *Adv. At. Mol. Phys.*, *19*, 345 (1983).]

collisions, sensitive to the orientation dependence of the intermolecular potential.

6.3.2 *Molecular beam studies of inelastic collisions*

In an inelastic collision not only the internal state but also the relative kinetic energy undergoes a change. By measuring the relative velocity change we can (via conservation of energy) determine the energetics of the collision. The analogy with Raman spectroscopy (where one measures the frequency of the emitted photons) is so striking that one sometimes speaks of "translational spectroscopy." As an example, consider an atom colliding with a diatomic molecule (Fig. 6.23). The most intense peak in the velocity distribution of the atom after the collision is due to elastic ("Rayleigh," in spectroscopy) scattering. On both sides of the elastic peak there can be weaker peaks corresponding to excitation of the molecule (and slowing down of the atom) and to deexcitation of the molecule (and acceleration of the atom). In Raman spectroscopy these will be the Stokes and anti-Stokes lines. The experimental results shown in Fig. 6.23 correspond primarily to rotational excitation, but contrast the behavior of the (isoelectronic) homonuclear diatomic molecule N_2 with that of CO.

Vibrational excitation is easier to resolve in a time-of-flight experiment since the energy spacings are wider (Figs. 6.14 and 6.24). Of course, at

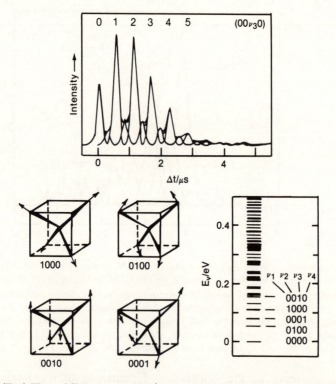

Fig. 6.24. (Top) Time-of-flight spectra of H^+ scattered by CF_4 at $E_T = 9.7$ eV, $\Theta = 10°$. Points: experimental curves, deconvoluted fits to peaks. The predominant peaks correspond to the vibrational excitation (of up to 6 quanta) of the ν_3 mode. (Bottom) Normal modes of CF_4 and the vibrational energy levels (with the four fundamentals designated). [Adapted from M. Faubel, *Adv. At. Mol. Phys.*, *19*, 345 (1983), based on results of U. Gierz, M. Noll, and J. P. Toennies, *J. Chem. Phys.*, *83*, 2259 (1985); see also T. Ellenbroek and J. P. Toennies, *Chem. Phys. 71*, 309 (1982).]

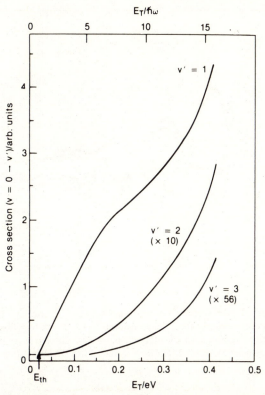

Fig. 6.25. Translational-energy dependence of the inelastic cross sections for the vibrational excitation of $I_2(v = 0 \rightarrow v' = 1, 2, 3)$ by collision with He. The magnitude of the (largest) cross section $(0 \rightarrow 1)$ at 0.4 eV is *ca.* 0.2 Å^2. The E_T dependence in the post-threshold region is linear for $v' = 1$, quadratic for $v' = 2$, and cubic for $v' = 3$. The threshold E_{th} is found to be $\hbar\omega = h\nu_{01}$, as expected. [Adapted from G. Hall, K. Liu, M. J. McAuliffe, C. F. Giese, and W. R. Gentry, *J. Chem. Phys.*, *81*, 5577 (1984).]

lower velocities, the vibrational transition probabilities are so much smaller that they are far harder to detect. Yet, Fig. 6.25 shows vibrational excitation cross sections determined in the immediate post-threshold region for the

$$He + I_2(v = 0) \rightarrow He + I_2(v' = 1, 2, 3)$$

collision. The simple observed post-threshold translational energy dependence, $\sigma \propto (E_T)^v$ is indicative of the impulsive mechanism, $\langle \Delta E_{vib} \rangle \propto E_T$ [*cf.* Eqs. (6.16) and (6.38)]. With $\bar{v} = \langle \Delta E_{vib} \rangle / h\nu \propto E_T$, the transition probability [Eq. (6.35)] has this expected E_T dependence.

6.3.3 Angular distributions

The advantage of the molecular beam method is its ability to provide angular information. Since the time-of-flight experiments can be carried out at a number of scattering angles, they can be combined to yield the c.m.

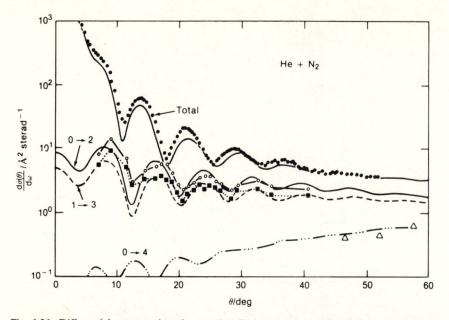

Fig. 6.26. Differential cross sections for rotationally inelastic scattering of $N_2(j \to j')$ by He at $E_T = 0.0273$ eV. Points: experimental, for transitions $0 \to 2$, $1 \to 3$, $0 \to 4$ and "total scattering," as indicated. Curves: theoretical calculations based on diffractive scattering from a rigid ellipsoid. [Adapted from M. Faubel, *J. Chem. Phys.*, *81*, 5559 (1984).]

angular distributions. The results, from the He–N_2 time-of-flight spectra shown in Fig. 6.23, are given in Fig. 6.26. Also shown for comparison are computational results for a realistic model potential (Fig. 6.27). As in optical spectroscopy proper, the link from the potential to the observation is firmly established. However, "inversion" of the data to the potential is nontrivial.

We have already mentioned the alternative route to such angular distributions using the Doppler shift. Of course one can also use the laser (as in laser-induced fluorescence) to probe the state of the scattered molecule in a crossed-beam arrangement (Fig. 5.5).

6.4 Collisions of molecules with surfaces

Just as for gas-phase molecular collisions, gas–surface encounters can be elastic, inelastic, or reactive in nature. A wide range of scattering behavior is observed depending upon the gas molecules, the composition, structure, and temperature of the surface, the translational and internal energies of the molecules; and even on the orientation of the colliding molecules with respect to the array of surface atoms.

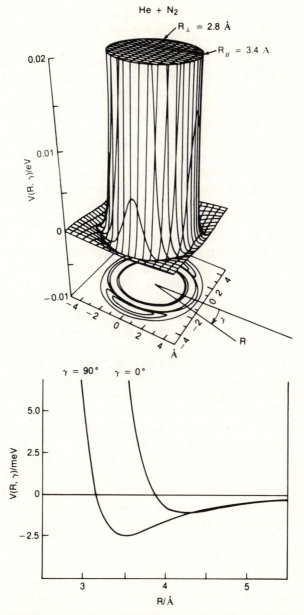

Fig. 6.27. A realistic potential function for He·N$_2$. (Top) Perspective view. The orientation angle γ is zero for the collinear configuration, 90° for a perpendicular approach. (Bottom) Cuts through the potential for $\gamma = 0°$ and 90°. (Adapted from same source as Fig. 6.26.)

6.4.1 Surface scattering

The processes that need to be considered include:

Two kinds of *elastic* scattering. In neither is there any energy loss, and in both cases the component of the projectile momentum in the direction perpendicular to the surface reverses its sign (without any change in magnitude) upon impact. For *specular* scattering, there is no change in the component of momentum parallel to the surface. The angle of scattering is thus equal to the incident angle. *Diffraction* scattering is characterized by a parallel momentum exchange.†

Three kinds of *inelastic* scattering. Two are of the direct mode, both involving energy transfers, but one type is the only possible direct inelastic process for atomic, "structureless" projectiles where the energy exchange is between the translation and the solid degrees of freedom (phonons and possibly also electronic excitation). Molecular projectiles offer the possibility of an additional type, namely, the excitation–deexcitation of the molecule, much as for gas-phase molecular inelastic scattering. Here, however, the energy balance need not be met entirely by the translation. The participation of the surface degrees of freedom is particularly evident in the third type of inelastic process, which proceeds via a temporary "trapping" of the projectile on the surface. This is reminiscent of the long-lived complexes in gas-phase collisions. Here, too, trapping is eventually followed by desorption but with a high probability of energy exchange having taken place.

Unique to surfaces is the possibility of "sticking" collisions, one type leading to the formation of a chemisorbed layer, the other involving dissociation of the projectile with retention on the surface of one or both fragments of the incident molecule. The molecules or fragments on the surface can be quite mobile. They can diffuse on the surface, but there can also be bulk penetration of the substrate lattice.

Several types of chemical reactions are possible and will be discussed in more detail in Section 7.5. Projectiles can react with surface atoms or with adsorbed molecules, but more complex processes involving interaction via diffusion of adsorbed species are also possible. All these reactive processes may involve desorption of products (and reagents).

6.4.2 Inelastic atom–surface collisions

We consider the simplest process, the collision of an atom (or an atomic ion) with a clean single crystal surface at low translational energies below

† The magnitude of the change in the component of projectile momentum parallel to the surface is determined by the "Bragg condition" familiar from X-ray structure studies of crystals. Here, however, the condition is determined by the atomic arrangement at the surface. Observing the angles at which diffraction scattering occurs provides an important surface diagnostic tool. As in the gas phase, the *intensity* of the diffraction scattering provides information on the projectile–surface potential.

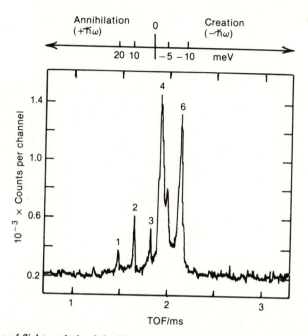

Fig. 6.28. Time-of-flight analysis of the He scattered from LiF(100). The incident translational energy of the He beam is $E_i = 0.019$ eV ($\pm 1.6\%$); the angle of incidence (with respect to the surface normal) is $\theta_i = 64.2°$; final (detected) angle (also with respect to the normal) is $\theta_f = 25.8°$. An analysis of the spectrum leads to the assignment of peaks 4 and 6 to creation and peak 1 to annihilation of single Rayleigh mode phonons, peak 5 to the creation of a bulk phonon, and peaks 2 and 3 to "understood" artifacts. From similar experiments at different angles, it has been possible to determine the "dispersion relation" for Rayleigh mode phonons of the surface, i.e., the dependence of $\omega(\equiv \Delta E/\hbar)$ versus k (the wave vector of the incident He beam). [Adapted from G. Brusdeylins, R. B. Doak, and J. P. Toennies, *Phys. Rev. Lett.*, **46**, 437 (1981).]

the threshold for electronic excitation. The change in the kinetic energy is due to surface phonon excitations (and deexcitations). Figure 6.28 shows results for He atoms scattered off the 100 face of a LiF crystal with the energy exchange interpreted as being due to single-phonon excitation. For heavier atoms (and molecules) and at higher incident momentum, the scattering reveals multiphonon excitations (Fig. 6.29). In addition, for heavier atoms, even the closed-shell ones such as Ar, Kr, or Xe, the attractive interaction with the surface atoms is sufficient to support the trapping–desorption inelastic process. As in the gas phase, the velocity and, in particular, angular distributions of the scattered molecules can be used in attempting to resolve the two concurrent processes of direct and trapping–desorption inelastic scattering. Figure 6.30 shows time-of-flight spectra for Xe atoms scattered off the 111 face of a Pt crystal at a surface temperature $T_s = 185$ K. The bimodal velocity distribution is deconvoluted into a contribution due to trapping–desorption and a direct inelastic component,

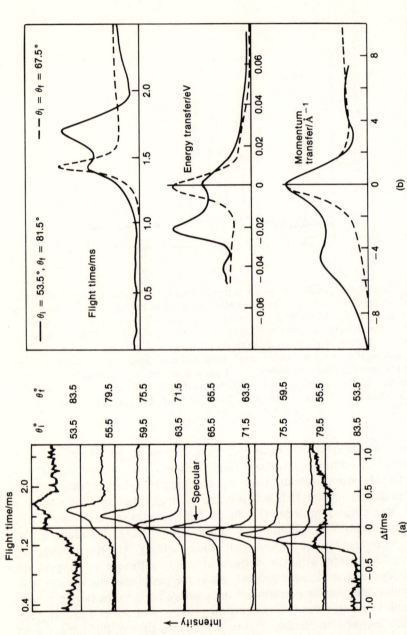

Fig. 6.29. (a) Time-of-flight spectra of Ne scattered from Ni(111). $E_i = 0.065$ eV, θ_i and θ_f as indicated. Note the shift in the Δt with respect to the specular conditions. (b) Representation of Ne scattering data, such as in (a), in terms of energy transfer and momentum transfer at the indicated specular and nonspecular conditions. [Adapted from B. Feuerbacher, in G. Benedek and U. Valbusa (1982).]

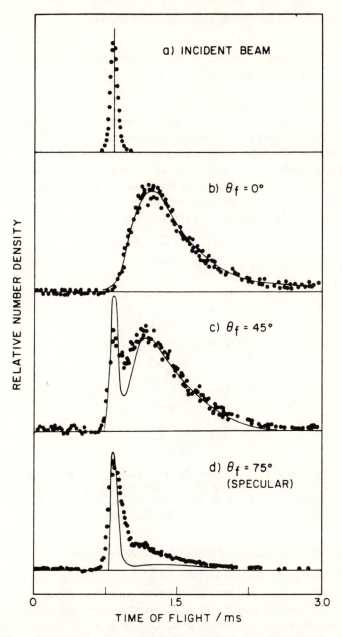

Fig. 6.30. Time-of-flight spectra for Xe scattered from Pt(111). $E_i = 0.145$ eV, $\theta_i = 75°$. Points: experimental; curves: calculated by model of direct inelastic and trapping–desorption scattering. (a) Incident Xe beam characterization; (b) Xe scattered at the normal ($\theta_f = 0°$); (c) $\theta_f = 45°$; (d) $\theta_f = \theta_i$ (=75°), i.e., specular scattering. Note the bimodal distribution in (c). The faster peak is attributed to direct inelastic scattering (most intense at the specular angle), the broad (slower) distribution from the trapping–desorption process. [Adapted from J. E. Hurst, C. A. Becker, J. P. Cowin, K. C. Janda, L. Wharton, and D. J. Auerbach, *Phys. Rev. Lett.*, *43*, 1175 (1979). Theoretical analysis by C. W. Muhlhausen, L. R. Williams and J. C. Tully, *J. Chem. Phys. 83*, 2594 (1985).]

the latter becoming dominant at the specular angle. The intensity of the trapping–desorption component varies according to the usual "cosine law" $[I(\theta) \propto \cos \theta]$ for "evaporation" from a surface, and the velocity distribution for the desorbing atoms can be fit by a thermal distribution at the surface temperature.† Additional experiments on the direct inelastic components indicate that the energy transfer is preferentially via the change in the component of the momentum normal to the surface.

Extensive studies of the translational accommodation of the heavier rare gases with various surfaces over a range of collision energies and surface temperatures relate the average translational energy of the scattered atoms to that of the incident atoms and the surface temperature (Fig. 6.31),

$$\langle E'_T \rangle = C_1 E_T + C_2 2kT_s \tag{6.53}$$

where $C_1 \equiv d\langle E'_T \rangle / dE_T \simeq 0.8$ and $C_2 < C_1$.

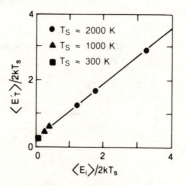

Fig. 6.31. Dependence of $\langle E'_T \rangle$ upon E_T (i.e., E_i) for Ar scattered by W, at surface temperatures $T_s = 300$, 1000, and 2000 K. [Adapted from K. C. Janda, J. E. Hurst, C. A. Becker, J. P. Cowin, D. J. Auerbach, and L. Wharton, *J. Chem. Phys.*, *72*, 2403 (1980). See also Janda et al., *J. Chem. Phys.*, *78*, 1559 (1983).]

6.4.3 Experimental studies of molecule–surface scattering

The principal experimental technique is the molecular beam method: A monoenergetic collimated molecular beam impinges on a well-characterized single crystal surface under conditions of ultrahigh vacuum. Mass spectrometric detection of the scattered molecules as a function of angle with time-of-flight velocity analysis is standard, as in the gas phase. However, additional complication of energy transfer to the surface makes the velocity analysis less definitive and calls for state-specific detection, sensitive to the

---

† Care should be exercised in generalizing these results to the desorption of molecules. If, for example, the molecules are formed by recombination on the surface (e.g., H_2 on transition metals) or if there are strong and anisotropic forces between the molecules and the surface, the distribution of the desorbed molecules need not reflect the accommodation of the absorbed species with the surface.

internal state of the molecules. These include laser-induced fluorescence, sometimes with polarized radiation, sometimes with narrow-band lasers for Doppler-shift measurements to ascertain recoil velocities, and laser multi-photon ionization (Sec. 7.3.4), which is also state specific and very sensitive.

Considerable attention has been given to the scattering of NO on Ag(111), for which the van der Waals binding energy is small ($\approx$0.2 eV) and on Pt(111), with a large binding energy (>1 eV). For the former system, a clear separation of the direct inelastic from the trapping–desorption process has been carried out via state-resolved detection as a function of angle. At incident energies above the well depth, direct inelastic scattering appears to dominate. For a cold NO beam scattered by Ag(111), the results show near-specular scattering for the low-j' states (Fig. 6.32). For larger j's, the peak in the scattered intensity shifts to forward angles. This can be understood if the parallel-to-the-surface momentum of NO is unchanged in the collision but the normal component is diminished because of the energetic requirement to reach these higher j' states with ever-larger energy spacings.

Figure 6.33 shows the rotational state distribution of NO for specular scattering off Ag(111), over a wide range of incident translational energy. The data suggest that for low j's, the final rotational distribution is thermal-like but with a rotational temperature lower than the surface temperature T_s. The rotational temperature of the scattered NO lags behind T_s more and more as T_s is increased for both Ag(111) and Pt(111) (Fig. 6.34). The rotational temperature does, however, linearly increase with the component of the initial translational energy normal to the surface.

Vibrational excitation has been observed only for NO scattering off Pt(111) in a lower-velocity range where trapping–desorption takes place, as suggested by a nearly cosine law angular distribution. The apparent vibrational temperature lags behind the surface temperature (Fig. 6.35).

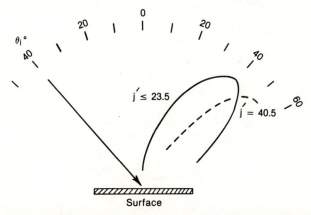

Fig. 6.32. Angular distribution (polar plot) of intensity of scattered NO from Ag(111), $T_s = 650$ K; $E_T = 1.1$ eV, $\theta_i = 40°$. Solid curve: average for all "low j'-states," i.e. $j' \leq 23.5$. Dashed curve: $j' = 40.5$. [Adapted from A. W. Kleyn, A. C. Luntz, and D. J. Auerbach, *Surf. Sci., 117*, 33 (1982).]

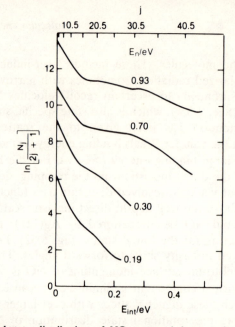

Fig. 6.33. Rotational-state distribution of NO scattered from Ag(111) at $T_s = 650$ K, with incident normal energies E_n ($= E_T \cos^2 \theta_i$) in the range 0.19–0.93 eV. The j's indicated are for the $^2\Pi_{1/2}$ state of the NO populations *via* laser-induced fluorescence of the scattered NO. (Top three curves) $\theta_i = \theta_f = 15°$; (bottom curve) $\theta_i = \theta_f = 40°$. Linear portion of graphs (at low j') characterized by a Boltzmann rotational temperature T_{rot}. [Adapted from A. W. Kleyn, A. C. Luntz, and D. J. Auerbach, *Phys. Rev. Lett.*, 47, 1169 (1981).]

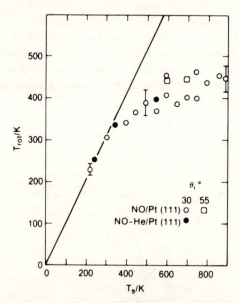

Fig. 6.34. Rotational temperature T_{rot} for NO scattered specularly from Pt(111) versus T_s (at two values of θ_i). Full accommodation would conform to the line of unit slope. [Adapted from J. Segner, H. Robota, W. Vielhaber, G. Ertl, F. Frenkel, J. Hager, W. Krieger, and H. Walther, *Surf. Sci.*, 131, 273 (1983). See also G. M. McClelland, G. D. Kubiak, H. G. Rennagel, and R. N. Zare, *Phys. Rev. Lett.*, 46, 831 (1981); and D. S. King, D. A. Mantell and R. R. Cavanagh, *J. Chem. Phys.*, 82, 1046 (1985).]

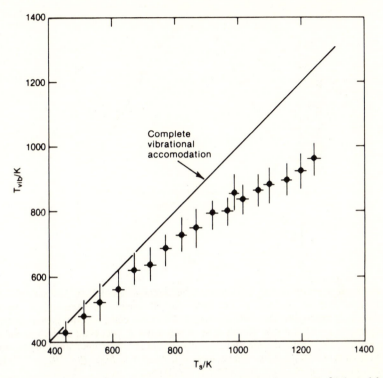

Fig. 6.35. Vibrational temperature for NO scattered from Pt(111) versus T_s. [Adapted from M. Asscher, W. L. Guthrie, T. H. Lin, and G. A. Somorjai, *J. Chem. Phys.*, **78**, 6992 (1983).]

The angular distribution of the scattered, vibrationally excited NO molecules very nearly follows a cosine law.

A very powerful development is the use of polarization data to infer the degree of alignment of the rotational angular momentum of the scattered molecules, i.e., the distribution of $m_{j'}$ for a given j'. In principle, it should be possible to measure the polarization of the laser-induced fluorescence as well as the dependence of the intensity of the fluorescence upon the polarization of the laser beam with respect to the surface normal.

Results for the latter type of measurement are shown in Fig. 6.36 for NO scattered off Ag(111) for an electronic transition dipole essentially parallel to $\mathbf{j}'$. To interpret the dependence of the laser-induced fluorescence intensity, $I(\theta_0)$, on the angle θ_0 of the laser polarization with respect to the surface normal, consider the so-called polarization anisotropy, P:

$$P = [I(0) - I(\pi/2)]/[I(0) + 2I(\pi/2)] \tag{6.54}$$

where $-\frac{1}{2} \leq P \leq 1$. The upper limit of $P = 1$ is only possible if $I(\pi/2) = 0$, that is, a perfect alignment of $\mathbf{j}'$ along the normal to the surface. The lower limit on P corresponds to a perfect perpendicular alignment. Figure 6.36

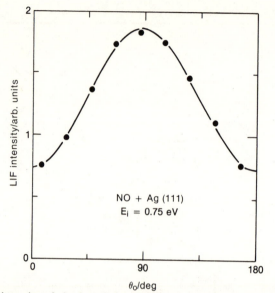

Fig. 6.36. Total intensity of the laser-induced fluorescence signal from NO scattered from Ag(111), $T_s = 650 \, \text{K}$ at $\theta_i = 40°$, $E_i = 0.75 \, \text{eV}$, plotted versus the angle θ_0 of the laser polarization with respect to the surface normal. The solid curve is a fit to the theoretical functionality, $I \propto 1 + aP_2(\cos \theta_0)$. The analysis of the result as presented in the text indicates that the scattered NO molecules are preferentially aligned in a direction normal to the surface. [Adapted from A. C. Luntz, A. W. Kleyn, and D. J. Auerbach, *Phys. Rev.*, *B25*, 4273 (1982); see also A. C. Luntz et al., *Surf. Sci.*, *152/153*, 99 (1985).]

immediately reveals that $I(0) < I(\pi/2)$, so that the alignment of $\mathbf{j}'$ is preferentially perpendicular to the surface normal.

The field of inelastic scattering of state-selected molecular beams by well-characterized surfaces is a growing one, with experimental progress being matched by theoretical and computational developments.

6.5 Bimolecular spectroscopy

Preparing an initial state or monitoring the final state provides definitive but indirect answers to our central question—What happens in the very act of the collision? The experimental realization that one can spectroscopically probe molecules *during* their collision opens up new vistas in the study of collision dynamics. In this section we consider the possibility of obtaining direct evidence, by spectroscopic means, of the "transition state."

6.5.1 The importance of collisions

Most of our attention thus far in this chapter has been devoted to collision phenomena that are explicitly inelastic. Yet there are other manifestations

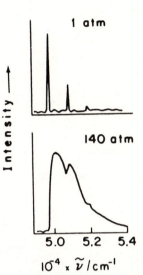

Fig. 6.37. Influence of foreign gas pressure upon the electronic absorption spectrum of CH_3I. This "pressure broadening" has its analog in the spectra of molecules in the condensed state. [Adapted from M. B. Robin and N. A. Kuebler, *J. Mol. Spectrosc.*, 33, 274 (1970).]

of energy-transferring collisions that are also very revealing of molecular dynamical behavior. Only at very low densities do spectroscopic observations refer to isolated atoms or molecules. At higher densities, collisions produce detectable perturbations. At the high densities characteristic of liquids and solids, such perturbations can be so severe that the spectrum is changed beyond recognition. Even a high gas density can change the spectrum markedly, as shown in Fig. 6.37.

In this section we are concerned with the low-density gas-phase limit where most collisions are bimolecular. Such collisions have diverse spectroscopic manifestations, as discussed below. The common theme is that the spectra depend upon the collision of the two molecules and hence serve as a useful probe of the dynamics of the collision.

6.5.2 Collision-induced light absorption

The most elementary bimolecular spectroscopic phenomenon is collision-induced absorption. In its simplest form it is exhibited as a broad-band absorption (usually in the far infrared) in mixtures of rare gases. The absorption of photons at the frequency v is found to depend upon the product of the densities of both gases,

$$\log[I(0)/I(l)] = l \, n_A n_B A(v) \tag{6.55}$$

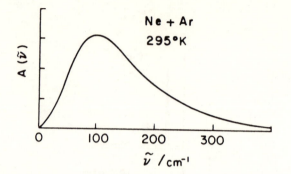

Fig. 6.38. "Bimolecular" absorption coefficient for the Ne–Ar system in the far infrared. This "collision-induced absorption" arises from the transition dipole moment associated with "fly-by" collisions of Ne with Ar atoms. [Adapted from D. R. Bosomworth and H. P. Gush, *Can. J. Phys., 43*, 751 (1965).]

where l is the path length. The binary absorption coefficient $A(v)$ is independent of the density, but it does depend upon the species present (Fig. 6.38). This absorption is not found in the pure rare gases and obviously disappears in the very low density limit. It is clearly a *bimolecular* phenomenon and requires the participation of *dissimilar* atoms.

The spectrum shown in Fig. 6.38 can be considered as the spectrum of the transient heteronuclear AB dimer formed during the A + B collision (a so-called fly-by dimer). This absorption is not found in the pure gases for the same reason that stable homonuclear, A_2, diatomics are infrared inactive. In classical language, light is absorbed or emitted when a dipole is changing with time. A heteronuclear diatom has a dipole moment, $\mu(R)$, that is a function of the internuclear separation and will have an infrared vibrational spectrum due to the oscillations of the AB distance. Similarly, a transient AB heteronuclear dimer has a dipole moment. During the collision the AB separation varies with time and so does the dipole moment, i.e., $\mu(R) = \mu[R(t)]$ (Fig. 6.39). This time dependence of μ is the mechanism whereby the transient dimer (existing only during the fly-by collision) can absorb light.

An estimate of the "lifetime" of such transient dimers is readily provided from the width of the absorption band via the Heisenberg *uncertainty principle*, $\tau = h/\Delta E$. These transient dimers have an absorption spectrum with a width $\Delta \bar{v}$ of about 200 cm^{-1} (Fig. 6.38). Hence the average duration of their existence, τ, is

$$\tau = h/hc \, \Delta\bar{v} = (c \, \Delta\bar{v})^{-1} \approx 2 \times 10^{-13} \, \text{s} \qquad (6.56)$$

which is comparable to a typical duration for a collision between two rare-gas atoms at ordinary temperatures (e.g., $a = 1$ Å, $v = 500 \, \text{ms}^{-1}$),

$$t_c \approx 10^{-10}/5 \times 10^2 = 2 \times 10^{-13} \, \text{s} \qquad (6.57)$$

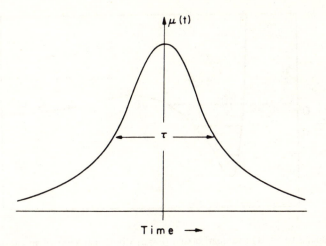

Fig. 6.39. Schematic representation of the time dependence of the (transient) dipole moment associated with a fly-by collision of unlike atoms, at a given b and E. The "width," τ, of the dipole-moment function can be related *via* the uncertainty principle to the frequency of the most intense emission or absorption of radiation.

6.5.3 Quasi-bound states and predissociation

The broad collision-induced absorption band is characteristic of the transient dimers formed during a direct collision (Sec. 3.3). What about the longer-lived dimers, formed during a "compound collision" (Sec. 4.3.6)? Because of their longer lifetimes, they should yield sharper absorption lines.

The simplest example is the case of the orbiting dimers that can be formed even in atom–atom collisions. To examine the origin of such states we consider the effective potential for the relative motion. At a given impact parameter b and initial translational energy E_T,

$$V_{\text{eff}}(R) = V(R) + E_T b^2 / R^2 \tag{2.32}$$

where $V(R)$ is the interatomic (or the central part of the intermolecular) potential (Fig. 6.40). For low values of b, the effective potential retains the inner well, characteristic of $V(R)$. Such a potential can support quasi-bound states, namely, states of positive energy that can energetically dissociate ("rotational predissociation") but [depending upon their depth below the top of the barrier of $V_{\text{eff}}(R)$] may have a long lifetime compared to the duration of a direct collision.

The quasibound or orbiting states of the effective potential are thus bound in the inner well by the centrifugal barrier. In classical mechanics such orbiting states would be permanently bound since to escape they would have to penetrate the "hump" inside of which their kinetic energy would be negative [*cf.* Eq. (2.33)]. Quantum mechanical tunneling allows the relative motion to cross the region of negative kinetic energy. Every time that the vibrational motion brings the relative separation to the classical turning

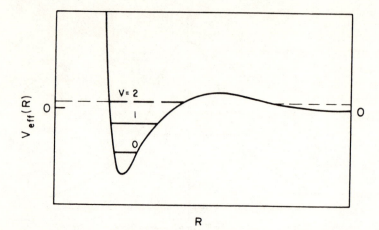

Fig. 6.40. Bound ($v = 0$, 1) and quasi-bound ($v = 2$) vibrational levels of an effective potential $V_{\text{eff}}(R)$ for a given angular momentum. This orbital angular momentum is identical to the rotational angular momentum of the bound (or quasi-bound) diatomic. The lifetime of the quasi-bound state is determined by the rate of (quantum mechanical) tunneling through the potential barrier. The light dashed line is the collision energy of the unbound atoms at which the quasi-bound state may be formed or into which the quasi-bound state may dissociate.

point [Eq. (2.34)] at the centrifugal barrier, a fraction, p, of the dimers dissociates by tunneling out.

If t_v is the vibrational period of the quasi-bound molecule in its particular state, the separation of the dimer reaches the turning point on the inner side of the barrier t_v^{-1} times per second, and, at each time, dissociation occurs with probability p. The rate of dissociation is thus p/t_v, and the lifetime τ of the dimer is $\tau = t_v/p$. Since t_v is of the order of the duration of a collision [Eq. (6.57)], quasi-bound states can be quite long-lived if their barrier penetration probability is small. In general, p decreases as the width of the barrier increases, as the level is further down below the barrier, and as the reduced mass of the dimer increases.

Quasi-bound states of the orbiting type have long been observed in ordinary absorption spectra of diatomic molecules. Transitions are observed from vibrational levels of the ground electronic state of the stable diatomic to a quasi-bound level in the excited electronic state. Since the latter state has a short lifetime, such transitions have a *broadened* absorption line [compared to the "sharp" lines for bound to (truly) bound state transitions]. In spectroscopy, the predissociation phenomenon has been studied extensively for hydrides (e.g., H_2, HgH, MgH, AlH), where the tunneling probability is large and the short quasi-bound lifetimes lead to larger widths of the absorption lines. From such studies one can learn about the *interatomic potentials* for the dissociated atoms.

In an atom–atom collision such quasi-bound states can be formed by tunneling in through the centrifugal barrier and can then dissociate by tunneling out. The scattered products will then be *delayed* with respect to the directly scattered wave. Since the energies of quasi-bound states at different l values will differ, then at a given energy the cross section will have a contribution from at most one quasi-bound state. Since the cross section is a sum of contributions from many partial waves [see Eq. (3.39)] and at most one partial wave is affected, the observed contribution to the total cross section will not be large, although it can have an effect on the angular distribution. This is where spectroscopy has a distinct advantage. By exciting a quasi-bound state of a bound molecule one is dealing with a single, particular l value.

In tri- and polyatomic molecules there is another type of predissociation, one that has already been mentioned several times. This is predissociation by vibration. Here the same total energy can be partitioned in several ways among two or more vibrational modes. For some of these the molecule is bound, but for others it is unbound. Excited van der Waals dimers (Fig. 6.10), are but one example. Another is a compound collision governed by a potential energy surface with a well in its middle (Sec. 4.3.6). What determines the lifetime is the rate of *intra*molecular vibrational energy transfer. When that rate is fast, we are in the regime where transition state theory can be used to compute the dissociation rate (see Sec. 7.2).

Finally, we mention trapping in atom– (or molecule–) surface collisions. In general, the potential energy in the direction perpendicular to the surface has an attractive component. Thus, the colliding particle can be trapped in the well, with the balance of the energy being taken up by the surface modes. Desorption is the predissociation of such surface-bound species.

6.5.4 Pressure broadening of spectral lines

An important spectroscopic manifestation of inelastic collisions is the phenomenon of pressure broadening of molecular absorption lines. Consider an excited vibrational state of the molecule that, in a very low density gas (i.e., in the absence of collisions), would lose its excitation by radiative decay. Inelastic $R–R'$ and $V–V'$ collisions change the state of the molecule and hence serve as a mechanism to quench the emission of the excited state. In the absence of collisions, the radiative lifetime is quite long (typically in the millisecond range for emission in the infrared). The quenching collisions can shorten the lifetime of the excited state considerably. According to the uncertainty principle, the absorption line will then be markedly broadened. The width of the line is (neglecting the radiative lifetime)

$$\Delta v = \tau^{-1} = nk \tag{6.58}$$

Here, k is the bimolecular rate constant for quenching the excited level and n is the number density ($n = P/\mathbf{k}T$) of the particular collision partner (see also Sec. 6.6.2). The slope of the pressure dependence of the linewidth provides a determination of the rate constant k for all inelastic collisions out

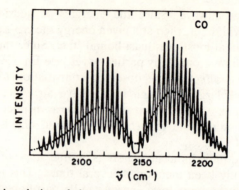

Fig. 6.41. Pressure broadening of the rotational structure in the fundamental vibrational absorption band of CO. The structure in the solid curve at the lower pressure (here *ca.* 10 atm) is "washed out" (dotted curve) at the high pressure (~90 atm), leaving only the envelope of the *P* and *R* branches. [Adapted from R. G. Gordon and R. P. McGinnis, *J. Chem. Phys., 58,* 4898 (1971).]

of the initial state. The very high rates of R–R' transfers were first determined in this way.

The infrared spectra of molecules contain many adjacent lines corresponding to different rotational–vibrational transitions.† As the pressure increases, these lines therefore begin to overlap. Ultimately, at high pressures they merge together and the detailed structure is lost (Fig. 6.41). The loss of detail due to pressure broadening is typical of spectra at high pressures (Fig. 6.37) and of molecules in the liquid phase.

Emission from electronically excited states is also pressure broadened. We invoke the Franck–Condon principle to interpret such broadening. Figure 6.42 shows the M—Na(3^2S) and M—Na(3^2P) interatomic potentials. The strong sodium D-line corresponds to the Na(3^2P) → Na(3^2S) transition. The frequency of the unperturbed D-line corresponds to the energy separation between the two potential curves in Fig. 6.42 at infinite M—Na separation. Say, however, the transition occurs when the perturbing M atom is but a finite distance R away. According to the Franck–Condon principle, the atoms will hardly move during the electronic transition. Hence the transition frequency will correspond to the energy separation between the two potential curves at the interatomic distance R. To each M—Na distance there can correspond a different frequency. If the two potential curves differ in their R-dependence, the D-line can acquire quite far-flung "wings." The experimental results are shown in Fig. 6.43.

---

† Pure rotational spectra are simpler. Because of the high resolution of microwave spectroscopy of low-pressure gases, one can measure much smaller line widths. This has proved to be a source of considerable insight into inelastic scattering and, thereby, of intermolecular forces.

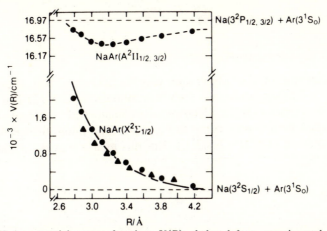

Fig. 6.42: NaAr potential energy functions $V(R)$, deduced from experimental data on Na D-line broadening by Ar. Circles: experimental results of G. York, R. Scheps, and A. Gallagher [*J. Chem. Phys.*, *63*, 1052 (1975)]. Triangles: experimental results on the (innermost, repulsive part of) the ground state ($X^2\Sigma_{1/2}$) potential via high-energy atomic beam scattering data [C. J. Malerich and R. J. Cross, Jr., *J. Chem. Phys.*, *52*, 386 (1970)]. The solid curve for the ground state is the pseudopotential calculation of J. Pascale and J. Vanderplague [*J. Chem. Phys.*, *60*, 2278 (1974)]. Recent beam-scattering experiments with excited Na* atoms have led to quantitative modifications and extensions of this early work. For results and a summary of accurate alkali-rare gas excited state potentials see R. Düren, E. Hasselbrink, and H. Tischer, *J. Chem. Phys.*, *77*, 3286 (1982) and R. Düren, E. Hasselbrink and G. Moritz, *Z. Phys.*, *A307*, 1 (1982).

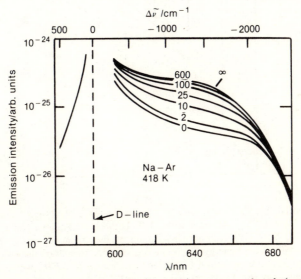

Fig. 6.43: "Normalized" emission intensity (log scale) versus wavelength (upper scale: wave number shift from the D-line) for Na(3^2P) at 418 K in the presence of Ar at the indicated pressures (in Torr). The blue wing is pressure independent (up to 10^3 Torr). The lowest curve is the extrapolated zero pressure result for the red wing. The zero pressure limiting lineshapes are attributable to bound–free transitions from the excited ($A^2\Pi_{1/2,3/2}$) state of NaAr to the repulsive part of the ground ($X^2\Sigma_{1/2}$) state, leading (within $\leq 10^{-3}$ s) to Na($^2S_{1/2}$) + Ar($^1S_0$). [Adapted from G. York, R. Scheps, and A. Gallagher, *J. Chem. Phys.*, *63*, 1052 (1975); analogous results based on the fluorescence of the Na* + N_2 system are reported by W. Kamke, B. Kamke, I. Hertel, and A. Gallagher, *J. Chem. Phys.*, *80*, 4879 (1984).]

6.5.5 Bimolecular emission spectroscopy

Upon excitation in the ultraviolet (say at 220 nm), NaI dissociates to $Na(3^2P) + I$. The dissociation is therefore followed by a Na D-line emission. En route to dissociation the electronically excited NaI* molecule moves on the upper potential energy curve in Fig. 6.44. The photodissociation of NaI is a half-collision of I with $Na(3^2P)$. But this means that the sodium D-line

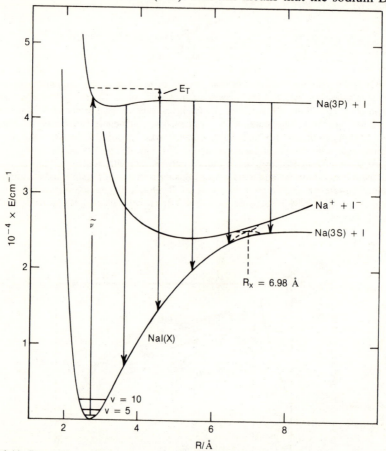

Fig. 6.44. Potential energy curves for the relevant states of NaI and "Franck–Condon" transitions $Na(^3P) \rightarrow Na(^3S)$. [Adapted from H.-J. Foth, J. C. Polanyi, and H. H. Telle, *J. Chem. Phys.*, **86**, 5027 (1982).] Photodissociation at the frequency $\bar{\nu}$ ($\lambda \sim 235$ nm) yields $Na(^3P) + I(5^2P)$ with a recoil energy E_T. In the course of the separation the system can radiate over a wide frequency range, shown by the arrows, from $\bar{\nu}$ to $\Delta E/h$, where ΔE is the Na D-line excitation energy (*ca.* 17,000 cm^{-1}, $\lambda \sim 590$ nm).

will be accompanied by a far-wing blue-shifted† emission. Of course, only those molecules that emit during the direct dissociation process will contribute to the wing. Taking a radiative lifetime of 10^{-8} s and a

---

† The shift is to the direction of shorter wavelengths (hence "blue") since at close distances the energy separation between the two potential curves is far higher than the asymptotic one.

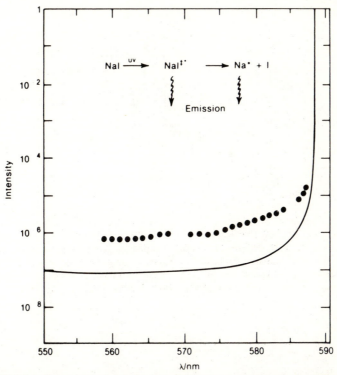

Fig. 6.45. Relative fluorescence intensity (log scale) versus wavelength from the photodissociating NaI*. Following uv (220–260 nm) excitation of NaI vapor, the Na D-line shows a blue wing in accord with expectations (shown as solid line) based on Fig. 6.44. [Adapted from H.-J. Foth, H. R. Mayne, R. A. Poirier, J. C. Polanyi, and H. H. Telle, *Laser Chem.*, **2**, 229 (1983).] The shortest wavelengths come from NaI* emitting early in the separation process, the longer wavelengths from the later stage.

half-collision duration of 10^{-13} s, only 1 in 10^5 NaI* molecules will emit on its way out. The quantum yield is thus low. The intensity at a given wavelength will thus be very weak, since the emission spans a very wide range in frequency (Fig. 6.45).

In principle, some of the NaI* molecules can emit into bound vibrational states of ground state NaI. The emission spectrum is thus of the free$\rightarrow$ bound type and should be structured. The small vibrational spacing in NaI and poor frequency resolution (due to the low quantum yield) smear out any expected structure in Fig. 6.45. Such a structure can, however, be seen in other experiments discussed below.

The electronically excited molecule need not be produced by excitation from the ground state. It can be formed in a chemical reaction. Figure 6.46 shows the wings of the Na D-line in the $F + Na_2$ reaction

$$F + Na_2 \rightarrow \begin{cases} NaF + Na(3^2P) \\ NaF + Na(3^2S) \end{cases}$$

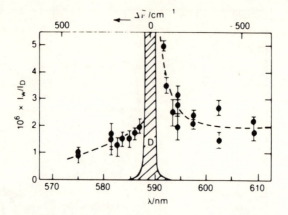

Fig. 6.46. Emission in the wings of the Na D-line from the FNaNa* transient from the chemiluminescent reaction $F + Na_2 \rightarrow FNaNa^* \rightarrow NaF + Na^*$. Both a blue and a red wing are observed. Points denote experimental results (with bars representing standard deviations), connected by "best" curves (dashed). Crosshatched region centered at the D-line, intense overall emission profile, scaled appropriately. [Adapted from P. Arrowsmith, S. H. Bly, P. E. Charters, and J. C. Polanyi, *J. Chem. Phys., 79,* 283 (1983).]

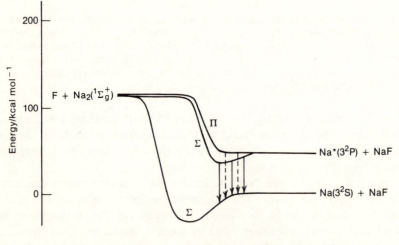

Fig. 6.47. Empirical potential energy curves for the collinear FNaNa system, intended to explain the red and blue wings observed in the emission (near the D-line) from the reaction of $F + Na_2$ (Fig. 6.46). They are based on the analogous FLi_2 curves of G. G. Balint-Kurti and M. Karplus, and are drawn in the adiabatic representation. The solid arrows from the excited Σ state downward to the ground (Σ) state denote transitions to the red of the Na D-line; the dashed arrows from the excited Π state, blue wing emission. [Adapted from Arrowsmith et al., 1983; see Fig. 6.46.]

which produces both ground and electronically excited Na* atoms. An estimate of the potential energy profiles (for a collinear configuration) for this reaction is shown in Fig. 6.47. As is suggested by this figure, one observes both blue and red wing emission. Of course, the FNa_2* molecule will be formed in a variety of configurations, and the main contributions to the emission will be from those configurations in which the system spends more time.

To obtain a better resolved emission spectrum it is desirable to limit the range of accessible configurations. One way of doing so is to make use of a half-collision. The molecule is optically excited to the upper electronic state in an initial configuration limited by Franck–Condon considerations. Figure 6.48 shows two contrasting possibilities for the subsequent emission. If the dissociation of the ABC* molecule occurs primarily through the repulsive A–B force (receding of A from BC, the situation essentially that of Fig. 6.44), the BC bond is not very perturbed.† Emission will be into the A–B excited modes of ABC. In other words, BC acts essentially like an atom. On the other hand, if ABC* is formed with a compressed BC bond, the dissociative separation of A from BC will have a high Franck–Condon overlap with the B–C excited modes of ABC. Two illustrations of such considerations are forthcoming. The first is the emission (Fig. 6.49) in the dissociation of CH_3I*. An exceedingly long progression of lines terminating in the overtones of the C–I stretch mode of ground state CH_3I is evident. This is the "diatomic-like" behavior. On the other hand, in O_3* (which dissociates by ultraviolet light via the B_2 excited state) the emission is primarily due to the symmetric stretching overtones of O_3 (with some mixing of even-quanta overtones of the asymmetric overtone) (Fig. 6.50). Since, by symmetry, there can be no bending force in the direction of the bending motion of the initial configuration of O_3*, no emission to the bending overtones of O_3 is observed.

6.5.6 Laser-assisted collision processes

Rather than excite a stable diatomic or triatomic molecule to an upper state, one can consider exciting a pair of atoms or an atom–molecule pair *during* their collision. In discussing such problems it is very suggestive to adopt a point of view suggested in part by considerations of the Franck–Condon principle. Consider an atomic collision along a potential energy curve (Fig. 6.44). The figure shows another, higher-energy potential energy curve. Even if the collision energy is sufficient, it is very unlikely that a translation-to-electronic energy transfer will take place—the required change in momentum of the relative motion will be too high. Now let the collision take place in a laser field but not necessarily at the frequency corresponding to the asymptotic energy separation of the two curves. But this is precisely

---

† Unless the translational energy is high enough so that intramolecular V–T coupling is effective.

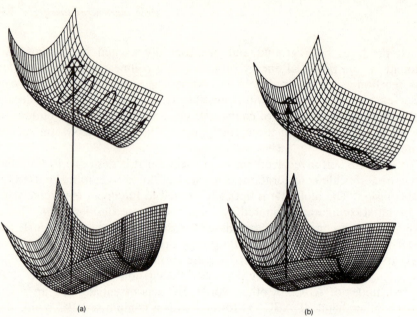

(a)   (b)

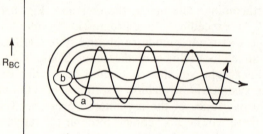

(c)

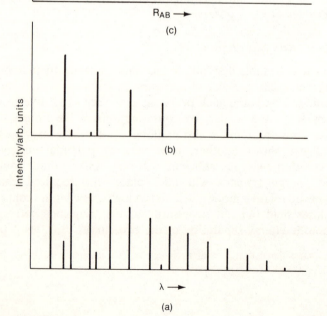

(b)

(a)

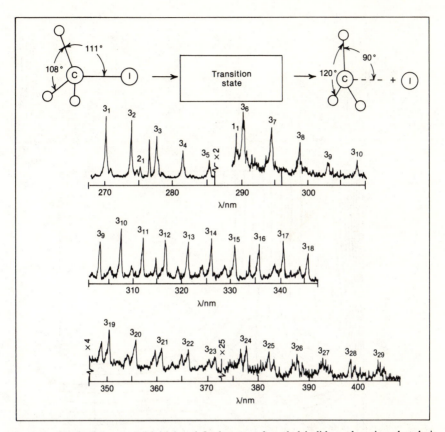

Fig. 6.49. (Top) Geometry of initial and final states of methyl iodide undergoing photolysis. For CH_3I, $R_{CH} = 1.11$ Å, $R_{CI} = 2.14$ Å, bond angles as shown; for $CH_3 + I$, $R_{CH} = 1.08$ Å, $R_{CI} = \infty$, angles indicated. (Bottom) Observed emission from CH_3I, i.e., Raman spectrum, using $\lambda = 266$ nm excitation. Grating orders indicated. The vibrational overtone progression is primarily based on the normal mode ν_3 (533 cm^{-1}), corresponding to the CI bond stretch. The labeling of the peaks is obvious (e.g., 3_9 means 9 quanta of ν_3). [Adapted from D. Imre, J. L. Kinsey, A. Sinha, and J. Krenos, *J. Phys. Chem.* **88**, 3956 (1984).]

Fig. 6.48. Wave-packet description of photofragmentation of a triatomic molecule, ABC. Photoexcitation from the ground electronic state surface, which then follows the path of steepest descent in the "half-collision" leading to separation of ABC* into A + BC. (a) ABC* formed with "compressed" BC bond leading to vibrationally excited BC†. (b) ABC* formed with normal BC bond; repulsion of BC by A leads to vibrationally cold BC fragment. (c) Qualitative features expected in emission spectra for cases (a) and (b) above. Case (a) produces a spectrum consisting mainly of the BC stretching frequency and its overtones, see trajectory above; case (b), very weak BC emission, mainly AB stretching frequency progression. [Adapted from E. J. Heller, *Acc. Chem. Res.* **14**, 368 (1981); E. J. Heller, R. Sundberg, and D. Tannor, *J. Phys. Chem.*, **86**, 1822 (1982).]

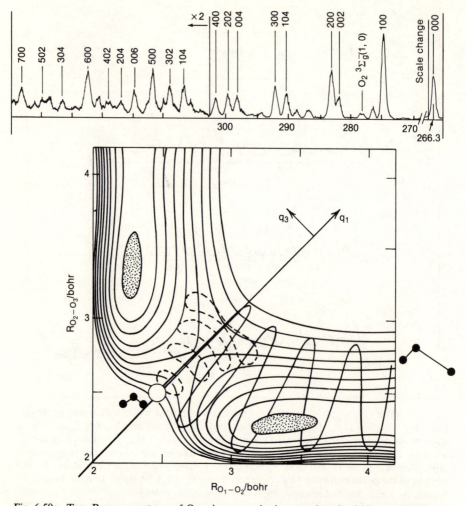

Fig. 6.50a. Top: Raman spectrum of O_3 using an excitation wavelength of 266 nm. Bands are labeled by the number of quanta of excitation in v_1 (symmetric stretch), v_2 (bend), and v_3 (asymmetric stretch). Note the predominance of the v_1 progression. Bottom: Calculated B_2 excited state potential-energy surface for ozone along the two stretching normal coordinates, q_1 and q_3; the bending coordinate is held fixed at its value in the ground state. The heavy line denotes a photodissociation trajectory; the dashed curves show the spreading wave packet. The shaded zones indicate quasi-bound regions (wells) in the exit valleys. [Adapted from D. Imre et al. (see Fig. 6.49), based on computations by P. J. Hay, R. T. Pack, R. B. Walker, and E. J. Heller, *J. Phys. Chem.* **86**, 862 (1982).]

the reverse of the description of the broadening of a downward electronic transition, i.e., emission. If the laser frequency matches the energy separation between the two potential energy curves at *some* relative

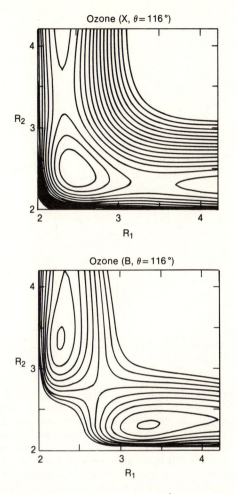

Fig. 6.50b. Comparison of potential surfaces for the $O_3(X^1A_1)$ ground state and ($^1B_2$) excited state at fixed bond angle. Separations in bohr. Semiempirical calculations adjusting *ab initio* computations to fit known energies and structures. Note the change from the deep "stable" well in the X state to the repulsive, bifurcating surface of the B state.

separation, then as the atoms reach that separation they can undergo an upward transition and the system can exit on the higher potential energy curve.

Transitions will be most probable when the laser frequency matches the potential-energy gap. Another way of stating this is to consider "dressed" potential energy curves, that is, curves shifted up in energy by $\Delta \equiv h\nu$, where ν is the laser frequency (Fig. 6.51).† At the aforementioned

† At very high fields the state can be dressed by more than one photon, so that integer multiples of $h\nu$ are also possible.

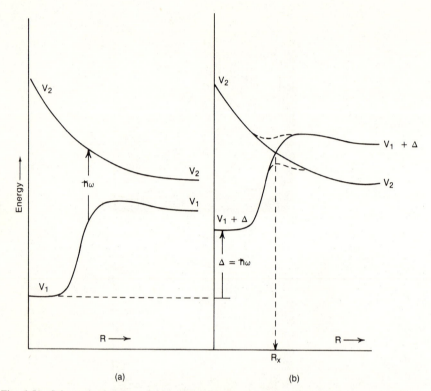

(a)   (b)

Fig. 6.51. Schematic drawing of two potential energy surface "profiles" along the reaction coordinate. (a) Field-free adiabatic potentials, both of the same symmetry, the lower one designated by V_1; upper, V_2. In the presence of an intense laser field at a frequency v (corresponding to an energy $\Delta = hv$, shown by the vertical arrow), the system may effect a transition from one adiabat to another. (b) So-called electronic-field surface (profiles) for the system, where there has been a shift by the photon energy Δ (shown here as an elevation of V_1 relative to V_2 by the amount Δ). V_1 is then the "dressed" potential; it now crosses V_2 at R_x. The dashed curves in the crossing region indicate the new adiabats, "avoided" crossing. See text for details. [Adapted from J.-M. Yuan and T. F. George, *J. Chem. Phys., 70*, 990 (1979); see also A. M. F. Lau, *Phys. Rev., A13*, 139 (1976), for early work on the subject.]

separation, the dressed lower curve *crosses* the undressed upper curve. Transitions preferentially occur at such curve crossings (see also Sec. 6.6) since at the transition there is no change in the relative kinetic energy.

A simple example of such a laser-induced process is as follows (see Fig. 6.52). First the $5s5p$ state of Sr is prepared by excitation at 460.7 nm from the ground $5s^2$ state. (This is a preliminary step.) The excited Sr* atom collides with a Ca atom in ground, $4s^2$, state *in the presence of a strong laser field* at a frequency in the neighborhood of 497 nm. This is "off" the asymptotic energy mismatch in the collision

$$Sr(5s5p) + Ca(4s^2) \rightarrow Sr(5s^2) + Ca(4p^2)$$

by 1954 cm^{-1}. Even so, emission from the Ca$(4p^2) \rightarrow$ Ca$(4s4p)$ transition is

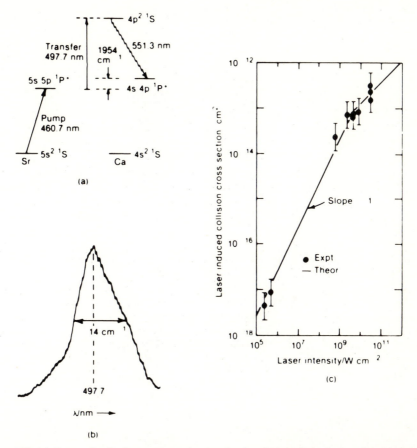

(a)

(b)

(c)

Fig. 6.52 (a) Energy level scheme appropriate to the description of the laser-assisted inelastic endoergic collision $Ca + Sr^* \rightarrow Ca^* + Sr$, $Q = -1954 \text{ cm}^{-1}$. The pump laser at 460.7 nm excites the ground state Sr to Sr*; a second laser is turned over a range of some 40 cm^{-1} centered at *ca.* 498 nm, and fluorescence at 551.3 nm from the Ca** is observed. (b) Intensity of Ca** fluorescence (at 551.3 nm) versus wavelength of the second laser; peaking at 497.7 nm. From the energetics shown in (a) note that this corresponds to the "transient excitation" of the Ca to the uppermost $4p^2$ ^1S state (Ca**), which fluoresces to the $4s4p^1P^0$ (Ca*) state (i.e., $20{,}092 \text{ cm}^{-1} \approx 18{,}139 + 1{,}954 \text{ cm}^{-1}$). (c) Log–log plot of laser-induced inelastic collision cross section versus laser power density. Since the transfer process involves one photon, the cross section is proportional to the first power of the laser intensity (unit slope), until "saturation" sets in, *via* reagent depletion, at high laser powers. [Adapted from W. R. Green, J. Lukasik, J. R. Wilson, M. D. Wright, J. F. Young, and S. E. Harris, *Phys. Rev. Lett.,* **42**, 970 (1979).]

observed and can be used to determine the dependence of the cross section on the laser power, as shown in Fig. 6.52.

The use of lasers is not limited to "bridging an energy gap"; they can also be used to cross barriers to reactive collisions. Figure 6.53 shows the potential energy profile in a collinear H_3 configuration for both the ground and an electronically excited (bound) state. Absorption to the upper state

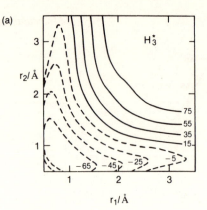

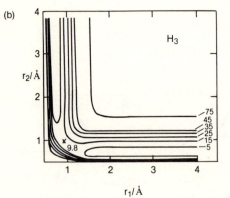

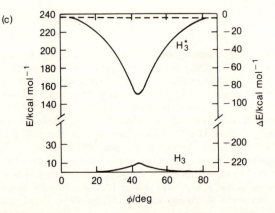

Fig. 6.53. (a) Semiempirically calculated potential energy surface for collinear $H_3^*(^2\Pi)$. (Note, however, that many H_3^* states are strongly bound, but with equilateral triangular geometry.) Energy contours in kcal mol^{-1}. (b) *Ab initio* ground state of H_3, i.e., the SLTH function. Contours in kcal mol^{-1}. (c) Profiles along the minimum-energy path of the SLTH surface; $\phi = 45°$ denotes the position of the transition state. [Adapted from H. R. Mayne, R. A. Piorier, and J. C. Polanyi, *J. Chem. Phys.*, *80*, 4025 (1984); for experiments on H_3^*, see G. Herzberg, *J. Chem. Phys.*, *70*, 4806 (1979); H. Figger, M. N. Dixit, R. Maier, W. Shrepp, and H. Walther, *Phys. Rev. Lett.*, *52*, 906 (1984); J. F. Garvey and A. Kuppermann, *Chem. Phys. Lett.*, *107*, 491 (1984).]

followed by emission at the other side of the barrier will carry the system across.

As a concrete example, consider the (bound) excited state of H_3^* that dissociates to $H(2p) + H_2$. Then we are really discussing the broadening of the Lyman-α (L_α) transition of a H atom in collision with H_2. Figure 6.54 shows the computed absorption wing at several collision energies, all below the barrier height. This is a free $\rightarrow$ bound transition and hence should show structure due to the bound vibrational levels of H_3^*. However, these are so numerous that the spectrum is not expected to show much more structure beyond that seen in the classical computation given in Fig. 6.54. The features that do survive in the figure represent those frequencies that are possible for a large number of distinct trajectories. They are associated with the turning points of the trajectories in the vicinity of the barrier.

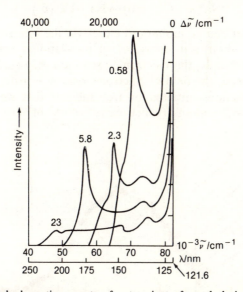

Fig. 6.54. Computed absorption spectra for transients formed during fly-by collisions of $H + H_2$ at indicated values of E_T (in kcal mol^{-1}). The ordinate is proportional to the time the trajectory spent at the geometry for which the difference potential (i.e., $V^* - V$) is $h\nu$. The top abscissa scale is the displacement $\Delta\tilde{\nu}$ (in cm^{-1}) relative to the position of the Lyman-α position (L_α, on the right ordinate scale). The lower abscissae are absolute values of ν and λ (in Å). [Adapted from Mayne et al., 1984; see Fig. 6.53.]

In bimolecular spectroscopy the laser is used not to prepare the reagents nor to probe the products but to influence the actual dynamics of the collision. The fields of absorption, emission and Raman spectroscopy of colliding–dissociating molecules are likely to grow in importance. In addition, the new research results have also forced us to recognize that there *are* excited electronic potential energy surfaces up there. Let us now switch off the lasers and examine *electronic* energy transfer in more detail.

6.6 Electronic energy transfer

6.6.1 Collisions of electronically excited species

We have thus far confined our attention mainly to V, R, and T transfer among molecules in their *ground electronic* states. There is, however, considerable interest in the collision dynamics of *electronically excited species*, not only due to their practical importance but also because such processes raise novel theoretical points. Since the electronic excitation energy is usually quite large (say, $\geq 1\,\text{eV}$), the adiabatic criterion would seem to rule out the efficient conversion of electronic to vibrational (or to translational) energy upon collision. Yet, such processes *do* occur, often with reasonable efficiency. Thus, an electronically excited sodium atom will induce $E-V$ transfer upon collision with diatomic molecule, for example,

$$\text{Na}^* + \text{CO}(v = 0) \rightarrow \text{Na} + \text{CO}(v')$$

The energy transfer can be monitored either via the infrared emission from the excited $\text{CO}^\dagger$ vibrational states or via translational spectroscopy (Fig. 6.55). For weaker bonds, the electronic excitation may suffice to dissociate the molecule. Indeed, the field of electronic energy transfer began with the observation of reactions initiated by free radicals that were produced by quenching of $\text{Hg}(6^3P)$ atoms (which carry $\approx 5\,\text{eV}$ of electronic excitation energy).

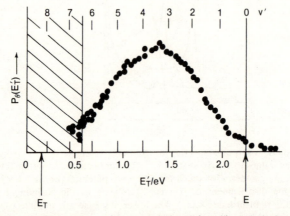

Fig. 6.55. Energy-transfer spectrum, from translational (recoil) analysis for Na* scattered by CO at 300 K at a laboratory angle of 10°; $E_T = 0.15\,\text{eV}$. Ordinate defined as $P_\theta(E_T') \equiv E_T' d^2\sigma/d\omega\,dE_T'$; lower, abscissa is E_T'. The maximum available c.m. energy for recoil $(E_T')_{\text{max}}$ is the total energy, $E = E_T + \Delta E_{\text{elec}}$ where ΔE_{elec} is the electronic excitation energy of the Na(3^2P) state (2.10 eV). In the shaded zone the inelastic data were obscured by overlapping, unresolved elastic scattering. The upper abscissa scale indicates the v' of the CO corresponding to the given E_T' (for $E_{\text{rot}}' = 0$). The distribution peaks in the neighbourhood of $v' = 3$–4. [Adapted from I. V. Hertel, in J. W. McGowan (1981).] Recent experiments with much improved angular and energy resolution (showing structure) have been published by W. Reiland, H. U. Tittes, I. V. Hertel, V. Bonacic-Koutecky, and M. Persico, *J. Chem. Phys.*, **77**, 1908 (1982).

The $E-V$ quenching process of Na* is but one example of a collision that does not take place on a single potential energy surface. We have already mentioned a variety of such processes, for example, collisional ionization in atom–molecule and molecule–molecule collisions, the harpoon mechanism (Sec. 4.2.5), and the formation of electronically excited products in highly exoergic reactions of ground-state reagents (Sec. 4.1.1).

Highly *endoergic* chemical reactions often proceed quite readily when the reagent is *electronically excited*. The reactions of excited oxygen or nitrogen atoms and molecules are particularly interesting because of their contribution to the chemistry of the atmosphere (aeronomy). Thus, the dominant contribution to the night glow in the infrared is the *chemiluminescent reaction*

$$H + O_3 \rightarrow O_2 + OH(v')$$

leading to vibrationally excited OH radicals. The source of H atoms for initiating the reaction[†] is believed to be the sequence

$$O_2 \xrightarrow{h\nu} O(^3P) + O(^1D)$$

$$O^* + H_2 \rightarrow OH + H$$

The (third-body-assisted) recombination of oxygen atoms leads to the formation of $O_2(A^3\Sigma_u^+)$, whose emission to the ground $O_2(X^3\Sigma_g^-)$ state is the dominant feature of the ultraviolet nightglow spectrum.

Electronically excited species can be formed in various ways, with the most common being absorption of light. Once formed, they can be detected by their fluorescence, which also serves to determine their concentration. For a so-called allowed transition, the lifetime for the unimolecular radiative decay of an electronically excited state is of the order of 10^{-8} s. On the other hand, for a forbidden transition, the lifetime can be much longer.[‡] Such excited states are often referred to as metastables. The radiative lifetime of the $a^1\Delta_g$ metastable state of O_2 is some 45 min! With their long radiative lifetimes, metastables are useful, energy-rich chemical reagents (e.g., Sec. 4.1.2).

6.6.2 Quenching processes

An excited state can decay either through a unimolecular fluorescence

$$Na^* \rightarrow Na + h\nu$$

[†] Once initiated, the reaction proceeds in a chain

$$OH(v') \rightarrow OH(v' - 1) + h\nu$$

$$OH + O \rightarrow O_2 + H$$

[‡] Strictly speaking, if the transition is forbidden, the excited state should be stable. The term "forbidden transition," however, is used to designate transitions with a low but finite probability.

or through a bimolecular transfer process, say, the $E-V$ process

$$\text{Na}^* + \text{H}_2(v = 0) \rightarrow \text{Na} + \text{H}_2(v')$$

Hence, the kinetic equation for the concentration of Na* atoms in H_2 is of the form

$$\frac{d[\text{Na}^*]}{dt} = -[\text{Na}^*](\tau_r^{-1} + k[\text{H}_2]) \tag{6.59}$$

Here, τ_r is the radiative lifetime and k is the bimolecular rate constant for the transfer process. The *effective lifetime* τ for the decay of the concentration of Na* atoms given by

$$\tau^{-1} = -[\text{Na}^*]^{-1}\frac{d[\text{Na}^*]}{dt} = \tau_r^{-1} + k[\text{H}_2] \tag{6.60}$$

will be shorter than the radiative lifetime τ_r. The intensity of the fluorescence, which is proportional to the concentration of Na* atoms, will decrease with time at a rate that increases with the pressure of H_2 (see Fig. 6.56). Bimolecular transfer processes that deplete the concentration of the excited species are referred to as *collisional quenching* of the fluorescence.

As in other examples (see Sec. 6.5.4), here too one can monitor the far wing emission from Na* as a probe of the upper and lower Na–H_2 potential energy surfaces.

Quenching processes can be divided into physical or chemical types. Physical quenching involves no chemical rearrangement and is a particular subclass of inelastic collisions. For atom–atom collisions, such processes can

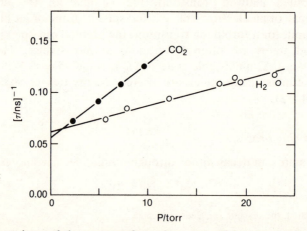

Fig. 6.56. Dependence of the apparent first-order rate coefficient for the disappearance of Na(3^2P) atoms upon the pressure of the quenching gas. Data are shown for both H_2 and CO_2. In principle, the intercepts should be the same, i.e., τ_r^{-1} for Na*. In this experiment the Na* was produced by the photodissociation of NaI using a laser at 225 nm, yielding translationally hot Na* atoms. [Adapted from J. R. Barker and R. W. Weston, Jr., *J. Chem. Phys.*, **65**, 1427 (1976).]

include

$$K^* + K \rightarrow K + K^*, \qquad \textit{symmetrical excitation transfer}$$
$$Hg^* + Tl \rightarrow Hg + Tl^*, \qquad \textit{asymmetric excitation transfer}$$
$$He^* + Ar \rightarrow He + Ar^+ + e^-, \quad \textit{Penning ionization}$$
$$Hg^* + Xe \rightarrow Hg^{*\prime} + Xe, \qquad \textit{E–T transfer}$$

The asymmetric excitation transfer was already discussed as a candidate for a laser-assisted process. Indeed, we shall shortly argue that, in general, transitions between two different potential energy curves (or surfaces) occur effectively only when these cross. If the bare curves (or surfaces) do not cross, we can achieve a crossing by "dressing" them by laser photons (Fig. 6.51). Quenching processes (and the inverses to them) can thus be *laser assisted*.

Chemical quenching can also occur in atom–atom collisions, for example,

$$Cs^* + Cs \rightarrow Cs_2^+ + e^-, \quad \textit{Associative ionization}$$

but it is more common in atom–molecule or molecule–molecule collisions. Examples include

$$Ar^* + NO \rightarrow ArNO^+ + e^-, \qquad \textit{Associative ionization}$$
$$Ne^* + H_2 \rightarrow NeH^+ + H + e^-, \quad \textit{Dissociative ionization}$$
$$Hg^* + H_2 \rightarrow Hg + H + H, \qquad \textit{Dissociative deexcitation}$$
$$Ar^* + OCS \rightarrow Ar + O + CS^\dagger, \quad \textit{Dissociative vibrational excitation}$$
$$C_3H_8 + I^* \rightarrow HI + C_3H_7, \qquad \textit{Atom abstraction}$$

The last three reactions illustrate the possible synthetic versatility of excited metastable atoms. A particularly interesting chemical quenching process is

$$Na^* + I_2 \rightarrow NaI + I$$

According to the harpoon model (Sec. 4.2.5), the cross section for the $Na + I_2$ reaction is determined primarily by the distance at which charge transfer can occur. Since the ionization potential of Na^* is lower than that of the ground-state atom, the crossing distance R_x [Eq. (4.5)] will be larger and the reaction (or quenching) cross section for the process should exceed the reaction cross section of $Na + I_2$, as is indeed the case.

6.6.3 The He–Ne *laser*

A physical quenching process with technological applications is the collisional quenching of the metastable $He(2^3S)$ atoms by ground-state Ne

$$He(2^3S) + Ne(1^1S_0) \rightarrow He(1^1S_0) + Ne(2s), \qquad Q \leq 0.15 \, \text{eV} \quad (6.61)$$

(The symbol 2s denotes the $2p^5 4s$ configuration.) The excited Ne atoms can radiate to one of the lower 2p levels (2p designating the $2p^5 3s$

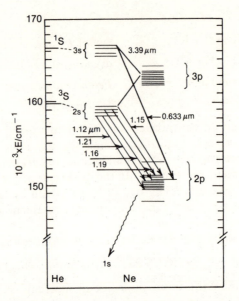

Fig. 6.57. He–Ne energy levels relevant to the He–Ne laser. The symbol 2s designates the Ne $2p^54s$ configuration; 2p, the $2p^53p$; and 1s is the $2p^53s$ configuration. An efficient energy transfer occurs between the He ^3S state at 159,850 cm^{-1} and the 2s group of Ne levels in the range 158,600 to 159,540 cm^{-1}, which can emit (and lase) at the indicated wavelengths from 1.118 to 1.207 nm to the 2p group. He ^1S can transfer its excitation to the 3s group of Ne, which will then lase at 632.8 nm to the $2p_4$ level, as shown by the heavy arrow. The commercial He–Ne "red" laser operates at this wavelength. (Adapted from Fig. 9.16 of B. A. Lengyel, *Lasers*, 2nd ed., Wiley, New York, 1971.)

configuration), as shown schematically in Fig. 6.57:

$$Ne(2s) \rightarrow Ne(2p) + h\nu$$

Here, the 0.7-eV energy difference is emitted as light; the radiative lifetime is 10^{-7} s. The transfer process (6.61) is surprisingly efficient† and can maintain an *inverted electronic population distribution* of Ne atoms and hence serves as a basis for the He–Ne laser. The excited He atoms can be formed in a discharge by electron impact,

$$He(1^1S_0) + e^- \rightarrow He(2^3S) + e^-$$

and the laser emission corresponds to the $2s \rightarrow 2p$ radiative decay. Ne atoms

---

† The translational release of the transfer process (0.15 eV, or 1210 cm^{-1}) is sufficiently large that a naive application of the adiabatic criterion,

$$v = \frac{a\,\Delta E}{h} \approx 3.6 \times 10^3 \,\text{m s}^{-1}$$

would suggest that at ordinary thermal velocities one should expect a more nearly adiabatic collision. It must therefore be that the *effective* change in the kinetic energy is somewhat lower than its nominal value based upon the exoergicity (see Sec. 6.6.6).

in the $2p$ levels *decay radiatively* to the $1s$ metastable state with a lifetime of 10^{-8} s, *much faster* than the $2s \rightarrow 2p$ radiative decay. The necessary conditions for *population inversion* are thereby satisfied.

6.6.4 Radiationless transitions

If we excite a large, polyatomic molecule, say, benzene, the electronically excited state can lose its excitation by emission, collisional transfer, and also by an intramolecular process. The electronic energy of the large molecule is redistributed among the internal vibrational levels. The vibrational modes that accept the electronic energy belong to a lower (or to the ground) electronic state (Fig. 6.58).

Intramolecular vibrations play the same role as the intermolecular translation in a physical quenching collision between simple systems. For *intra*molecular quenching the adiabaticity parameter has therefore to be

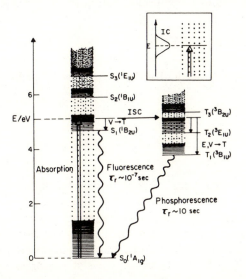

Fig. 6.58. Energy-level diagram showing intermolecular and intramolecular energy-transfer processes in benzene photoexcited to a particular vibrational level of the first excited singlet (S_1). The first few vibrational levels are shown for all the electronically excited states. For the ground state the vibrational level density at energies >1 eV is so great that they are shown as a continuum. The following processes are indicated: (1) Internal conversion (IC): an intramolecular $E \rightarrow V$ transfer from S_1 to S_0. The insert shows the "spreading" of the excitation of the first excited electronic singlet state S_1 among the (near continuum of) vibrational levels of the ground singlet state. (2) Unimolecular fluorescence of the excited state. (3) Bimolecular, $V \rightarrow V'$, T transfer from the excited benzene molecule. (4) Intersystem crossing (ISC): an intramolecular $E \rightarrow V$ transfer accompanied by a change in the spin multiplicity (singlet $\rightarrow$ triplet transfer). All four processes are also possible for the triplet state formed *via* ISC. Moreover, the highly vibrationally excited S_0 state can lose energy by $V \rightarrow V'$, T transfer. The possibilities multiply when unimolecular rearrangements or bimolecular reactions are also feasible.

written as t_v/t, where t_v is the period of the vibration that "accepts" the electronic energy. If ΔE is the amount of electronic energy converted to vibration, we can make the identification $t = h/\Delta E$, so that

$$\xi = \Delta E/h\nu \tag{6.62}$$

where $t_v = \nu^{-1}$. The most efficient vibrations for intramolecular quenching are those with the highest vibration frequencies. In a large organic molecule, these are usually the C—H vibrations. (Indeed, the efficiency of intramolecular quenching is much reduced by replacing H atoms with D atoms.)

Experimentally, radiationless transitions are manifested by the failure of the ordinary physical quenching mechanism summarized by Eq. (6.59). Even at very low pressures of a buffer gas, not all the absorbed photons are reemitted; the unimolecular decay rate of the electronically excited state is higher than that for fluorescence. The increase in the rate is due to *intra*molecular quenching.† The rate equation (6.59) is replaced by

$$X^* \xrightarrow{\tau_r} X + h\nu \text{ radiative decay}$$

$$X^* \xrightarrow{\tau_{nr}} X^\dagger \text{ nonradiative decay or intramolecular quenching}$$

$$X^* + M \xrightarrow{k} X + M \text{ collisional quenching}$$

$$\frac{d[X^*]}{dt} = -[X^*](\tau_r^{-1} + \tau_{nr}^{-1} + k[M]) \tag{6.63}$$

The rate constant for nonradiative, intramolecular decay is very small unless the density of vibrational final states is very large (recall Sec. 4.4.10). Hence intramolecular quenching is manifested experimentally for large molecules. A direct measure of its efficiency is the quantum yield or the fraction of electronically excited molecules that decay by photon emission. From (6.63) we obtain for the quantum yield at low pressures ($[M] \to 0$)

$$\Phi = \tau_r^{-1}/(\tau_r^{-1} + \tau_{nr}^{-1}) \tag{6.64}$$

For small molecules for which $\tau_{nr}^{-1} \to 0$, $\Phi \simeq 1$.

An example of the broadening of an electronic transition due to coupling between the excited and ground manifolds is shown in Fig. 6.59 for jet-cooled azulene. The cooling reduces the inhomogeneous broadening due to absorption of molecules that are not in the ground rovibrational state, so that the observed width yields the lifetime for electronic relaxation from the S_1 manifold (here ≤ 1 ps).

---

† Note that intramolecular vibrational energy redistribution (Fig. 6.5) does eventually result in fluorescence, except that it is red shifted.

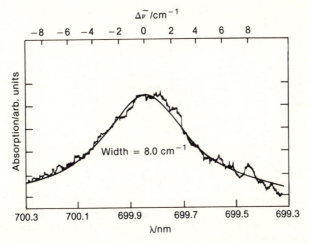

Fig. 6.59. Absorption spectrum ($S_0 \to S_1$ electronic band origin) of azulene, jet-cooled to a few degrees Kelvin by expansion with an Ar carrier. The solid curve is a Lorentzian fit to the line shape; the width parameter, Δ, exceeds the spectral resolution by a factor of 5. After corrections for resolution and a small degree of rotational broadening, the Lorentzian width, Γ, is found to be $9.6\,\text{cm}^{-1}$. Thus, the lifetime of the S_1 (vibrationless) state with respect to interstate electronic relaxation is calculated to be $\tau = (2\pi c \Gamma)^{-1} = 0.55\,\text{ps}$. [Adapted from A. Amirav and J. Jortner, *J. Chem. Phys.* **81**, 4200 (1984).]

6.6.5 Curve crossing

We have thus far encountered several electronic energy transfer collisions that are more efficient than their high nominal translational exoergicity would suggest on the basis of the adiabaticity criterion. An additional example is the process of *charge neutralization*, for example,

$$K^+ + I^- \to K + I$$

Figure 6.60 shows the interaction potentials for the KI system. The nominal energy release in the neutralization process is

$$\Delta E = \text{I.P.}(K) - \text{E.A.}(I)$$
$$= 4.34 - 3.06 = 1.28\,\text{eV} \qquad (6.65)$$

It is evident, however, from the figure that $|\Delta E|$ is the energy released when the atoms are far apart. At closer separation there exists a distance R_x, where the electronic energy of K^+I^- equals that of KI. Near the separation R_x the system can change its electronic state with only a very slight change in the kinetic energy of the nuclei.† Such a "switch" corresponds to a change from the electronic state that dissociates to $K^+ + I^-$ to the electronic

---

† The kinetic energy of the relative motion of the nuclei at the separation R is $E - V_{\text{eff}}(R)$. Near R_x, $V_{\text{eff}}(R)$ is the same for both electronic states.

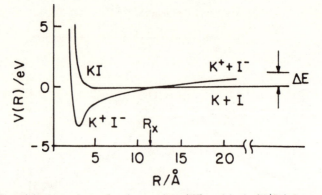

Fig. 6.60. Potential-energy curves for the covalent (KI) and ionic (K^+I^-) forms of potassium iodide. ΔE is the nominal (i.e., asymptotic) energy gap between the two states, Eq. (6.65). Thus, from $\Delta E = e^2/R_x$, $R_x \simeq 11.3$ Å. It is known that for this system the electronic energy gap (Sec. 6.6.6) is $\Delta E_x \simeq 2.5 \times 10^{-3}$ eV.

state that dissociates to $K + I$. Such a switch in the electronic state, accompanied by only a small change in the kinetic energy of the nuclei, is precisely what is needed to make the charge neutralization process efficient!

For the $Na^* + CO$ collision, the $E \to V$ transfer probably occurs through two such changes. The first switch is a "harpoon" type (Sec. 4.2.5) change, $Na^* + CO \to Na^+ + CO^-$, followed† by a "neutralization" switch $Na^+ + CO^- \to Na + CO$. If this is the case, the quenching cross sections for $Na^* + M$ collisions should depend on the nature of the colliding partner M. For such diatomics as NO, CO, or N_2, the effective cross section can be up to 50 Å^2, while it is less than 1 Å^2 for quenching by the rare gases. This is as expected on the basis of a harpoon mechanism, where R_x [see Eq. (4.5)] would increase with increasing electron affinity of M.

The convergence of two potential-energy curves so as to allow an efficient switch of the nuclei between them is known as curve crossing.‡

6.6.6 The adiabaticity parameter

When a switch between two potential curves is possible there is always a finite energy gap, $\Delta E(R_x)$, between them.§ The adiabaticity parameter is

---

† Of course, both of these changes occur during a single $Na^* + CO$ collision; they occur with high efficiency only when the Na—CO separation is such that the corresponding electronic states have roughly comparable electronic energies.

‡ Strictly speaking, this is a confusing terminology. The curves do not cross, they only converge, and Fig. 6.60 is only an approximation to the true state of affairs. It is, however, a very useful approximation.

A similar mechanism can operate also in polyatomic systems, where we need to consider "surface" intersection rather than curve crossing. Collisions where considerable electronic energy is released probably proceed through a surface intersection mechanism.

§ A simple way to regard this energy gap is from the uncertainty principle. The electronic rearrangment takes time, and $h/\Delta E(R_x)$ is a measure of how long it takes.

thus defined in terms of the time t required to traverse the crossing region,

$$\xi = t \, \Delta E(R_x)/h \tag{6.66}$$

where $\Delta E(R_x)$ is the change in the kinetic energy of the nuclei at the "crossing" point. This value will generally be much smaller than the asymptotic (or nominal) value of ΔE. Electronic energy transfer can thus be quite efficient, even for a large nominal ΔE, if curve crossing can take place, leading to a smaller "effective" value of ΔE, namely, $\Delta E(R_x)$. An adiabatic passage will be one for which $\xi > 1$. Note, however, that in the adiabatic limit the electrons *do* have enough time to readjust. In other words, there is a change in the character of the state if $\xi > 1$. This is further discussed following Eq. (6.69) below.

To estimate ξ, we examine the example of a $K + I$ collision where the switch is from a covalent KI potential for $R > R_x$ to an ionic K^+I^- form for $R < R_x$ (Fig. 6.60). It is evident from the figure that the region where the two potential curves converge and where ξ [Eq. (6.66)] is a minimum is a very localized region in R, say, a, about R_x. Efficient transitions between the two curves will occur only during the time spent by the system in the vicinity of R_x. The duration t in Eq. (6.66) is thus the time spent in this localized region of R, namely, $t = a/v$. v is the radial velocity at R_x and hence (Sec. 2.2) is b dependent.

An estimate of a is provided by regarding the two potential curves as linear functions of R, in the vicinity of R_x. The two curves diverge because of their different slopes. Hence, if we expand each potential in a Taylor series about R_x, we can estimate a from

$$\Delta E_x \equiv \Delta E(R_x) = a \left| \frac{dV_1}{dR} - \frac{dV_2}{dR} \right|_{R_x} \equiv a \, |\Delta F_x| \tag{6.67}$$

Here, a is the range about R_x over which the gap doubles and $|\Delta F_x|$ is the absolute value of the difference in slopes at R_x. Combining (6.67) and (6.66), we obtain

$$\xi = \Delta E_x^2/hv \, |\Delta F_x| \tag{6.68}$$

6.6.7 The Landau–Zener transition probability

We require an approximation for the probability of an overall change in electronic state induced by the collision. Given this probability (as a function of the impact parameter), we could obtain the cross section for the collisional ionization process, for example. Of course, our discussion assumes that the total energy exceeds the endoergicity ΔE [Eq. (6.65)] of the collision.

The classical path method (Sec. 5.6.3) yields the approximation

$$P = \exp(-\pi^2 \xi) \tag{6.69}$$

for the probability that there is no change in electronic state during the passage of the system through the critical region about R_x. Similarly $1 - P$ is the probability of a switchover, i.e., of a curve crossing. For a very slow collision, when ξ is very large ($P \to 0$), a change in the electronic state will occur with a high probability upon passage via R_x. We have already encountered this phenomenon in the harpoon mechanism (Sec. 4.2.5), where at about R_x, the charge transfer occurs with essentially unit probability for reactants with thermal velocities.

Consider now the collision of K and I atoms. As the two atoms approach, a curve crossing can occur in the region near R_x with the probability $1 - P$. The two newly formed ions will continue to approach, governed by the ionic potential, until they reach the turning point of their motion. They will then start receding from each other. If the two particles are to separate as ions then, as they recede across R_x to larger separations, curve crossing must *not* occur. The probability of forming $K^+ + I^-$ in this way is $P(1 - P)$. Alternatively, if curve crossing did not occur on the way in (probability P), then, if ions are to be formed, it must occur on the way out. The overall probability of collisional ionization is then† $Q \equiv 2P(1 - P)$.

In the particular case of the K + I collision where the covalent potential is only weakly R-dependent and $V_2 = -e^2/R$ (at large R; Fig. 6.60),

$$|\Delta F_x| = e^2/R_x^2 \tag{6.70}$$

and

$$\xi = R_x^2 |\Delta E_x|^2/e^2 h v \tag{6.71}$$

The probability of producing ions, Q cannot exceed $\frac{1}{2}$ regardless of the magnitude of v. When v is small (large ξ, adiabatic collision), P is small and so is Q. When v is large (low ξ), $1 - P$ is small and so is Q. Thus, there will be a maximum in the energy dependence of the cross section for collisional ionization, for some v, say v_{max} (Fig. 6.61). From the position of the maximum one can estimate $\Delta E(R_x)$.

6.6.8 Angular distribution in the curve-crossing problem

As discussed above, curve crossing can occur or fail to occur, both on the way in and on the way out. Hence at a given impact parameter, there can be four distinct classical trajectories (Fig. 6.62). Two of these will exit on one potential energy surface and two on the other. The result is that a new interference pattern in the angular distribution (over and above the one due to pure elastic scattering) can be manifest.

When we consider ionization in atom–molecule collisions (e.g., Sec. 2.3.1), a new feature comes into play. One way of exhibiting it is shown in

† Similarly, if the collision does not lead to ions, then either curve crossing failed to occur both on the way in and on the way out (probability P^2) or curve crossing occurs both on the way in and on the way out [(probability $(1 - P)^2$].

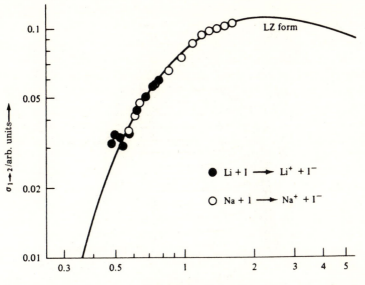

Fig. 6.61. Fit of experimental charge-transfer cross sections (for the indicated reactions) to the Landau–Zener cross-section function. The ordinate (log scale) is the cross section in units of $4\pi R_x^2$; the abscissa (log scale) is the reduced speed $\xi^{-1} = v/K$; see Eq. (6.71). Thus, the fit for a given reaction enables the value of K, and thus of $\Delta E(R_x)$, to be determined. [Adapted from A. M. Moutinho, J. A. Aten, and J. Los, *Physica*, *53*, 471 (1971); see also A. P. Baede, *Adv. Chem. Phys.*, *30*, 462 (1975).]

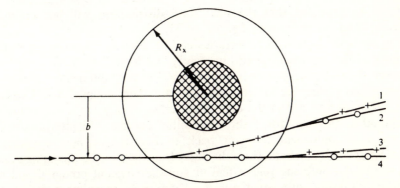

Fig. 6.62. Geometric interpretation of the four possible outcomes of a collision of a fast alkali atom with a fixed target halogen atom. The crossing point (separation), R_x, fixes the radius of the outer circle (the inner circle corresponds to the repulsive core). The open symbol denotes neutral; the plus sign indicates a trajectory in which an ion pair has been formed. If an electron jump occurs at the first crossing *and* the system remains ionized at the second, the result for the ionic trajectory corresponds to wide-angle scattering (labeled 1). For trajectory 2, a second electron jump has occurred at the second crossing, resulting in wide-angle neutral scattering. Trajectories 3 and 4 have obvious implications. [Adapted from J. Los and A. W. Kleyn, in P. Davidovits and D. L. McFadden (1979).]

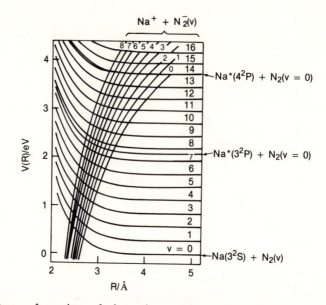

Fig. 6.63. Array of crossings of electronic potential curves appropriate to the ion-pair formation (charge-transfer) process $Na + N_2(v) \rightarrow Na^+ + N_2^-(v')$. Each curve is labeled with v or v' as appropriate. The zero of energy corresponds to the vibrationless state of $N_2 + Na(3^2S)$. [Adapted from E. Bauer, E. R. Fisher, and F. R. Gilmore, *J. Chem. Phys., 51,* 4173 (1969).]

Fig. 6.63. Each final $X_2^-(v)$ vibrational state of the diatom corresponds to a different crossing radius (and the same is true for the initial vibrational states of X_2). The result is that there is a whole maze of crossings. It is then only to be expected that the final state distribution will not be much different from the prior one (e.g., Fig. 5.33).

The method of classical trajectories can be adapted to allow for surface hopping by making each classical trajectory "bifurcate" at the crossing seam between the two potential energy surfaces. The probabilities for remaining on the same surface or for crossing over are then assigned *à la* Landau–Zener. Figure 6.64 shows the differential cross section for $K + I_2$ collisions computed in this way for a series of initial energies. In interpreting the results it is convenient to refer back to Fig. 6.62. Those trajectories undergoing a *non*adiabatic (i.e., diabatic) transition in the incoming region and not in the outgoing region feel mostly the covalent potential and will lead to scattering into small angles. Those trajectories that follow the adiabatic path in the entrance channel sample the strong "ionic" forces and give rise to the rainbow-type structure and wider scattering angles.

6.6.9 Steric effects in collisions of electronically excited reagents

Reactions using electronically excited reagents† are expected to show an additional dependence on the orientation of the reagents. The reason is the

---

† Or, by similar consideration, reactions producing electronically excited products.

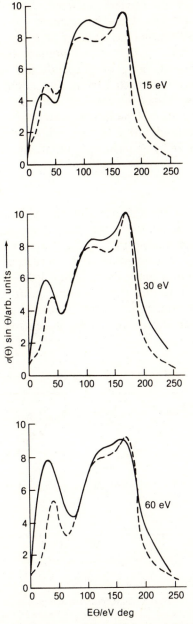

Fig. 6.64. Differential polar cross sections $\sigma(\Theta) \sin \Theta$ (in the laboratory system) for ion-pair production from the collision of fast K atoms by I_2: $K + I_2 \rightarrow K^+ + I_2^-$, at laboratory translational energies of 15, 30, and 45 eV. Θ is the laboratory scattering angle with respect to the incident K direction. Experimental curves *vs.* theoretical computations using the trajectory surface-hopping method. In the high-energy approximation, $E_T\theta$ is closely related to the impact parameter (Sec. 3.1.5) (and $E\Theta \propto E_T\theta$), so the principal effect of E_T is removed and the rainbow structure should be similar at each energy. [Adapted from C. Evers, *Chem. Phys. 21*, 355 (1977).]

spatial degeneracy of an excited electronic state, a degeneracy that is removed during a collision. Consider a beam–gas experiment with a fast Na^+ beam and $Na(3^2S)$ target gas. A laser polarized parallel or perpendicular to the relative velocity vector can prepare $Na(3^2P)$ atoms with the atomic p orbital aligned along or perpendicular to the relative velocity vector. The interaction potential with any colliding partner would be different for these two cases. Figure 6.65 summarizes the observed results, which show that an initial Π molecular state (prepared by polarization at 90°) quenches preferentially to a $3s$ state, while an initial σ molecular state (polarization at 0° or 180°) quenches preferentially to a $3d$ state.

6.6.10 Back to chemistry

We are now ready to resume our concern with *reaction* dynamics. Our intermezzo on inelastic collisions was not, however, in vain. We need to understand the process of energy transfer if we are to make full use of the selectivity of energy consumption and specificity of energy release of chemical reactions. On a more fundamental level, we must recognize that inelastic collisions bridge the gap between elastic and reactive collisions. The internal deformation during vibrational excitation naturally goes over into the atomic rearrangement during a reactive collision. Rotational

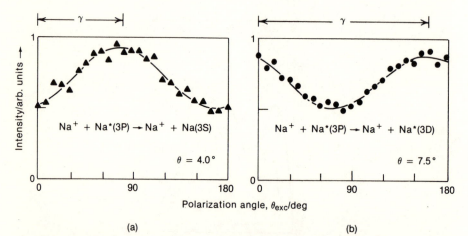

Fig. 6.65. Experimental data on the polarization angle dependence of the cross sections for the inelastic (quenching) processes (a) $Na^+ + Na(3P) \rightarrow Na^+ + Na(3S)$ and (b) $Na^+ + Na(3P) \rightarrow Na^+ + Na(3D)$ at $E_T = 45\ eV$. Center-of-mass scattering angles θ as indicated. The polarization angle θ_{exc} refers to the exciting laser used to prepare the $Na(3P)$; θ_{exc} of zero implies laser polarization parallel to z axis (initial relative velocity vector), $\pi/2$ perpendicular. The optimal alignment angle γ shown is close to 90° for process (a), which is presumed to go via a σ state for the transient molecular ion, and near 0° (180°) for process (b), $Na(3D)$ formation via a Π state. [Adapted from I. V. Hertel, H. Schmidt, A. Bähring, and E. Meyer, *Rep. Prog. Phys.*, **48**, 375 (1985).]

excitation is the simplest manifestation of a steric effect. Ultimately, our discussion of electronic energy transfer allows us to understand why at low collision energies chemical reactions appear to proceed on a "single" potential energy surface and suggests that this will no longer be the case for fast collisions of the same reagents.

In the next chapter we shall probe deeper in our attempt to achieve a detailed understanding of the *dynamics of chemical reaction*.

Appendix 6A: Stimulated emission, laser action, and molecular lasers

In Section 6.1 we discussed the return to equilibrium by energy-transfer collisions. Here we consider an additional mechanism, namely, photon emission. As we shall see, this mechanism becomes more important at shorter wavelengths and/or in the presence of photons that can stimulate the transition. The "detailed" in detailed balance means that each mechanism can be considered on its own. In a real experiment, both mechanisms are operative and compete with one another. The rate equations [of which (6.2) is a very special case] need then include both collisional and radiative population transfer. Here we consider only the latter.

Having displaced the equilibrium so that one (or more) excited state(s) is preferentially populated, we wish to determine the rate of depopulation by photon emission.

The radiative transition $f \rightarrow i$ can take place in two ways. One is a spontaneous (or "unimolecular") photon emission with a rate $A_{fi}n_f$. Here, A_{fi} is a "first-order" rate constant such that the lifetime τ of the state f, due to spontaneous emission, is $1/A_{fi}$. The spontaneously emitted photons are emitted into all spatial directions. There is also a "second order" mechanism for photon emission whose rate is proportional to both the density† ρ of the radiation present and to the concentration (i.e., number density) n_f of the state f. The rate of this so-called "stimulated emission" is then $B_{fi}\rho n_f$. Here, B_{fi} is a second-order rate constant. The combined rate of emission is $k_{fi}n_f$, with

$$k_{fi} = A_{fi} + B_{fi}\rho \qquad (6A.1)$$

Of course, when photons (at the frequency ν) are present they will be partially absorbed, due to $i \rightarrow f$ transitions, at the rate $k_{if}n_i$, where

$$k_{if} = B_{if}\rho \qquad (6A.2)$$

Absorption, like stimulated emission, is a second-order process whose rate is proportional to the density of the photons. The two rate coefficients are known as the Einstein A and B coefficients.

---

† $\rho = \rho(\nu)$ is the radiation energy per unit volume and unit frequency. Thus $\rho(\nu)\,d\nu$ is the energy per unit volume of the photons in the frequency interval ν to $\nu + d\nu$.

When radiation is incident on the system, those photons emitted through stimulated emission have the same phase as the external radiation. These photons therefore reinforce the stimulating beam; the radiation emitted via stimulation is coherent with the beam. This is in contrast to the spontaneously emitted photons, which are not coherent.

In order to achieve amplification of an external light beam it is thus not sufficient to achieve a net excess of emission over absorption. The mere introduction of some external pumping mechanism that raises the concentration n_f of state f to such a level that

$$k_{fi}n_f > k_{if}n_i \tag{6A.3}$$

is not enough. Condition (6A.3) is the requirement of net emission of photons. To achieve *amplification*, the rate of *stimulated* emission must exceed the rate of absorption, i.e., $\rho B_{fi}n_f > k_{if}n_i$, or

$$B_{fi}n_f > B_{if}n_i \tag{6A.4}$$

The amplification condition (6A.4) is more stringent than the luminescence condition (6A.3), which can be written as

$$[B_{fi} + (A_{fi}/\rho)]n_f > B_{if}n_i \tag{6A.5}$$

In order to obtain a quantitative comparison and to express (6A.4) as an explicit condition on the populations, we need to determine the ratios $B_{fi}\rho/A_{fi}$ and B_{fi}/B_{if}. We shall do so using the principle of *detailed balance*.

In a real system there can be many mechanisms for the transfer of molecules from state i to state f and vice versa. The absorption and emission of photons of frequency v is just one of those possible mechanisms. At equilibrium the overall rate of transfer from i to f is necessarily equal to the reverse rate. The principle of detailed balance states that at equilibrium not only are the overall forward and reverse rates equal but the same applies to each particular mechanism. At equilibrium, the rate of $i \rightarrow f$ transitions by photon absorption equals the rate of $f \rightarrow i$ transitions by photon emission, irrespective of any other processes that can take place.

Consider now our system *at equilibrium with radiation* of density ρ at the temperature T. Then, by detailed balance, $k_{fi}n_f = k_{if}n_i$ or

$$(k_{if}/k_{fi}) = (n_f/n_i)_{eq} = \exp(-hv/kT) \tag{6A.6}$$

Substituting into (6A.2) and (6A.1)

$$B_{if}\rho(v) = \exp(-hv/kT)[A_{fi} + B_{fi}\rho(v)] \tag{6A.7}$$

Solving for the equilibrium density $\rho(v)$ of the radiation, we obtain

$$\rho(v) = A_{fi}/[B_{if}\exp(hv/kT) - B_{fi}] \tag{6A.8}$$

The equilibrium density of radiation at the temperature T is given by Planck's equation,

$$\rho(v) = (8\pi hv^3/c^3)[\exp(hv/kT) - 1]^{-1} \tag{6A.9}$$

Comparing (6A.8) and (6A.9), we see that

$$B_{if} = B_{fi} = (c^3/8\pi h\nu^3)A_{fi} \tag{6A.10}$$

These results relate the three rates. They show that under equilibrium conditions stimulated emission is of negligible importance compared with spontaneous emission,

$$[B_{fi}\rho(\nu)/A_{fi}]_{eq} = c^3\rho(\nu)/8\pi h\nu^3 = [\exp(h\nu/\mathbf{k}T) - 1]^{-1} \tag{6A.11}$$

Only for very low frequencies, $h\nu \sim \mathbf{k}T$, do the two emission processes have comparable rates. At even lower frequencies, $h\nu \ll \mathbf{k}T$, the ratio can be approximated by $\mathbf{k}T/h\nu$. Under ordinary conditions, the lifetime for photon emission will be essentially that of spontaneous emission, $\tau = (1/A_{fi}) \propto \nu^3$. The lifetime for spontaneous emission will be much shorter (typically 10^{-8} s) for transitions in the visible region (i.e., changes in the electronic state) compared with transitions in the infrared (as slow as 10^{-1} s for the far infrared), which are typically associated with vibrotational and pure rotational changes. Quantum-mechanical analysis provides an explicit formula† for A_{fi} in terms of μ_{fi}, the "transition dipole moment" for the $i \to f$ transition:‡

$$A_{fi} = (64\pi^4/3hc^3)\,|\mathbf{\mu}_{fi}|^2\,\nu^3 \tag{6A.12}$$

Moreover, since $B_{fi} = B_{if}$, we can never hope to achieve amplification [i.e., Eq. (6A.4)] merely by using an external light beam with a high (nonequilibrium) photon density at the frequency ν. No matter how high $\rho(\nu)$ is, the most that we can achieve is $n_f = n_i$ [when $B_{fi} \gg (A_{fi}/\rho)$; cf. Eq. (6A.5)].

The condition for amplification (6A.4), combined with the relation between the Einstein coefficients (6A.10), can be now expressed as the condition

$$n_f > n_i \tag{6A.13}$$

on the populations of the two states. Amplification requires that the

---

† In practical units, $A_{fi}/s^{-1} \approx 10^{-38}\nu^3\mu^2$, with ν in s^{-1} and μ_{fi} in Debyes [1 Debye (D) $= 10^{-18}$ esu cm]. For $\mu_{fi} \approx 0.1$ D and $\nu \approx 10^{14}$ s^{-1} (corresponding approximately to the $\nu = 1 \to 0$ transition in HF), $\tau = A^{-1} \approx 10^{-2}$ s. Note the difference between the transition dipole $\mu_{if}^{(f\neq i)}$ and the permanent dipole moment μ_{ii} (for a molecule in state i); e.g. the vibrational transition dipole for the HF molecule is $\mu_{10} \approx 9.85 \times 10^{-2}$ D, whereas $\mu_{ii} \approx 1.82$ D.

‡ The magnitude of the transition dipole moment determines the rate of spontaneous emission, stimulated emission, and absorption [see (6A.12) and (6A.11)]. Equivalently, $\mu_{if} = |\mathbf{\mu}_{if}|$ measures the response of the molecular system to the radiation field. Transitions with smaller μ_{if} (e.g., the "overtone", $\nu = 0 \to 2$, transition of HF, $\mu_{02} \approx 10^{-2}$ D) are termed (nearly) forbidden, while those with large μ_{if} (e.g., the $\nu = 0 \to 1$, transition, $\mu_{01} \approx 0.1$ D) are termed (strongly) allowed. Based on the classical description of the interaction between atoms and radiation, where the electron is regarded as an oscillating dipole, the distinction between allowed and forbidden transitions is sometimes expressed in terms of the (dimensionless) "oscillator strength" of the transition $f_{fi} \approx (8\pi^2 m\nu/3he^2)\mu_{fi}^2$, where m and e are the electron mass and charge, respectively. Strongly allowed (forbidden) transitions are said to carry (not to carry) oscillator strength.

population of the upper state exceed that of the lower state. This represents such an extreme deviation from the common, equilibrium, situation that it is characterized by a special term, "population inversion." The normal order of occupancy is reversed. By way of comparison, the condition for net emission of light (6A.5)—due to auxiliary pumping, chemical or otherwise—can be expressed, using (6A.10), as

$$n_f/n_i > (n_f/n_i)_{eq} = \exp(-h\nu/\mathbf{k}T) \qquad (6A.14)$$

For most transitions, (6A.13) is much harder to realize that (6A.14), unless ν is quite low or T is high. Thus, while any preferential pumping will suffice to disturb the equilibrium in the direction required by (6A.14), amplification (i.e., population inversion) imposes special conditions on the pumping process. It needs to maintain a strict excess ($n_f > n_i$) of molecules in the upper state f.

Thus far we have assumed that there is just a single quantum state f at the energy $h\nu$ above the state i. If there are several, say, g_m, such states of the same energy, one refers to the "level" m with a degeneracy g_m. In this case, (6A.10) becomes

$$g_m B_{mn} = g_n B_{nm} \qquad (6A.15)$$

and the condition for amplification now reads

$$(n_m/g_m) > (n_n/g_n) \qquad (6A.16)$$

By way of contrast, at equilibrium

$$(n_m/g_m)_{eq} = (n_n/g_n)_{eq} \exp[-(E_m - E_n)/\mathbf{k}T] \qquad (6A.17)$$

As a specific example, consider an HCl chemical laser where the necessary displacement from equilibrium is produced by a rapid exoergic chemical reaction that preferentially produces HCl molecules in excited vibrotational states. For such molecules to lase on a $\nu, J \to \nu - 1, J'$ transition it is necessary [see (6A.16)] that

$$n_{\nu,J} - (g_J/g_{J'})n_{\nu-1,J'} > 0 \qquad (6A.18)$$

where $g_J = 2J + 1$ is the degeneracy of the vibrotational level.

The rapid rate (Fig. 6.1) of rotational energy transfer implies that within a given vibrational manifold of states, the rotational states rapidly reach thermal equilibrium (caveat: see Fig. 6.22). Hence, the population density $n_{\nu,J}$ rapidly reaches the form

$$n_{\nu,J} = n_\nu g_J \exp(-E_J(\nu)/\mathbf{k}T)/Q_R(\nu) \qquad (6A.19)$$

$$Q_R(\nu) = \sum_J (2J + 1) \exp(-E_J(\nu)/\mathbf{k}T)$$

$Q_R(\nu)$ is the rotational partition function for the νth vibrational manifold of states. $E_J(\nu)$ is the energy of rotational level J in the ν-manifold and is independent of ν in the rigid-rotor approximation, $E_J = BJ(J + 1)$. n_ν is the total number of molecules in the v vibrational manifold of states. For time intervals short compared to τ_{V-T}, n_v is not of the Boltzmann form.

In the rigid-rotor approximation, the necessary condition for lasing on the $v, J \rightarrow v - 1, J'$ transition is, from (6A.18) and (6A.19),

$$(n_v/n_{v-1}) > \begin{cases} \exp[-2B(J+1)/\mathbf{k}T], & J' = J+1 \\ \exp[2BJ/\mathbf{k}T], & J' = J-1 \end{cases} \quad (6A.20)$$

Lasing on a P-branch $(J' = J + 1)$ transition is thus possible even if $n_v < n_{v-1}$. When we speak of a "population inversion" in a strict technical sense what we really mean is (6A.18).

It should be emphasized that (6A.20) is an example where rotational relaxation plays a doubly beneficial role. The true nascent vibrotational distribution often populates J states that are quite high. Then condition (6A.18) is hard to meet. It is the rotational relaxation that insures that there are fewer $J' = J + 1$ quantum states than J states. The second role of rotational relaxation is to drain the final, $v - 1$, $J + 1$ level. In the absence of relaxation, lasing may start but rapidly cease since the lower level is populated by those molecules that emitted. The fast relaxation transfers this population to other rotational states of the $v - 1$ manifold.

Suggested reading

Section 6.1

Books

Cottrell, T. L., *Dynamic Aspects of Molecular Energy States*, Wiley, New York, 1965.

Flygare, W. H., *Molecular Structure and Dynamics*, Prentice-Hall, Englewood Cliffs, NJ, 1978.

Lambert, J. D., *Vibrational and Rotational Relaxation in Gases*, Clarendon Press, Oxford, U.K., 1977.

Levine, R. D., and Jortner, J. (eds.), *Molecular Energy Transfer*, Wiley, New York, 1976.

Stevens, B., *Collisional Activation in Gases*, Pergamon, Oxford, UK, 1967.

Yardley, J. T., *Introduction to Molecular Energy Transfer*, Academic, New York, 1980.

Review Articles

Amme, R. C., "Vibrational and Rotational Excitation in Gaseous Collisions," *Adv. Chem. Phys.*, *28*, 171 (1975).

Bailey, R. T., and Cruickshank, F. R., "Laser Studies of Vibrational, Rotational and Translational Energy Transfer," *Spec. Per. Rep.*, *3*, 109 (1978).

Beenakker, J. J. M., "The Internal Degrees of Freedom and the Transport Properties of Rotating Molecules," in Levine and Jortner (1976).

Ben-Shaul, A., "Chemical Laser Kinetics," *Adv. Chem. Phys.*, *47* (Pt. 2), 55 (1981).
Buck, U., "Rotationally Inelastic Scattering of Hydrogen Molecules and the Non-Spherical Interaction," *Faraday Discuss. Chem. Soc.*, *73*, 187 (1982).
Callear, A. B., "An Overview of Molecular Energy Transfer in Gases," *Spec. Per. Rep.*, *3*, 82 (1978).
Diestler, D., "Theoretical Studies of Vibrational Relaxation of Small Molecules in Dense Media," *Adv. Chem. Phys.*, *42*, 305 (1980).
Earl, B. L., Gamss, L. A., and Ronn, A. M., "Laser-Induced Vibrational Energy Transfer Kinetics: Methyl and Methyl-d_3 Halides," *Acc. Chem. Res.*, *11*, 183 (1978).
Fischer, S. F., "Intramolecular Vibrational Relaxation of Polyatomic Molecules," in Jortner and Pullman (1982).
Flygare, W. H., and Schmalz, T. G., "Transient Experiments and Relaxation Processes Involving Rotational States," *Acc. Chem. Res.*, *9*, 385 (1976).
Flynn, G. W., "Energy Flow in Polyatomic Molecules," in Moore, (1974).
Gole, J. L., "Probing Ultrafast Energy Transfer Among the Excited States of High Temperature Molecules," in Fontijn (1985).
Gordon, R. G., Klemperer, W. A., and Steinfeld, J. I., "Vibrational and Rotational Relaxation," *Annu. Rev. Phys. Chem.*, *19*, 215 (1968).
Gordon, R. G., and Steinfeld, J. I., "Spectroscopic Measurements of Energy Transfer by Fluorescence and Double Resonance," in Levine and Jortner (1976).
Heilweil, E. J., Moore, R., Rothenberger, G., Velsko, S., and Hochstrasser, R. M., "Picosecond Processes in Chemical Systems: Vibrational Relaxation," *Laser Chem.*, *3*, 109 (1983).
Laubereau, A., and Kaiser, W., "Picosecond Investigations of Dynamics Processes in Polyatomic Molecules in Liquids," in Moore (1974).
Lemont, S., and Flynn, G. W., "Vibrational State Analysis of Electronic to Vibrational Energy Transfer," *Annu. Rev. Phys. Chem.*, *28*, 261 (1977).
McCaffery, A. J., "Reorientation by Elastic and Rotationally Inelastic Transitions," *Spec. Per. Rep.*, *4*, 47 (1981).
McGurk, J. C., Schmalz, T. G., and Flygare, W. H., "A Density Matrix, Bloch Equation Description of Infrared and Microwave Transient Phenomena," *Adv. Chem. Phys.*, *25*, 1 (1974).
Moore, C. B., "Vibration→ Vibration Energy Transfer," *Adv. Chem. Phys. 23*, 41 (1973).
Moore, C. B., and Zittel, P. F., "State Selected Kinetics from Laser-Excited Fluorescence." Science, *182*, 541 (1973).
Phillips, L. F., "Mercury-Sensitized Luminescence," *Acc. Chem. Res.*, *7*, 135 (1974).
Rice, S. A., "Collision Induced Intramolecular Energy Transfer in Electronically Excited Polyatomic Molecules," *Adv. Chem. Phys.*, *47* (Pt. 2), 237 (1981).
Smith, I. W. M., "Vibrational Relaxation in Small Molecules," in Levine and Jortner (1976).
Smith, I. W. M., "Relaxation in Collisions of Vibrationally Excited Molecules with Potentially Reactive Atoms," *Acc. Chem. Res.*, *9*, 161 (1976).
Steinfeld, J. I., "Energy-Transfer Processes," *Int. Rev. Sci. Phys. Chem.*, *9*, 247 (1972).
Weitz, E., and Flynn, G. W., "Laser Studies of Vibrational and Rotational Relaxation in Small Molecules," *Annu. Rev. Phys. Chem.*, *25*, 275 (1974).

Yang, K., "The Mechanisms of Electronic Energy Transfer Between Excited Mercury ($^3P_1$) Atoms and Gaseous Paraffins," *Adv. Chem. Phys.*, *21*, 187 (1971).

Yardley, J. T., "Dynamic Properties of Electronically Excited Molecules," in Moore, (1974).

Section 6.2

Review Articles

Barker, J. R., "Direct Measurements of Energy Transfer Involving Large Molecules in the Electronic Ground State," *J. Phys. Chem.*, *88*, 11 (1984).

Clark, A. P., Dickinson, A. S., and Richards, D., "The Correspondence Principle in Heavy-Particle Collisions," *Adv. Chem. Phys.*, *36*, 63 (1977).

DePristo, A. E., and Rabitz, H., "Vibrational and Rotational Collision Processes," *Adv. Chem. Phys.*, *42*, 271 (1980).

Dickinson, A. S., "Non-Reactive Heavy Particle Collision Calculations," *Comp. Phys. Commun.*, *17*, 51 (1979).

Ewing, G. E., "Relaxation Channels of Vibrationally Excited van der Waals Molecules," *Faraday Discuss. Chem. Soc.*, *73*, 325 (1982).

Ferguson, E. E., "Vibrational Quenching of Small Molecular Ions in Neutral Collisions," *J. Phys. Chem.*, *90*, 731 (1986).

Fisk, G. A., and Crim, F. F., "Single Collision Studies of Vibrational Energy Transfer Mechanisms," *Acc. Chem. Res.*, *10*, 73 (1977).

Flynn, G. W., "Collision-Induced Energy Flow Between Vibrational Modes of Small Polyatomic Molecules," *Acc. Chem. Res.*, *14*, 334 (1981).

Gentry, W. R., "Vibrational Excitation: Classical and Semiclassical Methods," in Bernstein (1979).

Helm, H., "Ion–Atom Interactions Probed by Photofragment Spectroscopy," in Eichler et al. (1984).

Janda, K. C., "Predissociation of Polyatomic van der Waals Molecules," *Adv. Chem. Phys.*, *60*, 201 (1984).

Kouri, D. J., "Rotational Excitation: Approximation Methods," in Bernstein (1979).

LeRoy, R. J., Corey, G. C., and Hutson, J. M., "Predissociation of Weak-Anisotropy van der Waals Molecules: Theory, Approximations and Practical Predictions," *Faraday Discuss. Chem. Soc.*, *73*, 339 (1982).

Oxtoby, D. W., "Vibrational Population Relaxation in Liquids," *Adv. Chem. Phys.*, *47* (Pt. 2), 487 (1981).

Oxtoby, D. W., "Vibrational Relaxation in Liquids," *Annu. Rev. Phys. Chem.*, *32*, 77 (1981).

Rapp, D., and Kassal, T., "The Theory of Vibrational Energy Transfer Between Simple Molecules in Nonreactive Collisions," *Chem. Rev.*, *69*, 61 (1969).

Secrest, D., "Theory of Rotational and Vibrational Energy Transfer in Molecules," *Annu. Rev. Phys. Chem.*, *24*, 379 (1973).

Shin, H. K., "Vibrational Energy Transfer," in Miller (1976).

Weitz, E., and Flynn, G. W., "Vibrational Energy Flow in the Ground Electronic States of Polyatomic Molecules," *Adv. Chem. Phys.*, *47* (Pt. 2), 185 (1981).

Section 6.3

Review Articles

Brunner, T. A., and Pritchard, D., "Fitting Laws for Rotationally Inelastic Collisions," *Adv. Chem. Phys., 50,* 589 (1982).

Faubel, M., "Vibrational and Rotational Excitation in Molecular Collisions," *Adv. At. Mol. Phys., 19,* 345 (1983).

Faubel, M., and Toennies, J. P., "Scattering Studies of Rotational and Vibrational Excitation of Molecules," *Adv. At. Mol. Phys., 13,* 229 (1977).

Gianturco, F. A., "Internal Energy Transfer in Molecular Collisions," in Gianturco (1982).

Gianturco, F. A., and Staemmler, V., "Selective Vibrational Inelasticity in Proton–Molecule Collisions," in Pullman (1981).

Loesch, H. J., "Scattering of Non-Spherical Molecules," *Adv. Chem. Phys., 42,* 421 (1980).

McCaffery, A. J., Proctor, M. J., and Whitaker, B. J., "Rotational Energy Transfer: Polarization and Scaling," *Annu. Rev. Phys. Chem., 37,* 223 (1986).

Rice, S. A., and Cerjan, C., "Very Low Energy Collision Induced Vibrational Relaxation: An Overview," *Laser Chem., 2,* 137 (1983).

Toennies, J. P., "The Calculation and Measurement of Cross Sections for Rotational and Vibrational Excitation," *Annu. Rev. Phys. Chem., 27,* 225 (1976).

Whitaker, B. J., and Brechignac, Ph., "The Physical Origin of Fitting Laws for Rotational Energy Transfer," *Laser Chem., 6,* 61 (1986).

Section 6.4

Books

Aussenberg, F. R., Leitner, A., and Lippitsch, M. E., *Surface Studies with Lasers,* Springer-Verlag, Berlin, 1983.

Benedek, G., and Valbusa, U. (eds.), *Dynamics of Gas–Surface Interaction,* Springer-Verlag, Berlin, 1982.

Gerber, R. B., and Nitzan, A. (eds.), *Dynamics of Molecule-Surface Interactions, Isr. J. Chem, 22, No.* 4 (1982).

Gomer, R. (ed.), *Interactions on Metal Surfaces,* Springer-Verlag, Heidelberg, 1975.

Goodman, F. O., and Wachman, H. Y., *Dynamics of Gas-Surface Scattering,* Academic, New York, 1976.

Pullman, B., Jortner, J., Nitzan, A., and Gerber, R. B. (eds.), *Dynamics on Surfaces,* Reidel, Boston, 1984..

Taglauer, E., and Heiland, W., *Inelastic Particle-Surface Collisions,* Springer-Verlag, Berlin, 1981.

Review Articles

Apkarian, V. A., Hamers, R., Houston, P. L., Misewitch, J., and Merrill, R. P., "Laser Studies of Vibrational Energy Exchange in Gas-Solid Collisions," in Pullman et al. (1984).

Auerbach, D. J., "Inelastic Scattering of Atoms and Molecules from Solid Surfaces," *Phys. Scripta, T6,* 122 (1983).

Barker, J. A., and Auerbach, D. J., "Gas-Surface Interactions and Dynamics: Thermal Energy Atomic and Molecular Beam Studies," *Surf. Sci. Rep., 4,* 1 (1985).

Benedek, G., "Probing Surface Vibrations by Molecular Beams: Experiment and Theory," in Pullman et al. (1984).

Cardillo, M. J., "Gas–Surface Interactions Studied with Molecular Beam Techniques," *Annu. Rev. Phys. Chem., 32,* 331 (1981).

Celli, V., and Evans, D., "Theory of Atom-Surface Scattering," in Benedek and Valbusa (1982).

Ceyer, S. T., and Somorjai, G. A., "Surface Scattering," *Annu. Rev. Phys. Chem., 28,* 477 (1977).

Chance, R. R., Prock, A., and Silbey, R., "Molecular Fluorescence and Energy Transfer Near Interfaces," *Adv. Chem. Phys., 37,* 1 (1978).

Chuang, T. J., "Laser-Induced Gas–Surface Interactions," *Surf. Sci. Rep., 3,* 1 (1984).

Cole, M. W., and Vidali, G., "Universal Laws of Physical Adsorption," in Benedek and Valbusa (1982).

Eichenauer, D., and Toennies, J. P., "Theory of One-Phonon Assisted Adsorption and Desorption of the Atoms from a LiF(001) Single Crystal Surface," in Pullman et al. (1984).

Garrison, B. J., "Classical Trajectory Studies of keV Ions Interacting with Solid Surfaces," in Truhlar (1981).

Kolodney, E., and Amirav, A., "Collision Induced Dissociation of Molecular Iodine on Single Crystal Surfaces," in Pullman et al. (1984).

Lapujoulade, J., Salanon, B., and Gorse, D., "The DIffraction of He, Ne and H_2 from Copper Surfaces," in Pullman et al. (1984).

Levi, A. C., "Atom Scattering from Overlayers," in Benedek and Valbusa (1982).

Micha, D. A., "Scattering of Ions by Polyatomics and Solid Surfaces: Multicenter Short-Range Interactions," in Truhlar (1981).

Proctor, T. R., and Kouri, D. J., "Magnetic Transitions in Heteronuclear and Homonuclear Molecule-Corrugated Surface Scattering," in Pullman et al. (1984).

Rabitz, H., "Dynamics and Kinetics on Surfaces," in Pullman et al. (1984).

Rosenblatt, G. M., "Translational and Internal Energy Accommodation of Molecular Gases with Solid Surfaces, *Acc. Chem. Res., 14,* 42 (1981).

Schinke, R., "Rainbows and Resonances in Molecule-Surface Scattering," in Pullman et al. (1984).

Tully, J. C., "Theories of the Dynamics of Inelastic and Reactive Processes at Surfaces," *Annu. Rev. Phys. Chem., 31,* 319 (1980).

Tully, J. C., "Interaction Potentials for Gas-Surface Dynamics," in Truhlar (1981).

Weare, J. H., "Atom-Surface Potential Information from Low-Energy Atom-Surface Scattering," in Truhlar (1981).

Weinberg, W. H., "Molecular Beam Scattering from Solid Surfaces," *Adv. Colloid Interface Sci., 4,* 301 (1975).

Wolken, G., Jr., "The Scattering of Atoms and Molecules from Solid Surfaces," in Miller (1976).

Section 6.5

Books

Birnbaum, G. (ed.), *Phenomena Induced by Intermolecular Interactions,* Plenum, New York, 1985.

Delone, N. B., and Krainov, V. P., *Atoms in Strong Light Fields,* Springer-Verlag, Berlin, 1985.

George, T. F. (ed.), *Theoretical Aspects of Laser Radiation and Its Interaction with Atomic and Molecular Systems,* University of Rochester Press, Rochester, N.Y., 1978.

Rahman, N. K., and Guidotti, C. (eds.), *Collisions and Half-Collisions with Lasers,* Harwood, Utrecht, 1984.

Rahman, N. K., and Guidotti, C. (eds.), *Photon-Assisted Collisions and Related Topics,* Harwood, New York, 1984.

Sobelman, I.I., Vainshtein, L. A., and Yukov, E. A., *Excitation of Atoms and Broadening of Spectral Lines,* Springer-Verlag, Berlin, 1981.

Review Articles

Andersen, N., "Laser Spectroscopy of Collision Complexes: A Case Study," in Ehlotzky (1985).

Ben-Reuven, A., "Spectral Line Shapes in Gases in the Binary-Collision Approximation," *Adv. Chem. Phys., 33,* 235 (1975).

Birnbaum, G., "Microwave Pressure Broadening and Its Application to Intermolecular Forces," *Adv. Chem. Phys., 12,* 487 (1976).

Birnbaum, G., Guillot, B., and Bratos, S., "Theory of Collision-Induced Line Shapes—Absorption and Light Scattering at Low Density," *Adv. Chem. Phys., 51,* 49 (1982).

Burnett, K., "Spectroscopy of Collision Complexes," in Eichler et al. (1984).

Foth, H. J., Polanyi, J. C., and Telle, H. H., "Emission from Molecules and Reaction Intermediates in the Process of Falling Apart," *J. Phys. Chem., 86,* 5027 (1982).

Frommhold, L., "Collision-Induced Scattering of Light and the Diatom Polarizabilities," *Adv. Chem. Phys., 46,* 1 (1981).

George, T. F., Zimmermann, I. H., Juan, J.-M., Laing, J. R., and DeVries, P. L., "A New Concept in Laser-Assisted Chemistry: Electronic-Field Representation," *Acc. Chem. Res., 10,* 449 (1977).

Hamilton, C. E., Kinsey, J. L., and Field, R. W., "Stimulated Emission Pumping: New Methods in Spectroscopy and Molecular Dynamics," *Annu. Rev. Phys. Chem., 37,* 493 (1986).

Imre, D., Kinsey, J. L., Sinha, A., and Krenos, J., "Chemical Dynamics Studied by Emission Spectroscopy of Dissociating Molecules," *J. Phys. Chem., 88,* 3956 (1984).

Lau, A. M. F., "The Photon-as-Catalyst Effect in Laser Induced Predissociation and Autoionization," *Adv. Chem. Phys., 50,* 191 (1982).

Mies, F., "Quantum Theory of Atomic Collisions in Intense Laser Fields," in Henderson (1981).

Mukamel, S., "Collisional Broadening of Spectral Line Shapes in Two-Photon and Multiphoton Processes," *Phys. Rep.*, *93*, 1 (1982).

Orel, A. E., "Laser-Induced Nonadiabatic Collision processes," in Truhlar (1981).

Rabitz, H., "Rotation and Rotation–Vibration Pressure-Broadened Spectral Line-shapes," *Annu. Rev. Phys. Chem.*, *25*, 155 (1974).

Roussel, F., "Laser-Assisted Atom–Atom Collisions," in Ehlotzky (1985).

Telle, H. H., "Stimulated Processes in a Half-Collision," *Laser Chem.*, *5*, 393 (1986).

Weiner, J., "Inelastic Collision Processes in the Presence of Intense Optical Fields," in Gloriuex et al. (1979).

Section 6.6

Books

Nikitin, E. E., and Umanskii, S. Ya., *Theory of Slow Atomic Collisions*, Springer-Verlag, Berlin, 1984.

Steinfeld, J. I. (ed.), *Electronic Transition Lasers*, MIT Press, Cambridge, Mass., 1976.

Wilson, L. E., Suchard, S. N., and Steinfeld, J. I. (eds.), *Electronic Transition Lasers II*, MIT Press, Cambridge, Mass., 1977

Review Articles

Alexander, M. H., "Pseudo-Quenching Model Studies of Spin-Orbit State Propensities in Reactions of Ca($4s4p\,^3P$) with Cl_2," in Fontijn (1985).

Aquilanti, V., Grossi, G., and Pirani, F., "Interference and Polarization in Low Energy Atom–Atom Collisions," in Eichler et al. (1984).

Avouris, Ph., Gelbart, W. M., and El-Sayed, M. A., "Nonradiative Electronic Relaxation Under Collision-Free Conditions," *Chem. Rev.*, *27*, 793 (1977).

Baede, A. P. M., "Charge Transfer Between Neutrals at Hyperthermal Energies," *Adv. Chem. Phys.*, *30*, 463 (1975).

Baer, M., "Quantum Mechanical Treatment of Electronic Transitions in Atom–Molecule Collisions," in Bowman (1983).

Baer, M., "Theory of Electronic Nonadiabatic Transitions in Chemical Reactions," in Baer (1985).

Bardsley, J. N., "Recombination Processes in Atomic and Molecular Physics," in Gianturco (1982).

Breckenridge, W. H., and Umemoto, H., "Collisional Quenching of Electronically Excited Metal Atoms," *Adv. Chem. Phys.*, *50*, 325 (1982).

Carrington, T., "The Geometry of Intersecting Potential Surfaces," *Acc. Chem. Res.*, *7*, 20 (1974).

Child, M. S., "Electronic Excitation: Nonadiabatic Transitions," in Bernstein (1979).

Desouter-Lecomte, M., Dehareng, D., Leyh-Nihant, B., Praet, M. Th., Lorquet, A. J., and Lorquet, J. C., "Nonadiabatic Unimolecular Reactions of Polyatomic Molecules," *J. Phys. Chem.*, *89*, 214 (1985).

Drukarev, G., "The Zero-Range Potential Model and Its Application in Atomic and Molecular Physics," *Adv. Quant. Chem.*, *11*, 251 (1978).

Dunning, F. B., and Stebbings, R. F., "Collisions of Rydberg Atoms with Molecules," *Annu. Rev. Phys. Chem.*, *33*, 173 (1982).

Freed, K. F., "Radiationless Transitions in Molecules," *Acc. Chem. Res.*, *11*, 74 (1978).

Freed, K. F., "Collisional Effects on Electronic Relaxation Processes," *Adv. Chem. Phys.*, *42*, 207 (1980).

Freed, K. F., "Collision Induced Intersystem Crossing," *Adv. Chem. Phys.*, *47*, (Pt. 2), 291 (1981).

Garrett, B. C., and Truhlar, D. G., "The Coupling of Electronically Adiabatic States in Atomic and Molecular Collisions," in Henderson (1981).

Hay, P. J., Wadt, W. R., and Dunning, T. H., Jr., "Theoretical Studies of Molecular Electronic Transition Lasers," *Annu. Rev. Phys. Chem.*, *30*, 311 (1979).

Hertel, I. V., "Collisional Energy-Transfer Spectroscopy with Laser-Excited Atoms in Crossed Atom Beams: A New Method for Investigating the Quenching of Electronically Excited Atoms by Molecules," *Adv. Chem. Phys.*, *45*, 341 (1981).

Hertel, I. V., "Progress in Electronic-to-Vibrational Energy Transfer," *Adv. Chem. Phys.*, *50*, 475 (1982).

Hertel, I. V., Schmidt, H., Bähring, A., and Meyer, E., "Angular Momentum Transfer and Charge Cloud Alignment in Atomic Collisions: Intuitive Concepts, Experimental Observations and Semiclassical Models," *Rep. Prog. Phys.*, *48*, 375 (1985).

Houston, P. L., "Electronic to Vibrational Energy Transfer from Excited Halogen Atoms," *Adv. Chem. Phys.*, *47*, (Pt. 2), 381 (1981).

Jaecks, D. H., "Molecules in Nonadiabatic Collisions," in Eichler et al. (1984).

Jortner, J., and Mukamel, S., "Radiationless Transitions," *Int. Rev. Sci. Phys. Chem.*, *1*, 329 (1975).

Kleyn, A. W., Los, J., and Gislason E. A., "Vibronic Coupling at Intersections of Covalent and Ionic States," *Phys. Rep.*, *90*, 1 (1982).

Köppel, H., Domcke, W., and Cederbaum, L. S., "Multimode Molecular Dynamics Beyond the Born–Oppenheimer Approximation," *Adv. Chem. Phys.*, *57*, 59 (1984).

Krause, L., "Sensitized Fluorescence and Quenching," *Adv. Chem. Phys.*, *28*, 267 (1975).

Leach, S., "Electronic Spectroscopy and Relaxation Processes in Small Molecules in the Resonance Limit," in Levine and Jortner (1976).

Leach, S., and Dujardin, G., "Ionic State Relaxation Processes in VUV-Excited Polyatomic Molecules," *Laser Chem.*, *2*, 285 (1983).

Mahan, B. H., "Recombination of Gaseous Ions," *Adv. Chem. Phys.*, *23*, 1 (1973).

Morgner, H., "Penning Ionization of Molecules," in Eichler et al. (1984).

Nakamura, H., "Unified Treatment of Nonadiabatic Transitions in the Rotating Frame of the Complex," in Eichler et al. (1984).

Niehaus, A., "Spontaneous Ionization in Slow Collisions," *Adv. Chem. Phys.*, *45*, 399 (1981).

Nikitin, E. E., "Theory of Nonadiabatic Collision Processes Including Excited Alkali Atoms," *Adv. Chem. Phys.*, *28*, 317 (1975).

Ovchinnikov, A. A., and Ovchinnikova, M. Ya., "Problems of Nonlinear Radiationless Processes in Chemistry," *Adv. Quant. Chem.*, *16*, 161 (1982).

Ozkan, I., and Goodman, L., "Coupling of Electronic and Vibrational Motions in Molecules," *Chem. Rev., 79,* 275 (1979).

Rebentrost, F., "Nonadiabatic Molecular Collisions," in Henderson (1981).

Rhodes, W., "Nonradiative Relaxation and Quantum Beats in the Radiative Decay Dynamics of Large Molecules," *J. Phys. Chem., 87,* 30 (1983).

Saha, H. P., Lam, K.-S., and George, T. F., "Recent Advances in the Theory of Chemi-Ionization," in Fontijn (1985).

Siebrand, W., "Nonradiative Processes in Molecular Systems," in Miller (1976).

Tramer, A., and Nitzan, A., "Collisional Effects in Electronic Relaxation," *Adv. Chem. Phys., 47* (Pt. 2), 337 (1981).

Tully, J. C., "Nonadiabatic Processes in Molecular Collisions," in Miller (1976).

Tully, J. C., "Collisions Involving Electronic Transitions," in Brooks and Hayes (1977).

Whetten, R. L., Ezra, G. S., and Grant, E. R., "Molecular Dynamics Beyond The Adiabatic Approximation: New Experiments and Theory," *Annu. Rev. Phys. Chem., 36,* 277 (1985).

Yencha, A. J., "Penning Ionization and Chemi-Ionization in Reactions of Excited Rare-Gas Atoms," in Fontijn (1985).

Zülicke, L., Zuhrt, Ch., and Umansky, S. Ya., "Some Problems of the Dynamics of Nonadiabatic Energy Processes," in Eichler et al. (1984).

7

Reaction dynamics and chemical reactivity

We are now ready to deal with state-of-the-art research in molecular reaction dynamics, for which experiment and theory are usually deeply intertwined and interactive. We begin with an elementary, well-studied reaction and proceed to more complex systems. At the end, we come back to the "structure" aspects of chemical reactivity: What have we learned about the steric requirements of reactions? The chapter concludes with a (subjective) overview of the frontiers of our field.

7.1 A case study of an elementary reaction

7.1.1 The $F + H_2$ reaction

One of the few elementary reactions that have received detailed chemical dynamical investigation by several of the experimental methods discussed in Chapter 5 is the F-atom exchange reaction,

$$F + H_2 \rightarrow HF(v', j') + H$$

and its isotopic analogs. Complementary theoretical studies have brought to bear all the tools of the trade in an attempt to account for the body of experimental results on this ("bellwether") reaction, showing the power (and the limitations) of the methods of chemical dynamics. It is perhaps of pedantic value to proceed in a very rough chronological order to unfold the experimental and theoretical story of the intimate details of this reaction.

7.1.2 Early experiments: "In the beginning. . ."

Atomic fluorine is not a "friendly" reagent. So it is not surprising that there has been difficulty in obtaining accurate, absolute bimolecular rate constants for the hydrogen abstraction reaction. However, there is no doubt that the reaction is "fast" and that the Arrhenius activation energy is small.† The best experimental value is $E_a \approx 4 \text{ kJ mol}^{-1}$.

† Since the reaction is so very fast and so very exoergic, care is necessary to ensure that the reactants are indeed maintained in thermal equilibrium.

Early chemical laser experiments with UF_6–H_2 (or D_2) mixtures showed that the reaction produced inverted vibrational populations, definitely establishing that the ratio of the nascent populations of the $v' = 2$ to $v' = 1$ levels exceeds unity.

Early infrared chemiluminescence experiments were carried out in a moderate pressure regime that allowed rotational but not vibrational relaxation. The results confirmed the population inversion (for the HF reaction) from the laser study. From the overall v' distribution it could be calculated that about 2/3 of the total available energy E is converted into vibrational energy of HF. The state $v' = 2$ was the most probable one.

This was the position until the first molecular-beam study of the reaction of $F + D_2$, discussed below.

7.1.3 A landmark: Crossed molecular beam study

The (laboratory) angular distribution of DF was measured using a beam of D_2 crossed with a beam of F atoms.† Even without velocity analysis, it was possible (for reasons discussed below) to estimate the center-of-mass (c.m.) product flux–velocity contour map. A more recent version of such a fiux–velocity map (obtained with time-of-flight velocity analysis) is shown in Fig. 7.1. The circles represent the highest possible E_T' consistent with a specified final vibrational level v'. The highest value of E_T' for a given v' obtains when the final rotational energy is zero. The different final rotational states with a given value of v' will all have final translational energies between zero and the maximal value, $E - E_{v'}$.

Although, in principle, states of different values of v' can have nearly equal values of E_T' (the balance being made up for by the rotational energy), this is not the case here. The range of final rotational energy of the DF product is quite small; only fairly low j' states are populated (see Appendix 4B). The possible final rotational states j' with different v' have sufficiently different E_T' values that they do not overlap in the intensity contour plot.

There are clear intensity peaks in Fig. 7.1 corresponding to the different manifolds of v' states. In general, there is a tendency for "backward scattering" of the DF product molecules. By integrating the intensity for a particular manifold of v' states over all angles and all final velocities, the relative probabilities of formation of each final vibrational state v' were estimated. The current results are listed in Table 7.1.

Using a variable-temperature nozzle for the supersonic D_2 beam, it is possible to change the collision energy over a fairly broad range. In general, we expect the entire angular distribution to shift gradually to the more forward direction. This is indeed the case, but one particular DF vibrational state ($v' = 4$) is found to be scattered especially much in the forward

---

† The fluorine atoms were derived from thermal dissociation of F_2.

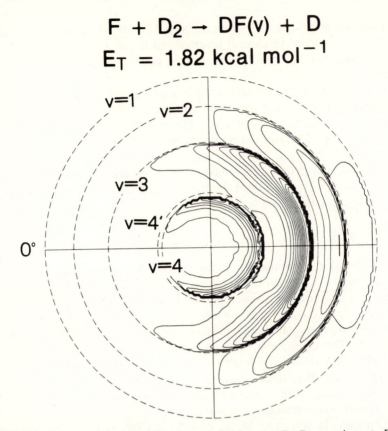

Fig. 7.1. Center-of-mass flux–velocity contour map for the $F + D_2$ reaction at $E_T =$ 1.82 kcal mol^{-1}. Dashed circles show DF c.m. velocity limits for $v = 1, 2, 3,$ and 4 as indicated. The circle labeled 4' is for $D_2(j = 2)$, constituting *ca.* 25% of the reagent D_2 beam. Lines are spaced linearly according to scattering intensity. [Reproduced with permission from D. M. Neumark, A. M. Wodtke, G. N. Robinson, C. C. Hayden, K. Shobatake, R. K. Sparks, T. P. Schafer, and Y. T. Lee, *J. Chem. Phys.*, *82*, 3067 (1985); see slso D. M. Neumark, A. M. Wodtke, G. N. Robinson, C. C. Hayden, and Y. T. Lee, *J. Chem. Phys.*, *82*, 3045 (1985) for the $F + H_2$ system. Extensive experimental details and discussion of results may be found in these articles.]

direction (see Fig. 7.11 below). It seems as though the behavior of the $v' = 4$ state is governed by different dynamical considerations than the other states. Indeed, theory suggested that this might be the case. What are these theoretical considerations?

7.1.4 Theoretical attacks on the dynamics

In the absence of an *ab initio* potential energy surface for the FH$_2$ system, the first attempts to account for the chemical dynamics of the system involved empirical LEPS surfaces (Sec. 4.2) that had been adjusted to have

Table 7.1. Vibrational-energy disposal: Summary of the relative populations $P(v')$ in the final vibrational state v' of DF from the reaction of $F + D_2$[a]

v'	Chemi-luminescence[b]	Chemical laser[c]	Molecular beam[d]	Trajectory calculations[e]	Quantal "sudden" computations[f]
0		0.04	0		
1	0.18	0.24	0.19	0	0.02
2	0.52	0.56	0.67	0.45	0.35
3	1.00	1.00	1.00	1.00	1.00
4	0.59	(0.4)	0.41	0.32	0.25

[a] All normalized to $P(v' = 3) = 1$.

[b] D. S. Perry and J. C. Polanyi, *Chem. Phys.*, **12**, 419 (1976).

[c] M. J. Berry, *J. Chem. Phys.*, **59**, 6229 (1973).

[d] D. M. Neumark, A. M. Wodtke, G. N. Robinson, C. C. Hayden, K. Shobotake, R. K. Sparks, T. P. Schafer, and Y. T. Lee, *J. Chem. Phys.*, **82**, 3067 (1985).

[e] J. T. Muckerman, in Henderson (1981).

[f] N. Abusalbi, C. L. Shoemaker, D. J. Kouri, J. Jellinek, and M. Baer, *J. Chem. Phys.*, **80**, 3219 (1984).

the approximately correct barrier height and location, such that the Arrhenius activation energy agreed with the existing experimental value and the average vibrational energy of the HF was approximately in accord with the population inversion data then available.

Classical trajectory calculations were used to simulate the experiments and thus test the surface. Though the results could account for the vibrational excitation of the HF (and DF), they showed a predominance of backward scattering, in accord with the early, lower-energy molecular-beam results.

An important new development was the completion of an extensive (and expensive!) *ab initio* computation of the potential energy surface for the FH_2 system (Fig. 4.3). As discussed in Section 4.2, the main features of this surface are similar to those of the empirical surface, including an early saddle point (before any significant change in the H–H distance) and a small barrier.

Early quantum mechanical computations for a collinear $F + H_2$ (or $F + D_2$) reaction on the LEPS potential energy surface showed that the distribution of final vibrational states is extremely sensitive to the collision energy (Fig. 7.2). Moreover, computations on reasonable but somewhat different potential energy surfaces led to markedly different results for the finer details of the final vibrational state distribution. The reason for this sensitivity is that the sharp maxima in the energy dependence of the state-to-state reaction probability reflect the contribution of quasi-bound FHH states or "resonances." These resonances are due to vibrational energy redistribution and are therefore a type of vibrational predissociation. The theoretical puzzle is that the potential energy surface does not seem to

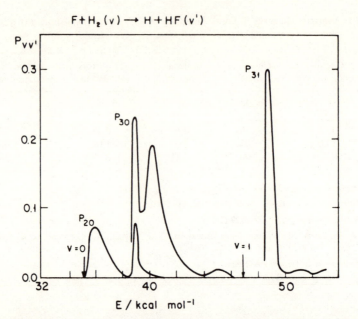

Fig. 7.2. Reaction probability, $P_{vv'}$ to specific final vibrational states of HF(v') in the collinear reaction $F + H_2(v) \rightarrow H + HF(v')$ as determined from a quantum-mechanical calculation. The arrows indicate the lowest value of E (i.e., the exoergicity) for the reaction from the $H_2(v)$ state $v = 0, 1$. [Adapted from S. F. Wu, B. R. Johnson, and R. D. Levine, *Mol. Phys.*, *25*, 839 (1973).]

have a well. Then how can there be quasi-bound states? To resolve this question let us take a somewhat simple (but valid) point of view. Consider the motion as we progress along the reaction path. Do not forget, however, the motion perpendicular to the path—the motion that begins as vibration of the H_2 and ends up as the vibration of HF. The vibrational frequency of this motion is often smaller near the saddle region than at either end. To see this unequivocally we backtrack to the simpler H_3 system.

Figure 7.3 shows the potential energy along the reaction coordinate when we freeze the quantum number v for the vibrational motion perpendicular to the reaction coordinate. By $v = 2$ the barrier along the reaction path has been compensated for by the decrease in the vibrational frequency in the H_3 region. Thus, even if the total energy E (Fig. 7.3) is below the threshold for $H + H_2(v = 2)$, the $v = 2$ state of the symmetric stretch motion can be reached *during* the collision. *If* that happens, the system moving along the reaction coordinate can find itself within the well shown in Fig. 7.3.

The intramolecular flow of vibrational energy into and out of the symmetric stretch gives rise to quasi-bound states in collinear $H + H_2$ collisions. In $F + H_2$ the mechanism is in essence the same. In both cases, however, experimental results refer to three-dimensional collisions. Because of the centrifugal barrier, the energy of the resonance will be somewhat

$[V_1(s) + E_v(s)]$ /kcal mol^{-1}

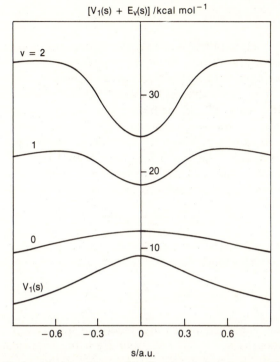

s/a.u.

Fig. 7.3. Adiabatic (static) channel potentials (for the $v = 0, 1, 2$ channels) for the Porter–Karplus potential-energy surface for $H + H_2$. Shown also is the profile of the surface along the s-reaction coordinate $[V_1(s)]$. Ordinate: energy in kcal mol^{-1}. [Adapted from R. D. Levine and S.-F. Wu, *Chem. Phys. Lett.*, **11**, 557 (1971).]

different for each value of the orbital angular momentum quantum number l. The effect of these resonances upon the energy dependence of the total cross section is thus unlikely to be dramatic. Not so in the angular distribution. The time delay due to the formation and decay of a quasi-bound state can make the duration of the collision comparable to a rotational period (Sec. 7.2). Consequently, the system can dissociate in all c.m. angular directions (Sec. 7.1.6). The participation of a resonance is signaled by the forward component of the reactive scattering (Fig. 7.11), not present for the products formed via the direct mode process.

The resonance phenomenon should be a very fine-tuned probe of the dynamics during a collision. It can rightly be compared to bimolecular spectroscopy (Sec. 6.5) for it is truly a translational spectroscopy of the (quasi) bound states of the colliding reagents.

7.1.5 Definitive product state distributions at last!

Using the technique of "arrested relaxation," it was possible to carry out an infrared chemiluminescence study that yielded nascent population distributions for both the rotational and vibrational states of the HF and DF, respectively.

The average fraction of the available energy disposed in product vibration was 66%, while 8% went into rotational excitation. The same results were roughly obtained for both isotopic reactions! However, the quantum-state distributions were quite different, owing to the different vibrotational spacings for HF versus DF. On the basis of the *energy* content in each of the modes, the results are essentially isotopically invariant. This can be understood in terms of the "classical" variable, f_v, the fraction of energy in vibration, as discussed in Section 5.5.2 (see also Fig. 5.31).

The relative populations of the different final rotational states arranged according to vibrational manifolds are shown in Fig. 7.4 for the reaction

$$F + D_2 \rightarrow D + DF(v', j')$$

The height of each "stick" is the relative probability $P(v', j')$ of the final v', j' state. The probabilities are normalized such that

$$P(v') = \sum_{j'} P(v', j') \tag{7.1}$$

The distribution is plotted versus the fraction of final energy in translation, $f_T' = E_T'/E$, where $E_T' = E - E_V' - E_R'$ and E_R' is the final rotational energy, as usual.

The rotational-state distribution for the different vibrational levels can be brought to a common reduced functional form by the introduction of the reduced rotational energy variable

$$g_R = E_R'/(E - E_V') = f_R'/(1 - f_v') \tag{7.2}$$

namely, the fractional energy in rotation within a given vibrational manifold. A plot of the normalized distribution

$$P(j' \mid v') \equiv P(j', v')/P(v') \tag{7.3}$$

versus g_R is often found to be nearly independent of v'.

A convenient summary of a detailed distribution such as Fig. 7.4 is provided, as in Section 5.4, by converting the discrete distribution to a continuum one and plotting the contours of equal probability. There are three variables that characterize any final state, the vibrational (E_V'), rotational (E_R'), and translational (E_T') energies. The three variables are not independent but are related by the conservation of energy

$$E = E_V' + E_R' + E_T' \tag{7.4}$$

A representation of the data of Fig. 7.4 as contours of equal probability is shown in Fig. 7.5. This provides a compact summary of the considerable specificity of the exoergicity partitioning among the three modes V, R, T. Shown also in Fig. 7.5 is a straight (dashed) line originating in the V apex.

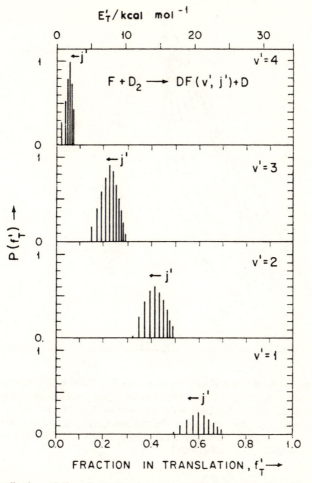

Fig. 7.4. Distribution of the vibrotational states of DF (arranged according to vibrational manifolds) from the $F + D_2$ reaction versus the fraction of energy in translation. The distribution within each manifold is limited to the high-f_T side by the conservation of energy (i.e., $f_T \leq 1 - f_v$). [Adapted from J. C. Polanyi and K. B. Woodall, *J. Chem. Phys.*, *57*, 1574 (1972).]

Along such a line g_R has a constant value.† The line drawn closely follows the "ridge" of the contour plot. This provides a simple indication that g_R is a useful reduced variable.

The population inversion, i.e., the excess population of the higher vibrational states of the product, is an expression of very specific energy partitioning. A measure of the deviation of the observed distribution $P(f_v')$

† $g_R = f_R'$ at $E_V' = 0$, hence, the value of g_R is determined by the intersection of the line with the R–T baseline.

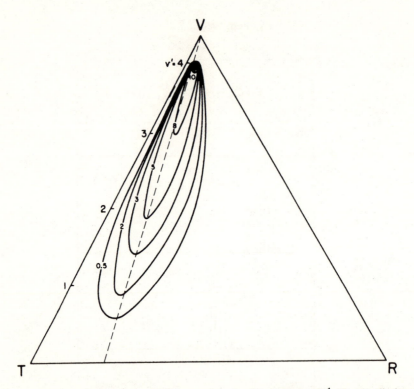

Fig. 7.5. Energy disposal in the $F + D_2$ reaction, for $E = 34\,\text{kcal mol}^{-1}$. Results displayed as relative population contours on a triangular coordinate system. All regions within the figure are allowed on energetic grounds, but most of the probability is found in the "upper" corner. Note also that the most probable intensity ("the mountain ridge") lies along the line $g_R = f_R/(1 - f_v) = \text{const.}$ [Adapted from the experimental results of J. C. Polanyi and K. B. Woodall, *J. Chem. Phys.*, *57*, 1574 (1972).]

from the distribution expected on pure statistical grounds (Sec. 5.5.1), $P^0(f_v')$, is known as the surprisal, $I(f_v')$, which has been defined [see (5.28)] as the logarithm of the ratio of populations

$$I(f_v) = -\ln[P(f_v)/P^0(f_v)] \tag{7.5}$$

(For simplicity we drop the primes denoting final states.) It is observed that $P(f_v)$ is nearly the same function of f_v for all four isotopic variants of the reaction. Since the same is true for $P^0(f_v)$ [see Eq. (5.29)], we conclude that the surprisal is (approximately) isotopically invariant and is a characteristic measure of the specificity of the energy release for the potential energy surface. If the energy disposal were merely that expected from the simple statistical considerations, our surprisal would vanish!

A differential measure of the specificity of the vibrational energy disposal is the vibrational surprisal parameter λ_v,

$$\lambda_v = dI(f_v)/df_v \tag{7.6}$$

We can thus represent the energy disposal compactly by the linear surprisal

$$I(f_v) = \lambda_0 + \lambda_v f_v \qquad (7.7)$$

or, in the more suggestive form (Fig. 7.6),

$$P(f_v) = P^0(f_v) \exp(-\lambda_v f_v)/Q \qquad (7.8)$$

Here, Q is a normalization constant

$$Q = \exp(\lambda_0) = \sum_v P^0(f_v) \exp(-\lambda_v f_v) \qquad (7.9)$$

Since $P^0(f_v)$ is a decreasing function of f_v (Fig. 7.6), it follows that the

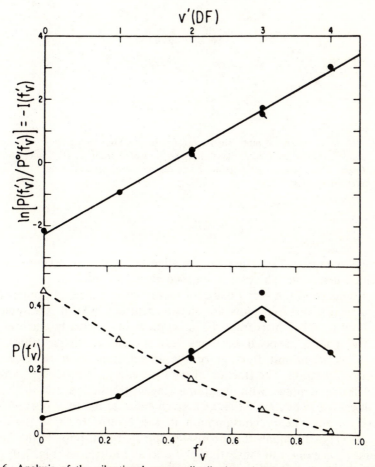

Fig. 7.6. Analysis of the vibrational energy distribution of DF from the $F + D_2$ reaction. (Bottom) Observed populations and *a priori* expected (statistical) populations of the vibrational states versus f_v (As in Fig. 5.32, the prior distribution is evaluated by discrete summation over rotational levels.) (Top) The surprisal $I(f_v)$ as a function of f_v, yielding $\lambda_v = -5.7$. [Experimental data: ●, M. J. Berry, *J. Chem. Phys., 59*, 6229 (1973); ◖, J. C. Polanyi and K. B. Woodall, *J. Chem. Phys., 57*, 1574 (1972).]

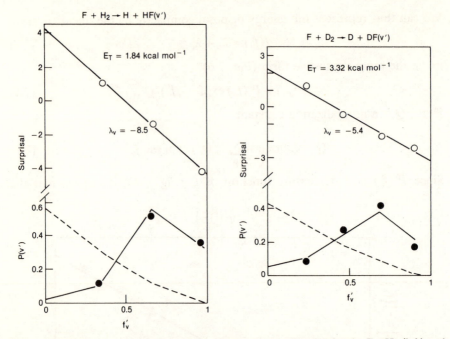

Fig. 7.7. Vibrational surprisal plots, similar presentation as Fig. 7.6 for the $F + H_2$ (left) and $F + D_2$ (right) reactions at the specified energies. [Adapted from Y. M. Engel and R. D. Levine, *Chem. Phys. Lett.*, *123*, 42 (1986), based on the experimental data of Y. T. Lee et al. (1985), references of Fig. 7.1.]

Lagrange parameter λ_v must be negative in order to account for the population inversion. The plot in Fig. 7.6 gives $\lambda_v = -5.7$.

The results shown in Fig. 7.6 are for thermally averaged reactants. In Fig. 7.7 two typical surprisal plots are shown, representative of the molecular beam results. As E increases, the magnitude of λ_v clearly declines (Fig. 7.8). The energy disposal becomes less specific at higher translational energies. A significant fraction of the initial translation is released as products' translation. The fraction of the total energy in products' vibration, $\langle f'_v \rangle$, will thus diminish with increasing reagent translation.

As discussed in Section 5.5.3, at a given value of the total energy E, the surprisal of the energy consumption in the inverse reaction [here $H + HF(v, j) \rightarrow H_2 + F$] equals the surprisal of the energy disposal in the forward reaction. Hence one can "invert" Fig. 7.5 to an illustration (Fig. 7.9) of the relative rates (at the same total energy) for different energy partitionings in the $H + HF$ reaction. Vibrational excitation is thus very effective in promoting this reaction. Rotation tends to diminish the rate, whereas changing the fraction of energy in translation has only a moderate effect.

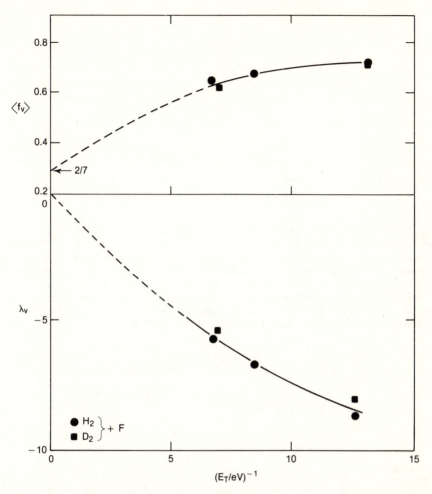

Fig. 7.8. Translational energy dependence of the vibrational population distributions for the HF and DF products of the F + H$_2$ and D$_2$ reactions. (Top) Plot of the average fraction of the available (total) energy E disposed in product vibration, where $f_v = E_v/E$ versus $1/E_T$. In the limit $E_T^{-1} \to 0$, $\langle f_v \rangle$ is extrapolated to its "prior" value of 2/7 (in the RRHO approximation). (Bottom) Plot of the vibrational surprisal parameter λ_v versus $1/E_T$, similarly extrapolated back to its limiting value (zero surprisal) as $E_T^{-1} \to 0$. The results are (nearly) isotopically invariant, in accord with earlier results. [Based on the experimental data of Y. T. Lee et al. (1985), references of Fig. 7.1.]

7.1.6 Angular distribution of products and the "resonances"

Center-of-mass flux–velocity contour maps, such as that of Fig. 7.1, for all four isotopic reactions F + H$_2 \to$ HF + H, F + D$_2 \to$ DF + D, F + HD $\to$ HF + D and F + DH $\to$ DF + H have been determined experimentally over a wide range of translational energy. By integrating a given map over all angles the product state distributions for that value of E_T are obtained, as

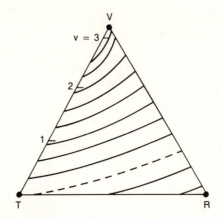

Fig. 7.9. Triangular contour plot (*cf.* Fig. 7.5) for the inverse of the $F + H_2$ reaction, i.e., $k(v, J \rightarrow$ all states; $E)/k(E)$ for the $H + HF(v, J)$ reaction at $E = 34\,\text{kcal mol}^{-1}$. Contours shown connect different partitionings of total energy E with the same reaction rate, with the uppermost contour (near the apex) having the highest rate, and successive contours smaller in k by successive factors of 2. The calculations are based on linear surprisals with $\lambda_v = -6.9$, $\theta_R = 1.75$. The dashed contour is the one of zero surprisal. The contour plots for the forward and inverse reactions at the same total E are related by detailed balance. [Adapted from A. Ben-Shaul and R. D. Levine, *Chem. Phys. Lett.*, *73*, 263 (1980).]

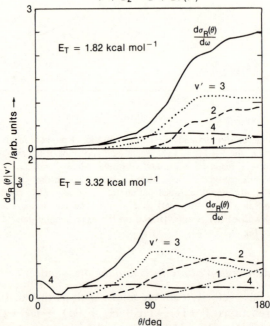

Fig. 7.10. Differential reaction cross sections for specified v' states of the DF product from the $F + D_2$ reaction at $E_T = 1.82$ (top) and 3.32 (bottom) kcal mol^{-1}. The total, summed over all v', is shown as the upper curve. Note the "forward" peak for $v' = 4$ at $E_T = 3.32\,\text{kcal mol}^{-1}$. (Adapted from same source as Fig. 7.1.)

discussed earlier, and the results are qualitatively consistent with (but more accurate than) infrared chemiluminescence- and chemical laser-derived population distributions (Table 7.1). But what can be learned from the *angular* dependence of the products, formed in different vibrotational states? This angular information turns out to be a sensitive probe of the "micro" reaction dynamics.

For these reactions, a coarse description of the angular distributions is that the fluoride products are strongly back-scattered with the peak of the differential reaction cross section near 180° (c.m.). Figure 7.10 shows typical results, for $F + D_2$ at two values of E_T.

However, closer inspection reveals that for the reaction at $3.32\,\text{kcal mol}^{-1}$, the $v' = 4$ state of the DF product has a significant "forward" component. This special feature is even more clearly evident from the flux–velocity contour map, shown in Fig. 7.11. Similar behavior is

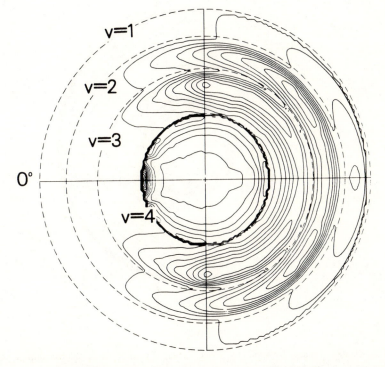

Fig. 7.11. Center-of-mass flux–velocity contour map for the $F + D_2$ reaction at $E_T = 3.32\,\text{kcal mol}^{-1}$. Compare with Fig. 7.1. Note the forward peak for $v = 4$. (Reproduced with permission from same source as Fig. 7.1.)

observed for the $v' = 3$ state of HF from the reaction of F with p-H_2, at an energy of $E_T = 1.84\,\text{kcal mol}^{-1}$ (Fig. 7.12). These "anomalies" have been interpreted, quite reasonably, as manifestations of long-lived resonances, occurring at specific energies, leading to wide-angle, sometimes nearly isotropic scattering of a particular vibrational state of the product. Theory and approximate quantal calculations based on the best available potential surface(s) confirm the origin of such resonant scattering, but quantitative shortcomings remain (Fig. 7.13).

We are forced to conclude this section with the admission that despite the intense experimental and theoretical efforts devoted to the study of this elementary reaction, a complete and quantitative description is not yet in

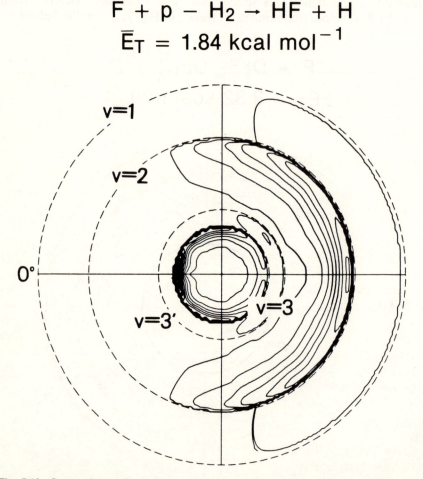

$$F + p - H_2 \rightarrow HF + H$$
$$\overline{E}_T = 1.84 \text{ kcal mol}^{-1}$$

Fig. 7.12. Center-of-mass flux–velocity contour map for the $F + p$-H_2 reaction at $E_T = 1.84\,\text{kcal mol}^{-1}$. Note the forward peak for $v = 3$. (Reproduced with permission from same source as Fig. 7.1.)

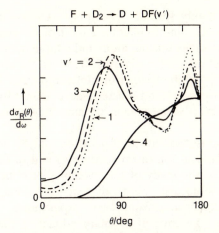

Fig. 7.13. Differential cross section into specific DF vibrational states for the $F + D_2 \rightarrow$ $DF(v') + D$ reaction, computed using the "quantal sudden" approximation at $E_T =$ $4.52 \, \text{kcal mol}^{-1}$. [Adapted from N. Abusalbi, C. L. Shoemaker, D. J. Kouri, J. Jellinek, and M. Baer. *J. Chem. Phys., 80*, 3219 (1984). See also R. E. Wyatt, J. F. McNutt, and M. J. Redmon, *Ber. Bunsenges. Phys. Chem., 86*, 437 (1982). For corresponding classical computations, see S. Ron, E. Pollak, and M. Baer, *J. Chem. Phys., 79*, 5204 (1983).]

hand. With "a little more work" on the *ab initio* potential surface, and more work on the quantal reactive scattering computations, it *should* be possible, in the near future, to account for the main observed reactive scattering behavior.

7.2 Collision complexes: Their formation and decay

We have concentrated much of our attention thus far on direct-mode elementary reactions (e.g., $H + H_2$, $X + H_2$, $X + HY$, etc.). There is, however, a still larger universe of reactions involving more complicated reagents with relatively strong potential wells (often with orientation-dependent barriers) that proceed by way of the formation of long-lived complexes. We have begun to examine such reactions in Chapter 4 and continue in this section.

7.2.1 A union of bimolecular–unimolecular concepts

By now we know the qualitative distinctions between direct (Sec. 3.3) and compound bimolecular reactions. We have learned that the archetype of the direct mode is the spectator-stripping reaction (Sec. 1.4), where the duration of the collision is $<10^{-13} \, \text{s}$, the momentum of the "spectator" is virtually unperturbed, the molecular product is strongly "forward", and the energy disposal is highly specific. The other extreme in the spectrum of "collision times" is the bimolecular formation of an almost stable adduct, e.g., when two large radicals or an atom and a large molecule combine. The entire range of "lifetimes" of complexes, from 10^{-13} s up, has by now been probed by the many different experimental techniques. Theories since the days of Rice, Ramsperger, and Kassel (more than fifty years ago) have provided us with ever-better means of analyzing rates of unimolecular reactions.

The new feature is the direct application of the molecular collision technique (molecular and ion beam scattering) and of laser excitation and detection methods to the study of the dynamics of both the (bimolecular) *formation* and (unimolecular) *decay* of reaction complexes. Several new kinds of information come from such experiments: (1) a qualitative statement as to the existence of a binary adduct of the reactants with a lifetime longer than, say a picosecond, and shorter than typical apparatus "transit times" of, say, a microsecond; (2) a quantitative measure of the branching ratio of the complex, i.e., the relative probabilities of decay into (various possible new sets of) products or back into the original pair of reactants; (3) angular distributions in the c.m. system of (all sets of) decay products of the complex; (4) energy disposal distributions of same, i.e., the translational distributions for the decay products or the complementary measurement of their internal energy distribution; (5) the cross section for complex formation, and (6) the dependence of all of the above upon reactant energy (internal and relative translational), and upon the mode of optical or chemical activation.

This renaissance in bimolecular–unimolecular experimentation has led to a renewed attack on the dynamic theory of the formation and decay of reaction complexes. We shall begin with the approach based on a configuration of no return—the RRKM approach. Then we will try to glimpse beyond the statistical limit. Such questions as the rate of intramolecular redistribution of vibrational energy and the mechanism thereof are then relevant. Finally, dealing with laser-pumped unimolecular processes in Section 7.3, we shall find that such considerations become of central importance.

7.2.2 A simple model of the angular distribution

As the simplest example of a long-lived complex, consider a collision of two structureless particles. Suppose the complex is formed when the two particles approach. The centrifugal energy of the relative motion is now the rotational energy of the complex, and it will rotate in the plane of the collision. By the time the complex has completed a few rotations it has lost all memory of the original direction of the relative motion of the colliding pair. Hence, if the lifetime of the complex exceeds a few periods t_R of

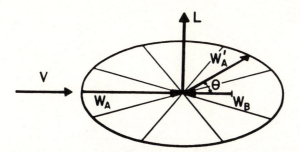

Fig. 7.14. Dynamics of a complex-mode collision between two structureless particles in the c.m. system. [The collision occurs in a plane normal to the (conserved) orbital angular momentum **L** (Sec. 2.2.5).] The diatomic complex must dissociate along its axis, and, when the complex is sufficiently long-lived, all final orientations θ of the relative velocity of the two particles will be equally probable.

rotation,† the subsequent dissociation of the complex will be isotropic in the plane of the collision (Fig. 7.14). The two particles will fly apart with equal probability in all directions θ in the plane.‡ The flux of products that appear in the angular range θ to $\theta + d\theta$ is thus independent of θ, i.e.,

$$\frac{d\dot{N}(\theta)}{d\theta} = \text{const.} \tag{7.10}$$

In terms of the solid-angle differential cross section, the angular distribution can be written§ (Sec. 3.1.3)

$$I(\theta) \propto (2\pi \sin \theta)^{-1} \frac{d\dot{N}(\theta)}{d\theta} \propto (\sin \theta)^{-1} \tag{7.11}$$

It is important to note that the angular distribution of products in space is not uniform. The physical reason is clear. The distribution is uniform in the *plane* of the collision and has (as usual) cylindrical symmetry about the direction of the initial relative velocity. When we generate the cones of equi-intensity at a given θ, and variable ϕ, (Fig. 7.15), there is much more intensity near the poles ($\theta \approx 0, \pi$) than about the equator ($\theta = \pi/2$) in

---

† The period is given by $t_R = \omega_R^{-1}$ where $\omega_R = (2E_R/I)^{1/2}$, where I is the moment of inertia of the complex and E_R the rotational energy. For the collision of structureless particles with an impact parameter b forming a complex at the separation d,

$$E_T = (E_T b^2/d^2) + V(d) = E_R + V(d)$$

and $I = \mu d^2$, $E_T = \mu v^2/2$. Thus, $t_R = (d/v)(1 - V(d)/E_T)^{-1/2}$. Note that d/v is often used as a measure of the duration of a direct collision for many purposes.

‡ Thus, in contrast to a direct reaction, the rotation of the collision complex destroys the one-to-one relation between b and the scattering angle θ that is typical of a direct collision (Sec. 3.3).

§ The unphysical divergence at $\theta = 0$ and π is smoothed out in the quantal treatment.

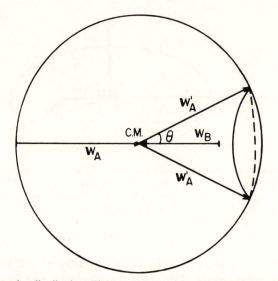

Fig. 7.15. Spatial angular distribution. The cone of equi-flux of product is generated by rotating the product velocity vector (in the c.m. system) about the initial relative velocity axis. The circle shown is the locus of all product molecules, scattered in the plane, with the same magnitude of final velocity, irrespective of scattering angle. The cone defines the locus of all product molecules scattered with the same magnitude of final velocity and the same scattering angle [see also Figs. 5.23(a)–(c)].

velocity space. Equation (7.11) implies that the c.m. angular distribution has a forward–backward symmetry.

Figure 7.16 shows flux–velocity contour maps for two examples of neutral–neutral reactions that proceed via complex formation and exhibit the forward–backward peaking as discussed above:

$$RbCl + Cs \rightarrow [RbClCs] \rightarrow Rb + CsCl$$

$$SF_6 + Rb \rightarrow [RbSF_6] \rightarrow RbF + SF_5$$

The exact form of (7.11) is not obtained in the general case of molecular (as contrasted with structureless) reactants since the rotational angular momenta of the molecules can couple to the orbital angular momentum, giving rise to "out-of-plane" angles of scattering. One can thus obtain a variety of shapes of $I(\theta)$, but all of them are characterized by forward–backward symmetry (i.e., symmetry about 90° in the c.m.).

The qualitative form of the angular distributions in realistic cases can be understood in terms of the conservation of the total angular momentum **J** (Appendix 4B). In Fig. 7.17 we consider a simple case when the reagents carry no rotational angular momentum. The products can be rotationally excited, however, and since $\mathbf{J} = \mathbf{L}' + \mathbf{j}'$ the final recoil velocity, $\mathbf{w}'$, which is in a plane perpendicular to $\mathbf{L}'$, need not be in a plane perpendicular to **J**. Let M' be the magnitude of the projection of **J** along the final recoil velocity $\mathbf{w}'$. For a given J and M' the products emerge in a cone of half-angle

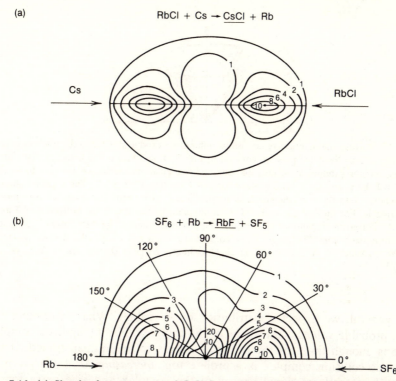

Fig. 7.16. (a) Sketch of contour map of CsCl flux–velocity distribution from the RbCl + Cs reaction. The forward–backward peaking of the angular distribution suggests a RbClCs complex which dissociates along its symmetry axis. (Adapted from unpublished results of W. B. Miller, S. A. Safron, G. A. Fisk, J. D. McDonald, and D. R. Herschbach.) (b) Detailed flux–velocity contour map for the Rb + SF$_6$ reaction. The RbF flux distribution is nearly symmetric (forward–backward peaking) but with a slight forward bias. Radial tic marks indicate recoil velocity intervals of 100 m s^{-1}. Peak intensity = 10. [Adapted from S. J. Riley and D. R. Herschbach, *J. Chem. Phys.*, *58*, 27 (1973). See also S. Stolte, A. E. Proctor, W. M. Pope, and R. B. Bernstein, *J. Chem. Phys.*, *66*, 3468 (1977).]

$\alpha' = \cos^{-1}(M'/J)$ about **J**. The direction of **J** is conserved throughout the collision, but since it is not selected, **J** is perpendicular to **v** but with all azimuthal orientations equally likely (as in the case for **L**; Fig. 7.14).

If the collision complex dissociates preferentially with high M' values, then the final relative velocity **v**' is almost in the direction of **J** and hence almost perpendicular to the initial relative velocity **v**. The products are then preferentially scattered about the equator. On the other hand, if low M' values predominate, **v**' and **v** are perpendicular to **J** and there is more products' intensity near the poles, just as in the case where **J** = **L**.

Quantitatively, it is seen from Fig. 7.17 that the scattering angle θ is, for a given M', limited to within the range α', $\pi - \alpha'$, or

$$I(\theta) \propto (\sin^2 \theta - \cos^2 \alpha')^{-1/2}, \quad \text{given } M' \qquad (7.12)$$

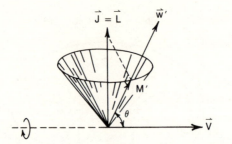

Fig. 7.17. Angular momentum relationships influencing products' velocity distribution follow-ing decay of long-lived complex. In contrast to Fig. 7.14, in which it was assumed that neither the reagent nor product molecules had rotational angular momenta, i.e., $j = j' = 0$, so that $L = L' = J$, here we take $j = 0$ but allow for $j' \neq 0$ (so that $J = L \neq L'$). For a given collision, with relative velocity v as shown, L is perpendicular to v, with all azimuthal orientations equiprobable. For the one shown $L(=J)$ is in the plane of the figure; the projection of J upon w' is M'. For given magnitudes of J and M' the products' angular (θ) distribution is generated by uniform precessions of w' about J (see cone) and J about v (see azimuthal arrow). [Adapted from W. B. Miller, S. A. Safron, and D. R. Herschbach, *Faraday Discuss. Chem. Soc., 44*, 108 (1967).]

In general, we expect a distribution of M' values. What determines the more probable values is the shape of the complex. For a long-lived complex there is enough time for the available energy to be equipartitioned. If the triatomic collision complex is a prolate top, *increasing M'* (at a given J) means increased rotational energy of the complex and vice versa for an oblate top. Hence a prolate top will dissociate preferentially towards the poles (as in Fig. 7.16), while an oblate top will result in sideways peaking. Figure 7.18 shows an example of a complex dissociating preferentially

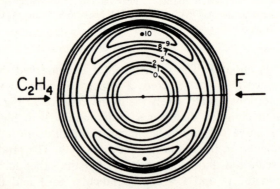

Fig. 7.18. Product flux–velocity contour map of C_2H_3F from the reaction of $F + C_2H_4$. The sideways angular distribution implies that the velocity of the departing H atom will be approximately perpendicular to the incident velocity of F, and hence to the C–C–F plane. [Adapted from J. M. Parson, K. Shobatake, Y. T. Lee, and S. A. Rice, *Faraday Discuss. Chem. Soc., 55*, 344 (1973).]

sideways for the reaction

$$C_2H_4 + F \rightarrow [C_2H_4F] \rightarrow HF + C_2H_3$$

Both (7.11) and (7.12) are classical approximations with the typical classical divergences at the boundaries. The quantal results are of course finite (but large) where the classical results diverge.

As the collision energy is increased there is a gradual transition from complex-mode to direct-mode of reaction. This is evident both in the observed angular distributions, which become asymmetric as E_T is increased, and in the decrease in the theoretically computed lifetime. The factors that determine whether a reaction proceeds via a long-lived complex or not thus include both the nature of the reactants and such dynamical variables as the collision energy and the internal energy of the reactants. Let us examine this in some detail.

7.2.3 *Qualitative criteria for complex formation: Structure and stability*

It is convenient to adopt a picture of the collision complex as an energy-rich molecule that would be stable if drained of its excess energy (Sec. 4.3.6). Hence the primary class of factors relevant to complex formation are the "static" ones, i.e., the nature of the reactants and their interaction. If a complex is to be formed, we expect a well in the energy profile along the reaction path (Sec. 4.3.6) leading from reactants to products (Fig. 7.19). We thus have our first criterion: The chemical structure of the possible collision complex must correspond to a hollow in the potential-energy surface. In the reaction

$$C_2H_4^+ + C_2H_4 \rightarrow [C_4H_8^+] \rightarrow C_3H_5^+ + CH_3$$

the ion $C_4H_8^+$ as the butene (or cyclobutane) cation satisfies this condition.

Another interesting example arises in connection with alkali halide systems. Stable dimers such as $(NaCl)_2$ are known gaseous species, so it is not surprising that the mutual scattering of alkali halide molecules should proceed via a long-lived complex. One of the reactions studied is

$$KCl + NaBr \rightarrow KBr + NaCl$$

Both experimental and trajectory-simulation scattering studies (see Fig. 4.31) confirm the complex mode for such reactions. More recently, supersonic expansion of a mixture containing alkali metal M and alkali halides $M'X$ has been used to prepare the alkali dimer–monohalide. These are the intermediates† in the reaction of Fig. 7.16(a). Photoionization measurements confirm the low ionization potential of these and show that they are stable but somewhat less strongly bound than the MX dimers.

---

† In the nozzle expansion process, the energy-rich MXM' triatomics do not fall apart since they are stabilized by "third body" collisions.

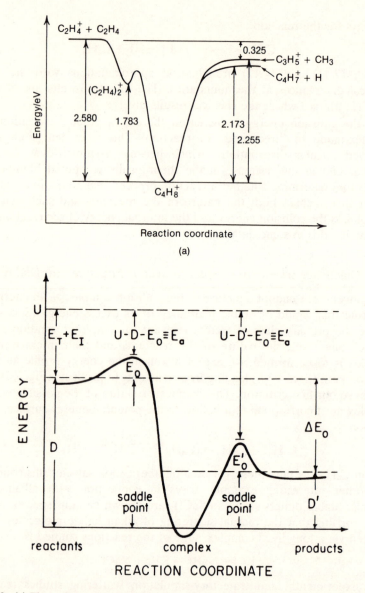

Fig. 7.19. (a) The main topographical features along the reaction coordinates for the exoergic reaction $C_2H_4^+ + C_2H_4 \rightarrow [C_4H_8^+] \rightarrow C_3H_5^+ + CH_3$ or $\rightarrow C_4H_7^+ + H$. Note the absence of a barrier along both exit valleys. [Adapted from W. J. Chesnavich, L. Bass, T. Su, and M. T. Bowers, *J. Chem. Phys.*, **74**, 2228 (1981).] (b) Schematic energy profile for the generic case. the symbols are defined and discussed in Sec. 7.2.4.

Other examples of long-lived complexes include

$$H_2^+ + H_2 \rightarrow [H_4^+] \rightarrow H_3^+ + H$$
$$H^+ + D_2 \rightarrow [HD_2^+] \rightarrow HD + D^+$$

where there is ample reason to expect the intermediate ions to be stable at low energies. In contrast, we recall that the analogous neutral intermediates are unstable, e.g., the ground electronic state H_3 surface has a barrier (saddle region), so that the reaction

$$H + D_2 \rightarrow HD + D$$

proceeds via a direct mode.†

The hollow in the potential surface must be such that the supermolecule is stable against *any* mode of dissociation (there can be no "leaks" in the bowl). Thus, the reaction of $K + I_2$ does not form a long-lived complex even though there is a steep drop in the potential surface "early downhill" along the reaction coordinate, apparently since the charge-transfer intermediate $K^+I_2^-$ quickly dissociates to $KI + I$.

Another interesting complication is exemplified by the hydrogen-abstraction reaction

$$O(^3P) + H_2 \rightarrow H + OH$$

which proceeds by the direct mode even though the HOH is a stable species. In contrast, the analogous $O(^1D)$ reaction goes via insertion to form an HOH complex that decays to $H + OH$. (Of course, these reactions proceed on different electronic state surfaces.) Another example is the ionic analog, for which it has been shown from quantum-chemistry considerations that H_2O^+ does not dissociate to OH^+, but rather to OH:

$$H_2O^+ \begin{cases} \nrightarrow OH^+ + H \\ \rightarrow OH + H^+ \end{cases}$$

i.e., the H_2O^+ well is on a different potential energy surface from that relevant to the formation of OH^+.

The presence or absence of a well is very manifest in the dynamics. Figure 7.20 contrasts the computed CO vibrational state distribution from the two possible product sets in the reaction

$$O(^3P) + CN \rightarrow \begin{cases} CO(v) + N(^4S) \\ [OCN] \rightarrow CO(v) + N(^2D) \end{cases}$$

Thus, unless the electronic state of the nitrogen atom is determined, the observed CO vibrational state distribution will be bimodal.

Note that it is not enough that a "stable" intermediate is possible; it is

---

† Both experiments and *ab initio* computations suggest that the barrier for the four center exchange reaction $H_2 + D_2 \rightarrow 2HD$ is not much lower than the H_2 bond energy.

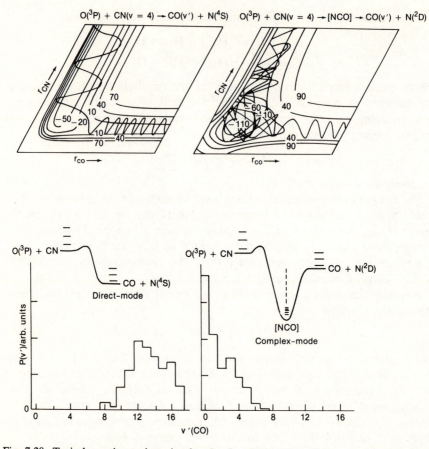

Fig. 7.20. Typical reactive trajectories for the $O + CN(v = 4)$ collision on potential energy surfaces corresponding to the formation of $N(^2D)$ and $N(^4S)$, respectively. The bottom panels show the resulting CO vibrational energy distribution (and, in the inset, the profile of potential energy along the reaction coordinate). [Adapted from J. Wolfrum, *Ber. Bunsenges. Phys. Chem.*, **81**, 114 (1977).]

necessary that it be energetically accessible, i.e., that the entrance barrier can be overcome (and, if there is to be decay into "products," that the exit barrier can be surmounted). We recognize also that the centrifugal barrier will serve to keep the particles outside the range of the necessary "close-in" separations for complex formation. Thus, there is always a value of b beyond which the distance of closest approach, R_0 ($\approx b$) is "too large" and no complex can form.

Our first criterion, that the energy-poor collision complex correspond to a stable entity, is not sufficient. If the energy-rich supermolecule formed in the actual experiment is to manifest itself as a collision complex, it must have a lifetime exceeding the typical (vibrational period) duration of the direct-mode reactions (*ca.* 10^{-13} s) and approaching the typical rotational period (*ca.* 10^{-12} s).

Thus, complexity of the reactants tends to favor complex formation since the collision energy can be well dissipated among the many vibrational modes, say *s*, characteristic of a large polyatomic molecule. It takes some time for this energy to be localized in a given bond (e.g., a fissile bond), so that the lifetime of the adduct is drastically lengthened with increasing *s*. An example is the reaction of $F + C_2H_4$, which, in contrast to the $F + H_2$ reaction, has enough degrees of freedom to equipartition the exothermicity. Another example is the pair of reactions

$$Cl + Br_2 \rightarrow ClBr + Br$$

and

$$Cl + \overset{\frown}{\underset{\diagdown Br}{\diagup}} \longrightarrow \overset{\frown}{\underset{\diagdown Cl}{\diagup}} + Br$$

where the first is direct but the second goes via complex formation.

Now let us turn to the dynamical factors that can be subject to experimental variation and that govern complex formation. Consider the effect of increasing the collision energy, for example. As the energy is increased, the local details of the potential-energy surface are of less and less importance and the result is the transition to the direct mode at higher energies.

Figure 7.21 contrasts the nearly symmetric scattering map at low energy with the strongly forward, anisotropic scattering at high energy observed for the reaction

$$C^+ + H_2O \rightarrow H + HCO^+$$

Another illustration of the tendency to direct scattering at higher energies is the $CH^+(A^1\Pi)$ vibrational energy distribution in the reaction

$$C^+(^3P) + H_2 \rightarrow CH^+(A^1\Pi) + H$$

Figure 7.22 shows the results of a surprisal analysis of the vibrational-state distribution as a function of the collision energy. At lower energies the observed distribution is essentially the prior one. As more energy is available, there is a preferential release of energy into the vibration, as in many other direct reactions.

7.2.4 Quantitative considerations: The available energy

Our need is to develop a simple model that would provide a quantitative account of the main features of complex formation and decay. In particular, we would like to estimate the lifetime of the complex against decay and the branching ratio, i.e., the ratio of the decay rates into products versus reactants. The details of the method, as well as the estimates of the cross section and the energy disposal, are given in Sections 4.4.10 and 7.2.5.

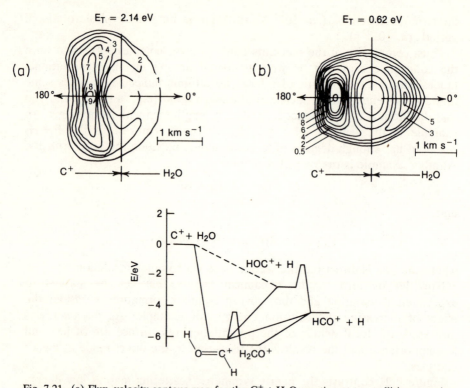

Fig. 7.21. (a) Flux–velocity contour map for the $C^+ + H_2O$ reaction at two collision energies. The ion detected is $m/z = 29$, corresponding to HCO^+ and/or HOC^+. At the lower E_T the angular distribution is much more nearly symmetric, showing near forward–backward symmetry. The slight asymmetry observed suggests an "osculating complex" collision, with the lifetime of the complex being somewhat shorter than its rotational period. The predominance of backward scattering at the higher energy and the analysis of the translational energy disposal suggests a contribution from a low-impact parameter, direct, "knockout" mechanism in which the incoming C^+ ion ejects an H atom into the forward direction. (b) Schematic potential energy profile along the reaction coordinate for the complex-mode reaction showing routes to both isomers of $m/z = 29$. The dashed line is for the knockout process. [Adapted from D. M. Sonnenfroh, R. A. Curtis, and J. M. Farrar, *J. Chem. Phys.*, **83**, 3958 (1985). See also B. H. Mahan, *Acc. Chem. res.*, **3**, 393 (1970).]

We consider an exoergic reaction with the schematic energy profile shown in Fig. 7.19(b). There is a barrier along the reaction path leading from reactants to the complex and along the path from the complex to the products. We will adopt the following assumption: The formation and decay of the complex is via a transition state (Sec. 4.4) located on the barrier in the reaction path.† Thus, for the reaction shown in Fig. 7.19(a) we envisage

---

† Note that this implies that the dissociation of the complex to form products is via a transition state located on the right barrier in Fig. 7.19(b), while the dissociation to the reactants (or formation) is via a different transition state located on the left barrier. Different decay modes have different transition states.

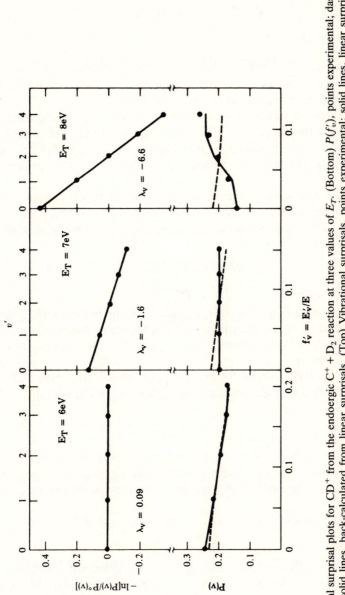

Fig. 7.22. Vibrational surprisal plots for CD$^+$ from the endoergic C$^+$ + D$_2$ reaction at three values of E_T. (Bottom) $P(f_v)$, points experimental; dashed lines, prior $P^0(f_v)$, solid lines, back-calculated from linear surprisals. (Top) Vibrational surprisals, points experimental; solid lines, linear surprisals. Abscissa scale, f_v' (upper scale shows v' values). [Adapted from E. Zamir, R. D. Levine, and R. B. Bernstein, *Chem. Phys.*, *55*, 57 (1981); experimental results of I. Kusunoki and C. Ottinger, *J. Chem. Phys.*, *71*, 4227 (1979). For a cautionary note, however, see I. Kusunoki and C. Ottinger, *Chem. Phys. Lett.*, *109*, 554 (1984).]

the formation stage as

$$C_2H_4^+ + C_2H_4 \rightarrow [C_4H_8^+]$$

and the decay to products as

$$[C_4H_8^+] \rightarrow [C_4H_8^+]' \rightarrow C_3H_5^+ + CH_3$$

Clearly, the transition-state configuration, located at the barrier, is not the same as that of the complex located over the hollow.

Having adopted a transition-state model,† we have the whole machinery of the statistical theory (Sec. 4.4.7) at our disposal. Every decay mode has its own transition state, and the rate of decay of the complex in a given mode is simply the rate of passage via the particular transition-state configuration.

The passage rate is determined, in the statistical theory, by the energy available to the transition-state configuration. The energetics have been shown in Fig. 7.19. U is the total energy *measured from the ground state of the complex*:

$$U = D + E_0 + E_a$$
$$= D' + E_0' + E_a' \tag{7.13}$$

Here, E_a is the energy available for partitioning between the internal degrees of freedom of the transition state and the translational motion along the reaction path at the entrance barrier; E_a' has the same meaning referring to the "exit" barrier.

Adopting the RRK model for the structure of the transition state (see Sec. 7.2.5), we find that the unimolecular decay rate of the complex to form products is

$$k' = \bar{v}(E_a'/U)^{s-1} \tag{7.14}$$

Here, $\bar{v}$ is an average vibrational frequency of the complex (typically, $\bar{v} \approx 10^{-13} \text{ s}^{-1}$) and s is the number of normal vibrational modes in the complex. The corresponding rate for decay to form back the reactants is $\bar{v}(E_a/U)^{s-1}$.

If the complex is to be long-lived, we thus require that $E_a' \ll U$ or

$$(U - D' - E_0')/U \ll 1 \tag{7.15}$$

and that $s \gg 1$. It is evident that these two conditions summarize the qualitative criteria discussed in Section 7.2.3. In particular, since $U =$

---

† Is a compound collision not a collision proceeding via a resonance (or a quasi-bound predissociating state in the terminology of Sec. 6.5)? Why then do we not write the rate constant as a sum over such contributions? We could, and the reason we do not is that for the processes under present considerations we are populating very, very many different quasi-bound states (e.g., in many different partial waves). The properties of any individual state are thus not of interest. This is in contrast to a spectroscopic experiment. The better we select the initial precollision state, the closer we come to a spectroscopic experiment.

$E_T + E_I + D$, only the total energy of the reactants should matter in determining the decay rate and branching ratio. Increasing the reactants' internal energy, E_I, has the same role as increasing the collision energy, E_T. Figure 7.23 gives the results (via a classical trajectories computation) for the decay rate, showing the near independence upon the different contributions

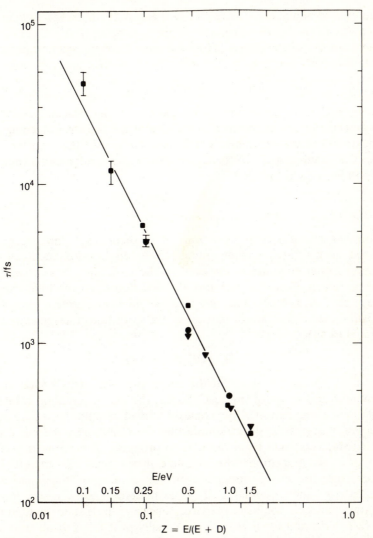

Fig. 7.23. Calculated lifetimes τ (in fs) for the decay of "strongly interacting" complexes of HD_2^+ from the reaction of $H^+ + D_2$ versus total energy, E. Different symbols denote different contributions to E, i.e., among $E_T(\blacksquare)$, E_V ($\blacktriangledown$) and E_R ($\bullet$). The results scatter around a "mean" curve of the RRK form: $\tau = \tau_0 \, z^{-s}$, where $z = E/(E + D)$, and $D = 4.58$ eV is the (assumed) well depth of HD_2^+ relative to $H^+ + D_2$. The fit yields $s = 1.97$ and $\tau_0 = 13.5$ fs. [Adapted from Ch. Schlier and U. Vix, *Chem. Phys.*, **95**, 401 (1985). See also D. Gerlich, U. Nowotny, Ch. Schlier, and E. Teloy, *Chem. Phys.*, **47**, 245 (1980).]

to E. Only the total energy matters. This, of course, is usually *not* the case for direct reactions.

The branching ratio, Γ, is independent of $\bar{v}$ and is simply a function of the available energy for the two decay modes:

$$\Gamma = k'/k = (E'_a/E_a)^{s-1} \tag{7.16}$$

For the case of an exoergic reaction, as in Fig. 7.19, $E'_a > E_a$, so that $\Gamma > 1$ and the decay favors the products. This is a general conclusion of the statistical theory, which always favors the most exoergic reaction path.†

In the simplest case, if $E_0 \approx E'_0 \ll E_a$, $E'_a \approx E_a - \Delta E_0$ and

$$\Gamma \approx [1 - \Delta E_0/(E_T + E_I)]^{s-1} \tag{7.17}$$

where $E_T + E_I = U - D$ is the initial energy of the reactants. Equation (7.17) emphasizes the importance of the sign of ΔE_0 on the branching ratio. An illustration of the qualitatively different behavior of the branching ratio for reactive–nonreactive collisions for exothermic and endothermic processes can be seen in Fig. 2.16.

7.2.5 The RRKM unimolecular reaction rate

We consider the theory for the rate of dissociation of an energy-rich, long-lived but ultimately unstable molecule. As discussed in Section 4.4, we shall assume that en route to dissociation the molecules cross a configuration of no return. Consider therefore the situation when the total energy is in the range $E, E + dE$ and the total angular momentum is J.‡ The dissociation rate is the rate of passage at the transition state, leading to [*cf.* (4.53) and (4.54)]

$$k_J(E)\, dE = N_J^{\ddagger}(E - E_0)\, dE/h \tag{7.18}$$

$N_J^{\ddagger}(E)$ is the number of states of the transition state, with a total angular momentum J and an energy up to E. So far, the only assumption is that of a configuration of no return. Now comes a second assumption, one that was not necessary in treating direct bimolecular reactions where there is a single transition state en route from reagents to products. There, the trajectories that cross at the transition state originate in the reactants' region. Here, however, the trajectories originate in the region of the quasi-bound molecule, over the potential well. It is therefore not obvious that the energy-rich molecule has been prepared such that all its states in the energy range $E, E + dE$ and for a given J are equally probable.§ If all such states

---

† Exceptions to this rule are often found for direct reactions. In general, direct reactions often favor a low translational release (Sec. 4.1).

‡ The conservation of the total angular momentum has not been enforced in the discussion in Section 4.4.

§ For direct reactions, if the reagents are in thermal equilibrium, then all reagent states in any narrow energy range and given J are equally probable.

are not equally probable, then (7.18) is not necessarily an exact result even when there are no recrossings. One therefore *assumes* that even if the initial preparation did not populate all the possible states uniformly, the subsequent rapid intramolecular energy redistribution insures a thorough mixing. We have already discussed experiments designed to examine the validity of this assumption (Sec. 4.4.11) and will do so again in Section 7.3.

Subject to both assumptions, the dissociation rate constant (per state of the energy-rich molecule) is

$$k_{JE} = N_J^{\ddagger}(E - E_0)/h\rho_J(E) \qquad (7.19)$$

where $\rho_J(E)\,dE$ is the number of states (over the well) in the energy range $E, E + dE$, i.e., the density of states with a total angular momentum J.

The main difficulty in computing the number of states at the transition state lies in those degrees of freedom of the complex that become the free rotations and the relative orbital motion of the products.

The RRKM theory explains the dependence of the unimolecular decay or the isomerization rate of an energy-rich polyatomic molecule, with few assumptions about the structure of the transition state. An example is shown in Fig. 7.24 for the isomerization of excited diphenylbutadiene.

7.2.6 Energy disposal and energy consumption in compound collisions

That a collision proceeds via a long-lived complex does not necessarily imply that it is not selective in its energy requirements or that it is not specific in its energy disposal. An already mentioned example is the elimination of HX molecules from energy-rich alkyl halides, which channels a considerable fraction of the energy into the HX vibration. Even if the rate of unimolecular dissociation is given by the RRKM considerations, one cannot, strictly speaking, predict the energy disposal. The reason is that the theory does not look past the transition state.† Under what conditions can we expect an equipartitioning of energy in the products (or no dependence on the distribution of energy in the reactants)?

The answer is based on physical considerations that go beyond the minimal RRKM assumption that energy is randomized in the complex itself. These additional considerations can fail without the RRKM assumption itself being invalid. As an example, consider a unimolecular bond rupture, as in several of the reactions already mentioned. Where will the transition state be? For a simple bond rupture the potential energy barrier, if any, will be small. The maximum in the effective potential will therefore be primarily due to the centrifugal barrier. It will therefore occur at larger relative separation. The transition state will be a "loose" one—that is, the two products have already nearly achieved their final geometry and are not

---

† In Section 4.4.10 we have provided a more detailed discussion, including an explicit result, Eq. (4.69), for the state-to-state cross section in the limit where the transition-state assumption and the *additional* RRKM assumption of energy randomization in the complex are exact.

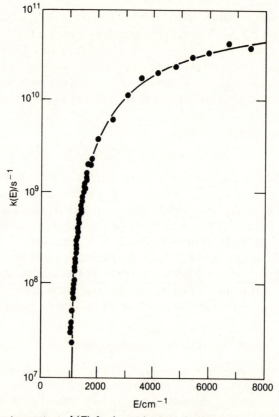

Fig. 7.24. Specific rate constants $k(E)$ for isomerization of excited diphenylbutadiene; Points experimental; solid line, optimized RRKM fit ($E_0 = 1100 \, \text{cm}^{-1}$, and a rigid transition state.) [From J. Troe, A. Amirav, and J. Jortner, *Chem. Phys. Lett.*, *115*, 245 (1985); experimental results of J. F. Shepanski, B. W. Keelan, and A. H. Zewail, *Chem. Phys. Lett*, *103*, 9 (1983); and A. Amirav, M. Sonnenschein, and J. Jortner, *Chem. Phys.*, *102*, 305 (1986).]

strongly interacting as they recede past the transition state. The RRKM assumption (plus the usual assumption of a configuration of no return) implies that energy is equipartitioned in the transition state and hence *de facto* in the products.

Another line of reasoning leading to the same conclusion is based on the inefficiency of $V-T$ transfer in the adiabatic regime. The argument begins in the same way as above. At the transition state, energy is equipartitioned. As the products separate beyond the transition state the coupling is weaker (since their separation is growing) and the motion is often slow enough for the transfer to be adiabatic.

These arguments are reasonable, but they can fail for a variety of realistic situations. For example, consider the four-center HX elimination. The potential energy barrier is very high (e.g., Fig. 4.20). The energy that is equipartitioned at the transition state is the excess energy above the barrier. Past the transition state en route to the products the entire barrier energy

($\approx 300\,\mathrm{kJ\,mol^{-1}}$ for CH_3CF_3) needs to be converted into internal and translational energies of the products. There is no reason for this energy to be equipartitioned, and indeed it is not. Three-center HX eliminations, which proceed via a much reduced barrier, lead to a much less specific HX vibrational-energy disposal.

Rotational-state distributions for diatomic molecules are more likely to distort en route from the transition state to the products. Indeed, even for reactions that proceed over a fairly deep well, the observed final distributions often deviate significantly from prior expectations. Table 7.2 is a summary of the rotational surprisal parameter for OH formed from several different elementary reactions.

Table 7.2. Surprisal analysis of rotational-energy disposal[a]

Reagents	E (kcal mol^{-1})	θ_R
$CO_2 + H$	18	2.8
	35	2.8
$O_2 + H$	6	1.2
	27	−0.7
	43	−1.2
$NO_2 + H$	31	−2.7
$O(^1D) + H_2$	45	−3.5
$O(^1D) + HCl$	46	−4.7
$O(^1D) + HCH_3$	45	−6.5

[a] Rotational surprisal parameter θ_R for nascent OH arising from decay of long-lived intermediates formed via several bimolecular reactions at specified values of (average) total energy E.

From K. Kleinermanns, E. Linnebach, and J. Wolfrum, *J. Phys. Chem.*, **89**, 2525 (1985).

Interactions past the transition state (en route to products) or prior to the transition state (en route from the reactants) are sometimes known as *exit* or *entrance valley interactions*. They can drastically modify the implications of energy randomization in the long-lived complex. Beyond that we should not overlook the fact that energy randomization is itself an assumption, as is the no recrossing of the transition state configuration. The latter requires, on physical grounds, that the barrier region be a real bottleneck for the motion. For a larger molecule, where energy is distributed over many degrees of freedom, this will often obtain up to fairly high total energies.

7.3 Multiphoton dissociation

When atoms or molecules are irradiated by a high-intensity laser beam, the rate of molecule–photon collisions can be very high, and so multiple photon processes can occur. Early reports that unimolecular dissociation or isomerization of molecules could be achieved via infrared multiphoton absorption were, however, accepted with some skepticism. Two features of the original results (based on CO_2 laser pumping of polyatomic molecules) were particularly puzzling. The first is related to the absorption mechanism. Laser radiation is very nearly monochromatic. Assuming the first infrared photon to be resonant with some vibrotational transition out of an initial level (thermally populated in the molecule), the inevitable vibrational anharmonicity means that the second photon will no longer be resonant with the next rung of the same vibrational ladder. The second part of the puzzle is the incredible efficiency. Almost all the molecules in a bulk gas sample can be dissociated by intense laser radiation at a particular wavelength. Yet in a mixture of two species it is possible to dissociate one of them highly selectively. This species selectivity is so great that one can achieve practical isotope separation by using a laser wavelength at which only one isotopic variant absorbs.

The same high species selectivity is also found to be the case for multiphoton absorption of visible or ultraviolet photons. Of course, at exceedingly high laser fields, the electric field of the radiation is strong enough to shift the molecular energy levels significantly, so species selectivity will be reduced. Yet "resonance enhanced" multiphoton absorption is a widespread phenomenon. The question is therefore how the second and subsequent photons get absorbed.

7.3.1 Intramolecular vibrational energy redistribution

To understand multiphoton absorption and a host of other problems in molecular dynamics we need to understand the level structure and dynamics of energy-rich polyatomic molecules. At low levels of excitation we have the limit made familiar by spectroscopy and depicted in Fig. 6.3: The motion of the entire molecule is a superposition of independent normal modes. The excitation energy in each normal mode remains constant in time. Yet even at the low energies shown in Fig. 6.3 there are telltale signs that all is not well. The 020 level is quite close in energy to the 100 level. Because of the anharmonicity of the potential, these levels will be coupled. As the energy is increased, overtones will tend to be increasingly in resonance. An example of such intramolecular energy exchange is vibrational predissociation.

At the other extreme, we have the energy-rich, long-lived quasi-bound molecules of Section 7.2; independently of the manner of their formation, by the time they dissociate their energy is equipartitioned. Spectroscopic experiments at energies that are high but remain below the threshold for dissociation are not easy to perform. Direct overtone excitation probabilities are very low, and the spectra often show quite broad lines, which are interpreted as broadening due to intramolecular energy transfer from the initially excited state to other modes. This is not to say that the bound

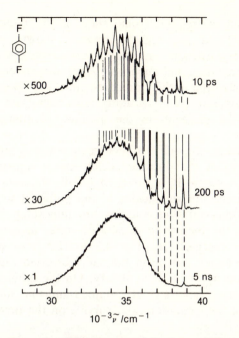

Fig. 7.25. Emission spectrum of p-difluorobenzene following excitation of the $3_0^2 5_0^1$ $S_1 \leftarrow S_0$ band with photons such that the electronically excited molecule "starts out" in a zero-order level with $E_v = 3100$ cm^{-1}. (Bottom) Spectrum corresponds to essentially collision-free emission (no added O_2) after 5 ns (*ca.* the radiative lifetime of the excited state). (Center) Spectrum with added O_2 (pressure 0.67 atm); reduction in fluorescence lifetime to *ca.* 200 ps. (Top) Spectrum, added O_2 (pressure 32.4 atm); lifetime *ca.* 10 ps. Multipliers (left) show loss in intensity (reduced fluorescence yield) at higher O_2 pressures. [Adapted from R. A. Coveleskie, D. A. Dolson, and C. S. Parmenter, *J. Phys. Chem.*, **89**, 645 (1985).]

molecule does not have stationary eigenstates that are solutions of the Schrödinger equation and whose energy does not vary with time. Of course such states exist, but at higher excitation they are no longer normal modes. Nor are such states easily accessed by optical or collisional excitation. Such excitations typically generate a superposition of such stationary states that evolve over time. Since each stationary state has a somewhat different energy, it acquires with the passage of time a somewhat different phase factor $[\exp(iEt/\hbar)]$ and the nature of the excited state changes.

One way of following the change in the character of the excited state is to monitor the emitted light, i.e., the fluorescence (see Fig. 6.5). How can we do a time-resolved fluorescence experiment on the rapid time scale of interest? An indirect but useful way is to observe the quenching of the fluorescence after a short time caused by the addition of an effective quencher molecule. Figure 7.25 shows the emission following excitation of a $3^2 5^1$ vibrational state of p-difluorobenzene (vibrational excitation,

$3310\,\text{cm}^{-1}$) in an excited electronic state in the presence of increasing pressures of O_2. At the highest pressure shown, where bimolecular collisions are very frequent, molecules survive on the average for about 10 ps between excitation and quenching. Such fluorescence that is observed is essentially due to the initially prepared state. At lower O_2 pressures, the emission is far less structured, as more states have been populated and can emit, whereas the emission in the absence of quencher is completely featureless.

Such experiments, as well as direct-frequency and time-resolved studies (see Fig. 6.5), verify that intramolecular energy redistribution is a quite general phenomenon, with, however, markedly different system-dependent details. Both direct anharmonicities of the potential well and rotationally mediated vibration–vibration coupling are important. In general, in any given molecule, the increasing density of states makes the redistribution faster at higher energies (see Fig. 4.39). The rates will therefore vary considerably, but they can be in the subpicosecond range at high excitations, as required by the RRKM theory. On the other hand, very refined studies show that not *all* states are coupled and that some memory of the initial manner of the excitation does survive on the time scale of chemical interest.

7.3.2 Infrared multiphoton dissociation

The thorough mixing of normal modes at high levels of excitation provides a key for the understanding of the extensive multiphoton up-pumping that is quite facile using infrared photons. Take the case of SF_6 where the ν_3 mode frequency is in the range of the CO_2 laser lines. The first absorbed photon is resonant with some particular rovibrational transition (Fig. 7.26). The ultimate matching is achieved by the "power broadening" of the levels by the electric field of the laser, which can reach, say, $\approx 10^{-2}\,\text{cm}^{-1}$ for SF_6 and a CO_2 laser at an intensity of $\approx 10\,\text{kW/cm}^2$. Suitable "rotational compensation" (Fig. 7.27) can also be achieved for the second and third photons, where the rotational excitation takes up the mismatch between the vibrational frequency and the laser photon. Once several photons have been absorbed, one reaches the "quasi-continuum" of high density of vibrational states where the ν_3 mode is mixed with many others, thus, transitions over a wider frequency range all have the ν_3 character and the pumping can therefore continue. The bottleneck is the first few photons. The intensity needs to be high enough to fine-tune the transitions via power broadening but not so high that another species will also absorb. Figure 7.28 shows that at intermediate powers it is indeed possible to dissociate $^{32}SF_6$ but not $^{34}SF_6$.

When the experiment is carried out in bulk, the initial absorbing rotational state depleted by the pumping is collisionally replenished owing to the large cross section for rotational energy transfer collisions. Of course, if the pressure is too high (>1 Torr for $^{34}SF_6$–$^{32}SF_6$ mixture), intermolecular

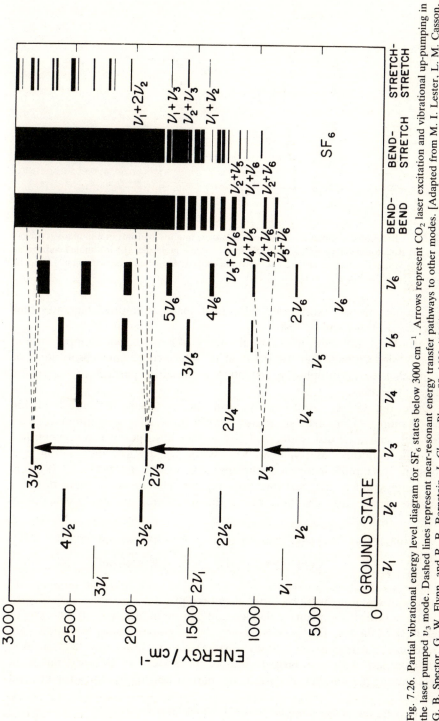

Fig. 7.26. Partial vibrational energy level diagram for SF$_6$ states below 3000 cm^{-1}. Arrows represent CO$_2$ laser excitation and vibrational up-pumping in the laser pumped ν_3 mode. Dashed lines represent near-resonant energy transfer pathways to other modes. [Adapted from M. I. Lester, L. M. Casson, G. B. Spector, G. W. Flynn, and R. B. Bernstein, *J. Chem. Phys.*, **80**, 1490 (1984), based on level diagram of R. S. McDowell and J. R. Ackerhalt (unpublished).]

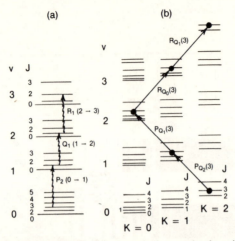

Fig. 7.27. Rotational compensation and power-broadening mechanisms in multiphoton excitation. (a) Rotational compensation principle in a diatomic molecule. (b) Rotational compensation in a symmetric top; note that J remains unchanged throughout the excitation. [Adapted from A. Ben-Shaul et al. (1981), Fig. 5.24.]

vibrational energy transfer will also take place at appreciable rates and species selectivity will be lost.

For a large molecule, the thermal energy content at room temperature is already high enough that the onset of the quasi-continuum absorption can be much lower. Multiphoton activation is therefore particularly suited for larger molecules.

Once the energy acquired by absorption exceeds the threshold for dissociation, the molecular energy levels form a true continuum. The molecule can then dissociate, and any further absorption is in competition with the unimolecular fragmentation. While the RRKM dissociation rate constant is a rapidly increasing function of the excess energy (Fig. 7.29), it is still quite small in the post-threshold regime. Hence, when there are two possible dissociation channels where the thresholds are not too different (Fig. 7.29), e.g.,

$$\begin{matrix} C_2H_5 \\ \quad\quad\quad\diagdown \\ \quad\quad\quad\quad O \\ \quad\quad\quad\diagup \\ C_2H_5 \end{matrix} \longrightarrow \begin{cases} C_2H_5O + C_2H_5, & E_a \approx 81 \text{ kcal mol}^{-1} \\ C_2H_5OH + C_2H_4, & E_a \approx 66 \text{ kcal mol}^{-1} \end{cases}$$

then both reactions will be observed to take place.

The nature of the primary dissociation can be interpreted on the basis of the RRKM theory. That is, it is possible to assign reasonable transition state structures such that the major dissociation pathway has a transition configuration with the largest number of states. A primary factor in determining the number of states is the barrier height† but the structure and

---

† Hence the rule of thumb that the weakest bond will break first.

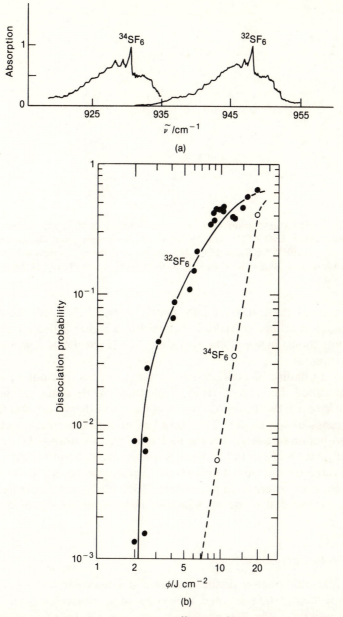

(a)

(b)

Fig. 7.28. (a) Low-resolution absorption spectra of $^{32}SF_6$ and $^{34}SF_6$ in the region of the ν_3 band (948 and 930 cm^{-1}, respectively). Note the easily observable isotopic shift. Based on this it is possible to dissociate one isotopic species selectively. [Adapted from R. V. Ambartzumian and V. S. Letokhov, *Acc. Chem. Res.*, *10*, 61 (1977).] (b) The multiphoton dissociation probability of $^{32}SF_6$ and $^{34}SF_6$ versus the fluence (pulse energy per unit area of the laser beam) of the CO_2 laser. [Adapted from W. Fuss and T. P. Cotter, *Appl. Phys.*, *12*, 265 (1977); and *Chem. Phys.*, *36*, 135 (1979).]

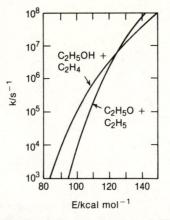

Fig. 7.29. RRKM unimolecular dissociation rate constant of $(C_2H_5)_2O$ versus energy. The dissociation into $C_2H_5O + C_2H_5$ has the higher activation energy but also higher frequency factor ($10^{16.3}$ vs. $10^{13.9}$ for $C_2H_5OH + C_2H_4$) and hence is the fast process at higher energies. [Adapted from L. J. Butler, R. J. Buss, R. J. Brudzynski, and Y. T. Lee, *J. Phys. Chem.*, **87**, 5106 (1983).]

frequencies of the transition state matter as well. Thus, for the $(C_2H_5)_2O$ dissociation reaction the radical channel has a higher activation barrier but softer vibrations; hence at energies above *ca.* 125 kcal mol^{-1}, it has a higher dissociation rate.

The very nature of multiphoton pumping implies that not all molecules have absorbed the same number of photons. While some molecules have already crossed the threshold for dissociation, others may still be in the lower rungs of the ladder. The measured unimolecular decay rates reflect the distribution of energies in the molecules past the threshold.† A thermal like distribution of the total energy is a reasonable, zeroth-order estimate.‡ For this reason one sometimes refers to multiphoton activation as "heating." However, as opposed to ordinary, "Bunsen burner," heating, we have here a method that is species selective, homogeneous, and of controllable rate.

7.3.3 Molecular beam studies

Proof that multiphoton dissociation is a unimolecular process was established by dissociating isolated molecules in a molecular beam.§ In this

---

† Where comparisons have been made, the decay rates are consistent with those measured for molecules with a sharper distribution of the energy content, as obtained via direct overtone excitation, for example.

‡ One region in which the estimate may be particularly poor is at low energies, where there may be bottlenecks. Of course, strictly speaking, the distribution is not continuous but discrete.

§ Of course, even in a supersonic beam there is some spread in the velocities of molecules, so that intrabeam collisions at very high impact parameters and very low relative velocities are still possible.

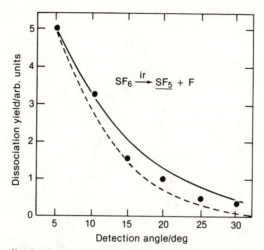

Fig. 7.30. Angular distribution of SF_5 from the multiphoton dissociation of SF_6 using a $5 \, J \, cm^{-2}$ CO_2 laser pulse of 60-ns duration (dots). The observed distribution is compared with predictions based on RRKM computed dissociation rate constants, assuming 12 (solid line) and 5 (dashed line) excess photons absorbed beyond the dissociation threshold. [Adapted from P. A. Schultz, A. S. Sudbo, E. R. Grant, Y. R. Shen, and Y. T. Lee, *J. Chem. Phys.* **72**, 4985 (1980).]

fashion one can not only establish the primary dissociation products, but also determine their angular and velocity (or state) distribution.

There is usually enough time between excitation and dissociation for the energy to be equipartitioned, as verified by the angular distribution of the dissociation products, shown for the archetypal process

$$SF_6 \rightarrow SF_5 + F$$

in Fig. 7.30. For a CO_2 laser at powers of $100 \, MW/cm^2$, the rate of up-pumping is *ca.* $10^9 \, s^{-1}$. Hence the molecules that do dissociate under such conditions have lifetimes greater than *ca.* $10^{-10} \, s$. This is ample time for many rotations and, evidently, also for energy equipartitioning.

The translational energy distribution in the products is consistent with previous studies of energy disposal in chemically activated or single-photon-activated molecules. For simple bond rupture, the energy is equipartitioned and consequently the fraction in the products' translation is low (Fig. 7.31). In a four-center elimination, e.g.,

$$CH_3CCl_3 \rightarrow CH_2CCl_2 + HCl$$

or, in general, whenever the reversed association reaction has a significant barrier [e.g., the molecular path in the dissociation of $(C_2H_5)_2O$], the fraction of energy in translation is higher (Fig. 7.32) and is not well accounted for on purely statistical grounds.

Molecular beam photofragmentation can also probe the absorption. Two variables are relevant. The intensity of the laser pulse and its fluence (the

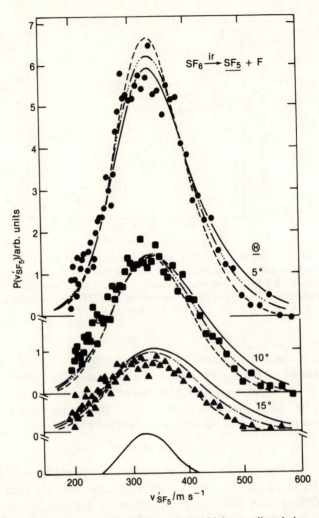

Fig. 7.31. Velocity distribution of SF_5 produced by multiphoton dissociation of SF_6 with $5 \, J \, cm^{-2}$ CO_2 laser pulses, compared with three RRKM calculated velocity distributions. The vertical axis plots $P(v')$, the number density of fragments, scaled to the observed angular distribution: (----) 5 excess photons, (---) 8 excess photons; (——) 12 excess photons. Bottom curve refers to SF_4, $\Theta > 15°$. [Adapted from P. A. Schultz, A. S. Sudbo, E. R. Grant, Y. R. Shen, and Y. T. Lee, *J. Chem. Phys.*, **72**, 4985 (1980).]

energy provided, i.e., the intensity integrated over the duration of the pulse). Intensity is required in the first few absorptions to overcome the bottlenecks to the quasi-continuum. Beyond this point, the important variable is fluence until the threshold for dissociation has been exceeded. Once the excess energy in the molecule is significant, dissociation is rapid and it competes with further up-pumping. At that point, unless the pulse is

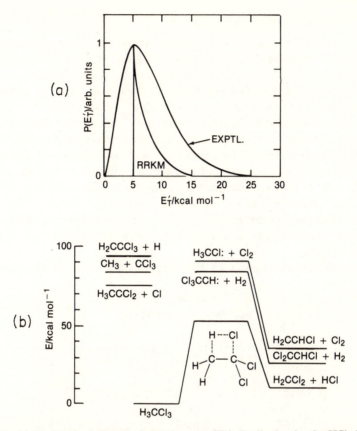

Fig. 7.32. (a) Products' relative translational energy (E_T') distribution for the HCl elimination reaction of CH_3CCl_3, in the c.m. frame of the parent molecule. The (four-center) elimination of HCl has been achieved by a 5–10 J cm^{-2} CO_2 laser pulse at 1073.3 cm^{-1} (the R12 line of the 9.6-μm CO_2 band), which accesses the 1075 cm^{-1} C—C stretching and 1084 cm^{-1} CH_3 rocking modes. The RRKM calculated distribution is scaled to agree with the peak but is not in accord with the experimental distribution. (b) Energy level diagram for the dissociation of CH_3CCl_3, showing the 4-center route. [Adapted from A. S. Sudbo, P. A. Schultz, Y. R. Shen, and Y. T. Lee, *J. Chem. Phys., 69*, 2312 (1978).]

very short, the laser intensity again matters. The higher the intensity, the further up the parent molecule can be pumped prior to dissociation.

The intensity-limited level of excitation can be observed experimentally in a number of ways. The most direct one is that the mean excess energy in the products should increase with laser intensity. For a given pulse but different molecules, the lower the unimolecular rate of dissociation, the further the molecule can be pumped up and the higher the energy content of the products. Indirect evidence is provided by the dissociation of the products. Using a high-power laser, the products will be quite energy-rich and hence in their vibrational quasi-continuum. They themselves can therefore absorb and undergo secondary multiphoton dissociation.

7.3.4 Multiphoton ionization and fragmentation

For most molecules, the first electronic excited state lies high above the ground state. However, past this first state the density of electronically excited states rapidly increases. Both valence states and Rydberg states are present. Because of selection rules, not all these states can be accessed by the absorption of a single ultraviolet or visible photon. Many of those that cannot, however, can be reached from the ground state by a coherent absorption of two (or, in general, n) ultraviolet or visible photons. It often turns out that the excited state reached by a resonance process can itself be readily ionized upon absorption of one (or m) additional photons. The rate of production of ions is then limited by the rate of absorption to form the intermediate resonant state. The overall process is termed "resonantly enhanced multiphoton ionization." Since ion collection is very efficient, this provides a sensitive probe for excited states reached by an n-photon resonant process.

The m-photon absorption of the intermediate state to the ionization

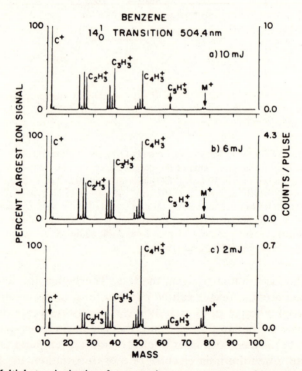

Fig. 7.33. Multiphoton ionization–fragmentation mass spectrum of benzene accessed on the 14_0^1 transition at 504.4 nm, recorded at several laser pulse energies (indicated). Note the increased fragmentation at higher pulse power levels. [Adapted from D. W. Squire, M. P. Barbalas, and R. B. Bernstein, *J. Phys. Chem.*, **87**, 1701 (1983); see also L. Zandee and R. B. Bernstein, *J. Chem. Phys.*, **71**, 1359 (1979), and K. R. Newton and R. B. Bernstein, *J. Phys. Chem.*, **87**, 2246 (1983) for further details.]

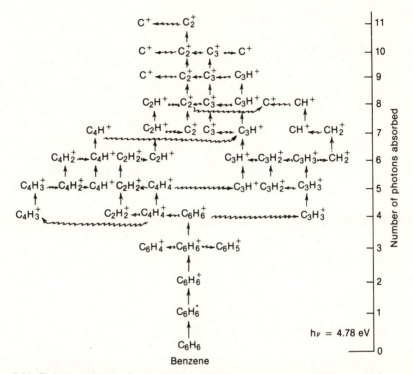

Fig. 7.34. Fragmentation pathways in the multiphoton ionization (MPI) of benzene with ultraviolet photons (259 nm, 4.78 eV), based on modeling calculations that fit observed fragmentation data. Vertical solid arrows denote absorption of one ultraviolet photon; wavy horizontal arrows indicate the fragmentation of an ion. [Adapted from W. Dietz, H. J. Neusser, U. Boesl, and E. W. Schlag, *Chem. Phys.*, **66**, 105 (1982).]

continuum is nonresonant and can be carried out with a second laser at a different wavelength. The resonance condition for the formation of the intermediate state does mean, however, that the ionization is species selective. Thus, isotopic separation can be achieved on the basis of differences in the initial absorption. Of course, for very intense lasers, power broadening is significant and molecules can also be ionized via a nonresonant process† thereby destroying the species selectivity.

If the molecular ion is stable it can be formed at high yields. Sometimes, due to the energetics of the $n + m$ photons, the molecular ion is formed at an energy above its dissociation threshold. Even if not, the ion is usually energy-rich, readily absorbs one or more additional photons, and dissociates. Typically, therefore, there is appreciable fragmentation of the parent ion (Fig. 7.33). The extent of fragmentation depends on the wavelength and

---

† Such multiphoton nonresonant absorption is a significant complication for laser-assisted processes (Sec. 6.5.6). Once the laser is intense enough to "dress" the ground state with one photon, it can also dress the excited state and thereby open an undesirable up-pumping route.

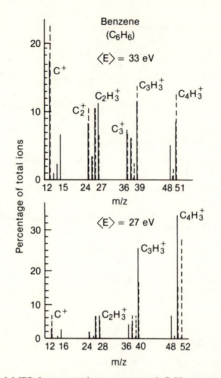

Fig. 7.35. Comparison of MPI fragmentation patterns of C_6H_6 computed via the statistical (maximal entropy) method (solid lines) with experimental results (dashed lines). (Top) Calculations assume an average energy absorbed per molecule $\langle E \rangle = 33$ eV, (Bottom) Same but for $\langle E \rangle = 27$ eV. [Adapted from J. Silberstein and R. D. Levine, *Chem. Phys. Lett.*, *74*, 6 (1980); experimental data from L. Zandee and R. B. Bernstein, *J. Chem. Phys.*, *71*, 1359 (1979), and U. Boesl, H. J. Neusser, and E. W. Schlag, *J. Chem. Phys.*, *72*, 4327 (1980), at two different laser wavelengths and pulse intensities. See also D. A. Lichtin, R. B. Bernstein, and K. R. Newton, *J. Chem. Phys.*, *75*, 5728 (1981), and J. Silberstein and R. D. Levine, *J. Chem. Phys.*, *75*, 5735 (1981).]

the laser pulse power, but the total yield of ions is essentially determined by the absorption cross section to the resonant intermediate state.

At higher laser powers, the fragmentation can be quite extensive, and small fragments (e.g., C^+) can be formed from larger molecules such as C_6H_6. It is not reasonable that such fragments are formed by direct fragmentation of a highly excited parent ion. The evidence favors a sequential mechanism where the parent ion absorbs one or more photons and dissociates; the secondary ions can also absorb and dissociate, and so on, all within the few nanoseconds of the laser pulse. The sequence of possible pathways readily multiplies, and Fig. 7.34 shows the more important steps in the fragmentation of the $C_6H_6^+$ ion.

As a first approximation, the distribution of ionic (and neutral) fragments can be computed by the procedure of maximal entropy (Sec. 5.5.4) using only the conservation laws as constraints. When an isolated molecule dissociates, the conserved quantities are energy, charge, and number of atoms of every element present. Such a simple approximation proves to be quite successful in accounting for the systematics of the multi-photon ionization–fragmentation patterns of polyatomic molecules (Fig. 7.35).

Appendix 7A: Beyond RRKM/QET

The RRKM or QET approach goes one step beyond transition-state theory. It assumes that in the energy-rich molecule the energy is equipartitioned. Yet we have seen that at low levels of excitation, where spectroscopy is usually applied, this is typically not the case. Equally, at very high levels of excitation (e.g., in a fast collision over a potential energy surface containing a well) this is also not the case. How can we understand this change in character of the dynamics for a given potential, as a function of the energy?

Mathematicians, disregarding the fact that spectroscopy preceded dynamics, pose the question the other way round. Since the true potential is anharmonic, how do we have normal modes without energy exchange between them? The important aspect is the latter. If we excite the symmetric stretch of OCS, then in the absence of collisions, the energy does not transfer to the asymmetric stretch.

Let us consider a system of two oscillators (e.g., a nonbending OCS molecule). There is always one good constant of the motion, namely, the total energy. This, in itself, does not preclude intramolecular energy transfer. The question is therefore whether there is a second constant of the motion. The analytical approach and extensive evidence from classical mechanics computations are that such second constants often exist for low energies of excitation just above the bottom of the well. However, as the energy is increased, the classical trajectory studies show that this second constant *gradually* disappears. To understand the idea of a gradual onset of energy exchange, recall that a classical trajectory for a system of N degrees of freedom requires the specification of $2N$ initial conditions (coordinates and momenta). For a given total energy E, this still leaves $2N - 1$ (i.e., 3 for two oscillators) initial conditions. If there is an additional constant of the motion, it imposes a relation among these $2N - 1$ variables, so only $2N - 2$ (i.e., 2 for two oscillators) are independent. Hence for the collinear triatomic the existence of another constant of motion makes the system behave like a single oscillator (2 independent initial conditions in classical mechanics). Whether this is or is not the case can be established by computational studies for model potential energy surfaces. What is found is that at some intermediate value of the energy, certain initial conditions give rise to a confined motion (i.e., like a single oscillator), whereas others do not. The fraction of initial conditions for which the motion shows the role of a second constant of motion diminishes as the energy is increased. How soon does the irregular or "chaotic" behavior set in? Two factors are clearly implicated. One is the magnitude of the anharmonicities.† The other, equally important, is how commensurate are the harmonic frequencies. That is, how large are the smallest integers n and m for which $n\omega_a \approx m\omega_b$? After all, van der Waals adducts like $I_2 \cdot He$ are very non-RRKM-like in their dissociation upon excitation, while an excited Ar_3 will shed off an Ar atom in near-RRKM-like fashion.

The widespread classical mechanical computational evidence for collinear triatomics is limited, however, in two important respects. For a system with two degrees of freedom, the presence of just one constant of motion (in addition to the ever-present total energy) confines the motion completely. Not so for realistic systems. There is a range of possibilities, from $N - 1$ to no such additional constants. Even in the presence of additional constants, as long as their number is below $N - 1$, some energy transfer is possible. The energy will not be equipartitioned, but it will not be confined to one mode either. The limited experimental evidence is that such a restricted intramolecular energy transfer is indeed common. Thus, the CH stretch modes in organic molecules (e.g., in C_6H_6), which are very different than the frequencies of the other modes, are coupled preferentially to the CH wags. The second problem is the validity of a classical mechanical simulation for a system that is essentially discrete.

As chemists we are interested not only in whether energy exchange does or does not ever take place but also in the rate of the transfer. Below the dissociation threshold, does it compete with the rate of multiphoton up-pumping? Above the threshold, is it fast enough to preclude preferential bond fission? Experimentally, there are examples of incomplete energy sharing on time scales of chemical interest. In a quantal language, the initially (optically) excited state is coupled to significantly fewer levels than the large number (Fig. 7A.1) energetically available. It remains, however, for such effects to be demonstrated as bond-selective processes in ordinary molecules.

7.4 Van der Waals molecules and clusters

The presence of equilibrium concentration of weakly bound van der Waals (vdW) "dimers" and higher "clusters" and their role in gas imperfection and condensation has been known for a long time. Direct experimental evidence included electron-impact ionization mass spectrometry and optical spectroscopy. The low abundance of such species (as sampled, say, in an effusive molecular beam) restricted the amount of attention given to their study.‡

---

† Alternatively stated, "the magnitude of the nonseparable terms in the Hamiltonian."
‡ Nevertheless, as mentioned in Chap. 3, vibrational bound states for several rare-gas dimers were obtained via optical spectroscopy. The potential wells determined in this fashion are in accord with the results from elastic scattering of rare-gas atomic beams.

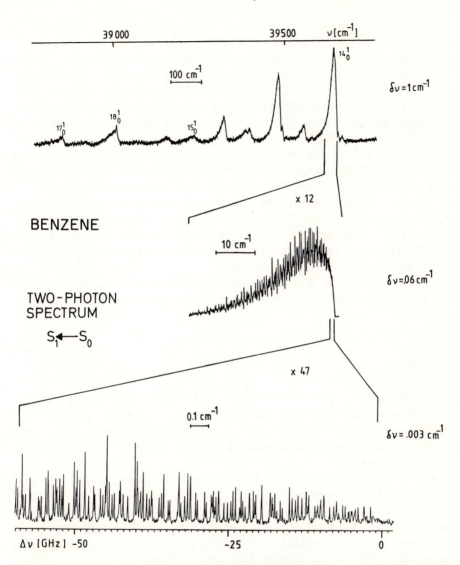

Fig. 7A.1. Fully resolved two-photon spectrum of benzene ($S_1 \leftarrow S_0$) via jet spectroscopy. (Top) Low (*ca.* 1 cm^{-1}) resolution spectrum showing the 14_0^1 and other bands. (Center) Medium (*ca.* 0.06 cm^{-1}) resolution of the 14_0^1 band. (Bottom) High (*ca.* 0.003 cm^{-1}) resolution spectrum of a 60 GHz region of the above. Every line has been assigned, and there are none calculated (from the energy level scheme) that are missing. [See S. H. Lin, Y. Fujimura, H. J. Neusser, and E. W. Schlag, *Multiphoton Spectroscopy of Molecules*, Academic Press, Orlando, FL (1984), Fig. 5.14.]

7.4.1 Beams of molecular clusters

With the development of supersonic nozzle sources (for intense, nearly monoenergetic, beams) and jet-expansion (for cold molecule spectroscopy), the presence of clusters in the beam (or jet) became so prevalent as to warrant considerable experimental and theoretical study. Limited only by the upper range of mass resolution of the mass spectrometer, ever larger vdW clusters are being investigated, e.g., Ar_{170} and many other examples to be discussed below. Somewhat different experimental methods (e.g., laser evaporation of the metal or nucleation of metal vapor in a cold carrier gas) are used to prepare strongly bound atomic clusters such as Cu_{29} or C_{60}.

There are three broad classes of studies of vdW molecules and clusters: (1) their formation, (2) their spectroscopy and structure, and (3) their reactivity and dynamics. Since these species are readily ionized by electron or photon impact, the neutral molecules provide a large number of "cluster ions" worthy of study in their own right.

The available evidence is that during their initial growth, the vdW molecules are formed via a third body (X) assisted (exoergic) association reaction:

$$M + M + X \rightarrow M_2 + X$$
$$M_2 + M + X \rightarrow M_3 + X$$
$$\vdots$$
$$M_{n-1} + M + X \rightarrow M_n + X$$

The need for the third body is less for the later growth steps since the exoergicity can be "soaked up" by the cluster molecule itself (and particularly so when many vibrational degrees of freedom are available to act as the sink).

The relative abundance of the various clusters depends on many factors, kinetic and thermodynamic. Typically, the expanded jet is not in equilibrium so that the relative stability of the clusters is an important consideration but not the only one. Kinetic factors are especially relevant in the expansion of a mixture of molecules where competition between like–like and like–unlike adducts is possible.

Understanding the systematics of cluster formation is hampered by the lack of a direct experimental method for the determination of the nascent size distribution of vdW molecules. Electron impact (even at lower energies) induces extensive fragmentation concurrent with ionization. Figure 7.36 shows the relative intensities of Cl_n^+ ions in the mass spectrum sampling an extensively clustered Cl_2 beam. Although the even-n ions, e.g., Cl_8^+ are more intense than neighboring odd-n ions, e.g., Cl_7^+ or Cl_9^+, it cannot be inferred that the Cl_8^+ ion derives only from the $(Cl_2)_4$ neutral parent.

In the mass spectrum of rare-gas clusters, peaks are often found at "magic numbers," e.g., $n = 13, 19, 55, 71$, etc. It is possible to argue on

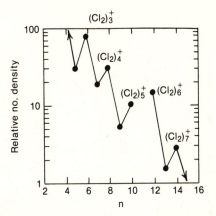

Fig. 7.36. Relative abundance of different ions $(Cl)_n^+$ in the electron-impact mass spectrum of a supersonic Cl_2 nozzle beam (semilog plot vs. number n of Cl atoms in the ion). Even-n peaks predominate; $(Cl_2)_m$ clusters molecules tend to lose molecular Cl_2 (rather than Cl atoms) upon ionization. [Adapted from R. Behrens, Jr., A. Freedman, R. R. Herm, and T. P. Parr, *J. Chem. Phys.*, *63*, 4622 (1975).]

theoretical grounds that neutral clusters of specific sizes will be more stable than their "neighbors." Another important factor is the higher stability of the corresponding *ions*. Direct experiments show that certain cluster ions are metastable and can eject a neutral to leave a more stable, lower, cluster ion, e.g.,

$$Ar_{20} + e^- \longrightarrow Ar_{20}^+ + 2e^-$$
$$\searrow$$
$$Ar_{19}^+ + Ar$$

Such cascade processes can distort the distribution of neutral vdW molecules as revealed through the ions.

For strongly bound (e.g., metal atom) clusters, fragmentation in the course of ionization is much less of a problem. Figure 7.37 shows, as an example, the multiphoton ionization mass spectrum of a beam containing C_n and $C_n La$ clusters. By far the most abundant pure carbon cluster is C_{60}. The most abundant of the La complexes as $C_{60}La$. It is tempting to interpret these results as follows: C_{60} can be a closed, hollow shell† made up of five- and six-membered "aromatic" rings. $C_{60}La$ would then be a La atom encapsulated within the carbon shell (Fig. 7.37). Of course, all that has actually been determined in the experiment shown in Fig. 7.37 is the mass of the cluster ion. We need much more before we can actually infer structures.

† Such 60-unit shells belong to the icosohedral point group.

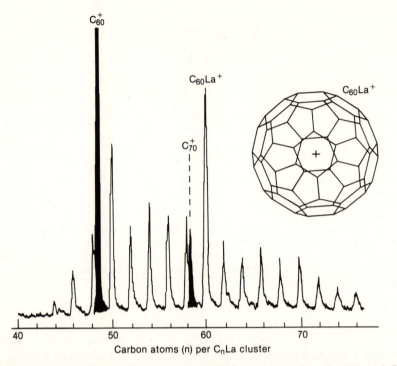

Fig. 7.37. Laser multiphoton ionization—time-of-flight mass spectrum of a beam of $C_n La$ cluster molecules and "bare" C_n clusters (blackened peaks; only $n = 60$ and 70 observed). The ionizing laser is an ArF excimer, $h\nu = 6.4$ eV; the vaporization laser is Nd:YAG, $\lambda = 532$ nm, irradiating a La-impregnated graphite sample. The most abundant La–C complexes are for even n, with $C_{60}La^+$ being the predominant one. A possible spheroidal structure for the $C_{60}La$ neutral precursor is shown, with the La bound within the spheroid. [Adapted from J. R. Heath, S. C. O'Brien, Q. Zhang, Y. Liu, R. F. Curl, H. W. Kroto, F. K. Tittel, and R. E. Smalley, *J. Am. Chem. Soc.*, *107*, 7779 (1985).]

7.4.2 Spectroscopic and structural studies

Spectroscopic studies of vdW molecules have led in some cases to real structure determination and in many cases to strong inferences. The most accurate results are obtained from molecular-beam electric resonance and from microwave spectroscopy, exemplified by the detailed Kr·HCl potential shown in Fig. 3.20. Figure 7.38(a) shows the equilibrium structure of the C_2H_2·HCN vdW molecule. Such a simple, hydrogen-bonded structure is also found for C_2H_2·HCl, but this is not general. The equilibrium structure of CH_2O·HF [Fig. 7.38(b)] has a nonlinear O—HF hydrogen bond, and, moreover, the molecule is not coplanar. Many other hydrogen-bonded dimers are known, but the vdW molecules HF·Cl$_2$ and HF·ClF are "anti-hydrogen bonded." Quantum chemical considerations help explain the bonding in such molecules, but simple rules for the structure of vdW molecules are not fully developed.

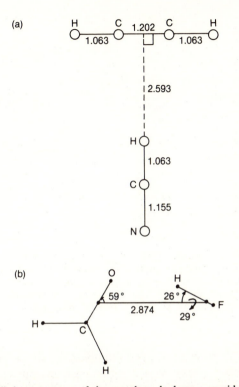

Fig. 7.38. (a) Equilibrium structure of the acetylene–hydrogen cyanide van der Waals (vdW) complex. The hydrogen bond is at a right angle to the HCCH. Distances are in Ångstroms. [Adapted from P. D. Aldrich, S. G. Kukolich, and E. J. Campbell, *J. Chem. Phys.*, *78*, 3521 (1983); for analogous results on C_2H_2–HF and –HCl, see J. A. Shea and W. H. Flygare, *J. Chem. Phys.*, *76*, 4857 (1982).] (b) Equilibrium structure of the vdW molecule $H_2CO \cdot HF$ (determined by molecular-beam electric-resonance spectroscopy) with the distance between the c.m.'s of H_2CO and HF indicated, as well as the orientation angles. Note that the HF is rotated out-of-plane by 29°. [Adapted from F. A. Baiocchi and W. Klemperer, *J. Chem. Phys.*, *78*, 3509 (1983).]

For larger vdW molecules most of the structural data comes from electronic spectroscopy (usually, laser-induced fluorescence, sometimes multiphoton ionization). Figure 7.39 shows results for the tetrazine (*s*-$C_2N_4H_2$) molecule, bare and vdW-bonded to one and to two He atoms. Accurate (± 0.01 Å) geometries can be obtained from the high-resolution spectral data.†

† The vdW bond length in the He and He_2 complexes are 3.28 and 3.32 Å, respectively. For the analogous H_2 and $(H_2)_2$ complexes, the vdW bond length is *ca.* 3.24 Å. Quite accurate geometries *and* force constants (that is, vibrational frequencies) for such vdW molecules can be computed using a potential energy hypersurface where the vdW bonding is represented as a sum of atom–atom vdW potential between the rare-gas atom(s) and all atoms of the molecule. Such energy hypersurfaces often have more than one minimum [for the location of the adduct atom(s)], showing that geometrical isomers are possible.

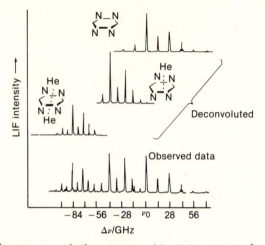

Fig. 7.39. Laser fluorescence excitation spectrum of the 0–0 band of the $^1B_{3u}$–$^1A_{1g}$ transition in s-tetrazine ($C_2N_4H_2$), say, S, and its two vdW complexes (with one and two He atoms, as shown). The observed data (bottom) have been deconvoluted to yield the three spectra above, for S, S–H, and H–S–H, as designated. Note the red shift for H–S–H (vs. S–H). [Adapted from R. E. Smalley, L. Wharton, D. H. Levy, and D. W. Chandler, *J. Chem. Phys.*, *68*, 2487 (1978).]

For still larger vdW molecules, even though structure determination is not achieved, considerable information is obtained from, for example, spectral shifts. Figure 7.40 shows the laser-induced total fluorescence intensity for anthracene seeded in Ar. The shifts in frequency from the 0–0 transition for "bare" anthracene are determined as a function of the number of Ar atoms

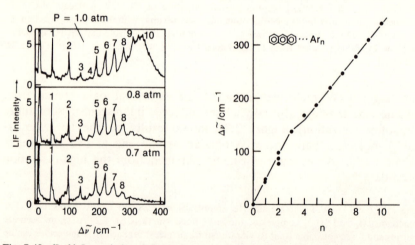

Fig. 7.40. (Left) Laser-induced fluorescence spectra of anthracene seeded in Ar at increasing pressures (indicated); numbers on peaks denote number n of Ar atoms complexed. (Right) Dependence of the wave number shift versus n for $n = 0$–10. Note the two series of Ar complexes: one for peaks 1–4 (not 3, nor 6), the second for $n = 5$–10. [Adapted from W. E. Henke, W. Yu, H. L. Selzle, and E. W. Schlag, *Chem. Phys.*, *92*, 187 (1985).]

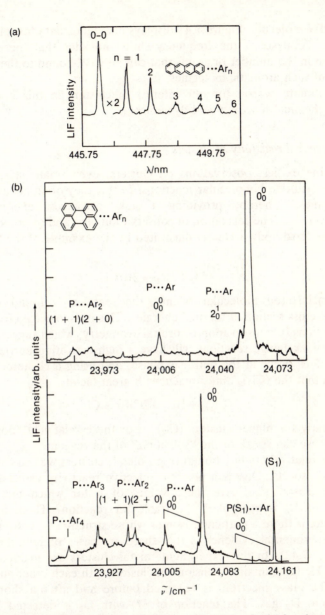

Fig. 7.41. (a) Laser-induced fluorescence of tetracene ($S_0 \rightarrow S_1$ band origin) and tetracene–Ar$_n$ vdW complexes. [Adapted from A. Amirav and J. Jortner, *Chem. Phys.*, **85**, 19 (1984).] (b) Similar: excitation spectrum ($S_0 \rightarrow S_1$) of perylene (P). Upper: $P_0 = 1.5$ atm; lower: $P_0 = 2.0$ atm. The band designated 0_0^0 is the origin of the bare perylene (S_1 is the first vibrational excitation of P). Assignments include isomers of P·Ar$_2$, i.e., the $1+1$ and $2+0$ Ar adducts. [Adapted from S. Leutwyler, *J. Chem. Phys.*, **81**, 5480 (1984).]

in the vdW molecule. Figure 7.41 displays analogous data for tetracene and perylene. Analysis of the frequency shifts indicates that there will be a saturation in the number of atoms that can be vdW-bound to the π-electron "faces" of such aromatic molecules.

The ultimate reason for our interest in clusters in this book is their unusual chemical reactivity and dynamics.

7.4.3 Chemical reactivity of clusters

One of the earliest observations on clusters, even small ones, was their ability to facilitate bimolecular reactions. In this they often act as "portable third bodies," thereby providing a sink for excess energy, angular momentum, etc. The relaxation of conservation laws and of propensity rules due to the third body is clearly illustrated by the example of the four-center reaction

$$Cl_2 + Br_2 \rightarrow 2BrCl$$

Using unclustered molecular beams, the reaction is found to have a negligible cross section at thermal energies. This is as expected from general considerations of conservation of orbital symmetry, which suggest that such four-center exchange reactions will have a high activation barrier. Yet the reaction does occur in the bulk gas phase, and using a clustered Cl_2 beam one finds that the $(Cl_2)_2$ dimer reacts with great facility,

$$(Cl_2)_2 + Br_2 \rightarrow 2ClBr + Cl_2$$

and similarly for higher clusters $(Cl_2)_n$. For these relatively "floppy" vdW molecules we can speak of the "solvation" of the reagent.

For the relatively tightly bound (e.g., metal) clusters we can consider two additional aspects. One is a possible selective reactivity associated with a preferred cluster size: Are there values of n for which the electronic structure of M_n is particularly favorable for reaction? Then, for a given cluster size, is there a particular isomer whose structure favors reactivity?

The experimental evidence is that there are cases of remarkable selectivity (Fig. 7.42). In this experiment, metal clusters are maintained in an excess of He (to minimize collisions of clusters with each other and with the walls). The mass spectrum is recorded before and after a short $(150\,\mu s)$ exposure to D_2 gas. The reaction of M_n with D_2 is detected both by a depletion of the M_n^+ ion current and by the appearance of ions of heavier mass corresponding to even numbers of added D atoms.[†] For Co, the monomer and dimer and n-mers with $n = 6$–10 are only slightly reactive. But the Co_3, Co_4, Co_5 clusters and all n-mers with $n > 10$ are extremely reactive, with the higher ones adding many D_2 molecules. $(Nb)_n$ clusters exhibit an even higher n-dependent selectivity. Clusters with $n = 8$, 10, and

---

† There is no ejection of D atoms, as expected on thermochemical grounds.

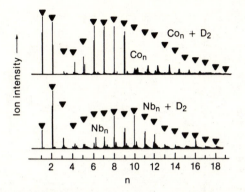

Fig. 7.42. Photoionization mass spectra of cobalt and niobium clusters reacted with D_2 (0.2 Torr) in an excess of He (*ca.* 100 Torr) at room temperature for 150 μs in a fast-flow gas reactor. The triangles shown are for the reagent cluster beam in He but without D_2. Reaction with D_2 is responsible for the reduction in the ion intensities (for those *n*-mers that are reactive) and the additional ion peaks in the mass spectrum. For $(Co)_n$, note the relative inertness of *n*-mers with $n = 1$ and 2 and 6–9, the great reactivity for $n = 3$–5, and especially $n \geq 10$. For $(Nb)_n$, there is an unusual inertness for $n = 8$, 10, and 16. [Adapted from M. E. Geusic, M. D. Morse, and R. E. Smalley, *J. Chem. Phys., 82,* 590 (1985).]

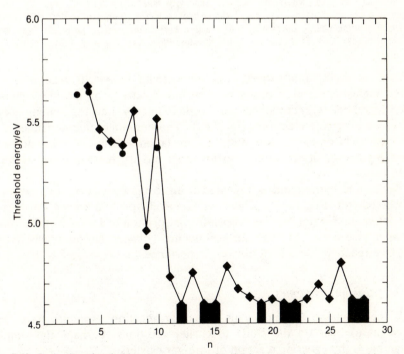

Fig. 7.43. Ionization threshold energy versus *n* for $(Nb)_n$ clusters (solid diamonds) and for $(Nb)_nO$ clusters (solid circles). Clusters with bars have thresholds below the top of the bar, i.e., ≤4.6 eV. Note the peaks in the electron binding energy for $n = 8, 10, 13, 16,$ and 26. Recall from Fig. 7.42 the inertness of $n = 8, 10,$ and 16. [Adapted from R. L. Whetten, M. R. Zakin, D. M. Cox, D. J. Trevor, and A. Kaldor, *J. Chem. Phys., 85,* 1697 (1986).]

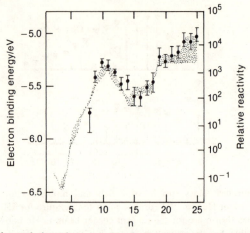

Fig. 7.44. Empirical correlation of measured ionization thresholds (the negative of the electron binding energies, plotted as ordinate) with relative reactivity of the n-mer from the reaction of Fe_n with D_2 (*cf.* Fig. 7.42 for similar reactivity data for Co_n and Nb_n with D_2). Shaded band shows range of uncertainty in the ionization threshold data; bars on points indicate uncertainties in relative data. The results are interpreted in terms of the equation $k_n/k_{n'} = \exp[\epsilon(I_{n'} - I_n)/kT]$, where the relative reactivity is $k_n(T)/k_{n'}(T)$, I_n and $I_{n'}$ are the ionization potential values, and ϵ is a dimensionless factor (found to be 0.2) that converts electron binding energy differences into reaction barrier differences. [Adapted from R. L. Whetten, D. M. Cox, D. J. Trevor, and A. Kaldor, *Phys. Rev. Lett.*, **54**, 1494 (1985).]

16 are essentially totally inert, while others (with $3 < n < 20$) react readily to yield highly deuterated products. The low reactivity of specific n-mers is well correlated with the ionization threshold energy, i.e., the binding energy of the highermost electron (Fig. 7.43). The nonmonotonic dependence of the reactivity on cluster size and the correlation of low ionization potential with high reactivity is found also for other metals, e.g., iron, as shown in Fig. 7.44.

When the experiment is repeated using copper clusters, no reaction is observed (for n up to 19). This is as expected from the known presence of an activation barrier for the dissociative adsorption of D_2 on metallic (bulk) copper (see Sec. 7.5). The correspondence between cluster reactivity and surface reactivity data is of obvious importance.

7.4.4 Reaction dynamics

All the available techniques of molecular reaction dynamics have been brought to bear on the study of cluster reactions. Perhaps the simplest example is the exoergic reaction of rare-gas clusters,

$$Xe + Ar_n \rightarrow XeAr_{n-1} + Ar$$

studied by mass spectrometric detection of products. Fig. 7.45 shows the laboratory angular distribution of $XeAr_7^+$ ions, which correspond to the

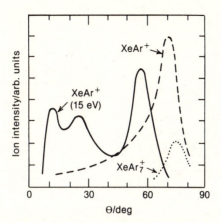

Fig. 7.45. Scattering angular distribution in the laboratory system: Xe beam at $0°$ scattered by Ar_n beam at $\theta = 90°$. Products detected by electron-impact ionization at 35 eV. Dotted curve, $XeAr_7^+$; dashed curve, $XeAr^+$; solid curve, $XeAr^+$ using 15 eV electrons, data $\times$ 100. Since the dashed curve's $XeAr^+$ peak comes at a high angle (*ca.* $80°$), inaccessible (kinematically) to a XeAr product, those ions must arise from fragmentation of larger ions, e.g., $XeAr_7^+$. [Adapted from D. R. Worsnop, S. J. Buelow, and D. R. Herschbach, *J. Phys. Chem.*, *88*, 4506 (1984).]

reaction products $XeAr_{7+m}$, $m \geq 0$ (since the parent ion may dissociate following ionization).

Direct evidence for the fragmentation of vdW clusters in the course of ionization (which is required for mass analysis) has been obtained in the crossed molecular beam scattering studies. In the experiment shown in Fig. 7.46 the *n*-mers are spatially separated (using a He projectile) and then their individual fragmentation patterns measured. It is found that Ar_3, for example, is essentially completely fragmented even for lower-energy electron-impact ionization, while Ar_2 is about 50% fragmented. For molecular clusters, fragmentation is accompanied by rearrangement. As an example, $(NH_3)_n$ clusters give rise preferentially to the ionic species $(NH_3)_{n-2} \cdot NH_4^+$.

Disposal of the reaction exoergicity is an example where the "solvation" of the reagent can make a significant difference, Figure 7.47 contrasts the observed BaI vibrational distribution from the reaction of Ba with monomeric and dimeric CF_3I:

$$Ba + CF_3I \rightarrow BaI + CF_3$$

$$Ba + (CF_3I)_2 \rightarrow BaI + CF_3 + CF_3I$$

The latter reaction produces BaI with significantly lower vibrational excitation. That the additional "sink" for energy is selective is shown by the Ba reaction with $CH_3I \cdot Ar$, on the one hand (where the BaI product is only slightly colder as compared to the neat CH_3I reaction), and with RI (R = higher alkyl groups), on the other. The longer alkyl group tail does not soak up much energy.

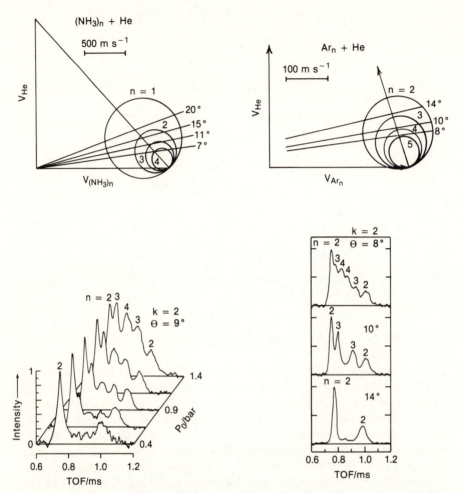

Fig. 7.46. Differential scattering analysis of beams of vdW molecules. (Left column) Scattering of $(NH_3)_n$ by He. Top: Newton diagram showing calculated loci of final velocity of elastically scattered individual n-mers; lines indicate laboratory angles at which time-of-flight analyses have been carried out. Bottom: Measured time-of-flight spectra at $\theta = 9°$, detected mass that of the dimer ion ($k = 2$), as a function of nozzle pressure P_0. Peaks corresponding to different neutral clusters ($n = 2$–4) are indicated. Their order is that of a cut at $\theta = 9°$ through the circles of the Newton diagram. [Adapted from U. Buck and H. Meyer, *Ber. Bunsenges. Phys. Chem.*, **88**, 254 (1984).] (Right column) Scattering of Ar_n by He. Top: Similar presentation as for $(NH_3)_n$. Bottom: Time-of-flight spectra at constant nozzle pressure but different laboratory scattering angles with detection set for $Ar_2{}^+$ ($k = 2$). Note the qualitative agreement with expectation based on the cuts through the Newton diagram. [Adapted from U. Buck and H. Meyer, *Phys. Rev. Lett.*, **52**, 109 (1984).]

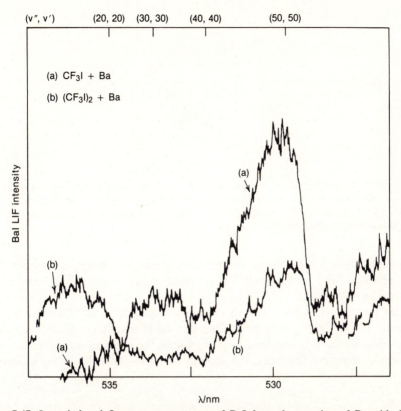

Fig. 7.47. Laser-induced fluorescence spectrum of BaI from the reaction of Ba with CF_3I monomer and dimer. (A quadrupole electric field is used to deflect the monomers out of the nozzle beam.) The monomeric reaction produces $BaI(v)$ with the most probable v of about 50, while the distribution from dimers peaks at about $v = 10$. This corresponds to a decrease of *ca.* 20 kcal mol^{-1} in the mean vibrational excitation of BaI. [Adapted from R. Naaman, *Laser Chem.*, *5*, 385 (1986).] When no attempt is made to separate the contributions from the two reactions, the BaI vibrational distribution will appear to be bimodal. [See T. Munakata and T. Kasuya, *J. Chem. Phys.*, *81*, 5608 (1984).]

The products' angular distribution is another probe for the dynamics of cluster reactions. The discussion in Section 7.2 suggests that if the collision energy is not high compared with the "solvation energy," cluster reactions will proceed through a clustered or solvated transition state that will be longer lived. Figure 7.48 shows the angular distribution of (electronically excited) NO_2 produced in the

$$O(^3P) + NO \cdot Ar \rightarrow NO_2^\dagger + Ar$$

reaction. There is forward peaking of NO_2 and a substantial fraction of the available energy is released as translation. The Ar atom almost acts as a spectator. Not so for the more strongly bound $(NO)_2$ reagent,

$$O(^3P) + (NO)_2 \rightarrow NO_2^\dagger + NO$$

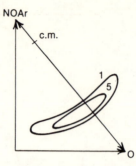

Fig. 7.48. Flux–velocity contour map for the $O(^3P) + NO \cdot Ar$ reaction. Note that the Ar atom is essentially a spectator since the available energy (about 9 kcal mol^{-1}) is high compared to the estimated (~1.5 kcal mol^{-1}) binding energy of NO·Ar. [Adapted from J. Nieman and R Naaman, *J. Chem. Phys., 84,* 3825 (1986).]

The NO_2 angular distribution is symmetric about 90°, as the reaction is far less exoergic [about 6 kcal mol^{-1} is required to dissociate $(NO)_2$ versus *ca.* 1.5 kcal mol^{-1} for NO·Ar].

Last but not least we come to orientation effects. A vdW adduct has a unique, albeit floppy, structure. One can take advantage of this relative orienation of the two constituents and explore the steric effects. Figure 7.49 shows the structure (assumed, by analogy to HCl— and HF—CO_2 adducts)

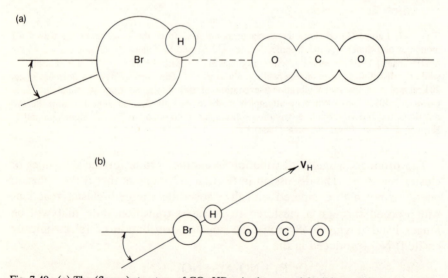

Fig. 7.49. (a) The (floppy) structure of $CO_2 \cdot HBr$. At the potential minimum, all the nuclei are along a straight line, but the restoring force is weak and the configuration shown is somewhat displaced from collinearity. The mean angle that the H atom makes with the line connecting the c.m.'s of the HBr and OCO is *ca.* 25°. (b) Similar to (a) but after photolysis of HBr, showing the orientation of the velocity of the hot H atom with respect to OCO axis. [Adapted from G. Radhakrishnan, S. Buelow, and C. Wittig, *J. Chem. Phys., 84,* 727 (1986).]

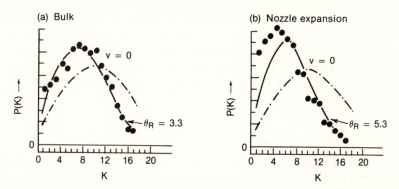

Fig. 7.50. Experimental (circles, from laser-induced OH fluorescence) OH ($v = 0$) rotational distributions versus the rotational state (K) for the hot H atom–CO_2 reaction for the bulk reaction (a) and for the photolysis of the $CO_2 \cdot HBr$ vdW molecule (b). The OH ($v = 0$) rotational excitation is clearly colder in the vdW "oriented" case. The dashed line is the prior distribution; both distributions are "colder" than the prior, but the vdW case is more so. The solid curves are calculated assuming a linear rotational surprisal with the specified rotational surprisal parameters θ_R. [Adapted from S. Buelow, M. Noble, G. Radhakrishnan, H. Reisler, C. Wittig, and G. Hancock, *J. Phys. Chem.*, **90**, 1015 (1986), and G. Radhakrishnan et al., ref. of Fig. 7.49. For related earlier work see C. Jouvet and B. Soep, *Laser Chem.*, **5**, 157 (1985).]

of $CO_2 \cdot HBr$. Photolysis of this vdW molecule in the region where HBr dissociates yields translationally hot H atoms whose velocity is more or less confined to a cone about the OCO axis (Fig. 7.49). In the bulk, hot H atoms react with CO_2 (see Fig. 4.37):

$$H + CO_2 \rightarrow OH + CO$$

Laser-induced fluorescence of OH can be used to compare the energy disposal for unoriented (bulk conditions) and for a more confined H atom approach to CO_2. The results (Fig. 7.50) show that the OH produced from the more collinearly oriented reagents is rotationally colder. Understanding such correlations between initial conditions and final outcome is a challenge for the theory and an opportunity for novel experiments. A new area, the molecular reaction dynamics of clusters is clearly on the horizon.

7.5 Molecular dynamics of gas–surface reactions

The field of heterogeneous catalysis has enormous practical applicability. The kinetics of heterogeneous chemical reactions have therefore been extensively studied. Our intent in this section, as in the rest of this volume, is to examine the molecular level description. We shall thus make no attempt to review the enormous literature on the macrolevel description but proceed immediately to the microlevel, considering first "prereactive" and then reactive collisions on well-defined surfaces. (The nonreactive processes have already been discussed in Section 6.4.) Many of the experimental and theoretical techniques are closely related to those used to study collisions in the gas phase, but both the experiments and the collision dynamics *per se* are more complicated here. Nevertheless, a great deal has been learned, and we are witnessing a microlevel "takeover" of the chemistry component of the field of surface science.

7.5.1 Adsorption and desorption

Scattering of molecules from clean metal surfaces is often "direct," showing incomplete energy accommodation and deviations from a cosine law angular distribution (Sec. 6.4). This is particularly so when the incident kinetic energy and surface temperature are high compared to the binding energy E_d of the molecules to the lattice. Otherwise, however, one finds a range of behavior with the short (~1 ps) interaction time, direct-mode collisions at one extreme through "compound mode" collisions with sojourn times in the pico- to nanosecond range [typically, physisorption process (Fig. 7.51)] to long-lived complexes with the surface (with still longer residence times).

The first step in characterizing the process of adsorption is to introduce the sticking probability S as the fraction of the incident flux that is completely accommodated with the solid. The motivation here is to isolate the primary step of adsorption from the myriad of possible subsequent processes (surface diffusion, trapping into a deeper well, dissociation, reaction, penetration into the bulk solid, desorption, etc.). The very act of adsorption implies an attractive interaction with the surface, with the component of the translational energy normal to the surface reduced below the well depth, E_d. As in the analogous process of "third body assisted" recombination in the gas phase, an energy sink is required to reduce the available energy (i.e., to soak up the exoergicity of bond formation). The sticking probability is thus critically dependent upon the coupling of the incident molecule with the lattice and will diminish with increasing surface temperature.

Experimental values of sticking probabilities of atoms and simple molecules with clean surfaces span a wide range from below the detectable limit up to near unity, depending on the nature of the interaction, the surface structure, and, of course, the temperature.

There is often a dramatic decrease of sticking probability with surface coverage. Figure 7.52 shows the results for the well-characterized collision of NO with Pt(111) (see Sec. 6.4). The sticking probability on the clean surface is quite high, indicating an efficient coupling of the NO translational energy to the lattice, but declines nonlinearly with increasing surface coverage by adsorbed NO. A model that explains the nonlinear behavior assumes that a molecule striking an occupied site is *not* promptly scattered back into the gas (as is assumed in the Langmuir adsorption model) but rather is temporarily vdW-bound to the existing adsorbed molecules. The

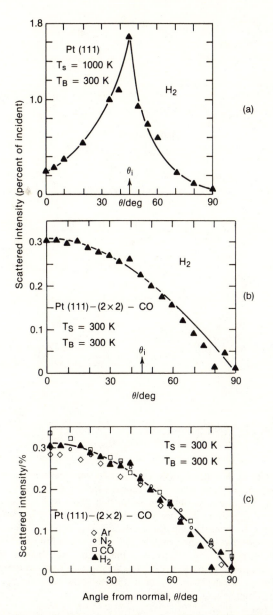

Fig. 7.51. (a) Angular distribution of scattering of a 300 K effusive H_2 beam from a clean Pt(111) surface, $T_s = 1000\,K$, with angle of incidence, $\theta_i = 45°$. Scattering angles θ measured from the surface normal. Ordinate is percentage of incident beam intensity detected by the (mass spectrometer) detector, of a given solid angle, set at the specified scattering angle. (b) Angular distribution of the H_2 beam from a CO-covered surface of Pt(111), $T_s = 300\,K$. Solid curve is $\cos\theta$ dependence. (c) Same as (b) but for beams of Ar, N_2, and CO, as well as H_2 showing cosine law scattering. [Adapted from S. L. Bernasek and G. A. Somorjai, *J. Chem. Phys.*, *60*, 4552 (1974), and G. A. Somorjai, *Chemistry in Two Dimensions: Surfaces*, Cornell University Press, Ithaca, NY (1981), p. 340.]

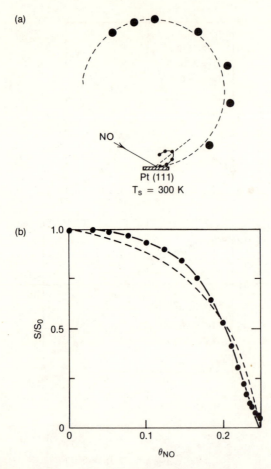

Fig. 7.52 (a) Polar plot of the intensity (vs. angle) of NO molecules scattered from a Pt(111) surface at $T_s = 300$ K. Small dots: Scattering from a bare (coverage, $\theta = 0$) surface showing specular scattering at low intensity, since a high fraction (0.85) of the incident molecules are adsorbed. Solid circles: Scattering from the same surface saturated with NO (corresponding to $\theta = 0.25$ monolayers). The scattered intensity is much higher and has essentially a cosine law angular distribution (shown as a dashed circle). (b) Dependence of the NO sticking probability S upon the coverage by NO, i.e., θ_{NO}. Here, S_0 is the sticking probability for the clean Pt(111) surface, $T_s = 300$ K, measured to be $S_0 = 0.85$. The solid curve is passed through experimental points. The dashed curve is calculated by a model invoking the existence of a precursor state. [Adapted from C. T. Campbell, G. Ertl, and J. Segner, *Surf. Sci., 115,* 309 (1982). Model of P. J. Kisliuk, *J. Phys. Chem. Solids, 3,* 94 (1957); *5,* 78 (1958).]

"floppy" nature of vdW binding enables this second-layer molecule to diffuse laterally rather freely so that it can locate a vacant site on which to adsorb. At high coverage, as the fraction of available unoccupied sites become small and their average separation large, the sticking probability rapidly approaches zero, and all incident molecules are only physisorbed and are scattered back with a cosine angular distribution.

In the limit of zero coverage and for adsorption without dissociation, the inverse process of desorption can be simply related to adsorption by detailed balance. If the sticking probability is low, so will be the rate of desorption. Experimentally such rate coefficients have a typical Arrhenius-like temperature dependence,

$$k_d = \nu_d \exp(-E_d/\mathbf{k}T) \qquad (7.20)$$

where $1/k_d$ is the mean surface residence time. While desorption can be thought of as a unimolecular process of an "evaporation" of a molecule from the surface, the frequency factor ν_d is often significantly higher $(10^{14}-10^{15} \text{ s}^{-1})$ than is typical for gas-phase bond fission. The transition state theory of Section 4.4 offers the following interpretation: At the configuration of no return for desorption the molecule is practically free, whereas in the adsorbed state its configuration is highly restricted. Many more states (or, classically, a higher volume in phase space) are available at the point of no return, and so the partition function ratio [see Eq. (4.56)] is atypically high. Typical values E_d range from a few meV to several eV [e.g., *ca.* 1.5 eV for NO on Pt(111)], mirroring the trend from vdW wells to chemical binding energies in gas–gas interactions. E_d is often higher than the activation energy for surface reactions (and, of course, for lateral surface diffusion), but when very rapid heating can be achieved, as in laser-induced desorption, the high frequency factor implies that evaporation can compete successfully with other processes.

7.5.2 Dissociative adsorption

Thus far we have discussed adsorption without bond rupture, i.e., the incident molecule maintains its chemical integrity during the interaction with the surface and is detected (after some mean residence time) as a scattered but intact molecule with a possibly different internal and kinetic energy. More relevant to chemistry is the process of dissociative adsorption wherein the adsorbed molecule undergoes bond rupture.

The energy required to break a bond in the incident molecule (typically 1–5 eV) is provided by the formation of new bonds with the (unsaturated) surface atoms. The dissociative adsorption of H_2 on clean copper can serve as an example:

$$
\begin{array}{ccccc}
H_2(g) & & H\text{–}H & & H \quad\quad H \\
+ & \longrightarrow & | \quad | & \longrightarrow & | \quad\quad | \\
Cu\cdots Cu\cdot & \cdots CuCuCu\cdots Cu\cdot & & CuCu\cdots CuCu\cdots
\end{array}
$$

Such processes are confirmed by observations of facile isotopic exchange, for example, detection of scattered HD when the incident beam consists of a mixture of H_2 and D_2 (to be discussed further in Sec. 7.5.3 below).

The mechanism of dissociative adsorption necessarily involves the crossing of (at least) two potential energy hypersurfaces (see Sec. 6.6) since both

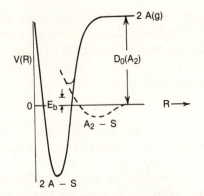

Fig. 7.53. Highly schematic illustration of the curve-crossing mechanism for dissociative chemisorption. The dashed curve with the shallow van der Waals well is the potential-energy curve for a diatomic molecule, say A_2, approaching an idealized surface (the abscissa R representing the separation of the c.m. of A_2 from the surface). The solid curve with the deep "chemical" well refers to the binding of 2 atoms of A with the surface. The curves cross and give rise to an adiabatic ground state potential with a barrier, E_b, as shown. [Adapted from G. Ertl, *Ber. Bunsenges. Phys. Chem.*, 86, 425 (1982).]

the undissociated and dissociated molecule can be at the (physical) surface. A simplified potential curve-crossing diagram is shown in Fig. 7.53 for a diatomic molecule approaching the (physical) surface. Given sufficient energy, the molecule can pass from the weaker (often, vdW) well, binding the undissociated molecule to the surface and over the barrier E_b to the region of a deep chemical well corresponding to the formation of two new bonds to the surface (at the expense of breaking the molecular bond).

The energy barrier, E_b, for dissociative (or "activated") adsorption can be measured as a translational energy threshold (E_{th}) for the dependence of the sticking probability S upon the component of kinetic energy normal to the surface, i.e., $E_\perp = E_T \cos^2 \theta_i$ (Fig. 7.54). The rapid rise in the dissociative sticking probability shown for H_2 on Cu as $E_\perp$ exceeds $4 \, \text{kcal mol}^{-1}$ suggests that $E_b \approx E_{th} \approx 4 \, \text{kcal mol}^{-1}$. However, as in the gas phase, the barrier height may well be different for different energies (e.g., translational vs. vibrational) of the incident molecule and will also depend on the orientation of the molecule with respect to the surface. So far, everything is much the same as in the gas phase. A new twist is that the barrier can also differ for different faces of the crystal (which differ in the detailed atomic configuration at the surface, as shown in Fig. 7.81 below).

Another indicator of the presence of the barrier is the energy disposal in the desorbed molecules. Thus, for a barrier in the exit valley we expect the barrier energy to appear as translational energy of the emitted molecules. The velocity distributions of H_2 (and D_2) molecules desorbing from Cu surfaces are indeed found to be peaked at high speeds. Time-of-flight distributions (Fig. 7.55) are characterized by $\langle E_T' \rangle \approx 15 \, \text{kcal mol}^{-1}$ and are narrower than for a Maxwellian velocity distribution at the surface

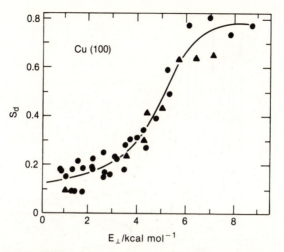

Fig. 7.54. Dissociative sticking probability S_d versus normal component of the translational energy of the incident H_2 molecules in a beam striking a Cu(100) surface. [Adapted from M. Balooch, M. J. Cardillo, D. R. Miller, and R. E. Stickney, *Surf. Sci.*, *46*, 358 (1974).]

temperature $T_s = 1000\,\text{K}$). In addition, the angular distribution is also narrow and is peaked at the normal. For the internal energy of the desorbing H_2 (and D_2) molecules it is found that the rotational distribution is only mildly deviant from a thermal one at T_s but the vibrational excitation is substantially higher than thermal.

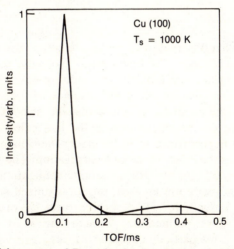

Fig. 7.55. Time-of-flight spectra of D_2 molecules desorbing in the normal direction from a Cu(100) surface at $T_s = 1000\,\text{K}$. The deuterium was supplied to the surface "from the rear" by permeation through the Cu crystal (as D atoms), recombining at the surface and desorbing into 2π steradians. From an analysis of the time-of-flight data the mean translational energy of the desorbing D_2 molecules is given by $\langle E_T'\rangle/2k = 3950 \pm 100\,\text{K}$ (four times larger than T_s). [Adapted from G. Comsa and R. David, *Surf. Sci.*, *117*, 77 (1982).]

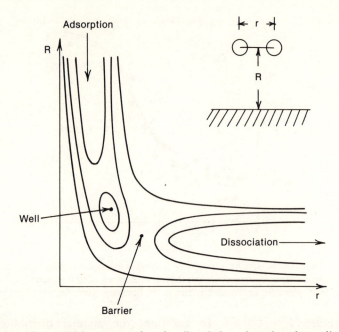

Fig. 7.56. Schematic potential energy surface for dissociative adsorption–desorption. In the drawing, the barrier to dissociation–recombination occurs when the molecule is already at the surface and is along the bond distance. This will lead to vibrationally excited desorbed molecules. Note also the precursor well along the approach to the (physical) surface. [Adapted from G. Ertl, *Ber. Bunsenges. Phys. Chem.*, *86*, 425 (1982).]

To explain all such observations, the simple curve-crossing picture of Fig. 7.53 is clearly inadequate. To begin with, we must consider at least a potential energy surface, showing the dependence not only on the distance from the surface but also on the bond distance of the diatomic molecule (Fig. 7.56). Then one needs to recognize that the interaction energy can be strongly dependent on the orientation of the molecule with respect to the surface, so that a hypersurface need be used. Then, unique to gas–surface interaction, there is the effect of surface corrugation: The interaction energy will depend (Fig. 7.57) on the precise atomic configuration of the surface. Finally, if all of this were not enough, the surface itself is neither static nor two-dimensional. Indeed, for H_2 desorbing from Cu or other metals, the recombination follows permeation through the solid, and the energy disposal is different for partially covered surfaces.

The many factors that come into play make it difficult to draw quantitative, micro-level conclusions from limited desorption data, but it is clear that the dynamics of these "prereactive" interactions must play an important role in reactive processes on the surface.

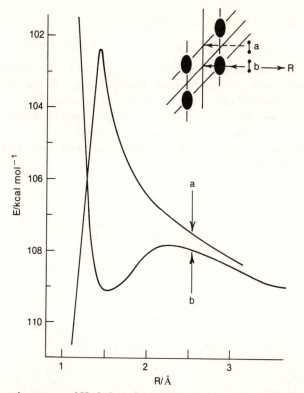

Fig. 7.57. Interaction energy of H_2 (oriented parallel to the surface) and Cu(100) as a function of distance from the surface for a semiempirical, LEPS-type (Sec. 4.2) potential energy surface. Curve (a) is the reaction path for dissociative adsorption. (The H_2 molecule is approaching towards a pair of adjacent Cu atoms.) In curve (b) the H_2 molecule is approaching a different site and the interaction is repulsive. [Adapted from A. Gelb and M. J. Cardillo, *Surf. Sci.*, 75, 199 (1978).]

7.5.3 Heterogeneous chemical reactivity

In a gas–surface reaction the surface atoms can be reagents and therefore can be consumed in the reaction or the surface can serve as a catalyst. In the latter case, one often distinguishes between the Eley–Rideal mechanism, where the molecules from the gas react with surface chemisorbed reagents, or the Langmuir–Hinshelwood mechanism, where both reagents are chemisorbed prior to reaction. Of course, in view of the often multistaged nature of the adsorption process, there can be intermediate situations as well.

Reactions of gaseous molecules with surface atoms are often of practical importance. They include the oxidation of graphite,

$$O_2 + C(gr) \rightarrow CO + O, \qquad E_a = 30 \, \text{kcal mol}^{-1}$$

and other, so-called corrosion reactions that lead to desorbing products at the temperature of the experiment, e.g.,

$$Cl_2 + Ni(s) \rightarrow NiCl(g) + Cl, \qquad E_a = 30 \, kcal \, mol^{-1}$$

The pre-exponential factors for such reactions can be estimated (to within one or two orders of magnitude) from transition-state theory adapted to the two-dimensional world of surface chemistry. In so doing, one must identify correctly the rate-determining step (whether reagent chemisorption or surface diffusion or reaction or product desorption).

Simple surface-catalyzed reactions that have been well studied include isotopic exchange reactions,

$$H_2 + D_2 \xrightarrow{\ \ \ \ } 2HD$$

$$^{14}N^{14}N + {}^{15}N^{15}N \xrightarrow{\ \ \ \ } 2\,^{14}N^{15}N$$

bimolecular reactions,

$$CO + O_2 \xrightarrow{\ \ \ \ } CO_2 + O$$

$$C_2H_4 + H_2 \xrightarrow{\ \ \ \ } C_2H_5 + H$$

and dissociation,

$$N_2O \xrightarrow{\ \ \ \ } N_2 + O$$

$$HCOOH \xrightarrow{\ \ \ \ } H_2 + CO_2$$

The exchange reactions are perhaps the most instructive since they require only dissociative adsorption followed by surface diffusion prior to recombination. Thus, when a beam containing a mixture of H_2 and D_2 is incident on a Pt surface, the integrated scattered yield of HD can be as high as 35% depending on the angle of incidence *and* the detailed atomic arrangement at the surface. Figure 7.58 compares the HD yield from a smooth (111) or stepped (332) Pt surface. The data imply that the exchange probability per unit surface is some seven-fold greater for "step" sites than for "terrace" sites and that this difference is due to a higher probability for dissociative adsorption on the stepped surface (recall Fig. 7.57). Subsequent processes are probably the same for both types of surface planes.

For N_2 on W(100), the yield of $^{15}N^{14}N$ is essentially identical with the directly measured sticking probability S (Fig. 7.59). The role of dissociative adsorption is also demonstrated in the absence of any detectable CO_2

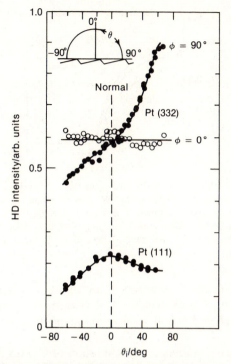

Fig. 7.58. HD formation as a function of the angle of incidence θ_i of a mixed H_2–D_2 effusive beam (300 K) scattered from clean surfaces of Pt, $T_s = 1100$ K. The lowest curve is for a smooth Pt(111) surface, for which the HD yield is relatively small and there is no dependence on azimuthal angle θ. The two upper data sets are for a stepped Pt(332) surface at the same T_s. As seen in the upper sketch, the view into the drawing defines the azimuthal angle ϕ to be zero. At $\phi = 0$ the HD yield is independent of θ_i. However, for $\phi = 90°$ the incident beam strikes the step edges; here a strong dependence of HD intensity upon θ_i is found, with greatly enhanced yields at large positive θ_i, decreased yields at negative θ_i (see sketch for visualization). [Adapted from G. A. Somorjai, *Chemistry in Two Dimensions: Surfaces*, Cornell University Press, Ithaca, NY (1981), p. 370; see references therein.]

formation when O_2 is incident on a Pt surface saturated with CO. The converse is not the case. However, since the O atom saturation coverage is low, the question of whether CO is adsorbed prior to reaction needs to be considered. By modulating the CO beam it is found that the mean residence time on the surface is very long. The reaction occurs between two adsorbed species (a Langmuir–Hinshelwood mechanism) to produce adsorbed CO_2 (Fig. 7.60). That the adsorbed species need not be in its gas-phase equilibrium configuration is revealed by measuring the energy disposal, as discussed next.

7.5.4 Dynamics of gas–surface reactions

The many relevant variables in gas–surface reactions make it almost imperative to probe the dynamics before even tentative conclusions on the

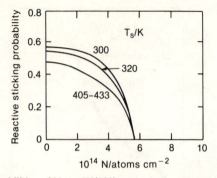

Fig. 7.59. Sticking probabilities of N_2 on W(100) at the indicated surface temperatures, T_s. The "reactive" sticking coefficient, measured by the $^{14}N^{15}N$ yield agrees with the directly determined sticking probability, as a function of the atomic nitrogen coverage in units of 10^{14} atoms cm^{-2}.) [Adapted from P. Alnot and D. A. King, *Surf. Sci.*, **126**, 359 (1983); see also D. A. King, *Surf. Sci.*, **126**, 359 (1983), and D. A. King and M. G. Wells, *Proc. R. Soc. London, Ser. A, 339*, 245 (1974).]

molecular mechanism can be reached. If anywhere, it is here that the (yet to come into general use) laser probing to determine the velocity and angular distribution for each individual final quantum state (Sec. 5.2) will prove decisive.

The beam scattering studies of the oxidation of CO on Pt in the presence of adsorbed O_2 have not only detected a long residence time prior to CO_2 desorption but have also shown that the emerging CO_2 molecules are translationally fairly hot ($\langle E'_T \rangle \approx 0.3 \, eV$) and vibrationally excited (as

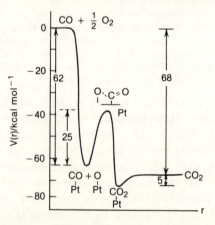

Fig. 7.60. Potential energy diagram along the reaction coordinate, intended to illustrate the mechanism of the Pt(111)-catalyzed reaction of CO with O_2 (energies in kcal mol^{-1}). Note the weak van der Waals well for the adsorption of CO_2. The Langmuir–Hinshelwood barrier of 25 kcal mol^{-1} to convert adsorbed CO and adsorbed oxygen atoms to adsorbed CO_2 has been derived from the measured activation energy. The adsorption energies of both CO and O have been directly measured. [Adapted from T. Engel and G. Ertl, *J. Chem. Phys.*, *69*, 1267 (1978); see discussion by M. P. D'Evelyn and R. J. Madix, *Surf. Sci. Rep.*, *3*, 413 (1984).]

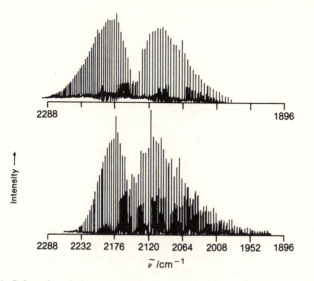

Fig. 7.61. Infrared emission spectrum of CO from the reaction of atomic oxygen with a monolayer of C on a Pt surface at $T_s = 1400\,\text{K}$ (bottom). (Top) Spectrum from a thermally accommodated beam of CO at 1400 K. The nascent CO from the reaction has a significantly higher mean vibrational energy than the Boltzmann CO at T_s. [Adapted from M. Kori and B. L. Halpern, *Chem. Phys. Lett.*, **98**, 32 (1983).]

revealed by infrared emission). The reaction of atomic oxygen with a carbonized surface at $T_s = 1400\,\text{K}$ also yields vibrationally hot CO (Fig. 7.61). The angular distribution of the desorbing CO_2 molecules is sharper than cosine, and this also is indicative of the forces operating in the exit valley. In contrast, the angular distribution of NO from the Ni(100)-catalyzed dissociation of N_2O is cosine, suggestive of accommodation of NO prior to desorption.

A simpler dissociation reaction that is more easily interpreted is the impact atomization of I_2 (Fig. 7.62). The dissociation probability of I_2 on MgO(100) measured as a function of the initial kinetic energy rises rapidly from the thermochemical threshold $(E_{\text{th}} \approx D_0(I_2) = 1.54\,\text{eV})$ and tends to level off for $E_T \gtrsim 5\,\text{eV}$. Although a rigid-surface model can account for this trend, time-of-flight data on the undissociated I_2 show a large average energy transfer to the solid requiring a nonrigid description of the solid. Such a model, which fits the inelasticity data also recovers the energy dependence of the dissociation probability (Fig. 7.62).

The theoretical approach that so far has been most effective in describing the dynamics of adsorption–desorption and of reactive gas–surface collisions is based on the method of classical trajectories. The essence of the problem is to provide a tractable yet realistic approach to the coupling of the molecular and surface (and bulk) degrees of freedom. In principle, one can introduce a (usually, semiempirical) potential energy, which is a function of the positions of all atoms, both those of the molecule and those

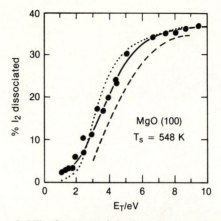

Fig. 7.62. Dissociation probability (percentage) for I_2 molecules striking a MgO(100) surface at $T_s = 548$ K as a function of their incident translational energy. Experimental data: points and solid curve. Theoretical calculations: dotted curve, rigid surface model; dashed curve, nonrigid. [Adapted from E. Kolodney, A. Amirav, R. Elber, and R. B. Gerber, *Chem. Phys. Lett.*, *111*, 366 (1984).]

of the surface. The classical equations of motion can then be solved. Since each atom of the solid is interacting with its neighbors, the number of coupled† differential equations that need to be solved is enormous. Moreover, the solution needs to be repeated many times with different initial conditions, so that, for example, thermal averaging for the solid's degrees of freedom can be carried out. This is computationally unrealistic and is also unreasonable on physical grounds: Surely the outcome of the surface reaction cannot depend on the precise details of the motion of an atom several layers below the surface.

The way out is provided by statistical mechanics: We average over the solid's degrees of freedom first and then integrate for the time evolution of the molecular degrees of freedom. This can be done in a manner that is formally exact and leads to the result that the presence of the surface introduces two additional terms in the equation of motion for the molecular degrees of freedom. One is a dissipative or frictional force leading to energy flow into (out of) the surface. The other is a random force reflecting the thermal fluctuations in the positions and momenta of the surface atoms. The resulting equation is often known as a generalized Langevin equation.

Evaluating the two additional terms exactly is as difficult as the original problem. However, because of their obvious physical interpretation it is possible to provide simple yet realistic approximations for them. In practice, one sometimes solves the equations of motion not only for the molecular degrees of freedom but also for those surface atoms to which they are directly coupled. Only the rest of the solid is averaged over. This structureless, "pillowlike" description of the environment has enabled the

† The equations are coupled since the force on any given atom depends on the instantaneous position of other atoms.

method of classical trajectories to be applied not only to reactions at the surface but also in solution.

7.5.5 Laser-induced processes

Finally, we consider the interaction of photons with adsorbed molecules and, specifically, laser-induced desorption and laser-induced photofragmentation processes. The potential practical applications of the former has resulted in considerable activity, but definitive generalizations on the dynamics are still not at hand. Photofragmentation at surfaces is still largely unexplored.

Laser desorption using CO_2 lasers for heating the substrates have been widely used to produce evaporated films, as well as beams ranging from metal atoms to complex organic polyatomic molecules (even for mass-spectrometer sample introduction, i.e., vaporization, purposes). This method has also been used to produce clusters of atoms or molecules for spectroscopic or kinetic study. Visible and ultraviolet lasers providing larger photon energies are sometimes preferred when attempting vaporization of refractory metals—some of the rare earths, transition metals, etc.

In the case of photodissociation, some dynamical information has resulted from a study of the CH_3Br molecule chemisorbed on a LiF(100) surface. Using an ultraviolet laser tuned to 222 nm, the time-of-flight velocity distribution of the CH_3Br photodesorbed was compared to that of the CH_3 radical from the photodissociation of the adsorbed CH_3Br (Fig. 7.63). The photodissociation results are as follows: a sharp translational energy distribution of the CH_3 normal to the surface, $P(E'_\perp)$ extending to the theoretical upper limit, assuming all the excess energy (i.e., the photon minus the H_3C—Br bond dissociation energy) goes into CH_3 recoil *and* that the CH_3 is recoiling from a more massive "particle" than Br, namely, the "near-infinite" mass of the surface. The inference is that the H_3C—Br bond of the chemisorbed CH_3Br is perpendicular to the surface (and that the Br is located over Li^+ ions, rather than F^-). Separate thermal-desorption studies indicate a binding energy $E_d \approx 0.3$ eV for the CH_3Br–surface bond, consistent with a Br—Li^+ affinity. Thus, it has been possible to deduce dynamical information at the microlevel on the process of photodissociation *concurrent* with photodesorption of the same molecule.

Clearly there are a host of new and exciting experiments, and an equal body of theoretical investigations, waiting to be done in the expanding new field of gas–surface reaction dynamics.

7.6 Stereospecific dynamics

The influence of the mutual orientation of the reagents upon their reaction probability is a subject that has been dear to the hearts of synthetic chemists for generations. Considerations of the "steric requirements" of the reaction

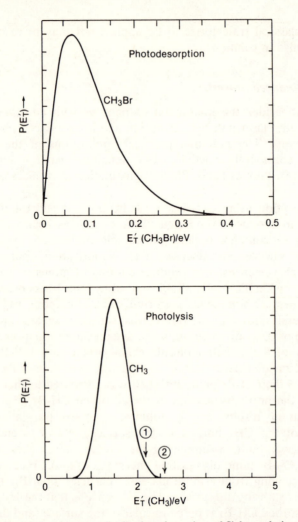

Fig. 7.63. (a) Translational energy distribution, from time-of-flight analysis, of photodesorption of CH$_3$Br adsorbed (submonolayer coverage) on a LiF(001) surface at $T_s = 115$ K. An excimer laser beam ($\lambda = 222$ nm, 2–3 mJ/pulse) was directed at the surface at 85° from the normal and photodesorbed CH$_3$Br molecules detected (mass spectrometrically) at 5° from the normal. The peak energy is 0.06 eV, with a width of 0.14 eV. (b) Same for the CH$_3$ radical produced by photolysis of the adsorbed CH$_3$Br. Here, the peak translational energy is 1.5 eV, with a width of 0.6 eV. Arrow 1 denotes the maximum thermochemically allowed recoil energy for CH$_3$ from a "free" CH$_3$Br molecule; arrow 2, the maximum allowed E'_T for CH$_3$ from a CH$_3$Br bound to a (massive) crystal. [Adapted from E. B. D. Bourdon, J. P. Cowin, I. Harrison, J. C. Polanyi, J. Segner, C. D. Stanners, and P. A. Young, *J. Phys. Chem., 88,* 6100 (1984).]

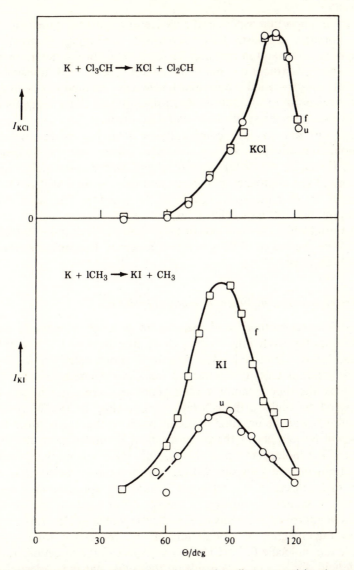

Fig. 7.64. Direct experimental observations of orientation effect on reactivity *via* crossed-beam scattering technique: K versus oriented molecules Cl_3CH (top) and ICH_3 (bottom). Laboratory angular distributions $I(\Theta)$ of the KI product from the methyl iodide reaction show directly a strong orientation effect; here f denotes favorable orientation (K–ICH$_3$); u, unfavorable (K–H$_3$CI). No such reactive asymmetry is observed for the analogous chloroform reaction. More detailed studies (*cf.* Fig. 2.18) show that there is a negligible reaction probability for head-on collisions in the u configuration. [Adapted from G. Marcelin and P. R. Brooks, *Faraday Discuss. Chem. Soc.*, *55*, 318 (1973).]

have been an important aspect in the choice of a synthetic procedure. More recently, organic chemists have developed computing programs using semiempirical inter- and intramolecular force fields for the determination of minimum-energy structures, minimum-energy paths, and transition-state geometries. This so-called "molecular mechanics" method has been applied to predict regio- and stereoselectivity in reactions of chiral reagents and to guide the stereospecific synthesis of chiral products.

A dominant role in such considerations is played by the short-range atom–atom repulsive forces. In addition, there will be longer-range attractive forces. If the latter are strong and highly anisotropic, they will tend to orient the approaching reagents and there is little that can be done "externally" to alter the steric course of the reaction. The relatively slow rotational motion of molecules does, however, suggest that it may be possible to influence the outcome of the collision, often quite dramatically (Fig. 7.64) by the use of initially oriented reagents. Dynamical stereochemistry has been born.

7.6.1 *Orientation and alignment of reactant molecules*

The experimental method used thus far to produce a beam of *oriented* molecules employed an inhomogeneous electric field (via a hexapole rod configuration) to effect a $|JKM\rangle$ rotational state† selection for a polar symmetric top molecule, followed by a weak homogeneous electric field $\mathbf{E}$ for orienting the dipole moment (and hence the molecular frame) of the state-selected molecule in the laboratory frame (Fig. 7.65). The co-reagent beam is directed at the oriented molecule beam so that the relative velocity vector $\mathbf{v}$ will be parallel to the orienting field $\mathbf{E}$ (and thus to the average direction of the molecular dipole moment $\boldsymbol{\mu}$). By reversing the direction of $\mathbf{E}$ and thereby inverting the orientation of $\boldsymbol{\mu}$, one can alternate between two opposite orientations of the molecule towards its approaching co-reagent (Fig. 7.66). A schematic drawing of the apparatus used for the $Rb + CH_3I$ reactive asymmetry experiments of Fig. 2.18 is shown in Fig. 7.67.

The average degree of orientation of the precessing (and spinning) symmetric top molecule (e.g., CH_3I) with respect to the direction, $\hat{\mathbf{E}}$, of the orienting field, i.e., $\langle \hat{\boldsymbol{\mu}} \cdot \hat{\mathbf{E}} \rangle$, depends on the rotational state selected. For a definite $|JKM\rangle$ state it is given by

$$\langle \hat{\boldsymbol{\mu}} \cdot \hat{\mathbf{E}} \rangle = \langle \cos \theta \rangle = KM/J(J+1) \tag{7.21}$$

---

† For a symmetric top molecule, in a weak electric field $\mathbf{E}$, the square of the total angular momentum is quantized with the eigenvalues $P^2 = J(J+1)\hbar^2$, where $J(J \geq 0)$ is the total angular momentum quantum number. Also quantized are two components of $\mathbf{P}$. One is P_z, the component of $\mathbf{P}$ along the figure axis (which we take as the z direction) $P_z = K\hbar$, where $K(=0, \pm 1, \ldots, \pm J)$ is the projection quantum number in the molecular frame of reference. The second is the component, $M\hbar$, of $\mathbf{P}$ along the direction $\hat{\mathbf{E}}$ of the electrical field. $M(=0, \pm 1, \ldots \pm J)$, is the projection quantum number in the laboratory frame.

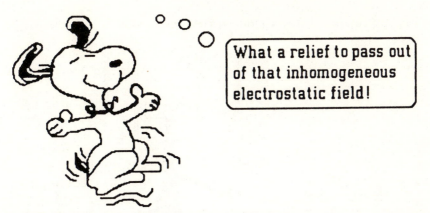

Fig. 7.65. On the need for an orienting field following state selection (*via* the inhomogeneous electrostatic field).

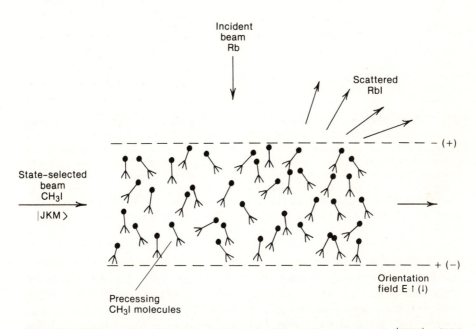

Fig. 7.66. Idealized, pictorial, representation of a beam of state-selected, $|JKM\rangle$, CH_3I molecules passing through the orientation field and serving as targets for the incident Rb atoms. The "snapshot" view of the CH_3I molecules undergoing precession but, on the average, with the I end pointing to the negative electrode. For the class of $|JKM\rangle$ states transmitted by the hexapole state-selector, the dipole moment $+ \rightarrow -$ orients parallel with the electrostatic (**E**) field. Reversing the polarity of **E** (shown in parentheses) causes the dipole moment (and thus the molecule) to reorient by 180°. [See K. H. Kramer and R. B. Bernstein, *J. Chem. Phys.*, *42*, 767 (1965), and R. J. Beuhler, Jr., R. B. Bernstein, and K. H. Kramer, *J. Am. Chem. Soc.*, *88*, 5331 (1966). For recent state-selection results, see S. R. Gandhi, T. J. Curtiss, Q.-X. Xu, S. E. Choi, and R. B. Bernstein, *Chem. Phys. Lett.*, *132*, 6 (1986).]

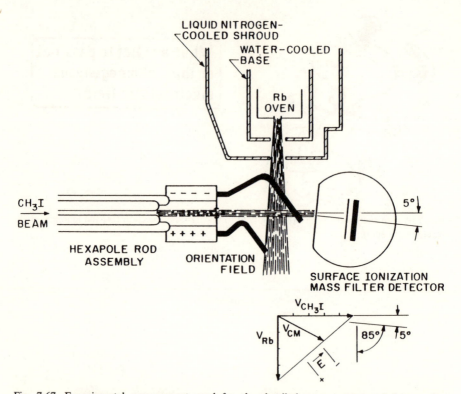

Fig. 7.67. Experimental arrangement used for the detailed measurements of the reactive asymmetry of the $CH_3I + Rb$ reaction (*cf.* Fig. 2.18). The CH_3I molecules pass through the hexapole field and are focused according to their $KM/(J^2 + J)$ value [determined by the hexapole rod voltage (V, $-V$, ... alternating rod to rod) and the CH_3I velocity v, its mass m, and dipole moment μ] into the scattering center, defined by the intersection with the Rb beam (shown). The orientation field is "tilted" at *ca* 45° to ensure that the $\hat{\mathbf{E}}$-field is parallel (or antiparallel) to the average relative velocity vector (see Newton diagram, below). [Adapted from D. H. Parker, K. K. Chakrovorty, and R. B. Bernstein, *Chem. Phys. Lett.*, **86**, 113 (1982).]

where θ is the angle between $\mathbf{\mu}$ and $\mathbf{E}$ and the "average" is quantum mechanical in origin, as a state with sharp values for J, M, and K does not have a definite "classical" orientation in space. States with $K \approx M \approx J$ are, however, only slightly precessing. When a CH_3I molecule is selected in such a state and allowed to collide with Rb, its dipole moment is essentially parallel (or antiparallel) to the incident relative velocity vector. By selecting other states, for example, those for which $K, M \ll J$, $\langle \cos \theta \rangle \approx 0$, the collisions with Rb will be essentially "broadside" rather than head or tail. The change in the reactivity upon reversing the direction of the orienting field $\mathbf{E}$ will now be much reduced. The dependence of the reaction probability (for backscattering, $b \approx 0$ collisions) upon the "initial angle of attack" of Rb on CH_3I as shown in Fig. 2.18 has been deduced from such experiments. The primary observation is that there is indeed "steric

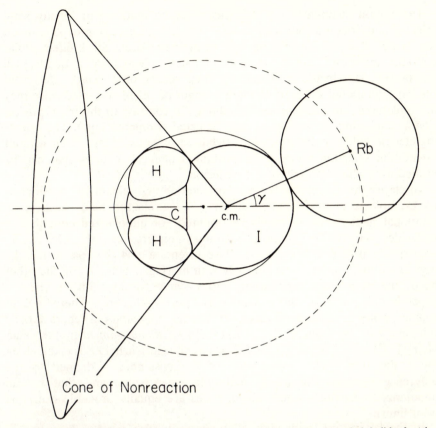

Cone of Nonreaction

Fig. 7.68. Pictorial representation of the "cone of nonreaction" of methyl iodide for the $Rb + ICH_3$ system. The C, H, I, and Rb atoms are sized according to conventional van der Waals radii. The "angle of attack," γ, is depicted. The geometrically defined cone of nonreaction (half-angle, 50°) is virtually identical to the best experimental value based on the so-called "cut-off" value of γ for reaction in the "unfavorable" orientation, based on a detailed analysis of the experiments of Fig. 2.18. [Adapted from S. E. Choi and R. B. Bernstein, *J. Chem. Phys., 83,* 4463 (1985).See also S. Stolte, K. K. Chakravorty, R. B. Bernstein, and D. H. Parker, *Chem. Phys., 71,* 353 (1982).]

hindrance," or a "cone of nonreaction," whose angular extent is determined essentially by the "size" of the atoms (Fig. 7.68). The classical chemical concept that an atom or group can "get in the way" of the reaction is thus borne out by the direct experiment. The new finding, already discussed in Section 2.4.5, is that when reaction is allowed, the reactivity is not constant but depends on the angle of attack. The "size" of the individual atoms is insufficient to account for this observation. Rather we need the potential energy surface and, specifically, the angle dependence of the barrier to reaction. Studies of other reactions, $K + CF_3I$, for example, also show the need for a more detailed (than merely "geometrical") interpretation of the results.

The reason symmetric top molecules can be readily rotationally state selected and oriented (but with a variable degree of precision depending on the specific J, K, M state) is because of their first-order Stark effect. Polar diatomic molecules can be rotationally state-selected by a quadrupole inhomogeneous electric field utilizing their second-order Stark effect, but the $|jm_j\rangle$ rotational states of diatomics cannot be readily oriented since they are undergoing an end-over-end tumbling (in contrast to the mere precession and spinning of rotationally state-selected symmetric tops). The orienting electric field for diatomics thus needs to be very strong. Even without orientation selection there are important dynamical features that can be probed using the *alignment* of the molecule.

To clarify the distinction between orientation and alignment, consider as an example a diatomic molecule (in a weak electric field) in a given rotational state j. For a "randomly oriented" or unpolarized ensemble of molecules, all the orientation quantum numbers, $m_j = 0$, $\pm 1, \ldots, \pm j$ are equally probable (Fig. 7.69). Pure orientation is the case when the population distribution varies linearly with m_j (Fig. 7.69b). If, on the other hand, the population distribution varies symmetrically with m_j (i.e., it depends only on m_j^2), the ensemble is "purely aligned." There are, of course, many intermediate cases. However, *orientation* is characterized completely by the value of $\langle P_1(\hat{\mathbf{j}} \cdot \hat{\mathbf{E}}) \rangle = \langle m_j \rangle / j$ and *alignment* by the value of $\langle P_2(\hat{\mathbf{j}} \cdot \hat{\mathbf{E}}) \rangle = (3/2)(\langle m_j^2 \rangle / j^2) - 1/2$, where the electric field $\mathbf{E}$ is taken along the z direction.† The *polarization* of the ensemble is defined as the difference between the actual multipole mean value and that for an unpolarized ensemble, where all m_j states are equally probable (i.e., the prior limit).

In the laboratory framework, diatomic molecules can be aligned using linearly polarized radiation of a laser. A molecule excited by such a laser will have its transition dipole moment $\boldsymbol{\mu}_{if}$ aligned parallel or antiparallel to the electric field of the laser radiation.‡ The excited molecules aligned in the laboratory frame can react with an incident beam of co-reagent molecules and the reactivity studied as a function of the angle between $\hat{\mathbf{E}}$ and the relative velocity vector $\mathbf{v}$ and thus of the angle between $\mathbf{v}$ and the molecular axis. The experimental arrangement used in the study of the reaction

$$\mathrm{HF}(v = 1, j, m_j) + \mathrm{Sr} \rightarrow \mathrm{H} + \mathrm{SrF}(v', j')$$

is shown in Fig. 7.70. By varying the polarization of the (HF chemical) laser that excites the HF molecules to the $v = 1$ level, they could be aligned perpendicular or parallel to the approach direction of the Sr atom. The

---

† The other possible cases require additional moments of $\hat{\mathbf{j}} \cdot \hat{\mathbf{E}}$ for their complete characterization.

‡ The excitation probability is proportional to $\cos^2 \theta = (\hat{\boldsymbol{\mu}}_{if} \cdot \hat{\mathbf{E}})^2$, where θ is the angle between $\boldsymbol{\mu}_{if}$ and $\mathbf{E}$ (see Sec. 5.2.4). For a diatomic molecule the transition dipole is either along or perpendicular to the molecular axis.

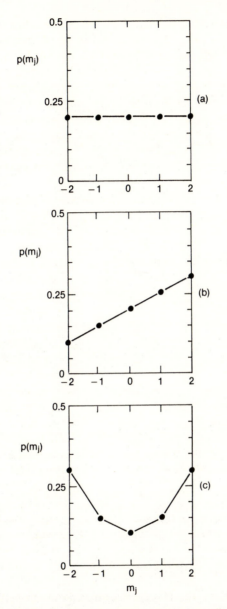

Fig. 7.69. Various possible distributions of m_j states, $p(m_j)$, for a diatomic molecule of $j = 2$ placed in a weak electric field defining the Z-direction. Note $\sum_{m_j=-2}^{2} P(m_j) = 1$ for each case. (a) Prior distribution (random orientation of **j**). (b) Distribution corresponding to "pure orientation." (c) Distribution corresponding to "pure alignment" (note symmetry with respect to $\pm Z$). [Adapted from C. H. Greene and R. N. Zare, *Annu. Rev. Phys. Chem.*, *33*, 119 (1982); see also C. H. Greene and R. N. Zare, *J. Chem. Phys.*, *78*, 6741 (1983).]

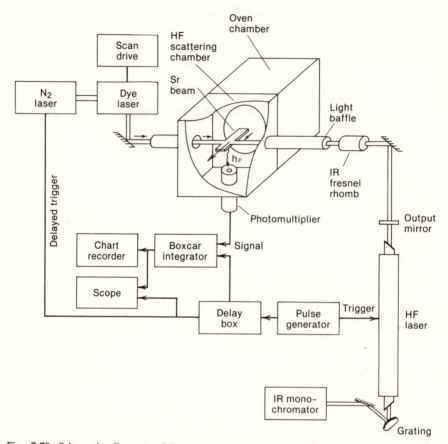

Fig. 7.70. Schematic diagram of laser–molecular beam apparatus for the study of the dependence of reactivity upon alignment (and vibrational state) of HF in collision with Sr. HF in the scattering chamber is resonantly excited by a polarized HF chemical laser and can then react with a beam of Sr. A tunable dye laser probes the nascent SrF by laser-induced fluorescence. [Adapted from Z. Karny, R. C. Estler, and R. N. Zare, *J. Chem. Phys.*, **69**, 5199 (1978).] Regarding the degree of alignment of laser-excited HF (taking into account the effect of nuclear spin), see R. Altkorn, R. N. Zare, and C. H. Greene, *Mol. Phys*, **55**, 1 (1985).

production of $SrF(v' = 2)$ is found to be enhanced for broadside attack (Fig. 2.19).

Both classical and quantum dynamics have been employed to analyze and predict steric requirements of chemical reactions. The classical trajectory computation needs to be redone for any different initial state and hence loses some of its computational advantage as compared to the quantum scattering theory, which generates the entire **S**-matrix in one go. The latter is often formulated in the so-called helicity representation. This consists in making a special choice for the axis of quantization of the angular momentum.† One possibility is to take it along the direction of relative

† The advantages of the helicity representation for inelastic collisions have been discussed in connection with Fig. 6.19.

motion. The orbital angular momentum **L** has a zero projection on that axis. (Recall that $\mathbf{L} = \mu\dot{\mathbf{R}} \times \mathbf{R}$.) Hence, $m_j = J_z$, where J is the total angular momentum. To see the advantage, consider a problem that we shall return to shortly: the photodissociation of a jet-cooled triatomic molecule. Since the rotation is cold, J is low and so is J_z ($|J_z| \leq J$). After the dissociation, the range of m_j values is thus low, which reduces the dimension of the computation. This will not be the case for other choices of the axis of quantization since j' can be large, even if J is low, provided $\mathbf{j}'$ and $\mathbf{L}'$ are in nearly opposite directions.

7.6.2 Orbital Control

The idea that the outermost electronic orbitals govern chemical reactivity is central to chemistry. The direct experimental examination of the role of the orientation of the valence orbital can be carried out using a polarized laser to pump into an aligned electronically excited state. Figure 7.71 outlines an experiment in which a $Ca(^1S_0)$ atom is laser-pumped (at 422.7 nm) to the $Ca(^1P_1)$ state, where one valence electron is in an aligned p-orbital. The alignment of $Ca(^1P_1)$ is monitored through its resonance fluorescence [Fig. 7.71(a)]. The reaction of $Ca(^1P_1)$ with chlorine-containing compounds yields CaCl in both the $B^2\Sigma^+$ state and the $A^2\Pi$ state. For the reaction of aligned $Ca(^1P_1)$ with HCl [Fig. 7.71(b)], the polarization of the fluorescence of the

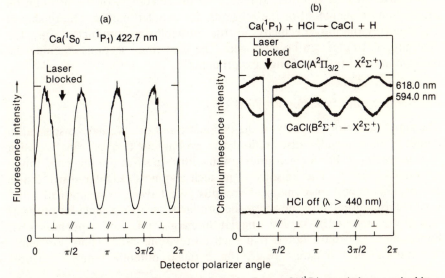

Fig. 7.71. (a) Variation of the fluorescence intensity from a $Ca(^1P_1)$ atomic beam excited by a polarized laser (from the ground, $^1S_0$, state at 422.67 nm) with the alignment of the laser electric field (i.e., the setting of the polarizer before the detector). (b) Similar to (a) except of the CaCl chemiluminescence intensity for the reaction of $Ca(^1P_1)$ with HCl. Note that the polarization for the $B^2\Sigma^+$ state is opposite to that of $Ca(^1P_1)$ or that of the $A^2\Pi$ state. [Adapted from C. T. Rettner and R. N. Zare, *J. Chem. Phys.*, **77**, 2416 (1982).]

two states is out of phase with one another. Those alignments of $Ca(^1P_1)$ that favor $CaCl(A^2\Pi)$ are least favorable towards $CaCl(B^2\Sigma^+)$ formation and vice versa. A quantitative measure is provided by the fluorescence intensity (I) sensitivity factor $S(-1 \leq S \leq 1)$

$$S = (I_\| - I_\perp)/(I_\| + I_\perp) \tag{7.22}$$

obtained when the laser field is parallel and perpendicular to the relative velocity vector. $S = 0$ implies no preferred alignment. If $S > 0$, the parallel alignment is more favorable; if $S < 0$, the perpendicular alignment is preferred. The S values for Fig. 7.71 are *ca.* -0.03 for $CaCl(A^2\Pi)$ and $+0.08$ for $CaCl(B^2\Sigma^+)$. An important observation, also evident from the figure, is that the selectivity is not large. This can be understood on the basis of the harpoon mechanism (Sec. 4.2.5): An electron is transferred from $Ca(^1P_1)$ to HCl as the reagents approach. If it is the electron in the p-orbital† that jumps over to yield HCl^-, then the initial alignment will not affect the outcome of the collision. If, however, it is the electron in the s-orbital that is used as the harpoon, then the Ca^+ ion has an aligned orbital. If the p-orbital on Ca^+ is along the direction of the relative approach, it will correlate with a $4p\sigma$ molecular orbital. A p-orbital perpendicular to the molecular axis will correlate better with a $4p\pi$ molecular orbital. Thus, parallel ("σ") alignment of the $Ca(^1P_1)$ orbital will favor the formation of $CaCl(B^2\Sigma^+)$, while a perpendicular ("π") alignment enhances the formation of $CaCl(A^2\Pi)$.

We have thus far centered attention on the selectivity of the precollision orientation or alignment of the reagents. By detailed balance, there should be a corresponding specificity in the postcollision products, e.g., the emerging molecules can be polarized. The next three sections examine the background and results of such experiments.

7.6.3 Alignment of reaction products

The simplest vector property of the products, and one we have discussed in detail, is their angular distribution: the probability that the final relative velocity vector $\mathbf{v}'$ has a given orientation with respect to the reagents' initial relative velocity $\mathbf{v}$. The anisotropy in such products' angular distributions can be understood on dynamical grounds. We have also mentioned other vector correlations, e.g., between the products' rotational angular momentum $\mathbf{j}'$ and the initial orbital angular momentum $\mathbf{L}$ (see Appendix 4B). Furthermore, we have noted that such correlations reflect both kinematic (e.g., mass) and dynamic (e.g., the detailed potential energy surface)

---

† The electron in the p-orbital of $Ca(^1P_1)$ has a lower ionization potential than that of the second valence electron, which is in an s-orbital. The crossing radius is thus higher for the p electron transfer. Most collisions will proceed via such a transfer, and the overall selectivity will be low.

effects. Are there additional vector correlations? Can we observe the orientation of $\mathbf{j}'$ with respect to $\mathbf{v}$ or, for that matter, with respect to $\mathbf{v}'$ or to the normal ($\mathbf{v} \times \mathbf{v}'$) to the scattering plane? Can we expect aligned reaction products from a beam–gas collision of randomly oriented reagents? The answers to these questions are provided in the following elementary account of the subject of polarization of nascent reaction products.

The study of polarization effects can be profitably traced back to the venerable observations and theory of light scattering by molecules, first for Rayleigh and then for Raman scattering (Fig. 7.72). Recall that for the

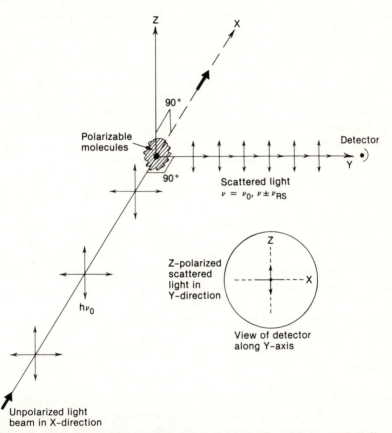

Fig. 7.72. Schematic arrangement for observation of polarization of light scattered (from a randomly oriented ensemble of molecules) at 90° from an incident unpolarized light beam, frequency ν_0 (directed along the X axis). The detector provided with a polarizer looks along the Y axis, and we note that the scattered light, both Rayleigh and Raman (frequency $\nu_0 \pm \nu_{RS}$, where RS stands for Raman shift) will be Z-polarized. If the "polarizability ellipsoid" of the target molecules is spherical (e.g., for S-state atoms or molecules such as CH_4 with T_d point-group symmetry), the 90° scattered light, both Rayleigh and Raman, will be *completely* polarized, $P = 1$. Regardless of the anisotropy of the polarizability ellipsoid the depolarization, say $d = I_\perp/I_\parallel$, can never exceed $\frac{1}{2}$ for Rayleigh and $\frac{6}{7}$ for Raman scattering (for unpolarized exciting light).

totally symmetric molecular vibrations (for which the polarizability tensor α is a scalar, so that the induced moment $\alpha\mathbf{E}$ is parallel to the incident electric field $\mathbf{E}$), the Rayleigh and Raman light scattered from an ensemble of randomly oriented molecules is totally polarized. The depolarization of the Raman lines occurs for those transitions due to asymmetric vibrations. In the case of molecular fluorescence, depolarization will be prevalent because of the relatively long lifetime of the electronically excited state as compared with the rotational period of the molecule. Reorientation of the molecular frame will destroy the memory of the *directionality* of the incident photon beam.

Let us turn now to elementary molecular reactions. The simplest case, that of photofragmentation, is discussed in Section 7.6.5. Here, we shall deal with bimolecular reactions. Specifically, consider a beam–target gas arrangement with a chemiluminescent reaction where the light emitted is monitored at right angles to the beam. That such light may be polarized can be anticipated on the basis of the discussion of Fig. 7.72, if the product molecule(s) remember that they were formed from a highly anisotropic (relative) velocity distribution of the reagents. This "memory" of the initial direction of approach arises through the conservation of angular momentum. For the simple chemiluminescent exchange reaction

$$A + BC \longrightarrow AB^* + C$$
$$\qquad\qquad\quad \Big\downarrow {\scriptstyle h\nu}$$
$$\qquad\qquad\quad AB$$

the argument is as follows. Initially, $\mathbf{j}$ (the angular momentum of BC) is randomly (i.e., spherically symmetrically) distributed, while $\mathbf{L}$ is randomly distributed in a plane perpendicular to the initial relative velocity vector $\mathbf{v}$ (e.g., the YZ plane in Fig. 7.72). For simplicity, consider a reaction where many partial waves contribute, so that $j \ll l$, and that AB^* is formed at high rotational excitation, so that $j' > l'$. Conservation of total angular momentum (see Appendix 4B), $\mathbf{j} + \mathbf{L} = \mathbf{j}' + \mathbf{L}'$, implies that for our assumed conditions

$$\mathbf{j}' \approx \mathbf{L}$$

Thus, $\mathbf{j}'$ will be nearly parallel to $\mathbf{L}$ and hence confined to a plane (the YZ plane of Fig. 7.72), so the emitted chemiluminescence will be polarized.

The degree of polarization is conveniently measured by P [cf. Eq. (7.22)]:

$$P = (I_\| - I_\perp)/(I_\| + I_\perp) \tag{7.23}$$

where $I_\|$ is I_Z of Fig. 7.72 and $I_\perp$ is I_X in that geometry. Depending upon whether the chemiluminescence arises from a parallel or perpendicular transition, its polarization will differ. Detailed analysis showed that the polarization of a perpendicular transition for the geometry discussed has a maximum value of 1/3 and that $0 \le P \le 1/3$, depending upon the degree to which $\mathbf{j}'$ deviates from confinement to the plane (YZ) defined by $\mathbf{L}$.

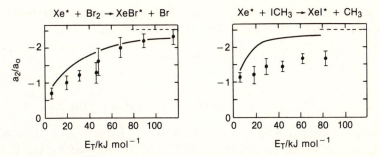

Fig. 7.73. Rotational alignment coefficient, a_2/a_0, of the diatomic products versus E_T for the reactions of $Xe^*(^3P_{2,0})$ with Br_2 [yielding $XeBr^*(B)$, left] and CH_3I [yielding $XeI^*(B, C)$, right]. Points are experimental; solid curve calculated from the classical DIPR model. The dashed horizontal lines represent the maximum theoretical limit on the alignment. The ordinate is a coefficient derived from the observable polarization, P, given, for the case of a parallel transition by $a_2/a_0 = 20 P/(P - 3)$. The maximum degree of polarization is $\frac{1}{3}$, for which $a_2/a_0 = -\frac{5}{2}$. [Adapted from R. J. Hennessy, Y. Ono, and J. P. Simons, *Mol. Phys.*, *43*, 181 (1981); for a recent related study see K. Johnson, R. Pease, J. P. Simons, P. A. Smith, and A. Kvaran, *J. Chem. Soc. Far. Trans.* II, *82*, 1281 (1986); see also C. D. Jonah, R. N. Zare, and C. Ottinger, *J. Chem. Phys.*, *56*, 271 (1972).]

For ground-state molecules, the observed polarization is often small ($P \leq 0.05$). High polarizations have, however, been measured for reactions of (metastable) electronically excited atoms. Figure 7.73 shows experimental results for the translational energy dependence of the polarization of the luminescence from the reactions of $Xe^*(^3P)$, where the ground state of the Xe halides is repulsive and where the $B \rightarrow X$ transition is known to be of the parallel type.

$$Xe^* + Br_2 \longrightarrow XeBr^*(B) + Br(^2P)$$

$$\downarrow hv$$

$$[XeBr(X)] \rightarrow Xe(^1S_0) + Br(^2P)$$

$$Xe^* + CH_3I \rightarrow XeI^*(B, C) + CH_3$$

$$\downarrow hv$$

$$[XeI(X)] \rightarrow Xe(^1S_0) + I(^2P)$$

For the Br_2 reaction at the highest E_T studied, the observed polarization is 0.3 (theoretical maximum, 0.33). In general, the higher the relative velocity, the better the expected alignment of BC^* angular momentum. Model computations verify this expectation, which can also be understood [see Eq. (4B.6)] in terms of the increasing role of kinematic constraints.†

---

† As E_T increases, so does the range of partial waves that contribute to the reaction (Sec. 2.4). If j is low, the term proportional to **L** in (4B.6) will then tend to dominate, barring special dynamic effects.

A related example is the laser-assisted photoassociation of ground state Xe and Br atoms to form the XeBr*(B) molecule (Fig. 7.74).

$$Xe(^1S_0) + Br(^2P) \xrightarrow{\ h\nu\ } XeBr^*(B)$$

$$\Big\downarrow h\nu$$

$$[XeBr(X)] \rightarrow Xe(^1S_0) + Br(^2P)$$

The polarized laser radiation at the frequency ν_L induces association of the colliding atoms.† The formation of XeBr*(B) is followed by the fluorescence, as before. Using horizontal polarization of the pump laser, the XeBr*($B \rightarrow X$) emission is found to be horizontally polarized (Fig. 7.74).

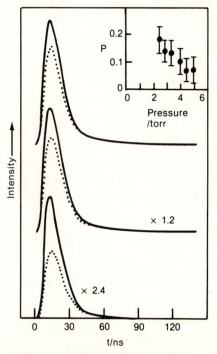

Fig. 7.74. Temporal dependence of the fluorescence of XeBr* (i.e., $B \rightarrow X$ radiation) from the polarized laser-assisted photoassociation reaction of Xe* with Br($^2P_{3/2}$) atoms, carried out in a gas cell at pressures of 2.5 Torr (bottom), 3.4 (center), and 4.5 Torr (top). The fluorescence is measured at 90° from the laser-beam direction. The solid curve is for fluorescence polarized parallel to the laser polarization; the dashed curve, perpendicular. The measured polarization P versus the pressure is plotted in the inset. Extrapolating to zero pressure, i.e., correcting for collisional depolarization, would yield a value of *ca.* 0.3, comparable to that of Fig. 7.73 for XeBr*. [Adapted from J. K. Ku, D. W. Setser, and D. Oba, *Chem. Phys. Lett.*, *109*, 429 (1984).]

† The photoassociation is induced by a pulsed dye laser in the wavelength range of 275–285 nm. Energies of about 1 mJ per pulse are already sufficient to promote the XeBr($X \rightarrow B$) excitation.

Extrapolation of the results to the zero pressure limit would yield P values comparable to those for the $Xe^* + Br_2$ reaction. The laser-induced photoassociation yields $XeBr^*(B)$ molecules with a definite propensity for their plane of rotation to lie in the plane of the laser polarization.

The alignment of electronically excited products is monitored through the polarization of their fluorescence. Reactions producing ground electronic state products can equally well lead to product alignment, but this requires a more elaborate detection. We have already mentioned (e.g., Fig. 6.36) the use of laser-induced fluorescence probing of the products for such purposes. The earlier studies employed the deflection of products in an inhomogeneous electric field, thereby resolving the different j, m_j states. In the more refined experiments, the deflection of the product molecule is measured for two orientations of the electric field, parallel to $\mathbf{v}$ and perpendicular to the scattering plane, i.e., parallel to $\mathbf{v} \times \mathbf{v'}$. From such data it is possible to determine not only the alignment of $\mathbf{j'}$ with respect to the initial relative velocity vector $\mathbf{v}$ (i.e., $\langle \hat{j}_z'^2 \rangle = \langle \cos^2(\hat{\mathbf{j}}' \cdot \hat{\mathbf{v}}) \rangle$), but also $\langle \hat{j}_y'^2 \rangle$ (and thus $\langle \hat{j}_x'^2 \rangle = 1 - \langle \hat{j}_z'^2 \rangle - \langle \hat{j}_y'^2 \rangle$) and to compare with the prior (isotropic) distribution $\langle \hat{j}_z'^2 \rangle = \langle \hat{j}_y'^2 \rangle = \langle \hat{j}_x'^2 \rangle = 1/3$. From a study of the reactions of alkali metals with halomethanes, Fig. 7.75 shows the results for the

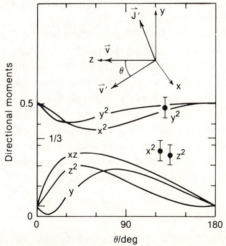

Fig. 7.75. Experimental determination of the preferred spatial alignment of the rotational angular momentum of the CsI product of the reaction of Cs with ICH_3. Results are from a crossed molecular beam experiment employing electric deflection analysis of the alkali halide product. Results are shown for two field orientations, (1) $\mathbf{E} \parallel \mathbf{v}$, the relative velocity (Z axis), and (2) $\mathbf{E} \perp \mathbf{v}$, $\mathbf{v'}$ plane (X axis). The first measurement yields $\langle \hat{J}_z^2 \rangle = 0.25$ and the second one $\langle \hat{J}_y^2 \rangle = 0.48$. [Here, $\langle \hat{J}_z^2 \rangle$ is the average of the square of the direction cosine between $\mathbf{J'}$ and $\mathbf{v}$, i.e., $\langle \cos^2(\hat{\mathbf{J}} \cdot \mathbf{v}) \rangle$, and similarly for $\langle \hat{J}_y^2 \rangle$.] Since the sum of these moments is unity, $\langle \hat{J}_x^2 \rangle = 0.27$. The experimental points are plotted at the approximate c.m. scattering angle of the measurements. The prior expectation for a random spatial distribution of $\mathbf{J'}$ would be $\frac{1}{3}$ for each of the X^2, Y^2, and Z^2 moments. The solid curves are model calculations. [Adapted from D. S. Hsu, G. M. McClelland, and D. R. Herschbach, *J. Chem. Phys.*, **61**, 4927 (1974); see also D. A. Case and D. R. Herschbach, *J. Chem. Phys.*, **64**, 4212 (1976), and D. A. Case, G. M. McClelland, and D. R. Herschbach, *Mol. Phys.*, **35**, 541 (1978).]

reaction

$$CH_3I + Cs \rightarrow CsI + CH_3$$

7.6.4 Repulsive energy release

We have emphasized the need to ascertain which steric effects reflect simple kinematic constraints and which reflect the intrinsic topography of the potential energy surface. An important aspect of the latter is the geometry and dynamics at the transition state. Consider, as an example, an exoergic

A + BC chemical reaction with a bent transition state $A \overset{\displaystyle B}{\diagup \diagdown} C$, where the energy is released repulsively, i.e., as C separates from AB. Such a repulsive release will clearly lead to a high rotational excitation in the AB product (see also Appendix 4C).

The kinematic confinement of $\mathbf{j}'$ parallel to $\mathbf{L}$ tends to dominate when $\sin^2 \beta$ is large [Eq. (4B.6)], that is, when the departing atom is light (e.g. Fig. 4.27). Dynamic effects are particularly important when the transferred atom, B, is lighter compared to the other two, so that $\cos^2 \beta$ is large. Of course, the potential energy surface must provide for a substantial repulsive energy release to cause a considerable spinning of the ejected AB molecule from the bent transition state.

These expectations are verified by trajectory calculations on model potential-energy surfaces exhibiting the required topographic characteristics. Figure 7.76 shows the remarkably strong correlation between $\mathbf{j}'$ and $\mathbf{L}'$. They tend to be antiparallel, as expected in general for dynamical effects on the basis of Eq. (4B.6), and as is illustrated for the special case under consideration in Fig. 7.76. The antialignment of $\mathbf{j}'$ and $\mathbf{L}'$ implies that the collision trajectory has a proclivity for coplanarity, or, in a more technical language, that only low helicities contribute.

More experimental results and dynamical computations and interpretations are clearly needed, and are forthcoming.

7.6.5 Photopolarization

The polarized laser-induced photofragmentation of simple polyatomic molecules, with polarized laser probing of the products bring us close to the ultimate experiment where the products' quantum state distribution is mapped as a function of initial conditions.

A case where both the "half-collision"

$$HONO(\tilde{X}) \xrightarrow[355 \text{ nm}]{h\nu} HONO(\tilde{A}) \rightarrow OH(X^2\Pi) + NO(X^2\Pi)$$

and the entire collision (Fig. 5.7)

$$H(^1S_0) + NO_2(\tilde{X}^2A_1) \rightarrow [HONO] \rightarrow OH(X^2\Pi) + NO(X^2\Pi)$$

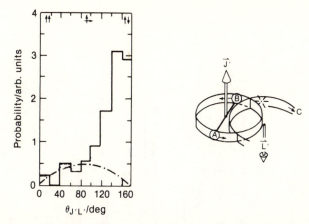

Fig. 7.76. Trajectory calculations for a model system $A + BC \rightarrow AB + C$ (with a large repulsive energy release) of the correlation between the directions of final angular momenta $\mathbf{j}'$ and $\mathbf{L}'$ (left). Shown in the form of a histogram is the probability of reaction with a given angle $\theta_{j'L'}$ (the angle between $\mathbf{j}'$ and $\mathbf{L}'$), peaked near 180°. The prior distribution, for random orientation of $\mathbf{j}'$ with respect to $\mathbf{L}'$, is $\sin \theta_{j'L'}$, maximized at 90° (the dashed curve, scaled to the histogram near 90°). At right is a pictorial representation of the sudden repulsive energy release for a bent ($A^B C$) transition state, forcing the product AB to rotate as shown, yielding $\mathbf{j}'$ "up," and C to recoil as sketched, with a resulting orbital angular momentum $\mathbf{L}'$ in the opposite direction. [Adapted from N. H. Hijazi and J. C. Polanyi, *J. Chem. Phys.*, **63**, 2249 (1975).]

were studied in detail is our first example. In the photolysis experiment, the dissociating and probe lasers are counterpropagating in a gas cell through which HONO flows at a low pressure. The OH is probed by measuring the undispersed laser-induced fluorescence for the $X \rightarrow A$ band excitation. The OH alignment is determined (Fig. 7.77) by using different polarizations of the laser. The detailed analysis of the results is beyond our scope, but even the raw data suffice to reveal the existence of a polarized OH nascent distribution.

The relative population of the OH vibrotational states as determined by laser-induced fluorescence shows the distribution to be relatively cold and hence translationally hot (see Fig. 5.7). We shall shortly come to the subject of Λ doublets, but we wish to mention here that both the full and the half-collisions lead to unequally populated OH Λ doublets.

For our finale in the discussion of product alignment we turn to the polarized laser-induced dissociation of H_2O from a repulsive upper electronic state:

$$H_2O(^1A_1) \xrightarrow[157 \text{ nm}]{h\nu} H_2O(^1B_1) \rightarrow H(^2S_{1/2}) + OH(^2\Pi^+, {}^2\Pi^-)$$

Laser-induced fluorescence from OH shows it to be translationally and vibrationally hot and rotationally warm, and the Λ-doublet populations are inverted, i.e., the lone unpaired electron of OH has its p-orbital oriented

355 nm Photolysis

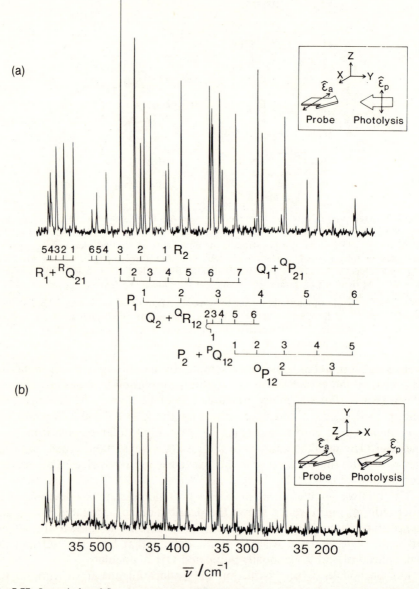

Fig. 7.77. Laser-induced fluorescence of the OH photofragment from the photolysis of HONO at $\lambda = 355$ nm. [At this wavelength the HONO(X) is excited to the so-called 2^2 level of the $\tilde{A}$ state, which then promptly dissociated to NO + OH.] The photolysis laser and the probe laser beams are counterpropagated. The total fluorescence intensity (detected along $\hat{\epsilon}_a$) is plotted versus the probe laser wave number in (a) for perpendicular polarization and in (b) for parallel polarization (see insets). The differences in the line intensities for the P and R versus the Q branches for configurations (a) versus (b) leads to information on the laboratory alignment of $\mathbf{j}_{OH}$, the OH rotational angular momentum. [From R. Vasudev, R. N. Zare, and R. N. Dixon, *J. Chem. Phys.*, *80*, 4863 (1984); with permission.]

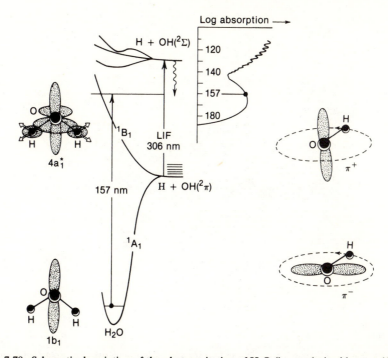

Fig. 7.78. Schematic description of the photoexcitation of H_2O (by a polarized laser at 157 nm) and its subsequent fragmentation to $H + OH$. The transition $H_2O(^1A_1 \rightarrow {}^1B_1)$ is believed to promote one electron from the nonbonding $1b_1$ orbital (pictured at lower left) to the $4a_1^*$ antibonding orbital, which is in the H_2O plane. One unpaired electron remains in the $^1b_1$ orbital, which is perpendicular to the H_2O plane (represented above left). The absorption spectrum of H_2O is shown (from 110 to 190 nm) at the top. The nascent $OH(^2\Pi)$ is probed by polarized laser-induced fluorescence from 306 to 320 nm to ascertain vibrational and rotational populations, Λ-doublet state ratios, the alignment of the OH product rotational angular momentum, and the spin distribution (ratio of abundances of $^2\Pi_{3/2}$, $^2\Pi_{1/2}$ states). Shown at right are idealized (classical) representations of the orientations of the unpaired electron with respect to the plane of rotation of the OH. [Adapted from P. Andresen, G. S. Ondrey, B. Titze, and E. W. Rothe, *J. Chem. Phys.*, *80*, 2548 (1984). See also: P. Andresen, V. Beushausen, D. Häusler, H. W. Lülf, and E. W. Rothe, *J. Chem. Phys.*, *83*, 1429 (1985), R. Schinke, V. Engel, P. Andresen, D. Häusler, and G. G. Balint-Kurti, *Phys. Rev. Lett.*, *55* 1180 (1985).]

preferentially perpendicular to the OH plane of rotation. It is further concluded from the OH alignment that the photofragmentation process is essentially coplanar. We shall now consider an interpretation of these findings.

Figure 7.78 shows cuts along the reaction coodinate through the relevant potential-energy surfaces. Upon photoexcitation, one electron is promoted from the doubly occupied, nonbonding $1b_1$ orbital of H_2O to the $4a_1^*$ antibonding orbital of a repulsive $^1B_1$ state of H_2O. The $4a_1^*$ orbital is perpendicular to the plane of the H_2O molecule (see Fig. 7.78). Also shown

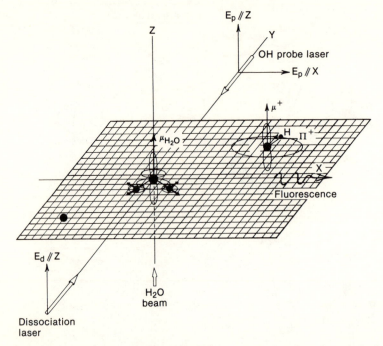

Fig. 7.79. Schematic outline of the OH polarization measurements from the polarized excimer laser dissociation of H_2O. [Adapted from same source as Fig. 7.78.]

are the two possible "classical" alignments of the p-orbital of OH, perpendicular to (Π^+) and within (Π^-) the OH plane of rotation. The Π^- state is somewhat lower in energy, yet the experiment shows that the Π^+ state is strongly preferred: The orientation of the orbital of the unpaired electron ($^1b_1$ in H_2O, $p\Pi$ in OH) is preserved with respect to the plane of the dissociation.†

The essence of the experiment is shown in Fig. 7.79. The transition dipole μ_{H_2O} (corresponding to a charge transfer from an orbital perpendicular to the plane to an in-plane orbital) is perpendicular to the H_2O plane. The linearly polarized photodissociation laser has its electric field $\mathbf{E}_D$ in the direction (defined as the Z axis) of the H_2O nozzle beam. Since absorption is proportional to $|\mu_{H_2O} \cdot \mathbf{E}_d|^2$, the $H_2O(^1B_1)$ molecules are preferentially aligned with the HOH molecular frame in the XY plane. The OH fragments tend therefore to rotate in the same XY plane. The probe laser is counterpropagating and can be polarized in the Z or X directions. The Q transitions of OH have a transition moment perpendicular to the OH rotation plane and are used for the laser-induced fluorescence probing. If

---

† The nozzle beam provides rotationally cold H_2O molecules. After dissociation the OH will rotate in the original H_2O plane since there are no forces that can take it out of that plane.

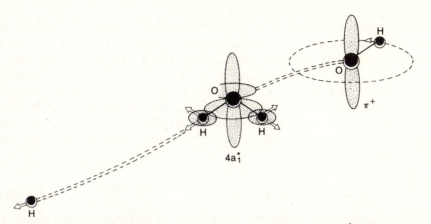

Fig. 7.80. Pictorial representation of the coplanar dissociation of $H_2O(^1B_1)$ on a single potential-energy surface. The $1b_1$ orbital of $H_2O(^1B_1)$ containing a single, unpaired electron is shown in the vertical direction and correlates with the (also vertical) $p\pi$-orbital of the OH radical, i.e., the Π^+ state is preferentially populated. [Adapted from same source as Fig. 7.78.]

$E_p \| Z$, the detection is then of OH radicals rotating primarily in the XY plane, while if $\mathbf{E}_p \| X$, the detection is for the YZ plane of rotation.

Measurements of the polarization of the laser-induced fluorescence signal have been carried out for both $\mathbf{E}_p \| \mathbf{E}_d$ and $\mathbf{E}_p \perp \mathbf{E}_d$ with the detector orthogonal to both lasers. The polarization ratios are large and approach the theoretical limit for high rotational states. The j dependence reflects a deviance of the OH transition dipole relative to the plane of rotation due to Λ-doublet mixing. The precession of $\boldsymbol{\mu}_{OH}$ about $\mathbf{j}$ decreases as j increases and reaches the "classical" limit of perfect orientation at high j's. (The high-j limit is shown in Fig. 7.78.) A summary of the results for this "direct" and coplanar photodissociation is provided in Fig. 7.80.

A byproduct of such detailed understanding is that we can now explain the origin of the interstellar OH maser emission, which requires Λ-doublet population inversion: The inversion itself, and hence the maser transition, is a direct consequence of the conservation of orbital symmetry in the course of the dissociation of $H_2O(^1B_1)$ excited state, formed via ultraviolet irradiation of H_2O molecules in the galactic environment.

7.7 New frontiers

We have by now achieved the ability to dissect the elementary chemical act at increasing levels of molecular complexity. In chemical dynamics we shall soon approach the current level of understanding of molecular structure. That is, we are no longer content to think of structure in static terms. The temporal and dynamical aspects of structure are being addressed. A central achievement of structural theory is the systematics that it provides. Spectra, structure and bonding of a given compound are no longer considered in isolation but as part of a unified approach encompassing many similar species. A recent example comes from the field of transition metal complexes. A sequences of theories of increasing sophistication provides successive frameworks for interpretation of an ever-wider set of observations. The more complex the theory (e.g., Do we include the role of π electrons on the ligands?), the more detailed and extensive are the systematics. Are we approaching a similar position with regard to chemical reactivity?

7.7.1 Understanding chemical reactivity

Can we now make qualitative predictions of systematics and trends in chemical reactivity? Can we, when necessary, refine such predictions to a quantitative level, all the way (if so desired) to full *ab initio* exactitude?

For reactions proceeding on a single potential-energy surface we have indeed covered much of the ground: Is there a barrier to reaction? Simple molecular orbital theory explains gross difference in activation energies and even finer details such as steric effects (i.e., the angle dependence of the barrier). Moreover, simple topographical aspects of the potential-energy surface (e.g., whether the energy will be released "early" or will be "late" and "repulsive") can come from such considerations. Correlations between features of the potential and modes of the dynamics derived from physical considerations (and validated by quantitative dynamical computations) can then be invoked to tell us much about the possible modes of reactivity. Next come the semiempirical theories of the potential-energy surface, which already yield a semiquantitative representation of the potential. Such a potential energy surface can be used as input for a full scattering calculation but also for less accurate (but all the same, realistic) approximations starting at the level of line-of-centers models. As always, it is only the finer details which warrant a corresponding detailed theory. The lower the available experimental resolution, the more we can expect from simpler theories. Thus, as our molecules become larger, we compensate for our ignorance of the potential and our inability to implement a full dynamical computation by the use of statistical methods. We have seen the power (and limitations) of the RRKM theory to interpret unimolecular processes, even in multiphoton dissociation.

For simple systems, *ab initio* potential energy surfaces of "chemical accuracy" are now available. In addition, benchmark quantal scattering computations have served to "calibrate" the various approximations (and quasi-classical trajectory simulations). The detailed experiments, on the one hand, and the detailed theory, on the other, are bringing us to a position where we have the ability to make physically realistic predictions (sometimes retrospective!) of the chemical reactivity at the desired level of detail. We are achieving an understanding of how the total energy of the system determines the reactivity, how different initial partitioning of this total energy affects the reactivity, and how the energy is disposed in the products. Equally, for the temporal aspects we have definite ideas (which are just approaching their experimental test via bimolecular spectroscopy) of how simple molecules react. Nor are our theories limited to energies accessible to ordinary thermal experiments. Stimulated by lasers (and by high temperatures via combustion and shock waves) we are able to compute not only reaction rates under extreme conditions but also more detailed attributes at higher energies. We are already at the stage of "feedback" where we are aiming for novel modes of reactivity (e.g., laser-assisted chemistry) based on our improved understanding.

It is for higher-resolution studies of larger systems (e.g., the understanding of intramolecular energy redistribution) or the full steric aspects of reactivity (e.g., product polarization versus scattering angle) that we require improved guidelines. Of course, and as in structure theory, there is only so much that a simple theory can do; exact predictions require a fully developed theory and exact computations. (Hence we can always use better potential energy surfaces!) Finally, we need to carry over our improved understanding into the condensed phase and onto interfaces. In other words, we need to make a full impact on the practice of real chemistry.

Even for nonadiabatic collisions, much of the understanding is at hand. A familiar example of systematics is the harpoon mechanism, which accounts for both successful electron-jump processes and for unsuccessful ones. On a semiquantitative level the role of curve crossing (and even surface-hopping) is well described by the Landau–Zener theory and its refinements. A variety of nonadiabatic processes, both inelastic and reactive, can thereby be treated. To the first approximation, the overall behavior of the system is governed by only one intrinsic parameter, the interaction matrix element responsible for the splitting $\Delta E(R_x)$, which can even be estimated semiempirically when not available from *ab initio* computations. For larger molecules, the theory of vibronic coupling is making rapid progress. Detailed understanding of complex photochemical processes is so advanced that now even the photosynthetic centers of leaves are subject to refined theoretical examination.

The practice of molecular reaction dynamics has indeed led to better understanding of chemical reactivity. Where, however, do we go from here?

7.7.2 *Future directions*

Inspection of our table of contents reveals that much has happened in this field since the publication of *Molecular Reaction Dynamics* some 13 years ago. Progress has by no means been limited to the acquisition of more data, on larger systems, with improved resolution. New phenomena not even considered and new questions not even asked then are now the subject of entire sections. Others, barely mentioned a dozen years ago, are already fully developed. These new topics are the growth areas of the next decade.

Consider the subject of dynamical stereochemistry. For many years, synthetic chemists maintained (without much controversy) that molecules have not merely a size but a shape. Their "molecular models" and their refinements ("molecular mechanics") have played a decisive role in the understanding of the factors that influence the course of their reactions and even in the development of new synthetic pathways. It is now becoming possible to understand steric effects not only on a static basis but on a dynamical one. This is so central to chemical thinking that we now find a detailed discussion of the subject already in Chapter 2. What were ground-breaking experiments at the time of our first book promises to provide gainful occupation for practitoners well into the future. While the organic chemists asks how steric considerations govern reactivity and yield of different products, the molecular dynamicist asks how the precollision conditions can be selected and how they influence not only reactivity but also the mutual orientation and energy disposal in the products.

Geometrical aspects of the dynamics are not limited to considerations of steric factors. Site- or regio-selective chemistry remains a favorite goal of the laser chemist. Can we selectively pump a molecule so as to create an energy-rich "hot spot" that will be the site of a chemical transformation? If the isolated molecule tends to dissipate the energy too rapidly, can we use an intense laser to dress the molecule during the collision so as to selectively induce a unimolecular process? Theory suggests that much more can be done than has so far been achieved.

Laser-induced, laser-assisted, and laser-catalyzed processes are important not only because of their potential practical applications. The "spectroscopy of the transition state" enables us to probe chemical reactions on a time scale comparable to (or shorter than) that of the bond reorganization process.

We are also far from exhausting the unique features of lasers. The applications based on the high selectivity provided by their near monochromaticity and high power permeate this entire volume. We are, however, just beginning to couple these with polarization and narrow linewidth towards the ideal of fully resolved (including angle) final state detection. Temporal resolution sufficient for the pump and probe study of bimolecular processes has already been introduced in Chapter 1, but we are just reaching the required subpicosecond pulse durations in the wavelengths of interest for the study of fast unimolecular processes. Finally, we are yet to take full advantage of the coherence of the laser radiation.

Another area where technological progress is making it possible to satisfy both practical and intellectual needs is the dynamics of processes occurring on surfaces. The need for providing a molecular basis for catalysis is obvious. We are finding that we need all of what we know from the gas phase, and then a considerable body of knowledge from solid state and surface science to unravel the myriad of processes occurring with molecules in the presence of a surface.

What are some of the prospects here? One aspect is an intrinsic steric

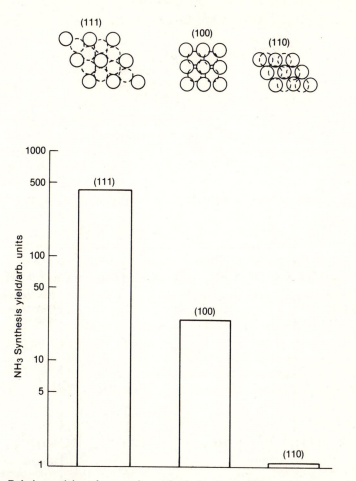

Fig. 7.81. Relative activity of ammonia synthesis for three different faces on a single iron crystal. The most efficient face, Fe(111), converts 2×10^{16} NH_3 molecules $cm^{-2} s^{-1}$ at $T_s = 870$ K and high pressure. The crystal faces shown above the reactivity bars represent the topmost Fe atoms as solid circles, while the dotted circles are atoms of the second layer. [Adapted from N. D. Spencer, R. C. Shoonmaker, and G. A. Somorjai, *J. Catal.*, *74*, 129 (1982).] Remarkably, the 11$\bar{2}$1 face of Re is almost 10-fold more active than Fe(111), and the structure sensitivity is even higher. [M. Asscher and G. A. Somorjai, *Surf. Sci.*, *143*, L389 (1984).]

phenomenon. An example of such an effect is found in the Haber ammonia synthesis, where the yield is quite different when carried out on different faces of a single crystal of iron (Fig. 7.81). Another aspect is reagent energy selectivity (e.g., translation and vibration can have quite different efficacies in overcoming the barrier to dissociative adsorption). Still another new area involves laser-assisted surface processes. It is possible to induce selective desorption by laser irradiation of an adlayer on a surface.

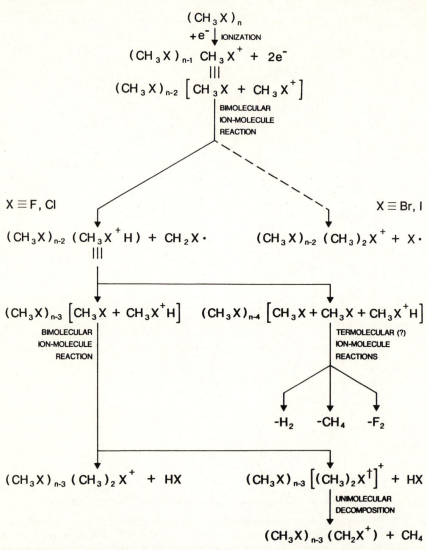

Fig. 7.82. Scheme of overall reaction mechanism for ionized methyl halide clusters. Symbols $\equiv$ and $\parallel$ denote equivalence of empirical formulas. [From J. F. Garvey and R. B. Bernstein, *J. Phys. Chem.*, *90*, 3577 (1986).]

Fig. 7.83. Snapshots from a computer simulation of surface coverage for a model reaction between two adsorbed species $A(ad) + B(ad) \rightarrow AB(gas)$. Initially, (a), 4000 A "atoms" are randomly distributed over 10,000 lattice sites. Because of lateral interactions, the A atoms tend to aggregate at a rate characterized by τ_A, the (mean) time required for an A atom to diffuse by one lattice spacing. Panels (b) and (c) show the "islands" formed when only A is present and diffusing. After (c), 4000 B atoms are randomly placed on vacant lattice sites. These B atoms diffuse freely and rapidly ($\tau_B \ll \tau_A$) and react with the adsorbed A atoms; AB desorbs immediately. Reaction is taken to be less favorable as the number of A-atom neighbors of the A(ad) atom increases. This results [stages (d)–(f)] in the rapid disappearance of isolated A(ad) atoms and the "smoothing" of the edges of the islands and a slower reaction rate than that expected for a random distribution of reagents. [Adapted from M. Silverberg, A. Ben-Shaul, and F. Rebentrost, *J. Chem. Phys.*, *83*, 650 (1985).]

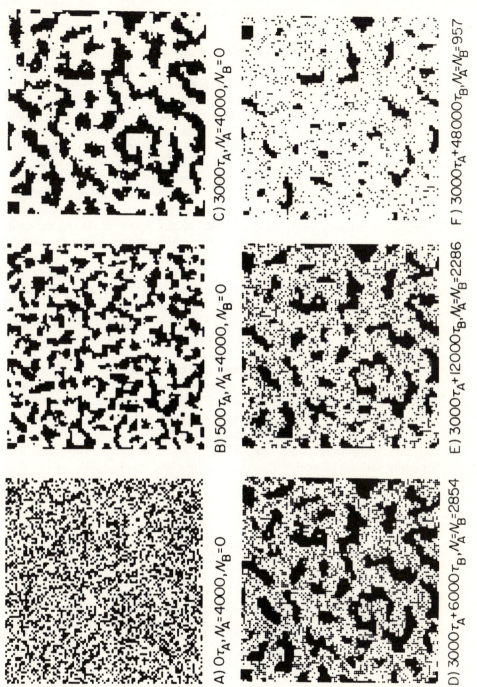

A) $0\tau_A$, $N_A=4000$, $N_B=0$

B) $500\tau_A$, $N_A=4000$, $N_B=0$

C) $3000\tau_A$, $N_A=4000$, $N_B=0$

D) $3000\tau_A+6000\tau_B$, $N_A=N_B=2854$

E) $3000\tau_A+12000\tau_B$, $N_A=N_B=2286$

F) $3000\tau_A+48000\tau_B$, $N_A=N_B=957$

Intermediate between the dynamics of gas phase and of surface processes are the dynamics of molecular clusters. Here we observe both intramolecular rearrangements and intermolecular reactions. The study of such processes is leading to a better understanding of processes in solution. The role of solvent molecules in "dressing" the transition state is likely to be much studied in the near future. Experiments on the ionization of molecular clusters indicate that extensive intramolecular rearrangements can take place, that commonly observed *bi*molecular ion–molecule reactions can occur *within* the cluster ion itself. The scheme for methyl halide clusters (Fig. 7.82) shows how a solvated CH_3F^+ ion, for example, reacts with its own solvent "shell" to form a protonated cluster ion (which then undergoes subsequent intramolecular reaction). Such cluster experiments will act as a bridge between gas-phase ion–molecule reaction dynamics and the chemistry of ions in solution.

Further possibilities are limited only by our imagination. How about, say, cluster reactions on surfaces? A cluster on a surface is what is usually called an island. When surface diffusion is rapid, and intramolecular attractive forces dominate over adsorptive attraction, the adsorbed species can aggregate into islands (Fig. 7.83), whose "shores" are then eroded by reaction.

Our goal has been and will continue to be a central theme of modern chemistry: the understanding of chemical reactivity. We have acquired from the real-world, practicing chemists the concept of structure and have tried to show how to forge the many successive links from the (static) structure to the (dynamic) reactivity. Ultimately, the field of molecular reaction dynamics will surely contribute to the practice of real-world chemistry, hopefully with dividends!

Suggested reading

Section 7.1

Book

Truhlar, D. G. (ed.), *Resonances in Electron Molecule Scattering, van der Waals Complexes, and Reactive Chemical Dynamics*, ACS Symp. Ser. No. 263 (1984).

Review Articles

Anderson, J. B., "The Reaction $F + H_2 \rightarrow HF + H$." *Adv. Chem. Phys.*, **41**, 229 (1980).
Bowman, J. M., Lee, K. T., Romanowski, H., and Harding, L. B., "Approximate Quantum Approaches to the Calculation of Resonances in Reactive and Nonreactive Scattering," in Truhlar (1984).

Hayes, E. F., and Walker, R. B., "Reactive Resonances and Angular Distributions in the Rotating Linear Model," in Truhlar (1984).

Kuppermann, A., "Reactive Scattering Resonances and Their Physical Interpretation: The Vibrational Structure of the Transition State," in Truhlar (1981).

Launay, J. M., and LeDourneuf, M., "Formation and Decay of Complexes: N_2^-, H_2^-, FH_2," in Eichler et al. (1984).

Lefebvre, R., "Complex Energy Quantization for Molecular Systems," in Truhlar (1984).

McCurdy, C. W., "Direct Variational Methods for Complex Resonance Energies," in Truhlar (1984).

Micha, D. A., and Kurudoglu, Z. C., "Atom–Diatom Resonances Within A Many-Body Approach to Reactive Scattering," in Truhlar (1984).

Neumark, D. M., Wodtke, A. M., Robinson, G. N., Hayden, C. C., and Lee, Y. T., "Dynamic Resonances in the Reaction of Fluorine Atoms with Hydrogen Molecules," in Truhlar (1984).

Polanyi, J. C., and Schreiber, J. L., "The Reaction of $F + H_2 \rightarrow HF + H$," *Faraday Discuss. Chem. Soc., 62,* 267 (1977).

Romelt, J., and Pollak, E., "Vibrationally Bonded Molecules: The Road from Resonances to a New Type of Chemical Bond," in Truhlar (1984).

Schaefer, H. F., III, "The $F + H_2$ Potential Energy Surface," *J. Phys. Chem., 89,* 5336 (1985).

Shoemaker, C. L., and Wyatt, R. E., "Feshbach Resonances in Chemical Reactions." *Adv. Quant. Chem., 14,* 169 (1981).

Zhang, Z. H., Abusalbi, N., Baer, M., Kouri, D. J., and Jellinek, J., "Resonance Phenomena in Quantal Reactive Infinite-Order Sudden Calculations," in Truhlar (1984).

Section 7.2

Books

Beynon, J. H., and Gilbert, J. R., *Application of Transition State Theory to Unimolecular Reactions,* Wiley, New York, 1984.

Hinze, J. (ed.), *Energy Storage and Redistribution in Molecules,* Plenum, New York, 1983.

Jortner, J., and Pullman, B. (eds.), *Intramolecular Dynamics,* Reidel, Dordrecht, 1982.

Pritchard, H. O., *The Quantum Theory of Unimolecular Reactions,* Cambridge University Press, Cambridge, U.K., 1984.

Review Articles

Bernstein, R. B. and Zewail, A. H., "Chemical Reaction Dynamics and Marcus' Contributions," *J. Phys. Chem., 90,* 3467 (1986).

Beynon, J. H., and Gilbert, J. R., "Energetics and Mechanisms of Unimolecular Reactions of Positive Ions: Mass Spectrometric Methods," in Bowers (1979).

Brumer, P., "Intramolecular Energy Transfer: Theories for the Onset of Statistical Behavior." *Adv. Chem. Phys., 47,* (Pt. 1), 201 (1981).

Davis, M. J., and Wagner, A. F., "The Intramolecular Dynamics of Highly Excited Carbonyl Sulfide (OCS)," in Truhlar (1984).

Frey, H. M., and Walsh, R., "Unimolecular Reactions," *Spec. Per. Rep.*, *3*, 1 (1978).

Gentry, W. R., "Vibrationally Excited States of Polyatomic van der Waals Molecules: Lifetimes and Decay Mechanisms," in Truhlar (1984).

Hase, W. L., "Dynamics of Unimolecular Reactions," in Miller (1976).

Hase, W. L., "Overview of Unimolecular Dynamics," in Truhlar (1981).

Hase, W. L., "Unimolecular and Intramolecular Dynamics; Relationship to Potential Energy Surface Properties," in El-Sayed (1986).

Hedges, R. M., Jr., Skodje, R. T., Borondo, F., and Reinhardt, W. P., "Classical, Semiclassical, and Quantum Dynamics of Long-Lived Highly Excited Vibrational States of Triatoms," in Truhlar (1984).

Holmer, B. K., and Certain, P. R., "Model Studies of Resonances and Unimolecular Decay of Triatomic van der Waals Molecules," *Faraday Discuss. Chem. Soc.*, *73*, 311 (1982).

Lifshitz, C., "Unimolecular Decomposition of Polyatomic Ions: Decay Rates and Energy Disposal." *Adv. Mass Spectrom.*, *7A*, 3 (1978).

Lifshitz, C., "Intramolecular Energy Redistribution in Polyatomics Ions." *J. Phys. Chem.*, *87*, 2304 (1983).

Parmenter, C. S., "Vibrational Redistribution Within Excited Electronic States of Polyatomic Molecules." *Faraday Discuss. Chem. Soc.*, *75*, 7 (1983).

Pritchard, H. O., "State-to-State Theory of Unimolecular Reactions," *J. Phys. Chem.*, *89*, 3970 (1985).

Reisler, H., and Wittig, C., "Photo-Initiated Unimolecular Reactions," *Annu. Rev. Phys. Chem.*, *37*, 307 (1986).

Rice, S. A., "Quasiperiodic and Stochastic Intramolecular Dynamics: The Nature of Intramolecular Energy Transfer," in Woolley (1980).

Sage, M. L., "The Dynamics of Intramolecular Coupled Local Modes," in Jortner and Pullman (1982).

Shapiro, M., "Dissociation and Intramolecular Dynamics," in Jortner and Pullman (1982).

Tabor, M., "The Onset of Chaotic Motion in Dynamical Systems." *Adv. Chem. Phys.*, *46*, 73 (1981).

Tardy, D. C., and Rabinovitch, B. S., "Intermolecular Vibrational Energy Transfer in Thermal Unimolecular Systems." *Chem. Rev.*, *77*, 369 (1977).

Wolfgang, R., "Energy and Chemical Reaction. Intermediate Complex vs. Direct Mechanisms." *Acc. Chem. Res.*, *3*, 48 (1970).

Section 7.3

Books

Cantrell, C. D. (ed.), *Multiple Photon Excitation and Dissociation of Polyatomic Molecules*, Springer-Verlag, Berlin, 1981, 1986.

Chin, S. L., and Lambropoulos, P. (eds.), *Multiphoton Ionization of Atoms*, Academic, New York, 1984.

Grunwald, E., Dever, D. F., and Keehn, P. M., *Megawatt Infrared Laser Chemistry*, Wiley, New York, 1978.

Haas, Y. (ed.), *Multiphoton Excitation, Isr. J. Chem., 24,* No. 3 (1984).

Lin, S. H., Fujimura, Y., Neusser, H. J., and Schlag, E. W., *Multiphoton Spectroscopy of Molecules,* Academic, New York, 1984.

Review Articles

Ambartzumian, R. V., and Letokhov, V. S., "Multiple Photon Infrared Laser Photochemistry," in Moore (1977).

Ambartzumian, R. V., and Letokhov, V. S., "Selective Dissociation of Polyatomic Molecules by Intense Infrared Laser Fields." *Acc. Chem. Res., 10,* 61 (1977).

Ashfold, M. N. R., and Hancock, G., "Infrared Multiple Photon Excitation and Dissociation: Reaction Kinetics and Radical Formation." *Spec. Per. Rep., 4,* 73 (1981).

Bagratashvili, V. N., Kuzmin, M. V., and Letokhov, V. S., "Chemical Radical Synthesis in Gas Mixtures Induced by Infrared Multiple-Photon Dissociation." *J. Phys. Chem., 88,* 5780 (1984).

Cantrell, C. D., Makarov, A. A., and Louisell, W. H., "Laser Excitation of SF_6: Spectroscopy and Coherent Pulse Propagation Effects." *Adv. Chem. Phys., 47* (Pt. 1), 583 (1981).

Chu, S.-I., "Recent Developments in Semiclassical Floquet Theories for Intense-Field Multiphoton Processes." *Adv. At. Mol. Phys., 21,* 197 (1985).

Danen, W. C., and Jang, J. C., "Multiphoton Infrared Excitation and Reaction of Organic Compounds," in Steinfeld (1981).

Davis, M. J., Wyatt, R. E., and Leforestier, C., "Classical and Quantum Mechanical Studies of Molecular Multiphoton Excitation and Dissociation," in Jortner and Pullman (1982).

Donovan, R. J., "Ultraviolet Multiphoton Excitation: Formation and Kinetic Studies of Electronically Excited Atoms and Free Radicals." *Spec. Per. Rep., 4,* 117 (1981).

Galbraith, H. W., and Ackerhalt, J. R., "Vibrational Excitation in Polyatomic Molecules," in Steinfeld (1985).

Gobeli, D. A., Yang, J. J., and M. A. El-Sayed, "Laser Multiphoton Ionization–Dissociation Mass Spectrometry." *Chem. Rev., 85,* 529 (985).

Golden, D. M., Rossi, M. J., Baldwin, A. C., and Barker, J. R., "Infrared Multiphoton Decomposition: Photochemistry and Photophysics." *Acc. Chem. Res., 14,* 56 (1981).

Haas, Y., "Electronically Excited Fragments Formed by Unimolecular Multiple Photon Dissociation." *Adv. Chem. Phys., 47,* (Pt. 1), 713 (1981).

Hall, R. B., Kaldor, A., Cox, D. M., Horsley, A., Rabinowitz, P., Kramer, G. M., Bray, R. G., and Maas, E. T., Jr., "Infrared Laser Chemistry of Complex Molecules." *Adv. Chem. Phys., 47,* (Pt. 1), 639 (1981).

Johnson, P. M., "Molecular Multiphoton Ionization Spectroscopy." *Acc. Chem. Res., 13,* 20 (1980).

Johnson, P. M., and Otis, C. E., "Molecular Multiphoton Spectroscopy with Ionization Detection." *Annu. Rev. Phys. Chem., 32,* 139 (1981).

Kaldor, A., Woodin, R. L., and Hall, R. B., "Recent Advances in IR Laser Chemistry." *Laser Chem., 2,* 335 (1983).

King, D. S., "Infrared Multiphoton Excitation and Dissociation." *Adv. Chem. Phys., 50,* 105 (1982).

Lee, Y. T., "Photodissociation of Polyatomic Molecules by Megawatt Infrared Lasers," in Glorieux et al. (1979).

Letokhov, V. S., "Laser Separation of Isotopes." *Annu. Rev. Phys. Chem.*, *28*, 133 (1977).

McAlpine, R. D., and Evans, D. K., "Laser Isotope Separation by the Selective Multiphoton Decomposition Process." *Adv. Chem. Phys.*, *60*, 31 (1980).

Mukamel, S., "Reduced Equations of Motion for Collisionless Molecular Multiphoton Processes." *Adv. Chem. Phys.*, *47* (Pt. 1), 509 (1981).

Parker, D. H., "Laser Ionization Spectroscopy and Mass Spectrometry," in Kliger (1983).

Quack, M., "The Role of Intramolecular Coupling and Relaxation in IR-Photochemistry," in Jortner and Pullman (1982).

Quack, M., "Reaction Dynamics and Statistical Mechanics of the Preparation of Highly Excited States by Intense Infrared Radiation." *Adv. Chem. Phys.*, *50*, 395 (1982).

Reisler, H., and Wittig, C., "Multiphoton Ionization of Gaseous Molecules." *Adv. Chem. Phys.*, *60*, 1 (1980).

Reisler, H., and Wittig, C., "Electronic Luminescence Resulting from Infrared Multiple Photon Excitation." *Adv. Chem. Phys.*, *47* (Pt. 1), 679 (1981).

Ronn, A. M., "Luminescence of Parent Molecule Induced by Multiphoton Infrared Excitation." *Adv. Chem. Phys.*, *47*, (Pt. 1), 661 (1981).

Schlag, E. W., and Neusser, H. J., "Multiphoton Mass Spectrometry." *Acc. Chem. Res.*, *16*, 355 (1983).

Schulz, P. A., Sudbo, Aa. S., Krajnovich, D. J., Kwok, H. S., Shen, Y. R., and Lee, Y. T., "Multiphoton Excitation and Dissociation of Polyatomic Molecules," in Cantrell (1981).

Thorne, L. R., and Beauchamp, J. L., "Infrared Photochemistry of Gas Phase Ions," in Bowers (1984).

Woodin, R. L., Bomse, D. S., and Beauchamp, J. L., "Multiphoton Dissociation of Gas-Phase Ions Using Low Intensity cw Laser Radiation," in Moore (1979).

Section 7.4

Books

Faraday Discussions of the Chemical Society, 73, van der Waals Molecules (1982).

Faraday Symposia of the Chemical Society *14, Diatomic Metals and Metallic Clusters* (1980).

Gole, J., and Stwalley, W. C. (eds.), *Metal Bonding and Interactions in High Temperature Systems,* American Chemical Society, Washington, D.C., 1982.

Review Articles

Beswick, J. A., and Jortner, J., "Intramolecular Dynamics of van der Waals Molecules." *Adv. Chem. Phys.*, *47* (Pt. 1), 363 (1981).

Blaney, B. L., and Ewing, G. E., "van der Waals Molecules." *Annu. Rev. Phys. Chem.*, *27*, 553 (1976).

Brady, J. W., Doll, J. D., and Thompson, D. L., "Classical Trajectory Studies of the Formation and Unimolecular Decay of Rare Gas Clusters," in Truhlar (1981).

Castleman, A. W., Jr., "Advances and Opportunities in Cluster Research," in Eichler et al. (1984).

Castleman, A. W., Jr., and Keesee, R. G., "Clusters: Properties and Formation," *Annu. Rev. Phys. Chem., 37,* 525 (1986).

Even, U., Amirav, A., Leutwyler, S., Ondrechen, M. J., Berkovitch-Yellin, Z., and Jortner, J., "Energetics and Dynamics of Large van der Waals Molecules." *Faraday Discuss. Chem. Soc., 73,* 153 (1982).

Ewing, G. E., "Structure and Properties of van der Waals Molecules." *Acc. Chem. Res., 8,* 185 (1975).

Haberland, H., "What Happens to a Rare-Gas Cluster When It Is Ionized," in Eichler et al. (1984).

Howard, B. J., "The Structure and Properties of van der Waals Molecules." *Int. Rev. Sci. Phys. Chem., 2,* 93 (1975).

Kappes, M. M., and Schumacher, E., "Preparation, Properties and Theory of Metal Clusters," in Eichler et al. (1984).

Klemperer, W., "Structure and Dynamics of Van der Waals Molecules," in Zewail (1978).

Levy, D. H., "Laser Spectroscopy of Cold Gas-Phase Molecules." *Annu. Rev. Phys. Chem., 31,* 197 (1980).

Levy, D. H., "Supersonic Molecular Beams and van der Waals Molecules," in Wooley (1980).

Levy, D. H., "van der Waals Molecules." *Adv. Chem. Phys., 47* (Pt. 1), 323 (1981).

Levy, D. H., Hayman, C. A., and Brumbaugh, D. V., "Spectroscopy and Photophysics of Organic Clusters," *Faraday Discuss. Chem. Soc., 73,* 137 (1982).

Levy, D. H., Haynam, C. A., Young, L., and Brumbaugh, D. V., "The Spectroscopy, Photophysics and Photochemistry of Organic Clusters," in Eichler et al. (1984).

Märk, T. D., and Castleman, A. W., Jr., "Experimental Studies on Cluster Ions." *Adv. At. Mol. Phys., 20,* 65 (1985).

Muetterties, E. L., Burch, R. R., and Stolzenberg, A. M., "Molecular Features of Metal Cluster Reactions," *Annu. Rev. Phys. Chem., 33,* 89 (1982).

Muetterties, E. L., Rhodin, T. N., Band, E., Brucker, C. F., and Pretzer, W. R., "Clusters and Surfaces." *Chem. Rev., 79,* 91 (1979).

Ng, C. Y., "Molecular Beam Photoionization Studies of Molecules and Clusters." *Adv. Chem. Phys., 52,* 265 (1983).

Sattler, K., "Microcluster Reearch: Between the Atomic and the Solid State," in Eichler et al. (1984).

Young, L., Hayman, C. A., and Levy, D. H., "Intramolecular Vibrational Relaxation and Photochemistry in Weakly Bound Organic Dimers," in Jortner and Pullman (1982).

Section 7.5

Books

Faraday Discussions of the Chemical Society, 72, Selectivity in Heterogeneous Catalysis (1981).

Gomer, R. (ed.), *Interactions on Metal Surfaces,* Springer-Verlag, Heidelberg, 1975.

King, D. A., and Woodruff, D. P. (eds.), *The Chemical Physics of Solid Surfaces and Heterogeneous Catalysis. Adsorption at Solid Surfaces,* Vol. 2, Elsevier, Amsterdam, 1983.

Rhodin, T. N., and Ertl, G. (eds.), *The Nature of the Surface Chemical Bond,* North-Holland, Amsterdam, 1979.

Somorjai, G. A., *Chemistry in Two Dimensions: Surfaces,* Cornell University Press, Ithaca, NY, 1981.

Review Articles

Adelman, S. A., "Generalized Langevin Equations and Many-Body Problems in Chemical Dynamics." *Adv. Chem. Phys., 44,* 143 (1980).

Adelman, S. A., "The Molecular Time Scale Generalized Langevin Equation to Problems in Condensed-Phase Chemical Reaction Dynamics." *J. Phys. Chem., 89,* 2213 (1985).

Adelman, S. A., and Brooks, C. L., III, "Generalized Langevin Models and Condensed-Phase Chemical Reaction Dynamics." *J. Phys. Chem., 86,* 1511 (1982).

Asscher, M., and Somorjai, G. A., "Energy Redistribution in Diatomic Molecules on Surfaces," in Pullman et al. (1984).

Bernasek, S. L., "Heterogeneous Reaction Dynamics." *Adv. Chem. Phys. 41,* 477 (1980).

Boudart, M., "Concepts in Heterogeneous Catalysis," in Gomer (1975).

Burden, A. G., Grant, J., Martos, J., Moyes, R. B., and Wells, P. B., "Variation of Catalyst Selectivity by Control of the Environment of Surface Sites." *Faraday Discuss. Chem. Soc., 72,* 95 (1981).

Canning, N. D. S., and Madix, R. J., "Toward an Organometallic Chemistry of Surfaces." *J. Phys. Chem., 88,* 2437 (1984).

Cardillo, M. J., and Tully, J. C., "Thermodynamic Implications of Desorption from Crystal Surfaces," in Pullman et al. (1984).

Chuang, T. J., and Hussla, I., "Molecule–Surface Interactions Stimulated by Laser Radiation," in Pullman et al. (1984).

Comsa, G., "The Dynamics Parameters of Desorbing Molecules," in Benedek and Valbusa (1982).

Comsa, G., and David, R., "Dynamical Parameters of Desorption." *Surf. Sci. 5,* 145 (1985).

D'Evelyn, M. P., and Madix, R. J., "Reactive Scattering from Solid Surfaces." *Surf. Sci. Rep., 3,* 413 (1984).

Ehrlich, G., and Stolt, K., "Surface Diffusion." *Annu. Rev. Phys. Chem., 31,* 603 (1980).

Ertl, G., "Chemical Dynamics in Surface Reactions." *Ber. Bunsenges. Phys. Chem., 86,* 425 (1982).

Ertl, G., "Reaction Mechanisms in Catalysis by Metals." *Crit. Rev. Solid State Mater. Sci., 10,* 349 (1982).

George, T. F., Lee, K. T., Murphy, W. C., Hutchinson, M., and Lee, H. W., "Theory of Reactions at a Solid Surface," in Baer (1985).

Gomer, R., "Electron Spectroscopy of Chemisorption on Metals." *Adv. Chem. Phys., 27,* 211 (1974).

Gomer, R., "Some Approaches to the Theory of Chemisorption." *Acc. Chem. Res., 8,* 420 (1975).

Greene, E. F., Keeley, J. T., and Stewart, D. K., "Alkali Atoms on Semiconductor Surfaces: The Dynamics of Desorption and of Surface Phase Transitions," in Pullman et al. (1984).

Halpern, B., and Kori, M., "Infrared Chemiluminescence from the Products of Exoergic Surface Catalyzed Reactions," in Fontijn (1985).

Halstead, J. A., Triggs, N., Chu, A.-L., and Reeves, R. R., "Creation of Electronically Excited States by Heterogeneous Catalysis," in Fontijn (1985).

Heidberg, J., Stein, H., Szilagyi, Z., Hoge, D., and Weiss, H., "Desorption by Resonant Laser-Adsorbate Vibrational Coupling," in Pullman et al. (1984).

Landman, U., and Rast, R. H., "On Energy Pathways in Surface Reactions," in Pullman et al. (1984).

Lin, M. C., and Ertl, G., "Laser Probing of Molecules Desorbing and Scattering from Solid Surfaces," *Annu. Rev. Phys. Chem.*, *37*, 587 (1986).

Lyo, S. K., and Gomer, R., "Theory of Chemisorption," in Gomer (1975).

Madix, R. J., "Surface Reactivity: Heterogeneous Reactions on Single Crystal Surfaces." *Acc. Chem. Res.*, *12*, 265 (1979).

Madix, R. J., and Benzinger, J., "Kinetic Processes on Metal Single-Crystal Surfaces." *Annu. Rev. Phys. Chem.*, *29*, 285 (1978).

Meek, J. T., Randolphlong, S., Opsal, R. B., and Reilly, J. P., "Laser Ionization Studies of Gas Phase and Surface Adsorbed Molecules." *Laser Chem.*, *3*, 3 (1983).

Menzel, D., "Desorption Phenomena," in Gomer (1975).

Ozin, G. A., "Spectroscopy, Chemistry and Catalysis of Metal Atoms, Metal Dimers and Metal Clusters." *Faraday Symp. Chem. Soc.*, *14*, 7 (1980).

Sachtler, W. M. H., "What Makes A Catalyst Selective?" *Faraday Discuss. Chem. Soc.*, *72*, 7 (1981).

Salem, L., "A Theoretical Approach to Heterogeneous Catalysis Using Large Finite Crystals." *J. Phys. Chem.*, *89*, 5576 (1985).

Schmidt, L. D., "Chemisorption: Aspects of the Experimental Situation," in Gomer (1975).

Selwyn, G. S., and Lin, M. C., "Laser Studies of Surface Chemistry," in Jackson and Harvey (1985).

Shustorovich, E., Baetzold, R. C., and Muetterties, E. L., "A Theoretical Model of Metal Surface Reactions." *J. Phys. Chem.*, *87*, 1100 (1983).

Simonetta, M., and Gavezzotti, A., "The Cluster Approach in Theoretical Study of Chemisorption." *Adv. Quant. Chem.*, *12*, 103 (1980).

Smalley, R. E., "Laser Studies of Metal Cluster Beams." *Laser Chem.*, *2*, 167 (1983).

Smith, J. R., "Theory of Electronic Properties of Surfaces," in Gomer (1975).

Somorjai, G. A., and Zaera, F., "Heterogeneous Catalysis on the Molecular Scale." *J. Phys. Chem.*, *86*, 3070 (1982).

Tully, J. C., Dynamics of Chemical Processes at Surfaces." *Acc. Chem. Res.*, *14*, 188 (1981).

Tully, J. C., and Cardillo, M., "Dynamics of Molecular Motion at Single Crystal Surfaces," *Science*, *223*, 445 (1984).

White, J. M., "Surface Interactions in Nonreactive Coadsorption: H_2 and CO on Transition-Metal Surfaces." *J. Phys. Chem.*, *87*, 915 (1983).

Woodruff, D. P., Wang, G. C., and Lu, T. M., "Surface Structure and Order–Disorder Phenomena," in King and Woodruff (1983).

Woodruff, D. P., Wang, G. C., and Lu, T. M., "Theory of Chemisorption," in King and Woodruff (1983).

Section 7.6

Review Articles

Bersohn, R., and Lin, S. H., "Orientation of Targets by Beam Excitation." *Adv. Chem. Phys.*, *16*, 67 (1969).
Dagdigian, P. J., "Spin–Orbit Effects in Chemiluminescent Reactions of State-Selected Ca($^3P_J^0$)," in Fontijn (1985).
Greene, C. H., and Zare, R. N., "Photofragment Alignment and Orientation." *Annu. Rev. Phys. Chem.*, *33*, 119 (1982).
Reuss, J., "Scattering from Oriented Molecules." *Adv. Chem. Phys.*, *30*, 389 (1975).
Sanders, W. R., and Miller, D. R., "Alignment of I_2 Rotation in a Seeded Molecular Beam," in El-Sayed (1984).
Stolte, S., "Scattering Experiments with State Selectors," in Scoles (1986).

Section 7.7

Book

Pimentel, G. C. (ed.), *Opportunities in Chemistry*, National Academy Press, Washington, D.C., 1985.

Review Articles

Adelman, S. A., "Chemical Reaction Dynamics in Liquid Solution." *Adv. Chem. Phys.*, *53*, 61 (1983).
Bado, P., Berens, P. H., Bergsma, J. P., Coladonato, M. H., Dupuy, C. G., Edelsten, P. M., Kahn, J. D., Wilson, K. R., and Fredkin, D. R., "Molecular Dynamics of Chemical Reactions in Solution." *Laser Chem.*, *3*, 231 (1983).
Connick, R. E., and Alder, B. J., "Computer Modeling of Rare Solvent Exchange." *J. Phys. Chem.*, *87*, 2764 (1983).
Frei, H., and Pimentel, G. C., "Infrared Induced Photochemical Processes in Matrices." *Annu. Rev. Phys. Chem.*, *36*, 491 (1985).
Hynes, J. T., "Theory of Reactions in Solutions," in Baer (1985).
Hynes, J. T., "Chemical Reaction Dynamics in Solution." *Annu. Rev. Phys. Chem.*, *36*, 573 (1985).
Kapral, R., "Kinetic Theory of Chemical Reactions in Liquids." *Adv. Chem. Phys.*, *48*, 71 (1981).
Oxtoby, D. W., "Vibrational Relaxation in Liquids: Quantum States in a Classical Bath." *J. Phys. Chem.*, *87*, 3028 (1983).
Warshel, A., "Dynamics of Reactions in Polar Solvents. Semiclassical Trajectory Studies of Electron Transfer and Proton-Transfer Reactions." *J. Phys. Chem.*, *86*, 2218 (1982).

General appendix

Table 1. Useful physical constants (rounded)

Designation of quantity	Symbol	Value	Units SI[a]	cgs
Avogadro's number	N_A	6.0221	$10^{26}\,\text{kmol}^{-1}$	$10^{23}\,\text{mol}^{-1}$
Atomic mass unit ($^{12}C = 12$)	amu	1.6606	$10^{-27}\,\text{kg}$	$10^{-24}\,\text{g}$
Electron charge/mass ratio	e/m_e	1.7588	$10^{11}\,\text{C}\,\text{kg}^{-1}$	$10^7\,\text{emu}\,\text{g}^{-1}$
Electron charge	e	1.6022	$10^{-19}\,\text{C}$	$10^{-20}\,\text{emu}$
Electron mass	m_e	9.1095	$10^{-31}\,\text{kg}$	$10^{-28}\,\text{g}$
Bohr radius (a.u.)	a_0	5.2918	$10^{-11}\,\text{m}$	$10^{-9}\,\text{cm}$
Rydberg constant	R_∞	1.09737	$10^7\,\text{m}^{-1}$	$10^5\,\text{cm}^{-1}$
Speed of light in a vacuum	c	2.99792	$10^8\,\text{m}\,\text{s}^{-1}$	$10^{10}\,\text{cm}\,\text{s}^{-1}$
Planck's constant	h	6.6262	$10^{-34}\,\text{J}\,\text{s}$	$10^{-27}\,\text{erg}\,\text{s}$
Dirac's $\hbar(h/2\pi)$	$\hbar$	1.05459	$10^{-34}\,\text{J}\,\text{s}$	$10^{-27}\,\text{erg}\,\text{s}$
Hartree (a.u.)	au	4.3598	$10^{-18}\,\text{J}$	$10^{-11}\,\text{erg}$
Gas constant	R	8.3144	$10^3\,\text{J}\,\text{kmol}^{-1}\,\text{K}^{-1}$	$10^7\,\text{erg}\,\text{mol}^{-1}\,\text{K}^{-1}$
Boltzmann's constant (R/N_A)	k	1.38066	$10^{-23}\,\text{J}\,\text{K}^{-1}$	$10^{-16}\,\text{erg}\,\text{K}^{-1}$

Adapted from *Pure Appl. Chem.*, *51*, 1 (1979).

[a] SI: Système International d'Unites (International System of Units), adopted in 1960. Special symbols for units: C, coulomb; J, joule; K, °K.

Table 2. Useful conversion factors[a]

Length
1 ångström (Å) $= 10^{-10}\,\text{m} = 10^{-1}\,\text{nm}$;
1 micron (μm) $= 10^{-6}\,\text{m}$

Force
1 newton (N) $= 1\,\text{kg}\,\text{m}\,\text{s}^{-2}\,[=10^5\,\text{dyn} = 10^5\,\text{g}\,\text{cm}\,\text{s}^{-2}]$

Pressure
1 pascal (Pa) $= 1\,\text{N}\,\text{m}^{-2} = 10^{-5}\,\text{bar}\,[=10\,\text{dyn}\,\text{cm}^{-2}]$;
$1.01325 \times 10^5\,\text{Pa}\,[=1\,\text{atm} = 1.01325 \times 10^6\,\text{dyn}\,\text{cm}^{-2} = 760\,\text{Torr}]$

Energy
1 joule (J) $= 1\,\text{kg}\,\text{m}^2\,\text{s}^{-2} = 10^7\,\text{erg}\,[=0.239006\,\text{cal}]$

[a] The familiar designations (units) enclosed in brackets are not part of the International System of Units (SI).

Table 3. Energy conversion factors (approximate)[a]

	erg	J	cal	eV	au	cm⁻¹	Hz	K	kJ mol⁻¹	kcal mol⁻¹
1 erg =	1	1.0000(−7)	2.390(−8)	6.241(+11)	2.294(+10)	5.034(+15)	1.5092(+26)	7.243(+15)	6.022(+13)	1.4393(+13)
1 joule (J) =	1.0000(+7)	1	2.390(−1)	6.241(+18)	2.294(+17)	5.034(+22)	1.5092(+33)	7.243(+22)	6.022(+20)	1.4393(+20)
1 cal =	4.1840(+7)	4.1840	1	2.611(+19)	9.597(+17)	2.106(+23)	6.315(+33)	3.031(+23)	2.520(+21)	6.022(+20)
1 eV =	1.6022(−12)	1.6022(−19)	3.829(−20)	1	3.675(−2)	8.065(+3)	2.418(+14)	1.1605(+4)	9.648(+1)	2.306(+1)
1 hartree (au) =	4.360(−11)	4.360(−18)	1.0420(−18)	2.721(+1)	1	2.195(+5)	6.580(+15)	3.158(+5)	2.626(+3)	6.275(+2)
1 cm⁻¹ =	1.9865(−16)	1.9865(−23)	4.748(−24)	1.2399(−4)	4.556(−6)	1	2.998(+10)	1.4388	1.1963(−2)	2.859(−3)
1 Hz =	6.626(−27)	6.626(−34)	1.5837(−34)	4.136(−15)	1.5198(−16)	3.336(−11)	1	4.799(−11)	3.990(−13)	9.537(−14)
1 °K (K) =	1.3807(−16)	1.3807(−23)	3.300(−24)	8.617(−5)	3.167(−6)	6.950(−1)	2.084(+10)	1	8.314(−3)	1.9871(−3)
1 kJ mol⁻¹ =	1.6606(−14)	1.6606(−21)	3.969(−22)	1.0364(−2)	3.809(−4)	8.359(+1)	2.506(+12)	1.2027(+2)	1	2.390(−1)
1 kcal mol⁻¹ =	6.948(−14)	6.948(−21)	1.6606(−21)	4.337(−2)	1.5936(−3)	3.498(+2)	1.0486(+13)	5.032(+2)	4.184	1

[a] Numbers in parentheses denote powers of 10 by which the entry is to be multiplied.

Subject Index

Absorption, 383–84
Absorption spectroscopy, 306
  fly-by collisions, 367
Abstraction, 214, 371
Action integral, 327
Activated complex theory of rate processes, 187–90
Activation energy, 181–82, 188. *See also* Arrhenius activation energy
  Tolman interpretation, 190
Addition-elimination, in chemical laser reactions, 214
Adiabaticity
  energy transfer reactions, 314–16, 318, 324, 376–77
  intramolecular quenching, 373–74
Adsorption, 470–71. *See also* Dissociative adsorption
Aeronomy, 369
Alignment, 480–95
Alkali halide system, formation of collision complex, 417
Allowed transition, 369
Ammonia synthesis, 499
Amplification of external light beam, 384–86
Angular deflection, 40, 68
Angular distribution, 11, 17, 91
  center-of-mass system, 70–73, 76
  collision complexes, 412–17
  crossed molecular beam scattering, 233–35
  curve-crossing problem, 378–80
  direct reactive collisions, 104–13
  elastic and reactive scattering, 17–18
  F-atom exchange reaction, 397, 401, 407–11
  gas-surface reactions, 461–62, 465, 471
  intermolecular potential and, 74–75
  laser probe experiments, 216
  molecular reactions, 47, 105, 208–9, 247–48, 258–59, 475, 484
  molecule-surface scattering, 345–47
  multiphoton dissociation, 437

nonreactively scattered molecules, 110
photofragmentation spectroscopy, 218
as probe of the potential, 95
rainbow effect, 86–87, 89
rare-gas systems, 95–96
rebound reactions, 111–13
rigid-sphere potential, 73
state-to-state inelastic collisions, 337–38
  in sudden limit, 328
van der Waals clusters, 454–55, 457–58
Angular momentum, 486, 492–93
  collision complexes, 414–16, 426–27
  elastic collisions, 41–42
  polyatomic system, 169–71
  quantal scattering theory, 278–79
  reactant molecules, 482–83
AO. *See* Atomic orbital
Aromatic endoperoxide, thermolysis of, 118
Arrested relaxation, in F-atom exchange reaction, 401
Arrhenius activation energy, 46*n*, 181, 399
Arrhenius-like cross-section functionality, 61
Associative ionization, 371
Asymmetric excitation transfer, 371
Asymptotic intermolecular potential, 103
Atmosphere, reactions of electronically excited species in, 369
Atom-atom interactions
  chemical quenching, 371
  dispersion constant, 103
  laser-assisted, 359
  physical quenching, 370–71
  potential, 97
  quasi-bound states, 351
Atom-atom repulsive force, 476
Atom-diatomic molecule
  long-range potential, 103
  velocity distribution, 335–36
Atom-diatom interaction
  chemical laser reactions, 214
  dispersion constant, 103

513

Author Index